PHYSICAL REALITY
AND MATHEMATICAL DESCRIPTION

PHYSICAL REALITY AND MATHEMATICAL DESCRIPTION

Edited by

CHARLES P. ENZ

Département de Physique Théorique, Université de Genève, Geneva, Switzerland

and

JAGDISH MEHRA

Instituts Internationaux de Physique et de Chimie (Solvay), Université Libre de Bruxelles, Brussels, Belgium

D. REIDEL PUBLISHING COMPANY

DORDRECHT-HOLLAND / BOSTON-U.S.A.

Library of Congress Catalog Card Number 74-81937

ISBN 90 277 0513 5

Published by D. Reidel Publishing Company,
P.O. Box 17, Dordrecht, Holland

Sold and distributed in the U.S.A., Canada, and Mexico
by D. Reidel Publishing Company, Inc.
306 Dartmouth Street, Boston,
Mass. 02116, U.S.A.

Printed in The Netherlands by D. Reidel, Dordrecht

Dedicated to

JOSEF MARIA JAUCH

on the Occasion of His Sixtieth Birthday

CONTENTS

J. M. Jauch

A PERSONAL INTRODUCTION

This collection of essays is intended as a tribute to Josef Maria Jauch on his sixtieth birthday. Through his scientific work Jauch has justly earned an honored name in the community of theoretical physicists. Through his teaching and a long line of distinguished collaborators he has put an imprint on modern mathematical physics.

A number of Jauch's scientific collaborators, friends and admirers have contributed to this collection, and these essays reflect to some extent Jauch's own wide interests in the vast domain of theoretical physics.

Josef Maria Jauch was born on 20 September 1914, the son of Josef Alois and Emma (née Conti) Jauch, in Lucerne, Switzerland. Love of science was aroused in him early in his youth. At the age of twelve he came upon a popular book on astronomy, and an example treated in this book mystified him. It was stated that if a planet travels around a centre of Newtonian attraction with a period T, and if that planet were stopped and left to fall into the centre from any point of the circular orbit, it would arrive at the centre in the time $T/\sqrt{32}$. Young Josef puzzled about this for several months until he made his first scientific discovery : that this result could be derived from Kepler's third law in a quite elementary way.

The youth was deeply impressed by the richness and order of the universe, and was thrilled by the fact that the human mind could absorb and understand the world in which it was placed, that there exist immutable laws which are not obvious but hidden behind mathematical structures of overwhelming beauty. Jauch was imbued with the desire to understand the world of natural phenomena with the help of mathematical concepts.

Mathematics he could pursue easily enough in the secondary school. But music competed with his interests in mathematics and natural science. Young Josef had made progress on the violin, which he studied first with his father and then with a professional teacher. He gave his first public concert in Lucerne at the age of twelve. At sixteen he discovered chamber music and founded a quartet which continued to play throughout his student years in Zurich. Jauch's music teacher wanted him to become a professional violinist, but he decided to remain an amateur musician.

Jauch's parents had died by the time he was fifteen years old. He had no money when he entered E.T.H. [the Swiss Federal Institute of Technology] in Zurich in 1933. Studies in Zurich were made possible with the help of generous friends from Lucerne who gave and loaned him the necessary funds. Jauch's decision to pursue scientific studies under such circumstances was a reckless one, possible only perhaps because he thought that he could not become more destitute than he was already.

The great intellectual shock of his early studies at E.T.H. was the encounter with Wolfgang Pauli in the fourth semester. Pauli's first course was on thermodynamics. It must have hurt Jauch's pride not to be able to understand Pauli's lectures. Jauch started taking stenographic notes and working them out at home. Slowly he caught on and found, as so many other students of Pauli did, that the effort was really worthwhile. Hidden beneath insufficient explanation, there were many gems and a logical structure in Pauli's lectures, which one could not find in the books. No doubt, these lectures by Pauli[1)] influenced Jauch's own style of teaching classical theoretical physics in later years.

There were other remarkable teachers as well at E.T.H. at that time, including F.Gonseth, Paul Scherrer, and G.Polya. Jauch took Polya's special courses on probability calculus and graph theory. From these he retained the method of generating functions in graph theory and in probability theory, a method to which he would return in his own research and teaching.

The deepest impression on Jauch was made by Heinz Hopf who was loved by all of his students. Jauch specially enjoyed Hopf's way of looking at mathematics, and this influence became lasting.

From Hopf he learned complex functions, Galois theory and topology, and these have remained favorite topics of pure mathematics with him.

Jauch wrote his Diplom Thesis under the supervision of Pauli in 1938. Pauli asked him to study Dirac's new theory[2)] of particles with higher spin, and to find out in particular whether one could define a current density and an energy tensor in this theory. This was obviously a very difficult task for the young diploma student. Indeed, Pauli was not known to spend much time with beginners and few wrote their diploma work under him. After several weeks of hard work, which sometimes brought poor Jauch to the edge of despair, he found the answer to Pauli's question. The answer was not trivial and he was very proud of himself. He sought an appointment with Pauli to explain his work. Pauli listened for a few minutes and said, "Das habe ich mir auch so gedacht". The interview was over and he had permission to write up the work,of which he gave an account at the Swiss Physical Society meeting in May 1938.[3)]

After the Diplom the future looked bleak. There were no assistantships available at E.T.H., and not a single position in sight in all of Switzerland other than part-time or substitute teaching in schools. Jauch accepted one such at Trogen (in Canton Appenzell). During his stay there he learned of his selection for an international exchange fellowship at the University of Minnesota. His Zurich teacher Polya had written in support of his application for the fellowship a recommendation which, as Polya joked later, was "much too good" to be refused.

Josef Jauch accepted the offer from Minnesota and started his work for a Ph.D. under E.L.Hill in 1938. The problem which they studied together concerned the higher symmetries of classical and quantum systems. Among the various special cases they studied, the isotropic harmonic oscillator in three dimensions became the prototype of the quark model in elementary particle physics since its internal symmetry group was shown to be SU_3. Jauch determined the eight generators of this group and the representations which occur in that system.[4)]

On the completion of his doctorate which he obtained in 1940 Jauch returned to Zurich where Pauli had offered him to become his research assistant. However, soon after his arrival in April 1940 Pauli left Zurich and settled in Princeton for the duration of the War. Jauch worked for two years under Gregor Wentzel who had taken over the direction of Pauli's students also.

Jauch returned to the United States in July 1942 and again collaborated with Pauli, serving at the same time as an assistant professor on the physics faculty of Princeton University from 1942 to 1945. Pauli was interested in strong coupling theory at that time. Jauch worked first on the λ-limiting process, especially regarding the value of the magnetic moment for the nucleon. The result was finite but of the wrong sign. This was the principal reason why Pauli subsequently abandoned this device. Pauli and Jauch then worked on the infrared divergence problem by applying Dirac's theory of field quantization and reported their work at the New York meeting of the American Physical Society in 1944.[5)] Again, the result was negative and served to eliminate this theory.

At the end of the War in 1945 students and professors were returning to their normal work at the universities. Jauch was not quite certain about the future direction of his work, and he felt the need of making new experiences. He joined the Bell Telephone Laboratory at Murray Hill, New Jersey, as a research scientist in 1945 for one year. There he studied the theories of luminescence in solids and wrote a review paper on the subject which was published as an internal report.

At Bell Telephone, Jauch became aware that he missed the students and wanted to return to a university. Teaching had, by that time, become an inseparable part of his activity and he missed its intellectual stimulation. Much of his subsequent research work has clearly had its origin in didactical problems.

In the fall of 1946 Josef Jauch joined the University of Iowa as a professor of theoretical physics. At Iowa he had the opportunity of developing his own style of teaching and research, and to choose his collaborators. Among the latter were Fritz Coester and Fritz Rohrlich. With Rohrlich he worked, from 1953 to 1955, on the ambitious project of "The Theory of Photons and Electrons", published in 1955, which laid the ground for a lasting friendship. This was the first book to appear containing the new renormalization theory developed by Schwinger and Feynman. Jauch duly sent a copy of the book to Pauli and later, in the summer of 1956, visited him. When he entered the office at the appointed time, Pauli looked at him in mock seriousness and said, "Ihr Buch ... ach ... Ihr Buch ... gefällt mir besser und besser." This was the highest praise that Pauli could accord, and it moved Jauch to tears of joy. He sent a telegramme the same day to his co-author, Fritz Rohrlich, informing him of Pauli's remark.

The authors of "The Theory of Photons and Electrons" had noted that "the study of the S-matrix forms the central part of this book". The S-matrix remained the central part of Jauch's research during the following years in Iowa which was devoted to scattering theory. In two fundamental papers[6)] Jauch led the way to a rigorous mathematical formulation of scattering theory which has attracted many mathematical physicists who brought this theory to a considerable state of perfection[7)].

Josef Jauch had always longed to return to his native Switzerland some day, and when the University of Geneva offered him, in 1959, the chair of theoretical physics he accepted it.

Before joining, however, Jauch spent a year with the U.S. Office of Naval Research, London Branch, as a scientific liaison officer. Beginning in August 1959 he wrote a series of technical reports for the ONR on the scientific organization and status of current research in physics in various European countries. Characteristically, the first report was on "Theoretical Physics in Switzerland and at CERN".[8)] In this report Jauch examined, diagnosed, and interpreted the state of the physics community in Switzerland before joining it himself. But he continued this exercise in perceptive reports on physics in Ireland and Scotland, the Scandinavian countries, Italy, Austria, Germany, France, England, and the Netherlands. Even after taking up his appointment in Geneva, Jauch continued in his role as a nuncius physicae by reporting on interesting summer schools and conferences.

The Fourth Centenary of Galileo's birth in 1964, and Jauch's report[9)] which covered the celebrations in Italy, brought into focus his own interest and scholarship in the life, science, and fate of Galileo Galilei, an absorption which has fulfilled Jauch's passionate interest in the historically human context and continuity of scientific creativity.

In Geneva, Josef Jauch was charged with reconstruction of the school of physics and the department of theoretical physics. Within four years he had brought the key people together and, eventually, resigned as the head of the physics institute in order to devote full attention to theoretical physics.

It was in Geneva that Jauch developed that other important research activity devoted to the foundations of quantum mechanics. As in his work on scattering theory the esthetic satisfaction of an adequate mathematical formulation was an important feature of this activity. In the preface to his book on the "Foundations of Quantum Mechanics" published in 1968 Jauch says : "Only beauty can lead to that « passionate sympathetic contemplation» of the marvels of the physical world which the ancient Greeks expressed with the orphic word « theory »". This second book of Jauch summarizes the efforts of axiomatization, mathematical formulation and interpretation made under his leadership by the "Geneva School"[10].

While the "Foundations of Quantum Mechanics" - dedicated to the memory of Wolfgang Pauli - was still rather anxious in tone (see its preface) Jauch has struck a more harmonious chord in his third book, "Are Quanta Real ? A Galilean Dialogue", published in 1973. In choosing the form of a dialogue Jauch allowed himself to develop freely his astonishing artistic imagination; and this quite different esthetic manifestation has given him probably more pleasure than many of the other things he has worked on[11].

Jauch's scientific activity is for him an intense human experience involving all his faculties. Science for him is not just, or even primarily, problem solving - it is that too -, but it involves a creative vision as in a work of art. To him the highest goal of science is the realization of beauty, for beauty also represents the deepest truth.

A scientist does his best work when he realizes to the fullest his creative potential, when he dares to be himself in the face of social pressure to conform his thinking to the leaders of the scientific community. Jauch has always sought to imbibe his collaborators, associates and friends with this spirit. In his "petit séminaire", a weekly meeting of all members of the department of theoretical physics at Geneva, people are invited to express their ideas and criticisms in a give and take of information and opinion that is hoped to be the source of new developments.

On behalf of the contributors to this volume and Josef Jauch's other friends and admirers not represented here, as well as on our own behalf, we take great pleasure in wishing Jauch a happy birthday and the continuing fullness of a rich and rewarding life.

Charles P. Enz, Jagdish Mehra

REFERENCES

1) Pauli Lectures in Physics, Vols.1 to 4, English translation edited by C.P.Enz (M.I.T. Press, Cambridge, Mass., 1973).

2) P.A.M.Dirac, Proc.Roy.Soc. (London) A155, 447 (1936).

3) J.M.Jauch, Helv.Phys.Acta 11, 374 (1938).

4) E.L.Hill and J.M.Jauch, Phys.Rev. 57, 641 (1940).

5) W.Pauli and J.M.Jauch, Phys.Rev. 65, 255-256 (1944).

6) J.M.Jauch, Helv.Phys.Acta 31, 127, 661 (1958).

7) For an up to date account of this development see the Introduction to the paper by W.O.Amrein, V.Georgescu and Ph.A.Martin in this volume.

8) Technical Report ONRL-61-59, 12 August 1959, American Embassy, London, England.

9) Technical Report ONRL-C-19-64, 7 December 1964, American Embassy, London, England.

10) For an up to date appreciation of these problems see the Introduction to the paper by B.Misra in this volume.

11) For an analysis of this Dialogue see the contribution by J.Mehra in this volume.

ACKNOWLEDGEMENTS

We take pleasure in thanking Prof.M.Guenin for editorial assistance and Mrs C.Brügger, Mrs D.Künzi, Miss M.-C.Bimbert, Miss M.Horisberger and Miss F.Mallmann for coping with the difficult task of typing the manuscript most pleasantly.-- The Editors.

PRINCIPAL PUBLICATIONS OF J.M. JAUCH

Articles

Über die Energie-Impuls-Tensoren und die Stromvektoren in der Theorie von Dirac für Teilchen mit Spin grösser als $\frac{1}{2}$ h. *Helv. Phys. Acta* 11, 374 (1938).

(J.M. Jauch and E.L. Hill). On the problem of degeneracy in quantum mechanics. *Phys. Rev.* 57, 641 (1940).

Die Streuung schneller Elektronen an Kernen. *Helv. Phys. Acta* 13, 451 (1940).

Über die Wechselwirkung schwerer Teilchen mit Elektronen. *Helv. Phys. Acta* 14, 465 (1941).

Kernkräfte in der Elektronenpaartheorie. *Helv. Phys. Acta* 15, 175 (1942).

Bemerkung zum Streuproblem in der Elektronenpaartheorie. *Helv. Phys. Acta* 15, 221 (1942).

Meson Theory of the Magnetic Moment of Proton and Neutron. *Phys. Rev.* 63, 334 (1943).

(J.M. Jauch and N. Hu). On the Mixed Meson Theory of Nuclear Forces. *Phys. Rev.* 65, 289 (1944).

(J.M. Jauch and J.L. Lopes). Scalar Meson Pair Theory of Nuclear Forces. *An. Acad. Brasil. Ciênc.* 16, 281 (1944).

Neutron-Proton Scattering and the Meson Theory of Nuclear Forces. *Phys. Rev.* 67, 124 (1945).

On the Inelastic Photo-Dissociation of the Deuteron. *Phys. Rev.* 69, 275 (1946).

The Hyperfine Structure and the Stark Effect of the Ammonia Inversion Spectrum. *Phys. Rev.* 72, 715 (1947).

(J.M. Jauch and K.M. Watson). Phenomenological Quantum-Electrodynamics. *Phys. Rev.* 74, 950 (1948).

(J.M. Jauch and K.M. Watson). Phenomenological Quantum Electrodynamics. Part II. Interaction of the Field with Charges. *Phys. Rev.* 74, 1485 (1948).

(K.M. Watson and J.M. Jauch). Phenomenological Quantum Electrodynamics. Part III. Dispersion. *Phys. Rev.* 75, 1249 (1949).

On the Relativistic Invariance of the Canonical Field Equations and the Location of Energy, Momentum and Angular Momentum in a Field. *An. Acad. Brasil. Ciênc.* 20, 353 (1948).

Cosmic Rays I and II. *Nucleonics* 4, 39 and 44 (1949).

(F. Coester and J.M. Jauch). On the Role of the Subsidiary Condition in Quantum Electrodynamics. *Phys. Rev.* 78, 149 (1950).

(P.R. Bell, J.M. Jauch and J.M. Cassidy). Transition Energy Determination for Orbital Capture. *Science* 115, 12 (1952).

(A. Simon, M.E. Rose and J.M. Jauch). Polarization and Alignment of Nuclei. *Phys. Rev.* 84, 1155 (1951).

Radiative Correction for the Collision Loss of Fast Particles. *Phys. Rev.*, 85, 950 (1952).

(F. Coester and J.M. Jauch). Theory of Angular Correlations. *Helv. Phys. Acta* 26, 3 (1953).

(E.J. Hellund and J.M. Jauch). The Sequence of γ-Emission in Triple Correlations. *Phys. Rev.*, 92, 203 (1953).

A Note Concerning the Quantization of Spinor Fields. *Helv. Phys. Acta* 27, 89 (1954).

(J.M. Jauch and F. Rohrlich). The infrared divergence. *Helv. Phys. Acta* 27, 613 (1954).

Covariant Hyperquantization. *Helv. Phys. Acta* 29, 287 (1956).

On the Relation Between Scattering Phase and Bound States. *Helv. Phys. Acta* 30, 143 (1957).

Theory of the scattering operator. *Helv. Phys. Acta* 31, 127 (1958).

Theory of the scattering operator II. Multichannel Scattering. *Helv. Phys. Acta* 31, 661 (1958).

(J.M. Jauch and I.I. Zinnes). The Asymptotic Condition for Simple Scattering Systems. *Nuovo Cimento* 11, 553 (1959).

(J.M. Jauch and Y. Yamaguchi). Eine Bemerkung zum Pionenzerfall. *Helv. Phys. Acta* 32, 251 (1959).

Systeme von Observablen in der Quantermechanik. *Helv. Phys. Acta* 32, 252 (1959).

(D. Finkelstein, J.M. Jauch and D. Speiser). Zur Frage der Ladungsquantisierung. *Helv. Phys. Acta* 32, 258 (1959).

On Pauli's Transformation. *Nuovo Cimento* 16, 1068 (1960).

Systems of Observables in Quantum Mechanics. *Helv. Phys. Acta* 33, 711 (1960).

The existence of the Hamiltonian for Causal Systems. In *Dispersion Relations*. *Oliver and Boyd, Edinburgh*, 1961.

(J.M. Jauch and B. Misra). Supersymmetries and Essential Observables. *Helv. Phys. Acta* 34, 699 (1961).

Les symétries dans la physique classique et moderne. *Archives des Sciences Genève* 14, 5 (1961).

Die Symmetrien in den Naturgesetzen. *Verh. Schweiz. Naturf. Gesellsch., Biel* (1961) S. 19-32.

(D. Finkelstein, J.M. Jauch, S. Schiminovich and D. Speiser). Foundation of Quaternion Quantum Mechanics. *J. Math. Phys.* 3, 207 (1962).

(D. Finkelstein, J.M. Jauch and D. Speiser). Quaternionic Representations of Compact Groups. *J. Math. Phys.* 4, 136 (1962).

(D. Finkelstein, J.M. Jauch, S. Schiminovich and D. Speiser). Principles of General Q-Covariance. *J. Math. Phys.* 4, 788 (1963).

(J.M. Jauch and C. Piron). Can Hidden Variables be Excluded in Quantum Mechanics. *Helv. Phys. Acta* 36, 827 (1964).

Gauge Invariance as a Consequence of Galilei-Invariance for Elementary Particles. *Helv. Phys. Acta* 37, 284 (1964).

The Problem of Measurement in Quantum Mechanics. *Helv. Phys. Acta* 37, 293 (1964).

(J.M. Jauch and B. Misra). The Spectral Representation. *Helv. Phys. Acta* 38, 30 (1965).

(G. Emch and J.M. Jauch). Structures logiques et mathématiques en physique quantique. *Dialectica* 19, 259 (1965).

A recent study on Copernicus, Kepler and Borelli. *Bibl. d'humanisme et renaissance* 28, 155 (1966).

(J.M. Jauch and J.-P. Marchand). The Delay Time Operator for Simple Scattering Systems. *Helv. Phys. Acta* 40, 217 (1967).

(J.M. Jauch and C. Piron). Generalized Localizability. *Helv. Phys. Acta* 40, 559 (1967).

(J.M. Jauch and C. Piron). Hidden Variables Revisited. *Rev. Mod. Phys.* 40, 229 (1968).

Projective Representations of the Poincaré Group in a Quaternionic Hilbert Space. In *Group Theory and its Applications*, E.M. Loebl, ed., *Academic Press, New York* (1968), p. 131-182.

(J.M. Jauch, B. Misra and A.G. Gibbson). On the Asymptotic Condition of Scattering Theory. *Helv. Phys. Acta* 41, 513 (1968).

(J.M. Jauch and C. Piron). On the Structure of Quantal Proposition Systems. *Helv. Phys. Acta* 42, 842 (1969).

Scattering Theory in General Quantum Mechanics. In *Analytical Methods in Mathematical Physics*, Proceedings, R.G. Newton and R.P. Gilbert,ed., *Gordon & Breach, New York* (1970), p. 185-205.

(J.M. Jauch and C. Piron). What is "Quantum Logic"? In *Quanta*, Essays in Theoretical Physics, dedicated to Gregor Wentzel, P.G.O. Freund, C.J. Goebel and Y. Nambu, ed., *University of Chicago Press, Chicago* (1970), p. 166-181.

Le procès de Galileo Galilei. *Arch. sc. Genève* 23, 543 (1970).

(W.O. Amrein, V. Georgescu and J.M. Jauch). Stationary States Scattering Theory. *Helv. Phys. Acta* 44, 407 (1971).

Foundation of Quantum Mechanics, IL Corso, Varenna Summer School *Academic Press , New York* (1971), p. 20-55.

On Bras and Kets. In *Aspects of Quantum Mechanics*. A. Salam and E.P. Wigner, ed., *Cambridge University Press* (1972), p. 137-167.

On a new Foundation of Equilibrium Thermodynamics. *Found. of Phys.* 2, 327 (1972).

(J.M. Jauch and J.G. Baron). Entropy, Information and Szilard's Paradox. *Helv. Phys. Acta* 45, 220 (1972).

(J.M. Jauch, R. Lavine and R.G. Newton). Scattering Into Cones. *Helv. Phys. Acta* 45, 325 (1972).

(J.M. Jauch, K.B. Sinha and B.N. Misra). Time Delay in Scattering Processes. *Helv. Phys. Acta* 45, 398 (1972).

(J.M. Jauch and K. Sinha). Scattering Systems with Finite Total Cross-Section. *Helv. Phys. Acta* 45, 580 (1972).

Determinism in Classical and Quantal Physics. *Dialectica* 27, 13 (1973).

The Problem of Measurement in Quantum Mechanics, in the Physicist's Conception of Nature, Jagdish Mehra, ed., *Reidel, Dordrecht* (1973), p. 684-686.

The Mathematical Structure of Elementary Quantum Mechanics. ibid. p. 300-319.

Books

(J.M. Jauch and F. Rohrlich). The Theory of Photons and Electrons. *Addison-Wesley, Cambridge, Mass.* (1955).

Foundations of Quantum Mechanics. *Addison-Wesley, Reading* (1968).

"Are Quanta Real?" A Galilean Dialogue. *Indiana University Press, Bloomington* (1973).

Die Wirklichkeit der Quanten. Ein zeitgenössischer galileischer Dialog. Translated from "Are Quanta Real?" by Friedel Bachwinkel and Marina Guenin. *Carl Hauser, München* (1973).

Part I

ART, HISTORY AND PHILOSOPHY

SCIENCE AND ART

André Mercier

Institut für theoretische Physik
Universität Bern

The choice of a subject-matter like this responds to a trend among quite a few contemporary scientists to enlarge their view on the correlations between their professional activities and other manifestations of the human mind. The author - who during his youth had learnt all the instrumental techniques and the theoretical background to become a musician - has spent a considerable part of his career investigating the connections which tie science, especially the exact sciences like physics, to both morals and the arts, and more specifically mathematics to ethics and aesthetics [see for instance: *De la science à l'art et à la morale* (Ed. du Griffon, Neuchâtel 1950), *Thought and Being, An Inquiry into the Nature of Knowledge* (Verlag für Recht und Gesellschaft, Basel 1959), *Erkenntnis und Wirklichkeit* (Francke Verlag, Bern und München 1968), *Science and Responsibility* (Filosofia, Torino 1969)]. He is happy to dedicate this article to a friend with whom he has, during a number of years, played many works of chamber music.

* * *

1. Too often, science and art are represented as being opposed on the ground of assertions which are not in agreement with facts. For instance, it is said that science (and science alone) rests on evidence not only capable of being repeated but which has been stated again and again identically, this being the demonstration of what might be called the exclusive publicness (or openness to all) of science which is said to establish its objectivity. Art on the other hand is said to be the rendering, by unique pieces, of purely personal feeling, no two individuals having the sa-

Enz/Mehra (eds.), Physical Reality and Mathematical Description. 3–21.

me feeling, in short to be the product of fantasy and even of illusion, being "therefore" subjective (in the pejorative sense of the word).

The whole reasoning behind these arguments is specious. For, objectivity does not rest upon publicness. Science is not only "public" in the above sense, it is also tightly connected with a personal concern of the individual (Bridgman); art is not only private in the above sense, it is as public as science (even more so socially speaking, else why would people buy tickets for concerts or visit exhibitions?), and the historical uniqueness of each single work of art has nothing to do with an alleged personal feeling.

Bridgman, the Nobel laureate and one of the founders of operationalism, has repeatedly contested the first part of the above argument and I agree with his critique; his writings are too numerous and well-known to need detailed quotation[1)]. The fallacy of the second part of the above argument is less well recognized.

In the scholastic tradition of medieval philosophy, the word subjective was used for what people call in our days objective; an observation due to Arthur Eddington, who knew his classics well. Actually - and we shall adopt the modern use of the words, rather than the scholastic one - objectivity and subjectivity do not constitute an opposition of the kind logicians think of when they talk, for instance about contradiction. Objectivity and subjectivity are two extreme modes of approach within the modality of knowledge which is to be called judicial.

If one is to understand it clearly, one has first to consider man in his primeval situation of cognitive activity as being separated or isolated from everything else and that isolation as being the cause of his need to reestablish the links with all that there is, as the source of both his scientific curiosity and artistic creativity[2)]. Two ways or possibilities at least are open to fulfil this need: (i) either to keep the greatest possible independence with regard to, even to keep continuous mastery over, the objects of his consideration, or (ii) to submit his self in the most acute possible degree to these same objects. The first of these attitudes of the human mind towards reality is the objective mode and shall be called objectivity; the second one is the subjective mode and shall be called subjectivity. For, keeping one's independence allows one to throw the things (ob-jects) onto the plane of his consideration, and this is exactly what is done by scientific investigation, drawing to one's help the kind of dialectics which exists between experimental activity and theoretical speculation. Science is always directed in the sense of an abstraction from concrete situations, producing intellectually conceived models of what appears universally to be the case within reality, whereas art goes just the other way and consists of moulding into reality the conceptions borne into one's mind by submitting one's self to that very reality, whence art always appears as proceeding from conceived ideas towards concrete realizations called works of art.

There cannot be any doubt that both ways are from the very beginning open to a choice and are therefore authentic modes of actualizing the link between man's cogitativeness and the reality of things. Therefore science as the cognitive enterprise of objective nature implies authentic knowledge not more and not less than art as the cognitive enterprise of subjective nature.

Subjectivity is wrongly described as an illusory attitude due to the refusal of "sound" critique. Subjectivity is as rigorous and efficient an attitude for the disciplines to which it applies as objectivity is for other disciplines, and it is endowed with a critical attitude of its own. Objectivity cannot be taken as a criterion to judge fields like the arts where it is not valid, and vice versa. Even a scientific investigation of art (by psychology for example) must in the long run end in a complete misunderstanding of the nature of the artistic enterprise, and an artistic view upon science will finally efface the genuine character of science.

Yet these last assertions must be read by keeping in mind that everything one says is either an exaggeration (as noted by Ortega y Gasset), or some sort of an approximation (as is well known to every theoretical physicists).

Specifically, in our case, it should not be understood that the position of (scientific[3)]) objectivity is absolute (however more or less obscure significance may be attached to the word absolute), for it would require a total independence of the scientist from the objects of his investigation. Such an independence was in the last century unconsciously believed to apply due to the success of classical physics since the time of Newtonian mechanics[4)]. However, we have since realized that a total independence would separate the subject from these objects and leave him in splendid isolation. Hence a certain lower limit is imposed upon objectivity. It is perhaps one of the great achievements of the 20th Century physics to have recognized (N. Bohr) such a relativity of objectivity, even measurable by Planck's constant, in a way similar to the relativity recognized by Einstein as measured by the velocity of light[5)].

Similarly, it cannot be that the subjective attitude be absolute, for a total submission of the subject would prevent him from retaining any initiative in the creation of the works from the ideas he received of possible conceptions of reality. In a way, a quasi-total submission is what is sought by people using drugs. Of course, no work of art worthy of the name need follow from that attitude.

As a first conclusion, therefore, I should like to call the attention of philosophers (and scientists) to the double correlation existing between science and art: they are both modes of knowledge belonging to the judicial modality[6)], but they are at the two extremes of that modality. (Incidentally, I believe that the French adage *"Les extrêmes se touchent"* applies here). In a first approximation, science is by definition objective and art is subjective, but on closer examination, science is not found free from some subjectivity and art free from some objectivity.

In that sense, it is understandable that within the controversies about the correct interpretation of quantum physics, the interpretation attributed to the so-called Copenhagen School (to which I admit to belonging ever since the days I spent at Bohr's Institute) is often called subjective. However, the sense given to the word subjective in these debates is wrongly identified with the pejorative and, to my mind, useless and unorthodox usage. The correct relation is that science (here physics and more precisely quantum physics) is not deprived of an artistic touch which could be said to endow it with a creativity of the very situation under consideration, whereas e.g. a Newtonian mechanics creates nothing, since it only describes a state of affairs as if the theoretician (or the astronomer) were a demi-god reigning on a throne outside the celestial vaults in which bodies move around one another.

In the same sense, but vice versa, it is understandable that so much talk and writing are spent upon the assumed abstractness of much contemporary art, though actually art is never abstract, as has been insisted upon by Picasso, since it must be actualized in concrete pieces of work realizing[7)] the ideas which lie behind them. (Even music must be played and is in that sense concrete[8)].) By abstractness of art, one means only the departure from a naive resemblance of a quasi-photographic (or quasi-phonographic etc.) type. Beethoven's *Pastorale* contains quasi-phonographic resemblances of the noises of storms and birds, but these are only means chosen to concretize the conceived idea, which is the joy within man's soul and has only as a secondary theme the happenings within an assumed country-life, or Virgil's *Georgics* concentrates on that one verse famous among quotations *"Fugit irreparabile tem-*

pus" more than it rests on a detailed description of Roman husbandry.

It is the motion from the idea to the concrete which makes works of art historically unique, and not such a thing as an assumed personal feeling of the artist. It is the process of abstraction from the concrete towards ideal forms which makes the scientific theories universal.

2. Pythagoras - if he did indeed live - was the first theoretical physicist. For he abstracted from the facts observed on the monochord the numerical relations valid in the harmony of musical tones. This, incidentally, established a link between music and physics. It was much later replaced by a different construction of the scale, unconsciously using a consideration from the group theory by inserting twelve "equal" half-tones within the octave. This latter construction is of purely mathematical character, it does not describe a physical phenomenon, whereas the Pythagorean researches dealt with physical data.[9)]

When physical data are left unconsidered, that which remains on the plane of objectivity is a set of rules about how to reason correctly. This is called mathematics. Mathematics is not concerned with the world as it is; it is, if you will, not concerned with that which is posited in the world. It offers ways of reasoning with certainty and is therefore not a positive science, not even a science at all (for all science is positive) because mathematics does not know anything, it proves, though it proves nothing about anything real. Much rather, it is a power (in Latin: *potentia*), not a science (*scientia*).

Arts are also positive like the sciences. Hence, similarly, when the positivity of the various arts is left out, there remains a procedure corresponding to mathematics and of comparable nature; it is aesthetics, consisting of techniques like counterpoint, or rules like the golden rule. Aesthetics is also a power, not an art.

It is interesting to notice how much these two powers condition each other, for, criteria of aesthetic satisfaction are always invoked in matters of mathematical concern (proofs are said to be elegant etc.) and criteria of mathematical origin are again and again used in aesthetics[10)]. Lichtenstein, Einstein, Dirac and others have repeatedly commented upon this. However, there remains a difference which reflects the distinction between the objective and the subjective. Mathematics can live a life of its own. It can be acquired[11)] and developed before it be applied to the elaboration and correct expression of scientific theories, whereas the evolution of aesthetics is always a consequence of the novelty of a work of art. Of course scientific discoveries can motivate new mathematics - e.g. mechanics exacting the invention of calculus by Newton. Furthermore an acquired piece of novel aesthetics like Debussy's harmony replacing that of the Romantics can, and even should, be then practised by students in the schools and conservatories before they become creative artists themselves.

3. The sciences throw light upon the workings of reality on the objective plane and the arts make them luminous on the subjective one. Reality is a generic name for all beings, and the general study of the being of things is called ontology. Sciences like arts fulfill an ontological task. If we think of a picture of a landscape, limited in breadth and height and painted by means of a small number of brushes from a few tubes of colour, the following characterization will easily be acceptable. Painting is the procedure of giving the impression and sense of the infinite by finite means; indeed, the horizon appears actually cut on the left and right sides of the frame, but if "well painted" it is "seen" as an infinite horizon. The heavens too are "visible" with an infinite depth etc. The same applies to less evident cases, say to a sculpture. One might object that there is no such thing as an horizon or anything supposedly infinite in a sculpture. I would reply: take a torso without a head, and the head is "seen" though it is failing. In this way, any good sculpture establishes the appurtenance of the body to the totality of reality. Or, a symphony may end, but when you go home, you have it still with you as transcending the limits of an hour of time.

Such limits (although they often look space-like) are, as a matter of fact, always of temporal nature, because time is essential if we are to grasp them. Now, what is concrete is rooted in matter; matter composes the world. If matter did not change in any way, never showing any other aspect, no one would ever be able to notice that it is. Conversely, if nothing called time spent itself like a stream in the course of which happenings are noticeable, nothing would be noticed. These not quite trivial remarks help us to understand that matter and time are the two necessary and complementary aspects of reality. They compose the being of things in time, their existence if you will, but they are at that primitive stage of our reasoning so entangled that not much can be said about them. If one is to say or do more about things in the world, he must first separate matter and time from each other by suitable ways of thought; the mind must invent "intellectual knives" to cut the knots which entangle them and pull time and matter apart. Such structures have been invented, and they are called *spaces*. For instance Euclidean three-dimensional space has been conceived and used for various purposes. The painters of the Renaissance used it for the representation of perspective (think of Raphael's *Academy*); physics has used it for a few centuries in mechanics and other branches. But the Chinese used other spaces to represent depth by painting, and physics also makes use of a number of other spaces (q-space, p-space, phase space, Hilbert space and others). So space has never the reality possessed by matter and time, not even the affine space of general relativity (which is not identical with "space-time", itself not a space, but a *time* endowed with a richer structure than the time analysed by means of Newtonian mechanics.[12])

The use of spaces for purposes of understanding the workings of reality is associated with the fact that we need first to see symmetries, and then to destroy them by the application of more elaborate symmetries if we are to foster our understanding of reality. This is so in science, especially in physics (as was explicitly recognized first by Pierre Curie) as well as in art (as has been practised since Antiquity). Of course, the word symmetry must be understood in a very general sense. We shall not dwell upon that.

The so-called creation of works of art can be said to reside in producing symmetries broken by further symmetries in concreto. If the face of a woman for instance were perfectly symmetric left and right of a plane through the mouth and the nose, her face would be dull. Feminine beauty is rendered in art by curving and the like, producing e.g. the *Venus de Milo*, the stone dancers of Khajuraho or the torso of the Hindu goddess now at the Boston Museum of Fine Arts[13]), ... all being ways of breaking symmetries in order to comprehend them in our attempt to (re-)cognize the beauty of the idea of a woman as revealed in matter (in stone).

There is no difference of principle between this and the use of broken symmetries in physics for the purpose of understanding fields and particles.

Repetition is also a means frequently applied in art or often noticed in physical phenomena, but very quickly becomes tedious or uninteresting unless broken in a suitable manner, like by the damping of an oscillation, or the bending of an arabesque. (In music, the tedious repetition of motifs is called *rosalies* in the composition class of a French music conservatory, and one learns there how to avoid it by introducing incidentals such as modulation, trills, etc.).

When a new piece of music is being played, a person sufficiently talented and trained is able to anticipate at every moment with an astonishing approximation what will be the next move of the piece, even though he has never heard it before. Any good musician will confirm this. It is by the way a proof that music is not a business of personal feeling and fantasy. But precisely the breaking of the symmetry used by the composer makes a total anticipation impossible and compels the person to make an effort to continue the procedure of anticipation to the end, i.e. to experience in his turn the disclosure of the value clad in the piece itself. Quite similarly, a trained scientist can re-make, or follow the exposition of, a piece of scientific work by anticipating nearly every phase of its development, except the turns which reveal the concealed values of truth which are marked as discoveries. (By the way, *this* is what makes science "public", i.e. public only among specialists, and actually no scientific finding is re-made identically by anyone. The assertion that every scientific step can be repeated by anyone is a myth that has turned into superstition.) The cognizance reached by the person who anticipates the whole step by step is in both cases absolutely authentic and renders familiar to him a piece of reality which had been previously alien.

Looked at from such view-points, science and art are very close in nature, *toutes choses égales d'ailleurs*. Both imply a moment of personal feeling on the part of each subject actively involved in them, which, by the way, is not only present with those who relive the experiences, but is strongest with those who experience them for the first time, namely the scientist-discoverers or artist-creators. For, even the prime discoverer of a scientific truth gets at this very truth along a path of anticipations broken again and again and finally crowned by the fulguration of evidences of the nature of categorical imperatives[14)], even more so than do his listeners when he tells them, or his followers when they try to verify his findings. The prime artist creating a work gets at its achievement along a similar path of anticipations, turns and evidences of the beautiful, sparing them and the sufferings they imply for the contemplator who experiences them only second hand.

4. In the argument above, it is said that on the one hand sciences are, like arts, necessarily positive and consequently of the nature of specific ontological enterprises, whereas on the other hand mathematics like aesthetics are not, since they are the powerful techniques by means of which sciences or arts are executed without formal incongruity.

Usually, ontology is considered an enterprise of its own, sometimes also called metaphysics. Used in the latter sense, metaphysics is then meant to be a "kind of science", which however is to be distinguished from the "other" (positive) sciences because its objects are not finite things like particles, atoms, crystals, stars, macromolecules, cells, plants, animals, societies and so on. The objects of metaphysics are always by their own nature infinite, not individual beings to be classified according to their similarities, but Being or Reality itself. It deals with attributes, powers and entities - like freedom, love ... God. So, if metaphysics is to be in some analogy treated as a science, it is one which implies difficulties of principle compared with the "other" sciences, for, firstly it does not allow for either experiment or observation on finite systems though it does have to go through the finitude of man, and secondly it cannot be expected to reduce, at least in all cases, to models of an assumed reality which could be clad in logico-mathematical language.

And yet, on the one hand, there can be no doubt that good metaphysics does rest upon a kind of experience, the experience of freedom, of love, ... of God. (People who deny this are simply bad metaphysicians, or they can be said to lack the capacity to make the experiences, just as there are people who are completely a-musical or totally incapable of understanding physics and its mathematical basis). The language used by metaphysicians need not be unsound as judged even by logicians, firstly because a good part of it still rests on logics, and even incidentally on mathematics[15)], secondly, because there are findings about the "last" entities which require a kind of reasoning not isomorphic to any system of logic developed so far, but a kind known as one of the forms given to dialectics. Hence metaphysics is wrongly despised by those who describe it as a mere construction of a phantasmal mind.

Now if there really is a parallelism between science and art, between the sciences and the arts, between mathematics and aesthetics, inspite of their being at the two extremes of judicial knowledge, and if metaphysics is an acceptable activity to be considered as the ontological approach to entities entailing the infinity of reality as opposed to its finitude[16)], then there should also be an ontological approach of the same kind on the side of the arts. Anyone familiar with these things will soon recognize that there is such an approach and that it is nothing but Poetry. In other words - since the distinction between science and art is the same as between authentic objectivity and authentic subjectivity - poetry is "metaphysics" (or "ontology" if you prefer) on the subjective plane. Indeed, poetry deals always, even if not necessarily at first sight, with love, freedom ... the last realities ... God.

One should in that respect understand that since there is a problem in trying to describe the infinite in terms of words evolved to describe the finite, the way such speech - dissertation or poem - is worded will necessarily refer to them by ways of showing how to escape or transcend the finitude of things. It has therefore a more or less concealed negative form or aspect, though it can and should act very positively. This may be one reason why many unmetaphysical or unpoetical minds confuse metaphysics and/or poetry with fantasy and illusion.

As a further conclusion at this stage, we might say that we have established a general analogy between the scientific enterprise and the artistic one, which includes both their formal and their intentional aspects.

The question, whether apart from metaphysics and poetry there is a further analogy of detail between the individual sciences (even within each individual one) and the individual arts, can be taken up now.

5. For this to be done, we must however first inquire into the exact nature of what is achieved by either the sciences or the arts.

If asked about such achievements, a rash person might answer: the discovery of the rotation of the Earth, the discovery of the electric battery, of atomic energy, of penicillin ..., or the Venus de Milo, the Joconda, the Ninth Symphony, ... But to my mind these are not the kind of advancements that should be called achievements. If they were, what could we say e.g. of the achievements of the art of dancing? For no doubt dance is an art, and a major one indeed. Or what should be said of e.g. natural selection? Actually, particular cases are not achievements within the cognitive experience, even if their disclosure may be recompensed by the Nobel prize or some other award.

An achievement is to be acknowledged where one has succeeded in establishing a structure or a monument endowed with the property of suspending the temporality which pervades all material beings.

In the field of science, this requirement is fulfilled by theories, e.g. Maxwell's equations, the classical form given to the theory of electromagnetism. They are valid and non-contradictory, they contain time as a mathematical parameter but precisely that means that their validity is timeless[17]. A theory such as that has the property of escaping the particular happenings of the world by describing the world as it seems to be in all cases in view (within some sort of an approximation); it is the expression of a *perfection*. Perfection is timeless. On the objective plane, perfection must eventually take a mathematical form, and its content of positivity is then freed from its original temporality.

Art does not develop theories. It develops styles which are very much like theories. It is sometimes said that styles are bound to historical periods. This is both true and untrue, according to what is meant. They are of course created at times which history recalls, but that does make them temporal only from the point of view of history, not of epistemology. Actually, a style is precisely that specific aspect of a work of art which makes it timeless. It can also be said (and often is) that each work of art worthy of that name is in itself timeless. This is also an acceptable proposition, for each piece can be said on the grounds of its uniqueness to deviate more or less from the macro-style to which it belongs and so to possess a micro-style of its own of which it is the sole representative.

The relation of theory to style is a peculiar one and could give rise to longer arguments.[18)]

At this point, rather, we shift over to the conclusion that since science and art produce a suspension of the temporality of actual happening each in its specific way, this means that from that point of view they accomplish a categorization of the notion of time.

I should like here to insist upon the difference between what should be called *notions* and what should be called *concepts*. A concept is the result of an abstraction by a process leading to the distinction of several elements of speech mutually connected by some sort of general principles recognized as working reasonably well among groups of such elements. So, they do not designate the entities of our immediate apprehension of the real. The latter is done by notions. Thus time and matter, ... love and freedom ... are notions, not concepts. A baby apprehends the real by a hunger every few hours, by the love of the mother ... but not space: for, in order to conceive of space, it must play with the details of its cradle and construct some mental scheme, which psychologists like Piaget elaborate upon ad infinitum. Abstraction assumes always a certain construction, so concepts are (mental) constructions. Spaces for instance are constructions, including the so-called three dimensional Euclidean, Newtonian space of classical mechanics. Newtonian space is devoid of immediate reality. The Ancient were not aware of it. It needed Euclid to give it the necessary mathematical structure. To the Egyptians of the Nile Valley space was two-dimensional, as abstracted from the work done on arable earth, the kind of things measured in "mornings"; but that is actually matter (earth) which, because of the Earth's attraction, appears horizontal in Egypt, and inclined in the Alps. In its abstraction, it leads to a two-dimensional *geo*-metry.

So spaces, again, though now said differently, are concepts used for the categorization of time (versus matter), and there is no such thing as Space, in the singular: Space *is not*, whereas God *may be*. And we draw this conclusion from the consideration of both the sciences and the arts.

For, the arts do a similar thing, though one does not speak there so often about spaces (yet some theoreticians of art do).

Now, a categorization of time can actually be recognized, first within the sciences.

Physics, to begin with, concerns itself with two main categories[19]: one working within "physics proper", the other within thermodynamics. The former is periodicity, which is a consequence of the reversibility of the dynamics involved: mechanics is time-reversible, so is electrodynamics, even in its quantum form; this is mathematically proved under certain assumptions, like assuming the Hamiltonian itself to be time-reversal invariant or certain potentials to reverse simultaneously with time. It is then possible to define clocks working according to the interactions involved. The generic name for a system is then that of a *mechanism* (P. Hertz). So physics proper deals with mechanisms, and all matter in the world appears to be made of mechanisms of a few different types according to the interaction reigning.

Thermodynamics does not deal qua thermodynamics with mechanisms, but with machines, i.e. in the hierarchy of natural systems with such as cannot escape the second principle of thermodynamics. A mechanism can qua mechanism in principle escape the second principle (e.g. planetary motion free from tides in the planets and sun); as such, it is not a machine. A machine must necessarily consist of one or more mechanisms. Statistical mechanics does not reduce thermodynamics to mechanics, since it assumes a non-mechanical, viz. a statistical principle (Pauli: *Schmier*-principle). So there is a second category involved which is usually called irreversibility. But it should be mentioned that this irreversibility is not contradictory to the reversibility of the constituent mechanisms, so it might perhaps be better to give it another name, like *dissipation*.

A categorial analysis like that can be refined still more, but we shall not attempt to do it here.

Not being a biologist, I shall refrain from developing in detail further categories from the field of biology. But surely we can say that grossly speaking one or a few such categories of time will characterize the sciences of life, as well as further sciences. These things have been developed in the literature[20]. It offers a classification of the sciences, different from the so-called positivistic one which can no longer be maintained[21].

Now, when one thinks of the arts, he will notice that they also allow for a categorization rooted in the aspects of temporality. What I have studied more carefully are the categories of rythm, of continuity, of motion and of others which are involved more specifically in music and dance and their interrelation. Such categories are rather similar to those with which physics is concerned. Physics and music especially offer the best analogies, as has been known for some time. These things, however, have not been worked out in detail by philosophers until now, because their study requires a simultaneous practical knowledge of both disciplines involved at the professional level. Such expertise remains rare, so what is said here is still rather programmatic.

Also the notion of time needs be analyzed not as has been the case till now: i.e. from the point of view of formal logic (Prior, Rescher, ...) from the traditional point of view of relativistic physics (innumerable authors), or even from the many usual points of view taken by either classical philosophers (Augustine, Pascal ...) or contemporary ones at large (Castelli, Heidegger, Jankélévitch, Pucelle ...). Attempts made so far are mostly not yet published[22].

Generally speaking, both the sciences and the arts can be said to aim at an apprehension of reality in its finitude. So we must conclude at this point that that very finitude is the temporal. This of course has been recognized by the philosophers of old, but since then one has usually given to this finding a religious content only, which it need not assume.

6. The analogy which we are examining on so many sides can be seen also between the couples of physics and metaphysics on the one hand, and music and poetry on the other.

It is commonplace to say that much poetry has been "put into" music in songs, chorals, ...operas. Many of the operas composed rest on very bad texts which cannot be said to be good poetry, therefore they do not seem to serve the purpose of our argument very well, but there are at least the ones by Richard Wagner who wrote his own text, those resulting from the collaboration between Hugo von Hofmannsthal and Richard Strauss, Debussy's *Pelléas et Mélisande* based on Maeterlinck's text, and a few others, ... and, after all, the great model from which Mozart wrote *Don Juan* goes back to Molière. So, for other operas composed on poor texts, it is the "plot" which supports the comparison, and the plot is a pattern of life mostly concerned with love, freedom or with morals, exacting an analysis rather of the relation between art and morals, and not between art and science. This analysis cannot be elaborated here, though it demonstrates very interesting analogies, just as is the case between science and morals.

The fact that poetry can be put in music at all is very significant, for it proves that the quality of the poetry can be tested by the means of music, in a process that could be described as its musicalizibility. We recall that poetry is the subjective form taken by metaphysics on the side of the arts.

A very similar process happens on the side of science. One could test good metaphysics by its physicalizibility: something like putting metaphysics into physics. Actually, such a process has always taken place. H. Margenau has maintained that all physics implies metaphysical presuppositions; before him, one had investigated such presuppositions in more or less casual chapters of physics (Poincaré,...). It seems to me, however, that it rather goes the other way round. I mean that (good) physics is rather apt to suggest (good) metaphysics, whereas vice versa - because of the inverse relation - (good) poetry is apt to inspire (good) music: this reciprocal comparison is in complete agreement with the whole analogy presented in this paper.

Music and physics are exceptionally *pure* in comparison with other positive activities. They are also extremely accessible to the powerful grip of mathematics in its double sense of mathematics proper and aesthetics. This is perhaps the explanation of their purity. This makes physics and music highly transparent with regard to the possibility of disclosing fundamentals about reality - fundamental meaning: approaching universality.[23] Of course, there is no objection to saying that the same ability could also be in a certain sense attributable to other sciences or arts. However, painting, for instance, is by nature much more bound to particular objects than is music, or even dance incidentally. In dance, the human body is to be understood as an instrument, as if it were better, more powerful than, say, a violin, and so more able to express a certain kind of "truth" (called beauty), and not as an object, even if at the same time dance shows what the body is, and is therefore a typically cognitive enterprise like all the arts.

For it should be clear by now that when I have talked about science, I always meant science in the making and not the achieved knowledge as written in the books which can be found on the shelves of a library. And when I have talked about art, I have meant exclusively art in the making, not the ready-made pieces exhibited in the galleries or the written poems already known from old, etc. In other words, I meant science as an act and art as an act, not as facts, science as a discovering activity and art as a creating one. The professors who do nothing but teach or the critics who do nothing but comment, did not retain my attention, even less did the students who learn or the amateurs who listen, watch or taste. For neither science nor art is made by the latter, who do not themselves disclose the values of truth and of beauty, even if they constitute (both teachers and disciples) the social condition for the development of each activity.

Keeping thus in mind the consideration of the two activities themselves, it will now lead us over to my final conclusion.

7. Truth is to science what beauty is to art. We could state this as well by saying that truth and beauty are the specific forms exhibited by the value when looked for along the objective or the subjective enterprise. So they are rooted in one and the same determination of reality. It is therefore not astonishing that we have recognized all the analogies shown above, and we may expect them to stand for quite comparable weights among the things which are dear to mankind.

At this point I should like to recall how extraordinarily close to each other are the values of truth and of beauty. Values are of empirical nature; for in the last instance they are paid for in real money when amateurs buy pictures or customers acquire facilities. However, the values are not the things themselves, though they are embodied in them. But those who search for them are neither amateurs nor customers, they are seekers and creators interested only in disclosing them for their own sake.

Indeed, the disclosure of beauty or truth is bound within the human self to an extraordinary feeling of joy. This joy is exactly the same whether related to truth or to beauty. Everyone who has experienced this joy will confirm that it is one of the great moments of life. Neither the English nor, for that matter, the French language has a separate word for that kind of experience;

in German it is called *Erlebnis*[24]), for it does give *life* a supplementary significance which it has not from the mere biotic point of view. It is the joy experienced in knowledge. It should not be confused with the personal component of either the scientific or the artistic enterprise as we have described it at the beginning of this paper. For it is not a component of an enterprise, it is the reward gained from the enterprise - a reward which repeats itself after each new step: the certainty or categorical evidence, of the valuable achieved. This certainty suggests that something like a proof has been given by either the scientific or the artistic progress. Yet, the proof of what?

To answer this question is very difficult, for it can hardly be done without leaving the field of scientific argument or the plane of artistic creation themselves. It requires an insight into the nature of man which goes beyond the two enterprises involved. Indeed, how could we pass a judgment on the joy produced by scientific activity by means of science itself, or how could we render the joy of artistic creation by artistic creation itself without running into a circle? This is the kind of questions with which another human activity concerns itself: philosophy. It should therefore be understood that philosophy is not reducible to the specific enterprises themselves, not even to metaphysics defined as ontology of the infinite being, or to poetry as the subjective grasp of that same being.

To maintain that these things are analysable by scientific methods like those of psychology is to misunderstand completely the situation; it is narrow-minded.

The "what" or *quod* which is "proven" by the appearance of that joy calls for a name escaping the vocabulary of science or art, for it is a quod in and of man making him capable of being scientist and artist, i.e. it is beyond science and art themselves. The best word I can find for it is to call it man's soul, psyche if you like, but no psychology will explain it since every science always destroys the objects of its consideration by cutting it into pieces (materially or intellectually) and is necessarily led to the impossibility of putting the broken "Humpty Dumpty" together again. Other names have been used too: mind, spirit etc. They may be used for more limited purposes, like mind which does not evoke anything like the receptacle for the joy of the disclosure of beauty and is therefore better suited for the experience of truth, or like spirit which is nearer that which is meant in relation to the experience of the divine values. Also the heart has been used, meaning not the muscle which pumps the blood, but precisely the heart which *'a ses raisons que la raison ne connaît pas'*.

It would not be appropriate to follow this up ad infinitum, since the purpose of this paper was to offer an analysis of the similitude in form, intent and content of science and art. But if a last thing may be said here, it would be that apart from the mere analysis I have outlined, this paper may finally serve as an example for, and hence as an apology of, philosophy: for what I have done is neither a piece of science nor of art. I have been philosophizing the whole time. So, if the reader has followed me up to the end, then he has not been narrow-minded but philosophically-minded.[25)]

NOTES

1) See for instance P. Bridgman, *Reflections of a Physicist* (2nd, enlarged edition, Philosophical Library, New York 1955).

2) It is also the source of one further mode within judicial knowledge, viz. morals, as well as the origin of the alternative modality, contemplative knowledge. These very foundations of the theory of knowledge have been expounded in detail elsewhere [see A. Mercier, *Mystik und Vernunft*, in *Mystik und Wissenschaftlichkeit* (Verlag Herbert Lang, Bern und Frankfurt/M. 1972, p. 9)].

3) Since objectivity is the very nature of science, calling science objective is expressing a pleonasm, like subjectivity being called artistic. (I have explained elsewhere that subjectivity, if well understood, does apply to nothing else than to the artistic enterprise.)

4) Even Kant as early as the end of the 18th Century believed that Newtonian mechanics was necessary, and he meant to have proved it apodictically.

5) However, there remain differences of epistemological nature between the two relativities. See e.g. A. Mercier, *L'inportanza filosofica dell'opera di Niels Bohr* (Filosofia, XV, p. 185, 1964) and *Two Giants - Albert Einstein and Niels Bohr* (Technion, 1, No.2, p.4, 1965).

6) The judicial modality is to be distinguished from the contemplative one as I have explained many times. (See e.g. *Erkenntnis und Wirklichkeit, loc. cit.* See also note 2)).

7) In the etymological sense of the word, of "making real", i.e. a thing.

8) We shall treat the case of Poetry further down.

9) Cf. A. Mercier, *Sur les opérations de la composition musicale* (Archives de Psychologie, XXVII, p. 186, 1939).

10) Not in the kind of aesthetics as developed by the men of literature, for what they do is not the same, it is something between literary critique and the philosophy of literature.

11) One does not learn mathematics, one acquires them, precisely because mathematics is not anything to be known, but something (a power) to be possessed. Sciences and arts are learnt.

12) This is explained in detail in a paper by H.J.Treder and myself to appear in the Anniversary Volume in honour of P.L. Bhatnagar.

13) Perhaps the most beautiful sculpture that has ever been carved!

14) See my paper *De l'évidence* in *Problèmes de la Théorie de la Connaissance, Entretiens 1970 de l'Institut International de Philosophie à Helsinki* (edited by G.H. von Wright, Martinus Nijhoff, Den Haag 1972, p. 32).

15) E.g. Frederick Fitch's "proof of God" using mathematical logic.

16) There are such as maintain that there is no such thing as infinite reality, that it makes no sense to speak about it (e.g. Hume); there are such as maintain that the only entities making good sense are the ones entailed with infinity (e.g. the Buddha). Never was any of these two groups able to prove its pretention or the falsity of the pretention of the other. E.g. the proof of God or the proof of non-God have never succeeded (without gratuitous inferences). See also my paper on *Contemplation versus Reason* in Religious Studies, Vol. 9, p. 49, 1973.

17) We leave it to the reader to develop that argument.

18) The reader may consult in this respect G.G. Granger's *Essai d'une philosophie du style* (Arm. Colin, Paris 1968) or J.J. Daetwyler's *Sciences et arts* (Ed. de la Baconnière, Neuchâtel 1972) where the subject matter is developed further.

19) We are using the word category in its generic sense as has been done by philosophers (Aristotle, ... Kant ...), but not the explicit lists of categories proposed by former authors (Aristotle's, ... Kant's categories...).

20) E.g., in the Contribution No. 17 to *Perspectives in Quantum Theory, Essays in Honor of Alfred Landé* (ed. by W. Yourgrau and A.v.d. Merwe, MIT-Press 1971, p. 244).

21) Cf. A. Mercier, *Die Zeit und die Relativität der Kategorien im Lichte der modernen Physik* in *Il Tempo* (Archivio di Filosofia, Roma 1958, p. 191).

22) See however *Petits prolégomènes à une étude sur le temps* (Theoria, XXIX, p. 277, 1963). One of the best books on Time in the recent literature is J.R. Lucas' *A Treatise on Time and Space* (Methuen & Co, London 1973).

23) During a discussion which took place at a Conference in Denver 1966, Prof. Quine, Prof. Polya and I agreed that one might in a popular though meaningful way characterize mathematics (including logic) by saying that it is searching for all the *universals*, whereas history would be looking for all the *particulars*; physics, then, would be the attempt to establish universals from known particulars, whereas "art" (yet rather in the sense of the Greek τεχνὴ) would attempt at producing particulars from recognized universals.

24) Actually, the true significance of experience (from the Latin: *Ex-* and *periri* = to go through) is that of the *Erlebnis*, though it has acquired the sense of knowledge or of expertness resulting from it.

25) Eventually, that is what we have learnt to do from Plato. Incidentally, we should remember that Plato (*Republic* 398 ss.) has also considered music, the arts of painting and sculpture, and poetry too, but not so much in their relation to science as in their connection with education and morals. However, he says, there is 'a music of the soul which answers to the harmony of the world'. (cf. Jowett's Analysis in *The Dialogues* of Plato, translated into English by B. Jowett, third ed. Oxford MDCCCXCII, Vol. III. The quotation is from Jowett's own *Introduction* p. xliv, itself referring to *Republic* 402).

LEONARD DE VINCI ET L'HYDRODYNAMIQUE

P. SPEZIALI

Département de Physique Théorique
Université de Genève
1211 Genève 4, Suisse

"Seele des Menschen,
Wie gleichst du dem Wasser!"
(Goethe)

L'un des traits caractéristiques de la Renaissance, celui qui nous frappe peut-être le plus fortement, pour la bonne raison qu'on ne le retrouve à aucune autre époque avec la même intensité, est l'ouverture d'esprit des peintres, des sculpteurs et des architectes à l'égard des disciplines scientifiques. Cet intérêt pour des domaines à première vue étrangers à leurs propres préoccupations, était particulièrement vif dans les ateliers florentins, où nul n'ignorait ce que Leon Battista Alberti avait écrit dans son Traité de la peinture, en 1435 : "Je souhaite que le peintre soit savant autant que possible dans tous les arts libéraux, mais je désire surtout qu'il soit versé dans la géométrie."

Léonard de Vinci, qui avait vingt ans à la mort d'Alberti, survenue à Florence en 1472, ira très loin dans la mise en pratique de ces recommandations. De tous ses contemporains, le génial fils naturel du notaire ser Piero da Vinci est certainement celui qui, la plupart du temps en autodidacte, s'est adonné avec le plus de succès à l'observation de la nature et à l'expérimentation, celui qui a inventé toute sorte de machines et en si grand nombre, qu'une maison de brevets d'inventions d'aujourd'hui en serait saturée de besogne pendant des années. Quel autre génie de cette époque a mis autant d'obstination à interroger la nature pour en pénétrer les secrets, qui s'est imposé autant de rigueur dans la recherche d'une méthode ?

Enz/Mehra (eds.), Physical Reality and Mathematical Description. 22–36.

Voilà des questions qui surgissent à la lecture de ses écrits et à l'examen de ses dessins. Et puis, comment échapper, au fur et à mesure que nous progressons dans cette lecture et dans cet examen et que nous avons l'illusion d'avoir trouvé une clef, comment échapper à l'envoûtement qui s'empare de nous et qui nous porte à voir en Léonard le précurseur de tant d'hommes célèbres et à le juger en avance d'au moins un siècle sur le sien ? Il faut être conscient des risques que l'on court dans cette démarche et savoir éviter les pièges d'un enthousiasme débordant, pourtant moins grave que l'excès contraire consistant à dénier à Léonard toute qualité de savant, une attitude qui serait injuste et pauvre en arguments.

Certes, quand nous lisons dans le 5e Cahier d'anatomie, au f° 125 r*) que "le Soleil est immobile", nous pensons à Copernic et à Galilée, mais nous aurions souhaité y trouver quelques lignes explicatives en guise de commentaire. C'est pourquoi nous lui préférons cet autre passage : "L'expérience n'est jamais en défaut. Seul l'est notre jugement, qui attend d'elle des choses étrangères à son pouvoir." (C.A., 154 r).

Dans la Vie de Léonard que Giorgio Vasari publia en 1550, on trouve plus d'une allusion à la technique et aux machines. En voici une. "Il inventait sans cesse des projets et des modèles pour aplanir facilement les montagnes, pour y percer un passage d'une plaine à l'autre, pour soulever, avec leviers, cabestans et engrenages, des poids énormes, des procédés pour curer les ports, des pompes pour faire monter l'eau." Mais le premier biographe de Léonard ne nous en dit pas plus que ce que devaient connaître et admirer les gens de Florence, de Milan et d'autres villes, sans en soupçonner la portée et encore moins les possibilités de ce que nous appelons une révolution industrielle. Quant aux carnets, remplis de notes personnelles griffonnées d'une écriture en miroir, personne n'eût été capable de les déchiffrer. Tant d'efforts, de découvertes et d'inventions géniales n'ont eu, en définitive, qu'une influence minime sur le progrès scientifique, et il faudra attendre longtemps avant que d'autres ne retrouvent les mêmes résultats. Lorsqu'en 1797, G.B. Venturi put examiner ces manuscrits à Paris (où ils avaient été envoyés comme butin de guerre par Bonaparte l'année précédente; le Codex Atlanticus sera restitué à la Bibliothèque Ambrosienne de Milan après la chute de Napoléon) et qu'il fit savoir que l'on retrouvait dans les papiers de Léonard bon nombre de lois de la mécanique moderne, la surprise a été très grande.

---

*) Voir Sources originales et notice bibliographique à la fin de l'article.

Depuis lors, on n'a cessé de s'intéresser à Léonard artiste-savant, en étudiant son oeuvre dans ce qu'il a de plus vaste et de plus particulier, et l'on s'est vite rendu compte qu'un seul de ses multiples aspects offrait assez de matière pour un travail d'approche, mais que, quel que soit le chemin suivi et la méthode adoptée et le degré de préparation du chercheur lui-même, il resterait toujours des zones d'ombre et de pénombre, ou, si l'on préfère, de clair-obscur.

* * *

Chez Léonard, tout est forme et mouvement, souvent aussi image, allégorie et symbole. Pour lui, dont la vie avait commencé près de l'Arno et s'était terminée sur les bords de la Loire, le fleuve est moins l'image d'une fuite inexorable des choses et du temps que celle de l'élément primordial, qui a été le berceau de la vie et qui continue à entretenir et à dispenser la vie, en fertilisant la terre. Léonard voit une analogie entre la circulation du sang et celle de l'eau, entre le coeur, les veines et les capillaires, et l'océan, les fleuves et les petits ruisseaux. Pour lui, l'eau est "le sang de la terre" et il existe un lien étroit entre l'homme et l'eau, un attrait irrésistible vers l'eau, un appel émouvant de celle-ci, comme le cri d'une mère pour son enfant égaré. C'est ainsi que d'aucuns ont voulu interpréter la signification de <u>Léda au cygne</u> : Léda est de la nature de l'eau et par l'intermédiaire de la dualité être humain - animal nous parvenons à la connaissance des secrets de l'univers. Mais laissons là ces jeux ambigus et lisons cette belle page que Léonard destinait à son <u>Traité de l'Eau</u> :

"Elle use les hautes cimes des monts. Elle déchausse et déplace les gros rochers. Elle chasse la mer de ses anciens rivages, en élevant le fond par des alluvions. Elle désarticule et détruit les hautes berges; on ne voit jamais rien de stable qu'elle n'altère aussitôt. Les fleuves cherchent la pente des vallées, où enlever et déposer à nouveau la terre. Aussi peut-on dire de bien des fleuves que l'élément tout entier les a traversés, et qu'ils ont rendu plusieurs fois la mer à la mer; aucune partie de la terre n'est si élevée que la mer n'ait atteint sa base, aucun fond marin n'est si bas que les hautes montagnes n'y aient leur fondation. Elle est tantôt âcre et forte, tantôt acide et amère, tantôt douce, tantôt épaisse ou légère, tantôt dangereuse et malsaine, tantôt salubre, tantôt empoisonnée.

On dirait qu'elle change de nature avec les lieux divers où elle passe. Comme le miroir adopte la couleur de son objet, elle adopte la nature du lieu où elle passe : salubre, dangereuse, laxative, astringente, sulfureuse, salée, sanguine, mélancolique, frénétique, colérique, rouge, jaune, verte, noire, bleue, onctueuse, grasse, maigre. Parfois elle entraîne le feu, parfois l'éteint; tantôt chaude, tantôt froide; elle emporte ou dépose, elle creuse ou élève, elle ruine ou affermit, elle remplit ou vide, elle monte ou s'enfonce, elle court ou se repose, elle est source de vie ou de mort, de fécondité ou de privation; tantôt elle nourrit et tantôt elle fait le contraire, tantôt elle a une saveur de sel, tantôt elle est insipide, tantôt ses déluges submergent les vastes vallées. Tout change avec le temps." (Arundel, 57 r).

Dans le MS A de la Bibliothèque de l'Institut de France, on trouve les premières études de Léonard en vue d'un Traité de l'Eau. Ces pages ont été écrites en 1492; douze ou treize ans plus tard, Léonard dressa une table des matières de l'ouvrage. Entretemps, une foule d'observations, de calculs, de dessins et de réalisations pratiques dans le domaine de l'hydraulique n'ont cessé de s'accumuler pour former les matériaux de ce traité, qui hélas ! ne verra pas le jour. Aujourd'hui, tout ce qui n'a pas été perdu de cet immense labeur, se trouve dispersé dans les précieux recueils conservés à Milan, Paris, Londres et Windsor. Il convient de noter qu'en 1643 le moine Luigi Maria Arconati avait pu examiner les manuscrits de Léonard et qu'il avait composé un livre intitulé Del moto e misura dell'acqua, dont le texte original appartient à la Collection Barberini, au Vatican. Le livre d'Arconati laisse quelque peu à désirer, car il est incomplet et contient des erreurs. Il sera à nouveau question de lui dans la notice bibliographique, à la fin de cet article.

Dans le second des deux volumes qui ont été découverts à Madrid en 1967, figure une carte géographique, dessinée par Léonard, de la vallée de l'Arno, de Florence à Pise, avec une indication concernant le canal projeté par lui et qui devait relier Florence à Prato et à Pistoia. Ajoutons qu'à la fin du recueil il y a une liste autographe de 116 ouvrages en sa possession : on y trouve les noms d'Albert de Saxe, d'Albert le Grand, de Scot, d'Aristote, d'Alcabitius, de Luca Pacioli pour son Arithmétique, de Ptolémée pour sa Cosmographie et de tant d'autres. Mais voyons, sans tarder, la table des matières du Traité de l'Eau :

Livre 1. De l'eau en soi.
2. De la mer.
3. Des sources.
4. Des rivières.
5. De la nature des grands fonds.
6. Des obstacles.

Livre 7. Des graviers.
8. De la surface de l'eau.
9. Des choses qu'elle remue.
10. De l'entretien des bords.
11. Des conduites.
12. Des canaux.
13. Des machines hydrauliques.
14. Des pompes à monter l'eau.
15. Des choses qu'elle détruit.

(Leicester, 15 v)

Voilà tout un programme qu'à défaut d'avoir développé pour sa rédaction définitive, Léonard a néanmoins, dans sa seconde partie, traduit dans la pratique. Sur le plan strictement théorique de l'hydrodynamique, il a même été plus loin que ne le laisse entendre cette table, ainsi que nous le verrons.

Si Léonard s'attarde longuement sur les différents cours d'eau, sur la genèse des remous et des tourbillons, sur l'observation de la dynamique de l'élément liquide, c'est dans le but d'asservir aux besoins de l'homme ce qu'il appelle le "voiturier de la nature".

"Si tu coordonnes tes notes sur la science des mouvements de l'eau, souviens-toi d'inscrire au-dessous de chaque proposition ses applications, pour que cette science ne demeure pas sans emploi." (MS F, 2v).

Ses projets de canaux d'irrigation sont fort nombreux et ils ont été, en grande partie, réalisés sous sa direction, ou, lui absent, d'après ses instructions. Ainsi, pour la région de Florence :

"Fais construire des écluses dans le Val de Chiana à Arezzo pour que, l'été, quand il y a pénurie d'eau dans l'Arno, le canal ne soit pas à sec; que ce canal ait 20 brasses de large au fond, 30 à la surface, et son niveau général sera de deux brasses ou 4, dont deux pour les moulins et les prés. La contrée en sera fertilisée, et Prato, Pistoia et Pise, ainsi que Florence, en tireront un revenu annuel de plus de 200.000 ducats et fourniront du travail et de l'argent pour cette utile entreprise." (C. A., 46 r)

Plusieurs notes nous apprennent comment il évalue le prix de revient de l'ouvrage, le salaire des ouvriers et la durée des travaux. D'autres nous montrent comment il se préoccupe d'assécher les marais proches de la mer; à cet effet, il invente une machine armée de deux grues. Puis, il dessine des écluses pour des voies d'eau navigables, il construit un type particulier de turbine et les premiers compteurs d'eau. De tels travaux d'ingénieur nous portent tout naturellement à établir un parallèle entre Léonard et Archimède : il est certain que le premier a rivalisé avec le second dans le domaine de la technique.

De 1517 à 1519, à la fin de sa vie, passée au château de Cloux à Amboise, Léonard a été ingénieur, architecte et ordonnateur des fêtes de François Ier. Après avoir dressé une carte hydrographique où figurent Montrichard, Romorantin, Tours, Amboise et Blois, il note : "Tu étudieras le niveau du canal [il s'agit d'un canal de navigation] qui doit conduire de la Loire à Romorantin au moyen d'un chenal large d'une brasse et profond d'une brasse", et il s'occupe également du tracé des conduites d'eau dans les jardins de Blois.

Pour terminer ce rapide coup d'oeil sur les travaux de navigation et d'irrigation de Léonard - dont le récit détaillé devrait être accompagné par la présentation des centaines d'admirables dessins qui le concernent -, je voudrais revenir à un travail antérieur, que Léonard avait entrepris dans sa patrie. Il est peut-être moins connu que les autres et les circonstances historiques qui l'entourent présentent un certain intérêt.

Il s'agit de la machine hydraulique que Léonard construisit, plusieurs années avant de partir pour la France, pour le compte du florentin Bernardo Rucellai, qui avait été l'ambassadeur de Florence auprès de Ludovic Sforza, à Milan, de 1481, une année donc avant l'arrivée de Léonard dans cette ville, jusqu'en 1486.

Les dessins de cette machine figurent dans le MS G de l'Institut de France et appartiennent aux années 1510-1511, époque à laquelle Léonard se trouvait en Lombardie, occupé par les études de navigation sur l'Adda, au service du roi de France Louis XII. Or, ces dessins n'ont pu être identifiés qu'en 1951, par Carlo Pedretti - un des meilleurs connaisseurs de Léonard et actuellement attaché à la Elmer Belt Library of Vinciana de Los Angeles - grâce à la découverte qu'il venait de faire à la Bibliothèque nationale de Saint-Marc à Venise d'un codex inédit de Benvenuto Lorenzo della Golpaja, où il était précisément question de la machine. Notons que l'existence de ce codex avait été signalée par Morelli en 1776, dans sa description des inventions léonardiennes.

En 1482, le dénommé Rucellai, homme de qualité, riche et cultivé, avait fait l'acquisition à Florence d'un vaste domaine et d'un palais, dont les plans avaient été dessinés par Leon Battista Alberti. Dans ce palais et dans ses jardins se réuniront les membres de la fameuse académie platonicienne, qui avait été fondée par Marsile Ficin, sous les auspices de Cosme des Médicis l'Ancien. Machiavel en sera l'un des membres actifs et y fera la lecture de ses Discours sur la première décade de Tite-Live.

Léonard rendit visite à Rucellai vers 1508, et ce fut à cette occasion que celui-ci lui demanda de réaliser l'irrigation de son domaine. Léonard accepta et, sans tarder, se mit au travail en donnant libre cours à son esprit inventif. L'ouvrage comporte un jeu fort complexe de roues dentées, de pompes, de balançoires, de déversoirs et de canalisations, avec une minutie de détails techniques et de mesures qui feraient l'admiration de nos ingénieurs : il y aurait là assez de matière pour écrire un livre sur le sujet. Léonard, qui avait même rédigé un mode d'emploi et donné des conseils pour le bon fonctionnement de l'installation, a terminé celle-ci probablement deux années plus tard. Il est permis de supposer que l'ouvrage a dû faire l'admiration des habitués du palais Rucellai et, en particulier, de Niccolò Machiavel.

* * *

La notion de loi physique, telle que nous l'entendons aujourd'hui, s'est élaborée très lentement et l'on s'est longtemps contenté de l'observation des phénomènes naturels. Il y a une loi, là où le semblable domine, dit-on souvent, mais encore faudrait-il individualiser tous les facteurs qui interviennent et préciser comment ces facteurs se comportent mutuellement. Quant à la formulation mathématique d'une loi physique, il a fallu attendre Galilée pour la voir déduite de l'expérience, de façon logique et satisfaisante.

C'est donc en ayant présent à l'esprit l'état de la science à la fin du XVe siècle que l'on pourra mieux apprécier la contribution de Léonard à l'étude théorique du mouvement des fluides. Dans ce qu'il a écrit à ce sujet, tout n'est pas correct, ni toujours clairement dit. Parfois on voit mal le lien entre certaines propositions; parfois on a de la peine à saisir le fondement de certaines hypothèses. Cependant, plus d'une fois, l'effort inlassable de l'auteur pour aboutir à un principe, à une loi porte ses fruits et nous surprend par sa vigueur. D'autre part, on ne peut certes pas faire grief à Léonard d'avoir négligé l'outil mathématique, malgré tout le respect qu'il lui portait, et encore moins de n'avoir su donner à l'hydrodynamique une structure et une systématisation, comme le feront seulement deux siècles et demi plus tard Daniel Bernoulli et d'Alembert.

Voici, exposés à l'aide de ses notes, quelques résultats remarquables relatifs au domaine qui nous intéresse et qui devraient permettre de porter un jugement serein et impartial sur la contribution léonardienne à la science, ou, plus exactement, sur ce qu'aurait été cette contribution si l'on avait pu lire plus tôt ces manuscrits.

Quelques mots, d'abord, à propos de la statique. Léonard note que "les surfaces des liquides au repos, qui communiquent entre eux, sont situées à la même hauteur" (C.A., 219 v.a). Ailleurs, il dit que cela est indépendant de la forme et de la dimension des récipients. Notons que le principe des vases communicants est attribué d'habitude à Simon Stevin, lequel est né en 1548. Un siècle et demi avant Pascal, Léonard trouve aussi le principe de l'égalité des pressions sur des éléments de surface égaux. Toujours à propos des pressions des liquides, on trouvera dans le Codex Atlanticus, 395 r.a, un grand dessin colorié représentant un bassin avec trois débits d'eau et, à côté, un autre dessin plus petit - ici reproduit -, avec deux débits. Léonard a certainement eu l'intuition du principe fondamental de l'hydrostatique, qui dit que la pression ne dépend que de la hauteur de la colonne d'eau, mais il n'est pas parvenu à l'énoncer.

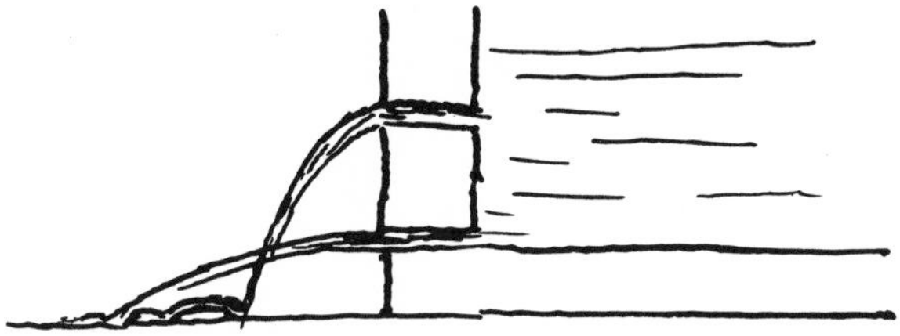

Léonard voit une analogie entre les mouvements de l'air et ceux de l'eau (A, 61 r) et, à propos de l'air, il dit qu'"un corps oppose autant de résistance à l'air, que l'air n'en oppose au corps" (C.A., 381 v.a). Nous voilà donc tout près du principe de l'action et de la réaction.

Les passages les plus nombreux et aussi les plus intéressants relatifs à l'hydrodynamique se trouvent dans le MS A et dans le Codex Atlanticus. Deux questions ont d'abord retenu plus particulièrement notre attention : le principe de continuité et l'analogie entre les ondes et les balles élastiques.

1° Il est permis d'accorder à Léonard la priorité dans la découverte du principe appelé aujourd'hui de continuité, qui peut s'énoncer ainsi, dans notre cas : à travers une section donnée d'une rivière il passe, pendant des temps égaux, la même quantité d'eau, quelle que soit la profondeur et la largeur. Ou encore : une rivière de profondeur constante s'écoulera plus rapidement là où elle est plus étroite et les vitesses seront dans le rapport inverse des largeurs. Léonard donne de ce principe plusieurs énoncés équivalents dans MS A 23 v, 57 v, C.A. 80 v.b, 80 r.b, 287 r et Leic. 6 v, 24 r. Celui de A 57 v est :

"Ogni movimento d'acqua d'eguale largheza e superfizie correrà tanto più forte 'n uno loco che nell'altro, quanto fia men profondo nell'uno loco che nell'altro." Le dessin qui l'accompagne est assez explicite, avec les quatre niveaux allant du simple au quadruple. A travers la section mn l'eau s'écoulera quatre fois plus vite qu'à travers ab.

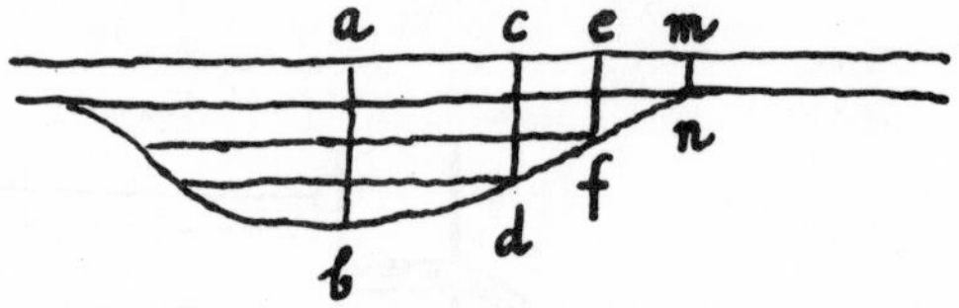

Il reste à savoir si Léonard est parvenu à ce principe par l'observation ou par le raisonnement. Notons qu'il sera retrouvé en 1625 par Benedetto Castelli, élève de Galilée.

2° Léonard voit une similitude entre les ondes qui viennent frapper les côtes et les balles qui rebondissent sur un mur. Ecoutons-le : "Si vede chiaramente e si conosce che le acque che percuotano l'argine de' fiumi, fanno a similitudine delle balle percosse ne' muri, le quali si partano da quelli per angoli simili a quelli della percussione e vanno a battere le contraposte pariete de' muri." (On voit clairement et l'on sait que les eaux qui frappent les bords des rivières se comportent comme les balles lancées contre les murs, qui en repartent en faisant des angles semblables à ceux de l'impact et s'en vont frapper la paroi opposée). (A, 63 v.)

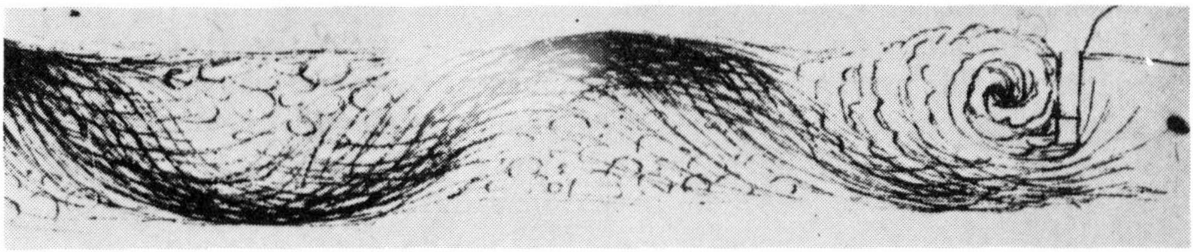

De plus, il voit dans les rebondissements d'une balle qui vont s'amortissant, une ressemblance avec certaines ondes qui en font autant. Il est tellement frappé par ce mouvement ondulatoire amorti, qu'il va jusqu'à émettre une hypothèse fort inattendue et nullement justifiée, d'ordre métrique, à propos des balles élastiques. Elle se trouve dans MS A, 24 r et dit que si l'on lance obliquement une balle vers le sol, le chemin total parcouru en

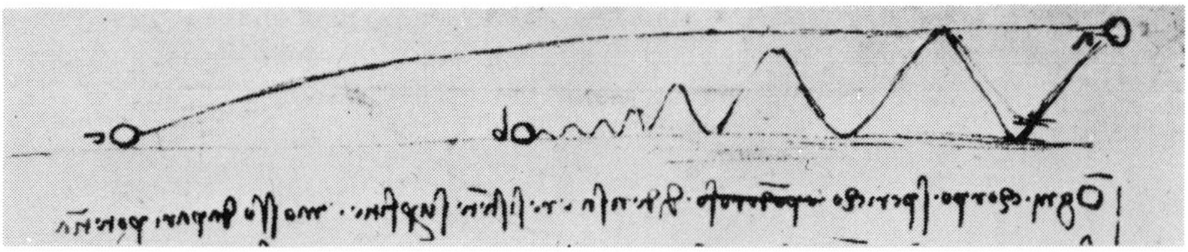

l'air jusqu'à l'arrêt est le même que celui de la balle lancée horizontalement du même point. Léonard suppose même que les temps sont égaux dans les deux cas.

Passons maintenant à d'autres problèmes concernant la dynamique de l'eau. Léonard constate la simultanéité de plusieurs trains d'onde (C.A., 84 v.a). Il constate aussi que l'eau des rivières coule plus lentement près des bords qu'au milieu et plus vite à la surface qu'en profondeur (C.A., 175 r.c).

A plusieurs reprises, il s'intéresse au travail d'érosion accompli par l'eau et il est capable de prévoir, selon la configuration du terrain et le cours de l'eau, quels seront les endroits les plus exposés et affectés par l'érosion (MS C, 23 v).

Le phénomène des remous et des tourbillons retient aussi son attention et il s'efforce d'en donner une explication. Lorsque l'eau rencontre un obstacle, elle ne peut brusquement arrêter sa course et il lui faut dépenser son énergie d'une manière ou d'une autre. "Universalmente, tutte le cose desiderano mantenersi in sua natura, onde il corso dell'acqua che si move, ne cerca mantere il [suo corso], secondo la potenza della sua cagione, e se trova contrastante opposizione, finisce la lunghezza del cominciato corso per movimenti circolari e retorti." (D'une manière générale, toutes choses ont la tendance à conserver leur nature, de sorte que le cours d'eau en mouvement essaie de garder le sien, selon la puissance de sa propre cause, et s'il rencontre une opposition qui le contrarie, il achève la longueur du trajet commencé par des mouvements circulaires et par des tourbillons). (MS A, 60 r).

Le début de ce passage mérite d'être souligné, car c'est une référence presque textuelle à Aristote, que Léonard cite ailleurs, dans C.A. 123, r.a, comme suit : "Dice Aristotile che ogni cosa desidera mantenere la sua natura."

Choisis parmi tant d'autres, ce sont là les textes qui nous ont semblé les plus dignes d'intérêt pour la science et son histoire. Pleins d'idées originales et de vues profondes, ils jettent, en tout cas, de belles lumières sur des problèmes qui ont obsédé Léonard presque autant que ceux du vol des oiseaux et de l'homme.

* * *

Comme l'eau a été le principe de la vie, elle en marquera la fin. Le cataclysme final qui effacera du globe toute trace de matière vivante sera un déchaînement conjugué des forces naturelles, parmi lesquelles l'eau jouera le principal rôle. Le spectacle d'un déluge universel, imaginé par Léonard, a quelque chose de grandiose et d'effrayant. Il ne s'agit pas, précisons-le, du déluge de la Bible, celui peint par Michel-Ange à la Chapelle Sixtine, mais bien du déluge du jugement dernier.

Cette vision de fin du monde figure dans une série de dessins célèbres et fort connus que possède la Bibliothèque de Windsor. La puissance créatrice de Léonard a peut-être atteint là ses sommets; l'impression de mouvement effréné qui s'en dégage est telle que l'on est pris de vertige. Comparées à la Joconde, par exemple, et à son sourire si calme et énigmatique, ces compositions d'artiste visionnaire présentent un contraste des plus frappants.

Mais laissons, pour conclure, à la plume de Léonard le soin de nous évoquer la scène.

"Il faut représenter d'abord le sommet d'une âpre montagne avec des vallées autour de sa base; on voit sur ses flancs l'écorce terrestre s'arracher avec les petites racines des arbrisseaux, en dénudant de grandes surfaces des rochers alentour. (...) Les montagnes, se dénudant, révèlent les failles profondes causées par les anciens tremblements de terre, leur pied est garni et en grande partie recouvert des débris de buissons arrachés aux cimes, mêlés à la boue, aux racines, aux branches, aux feuilles confondues avec la boue, la terre et les pierres. (...) L'eau gonflée tourbillonne dans le bassin qui l'enferme, frappe en remous vertigineux les différents obstacles, bondit en l'air avec une écume boueuse, puis retombe en faisant rejaillir l'eau sous le choc. Les ondes circulaires qui s'échappent du lieu de rencontre, se jettent dans leur élan au travers d'autres ondes circulaires qui viennent en sens inverse; après le choc, elles jaillissent dans l'air sans se séparer de leur base. (...)

Les campagnes submergées montraient des ondes généralement couvertes de tables, de bois de lits, de barques et de tous les expédients inspirés par le besoin et la peur de la mort; ils étaient chargés d'hommes et de femmes avec leurs enfants, au milieu de lamentations et de gémissements, pleins d'épouvante devant l'ouragan qui roulait les eaux en tempête, avec les cadavres noyés. Tout ce qui pouvait flotter était couvert d'animaux variés, réconciliés et groupés en tas apeurés : loups, renards, serpents, créatures de toute sorte, essayant de fuir la mort. Les eaux frappant leurs bords, les flagellaient avec les cadavres noyés, dont les corps achevaient ce qu'il leur restait de vie." (MS Windsor, 12665 r et v).

* * *

Sources originales et notice bibliographique

Les textes originaux concernant le mouvement de l'eau se trouvent dans : le Codex Atlanticus de la Bibliothèque Ambrosienne de Milan; les manuscrits (mss) A à M de l'Institut de France; les mss. Arundel du British Museum; Forster, du Victoria and Albert Museum de Londres; Leicester, à Holkham Hall; les Cahiers d'anatomie et les dessins de la Royal Library de Windsor, et dans quelques mss. du Castello Sforzesco, à Milan, et de la Collection Barberini, au Vatican. Leur contenu est facilement accessible aujourd'hui grâce surtout aux magnifiques éditions de la R. Accademia dei Lincei, de la Reale Commissione Vinciana, de Ravaisson-

Mollien, qui reproduisent ces mss. en fac-similés, avec transcription et commentaires. Le ms. A de l'Institut de France a connu l'année passée (1973) une nouvelle édition en 3 volumes, par les soins de la maison Il Polifilo de Milan.

Une description détaillée des mss., de leur histoire et des éditions en fac-similé a été donnée par Elmer BELT, dans Manuscripts of L. da V., Los Angeles, 1948. Carlo PEDRETTI consacre lui aussi quelques pages à leur histoire et dresse un tableau synoptique fort complet dans son gros volume L. da V. on Painting. A Lost Book (Libro A), University of Calif. Press, 1964, pp. 252-259.

Luigi Maria ARCONATI, fils de Galeazzo (qui avait fait don, en 1637, des mss. en sa possession à la Bibliothèque Ambrosienne), a été le premier chercheur qui a rassemblé les textes concernant le mouvement de l'eau. Son travail, Del Moto e Misura dell'Acqua, est resté inédit jusqu'en 1826, date à laquelle Francesco CARDINALI, un spécialiste bolonais en hydraulique, le publia dans sa Raccolta d'autori italiani che trattano del moto dell'acqua. Le texte d'Arconati s'y trouve aux pp. 270-450 et contient 566 passages sur l'eau.

En 1923, E. CARUSI et A. FAVARO, à Bologne, sont revenus aux sources d'Arconati avec leur L. da V. Del moto e misura dell'acqua. Libri nove ordinati da F. Luigi Maria Arconati, editi sul codice archetipo barberiniano.

Nando da TONI, L'Idraulica in L. da V. (Frammenti Vinciani, III-IX), Brescia, 1934-35. L'A. s'est limité aux mss. de l'Institut de France et y a trouvé 990 passages traitant de l'eau.

A propos de B. della Golpaja et de la machine hydraulique de Rucellai, cf. dans le No hors-série de la revue "Sapere" (Hoepli, Milan, 15 avril 1952), sorti à l'occasion du 5e centenaire de la naissance de L., les deux articles de C. PEDRETTI. Du même auteur, et sur les mêmes questions, cf. Raccolta Vinciana, XVII (1954), 177-215, et ses Studi Vinciani, Droz, Genève, 1957, p. 34-42. Dans son ouvrage déjà cité sur la peinture, il y a des textes importants sur l'hydraulique provenant des mss. Leicester. Cf. aussi Kate TRAUMAN STEINITZ, L. da V. Trattato della Pittura, Copenhague, 1958, pour sa riche bibliographie.

Les trois volumes de Pierre DUHEM, Etudes sur L. de V. Ceux qu'il a lus et ceux qui l'ont lu, Paris, 1906, 1909, 1913, sont fort bien documentés et toujours instructifs. Dans les deux premiers, on peut suivre la filière qui va d'Aristote à L. de V., en passant par Albert de Saxe; le 3e traite des précurseurs parisiens de Galilée.

Roberto MARCOLONGO, L. da V. Artista-scienziato, Milan, 1939, 3e éd. 1950. Cf. le chap. VIII, consacré à la mécanique et aux sciences physico-mathématiques.

L. de V. par lui-même, textes choisis, traduits et présentés par André CHASTEL, Paris, 1952. Historien de l'art, l'A. est un grand connaisseur de L. peintre. Les deux traductions les plus longues, qui figurent dans cet article, nous les avons empruntées à son livre.

L. de V. et l'expérience scientifique au XVIe siècle. Colloques internationaux du C.N.R.S., Paris, 1953. Cf. la communication de René DUGAS, "L. de V. dans l'histoire de la mécanique", aux pp. 89-114.

Vassili P. ZUBOV (1899-1963), historien des sciences soviétique, a traité de l'hydraulique de L. dans son imposante publication de textes scientifiques léonardiens (Moscou, 1955).

Marcel BRION, L. de V., 2 vol., Paris, 1959. Cf. dans le premier volume, le beau chapitre "Louange de l'Eau", pp. 245-285.

Clifford A. TRUESDELL, Essays in the History of Mechanics, Springer, Berlin, 1968, 384 p. Le premier chapitre (pp. 1-83) est consacré à la mécanique de Léonard. Cf., en particulier, les paragraphes 15-17 sur le mouvement des fluides, les ondes, les tourbillons et le principe de continuité.

Leonardo's Legacy. An International Symposium. Edited by C.D. O' MALLEY. Berkeley and Los Angeles, 1969. Cf. l'article de E.H. GOMBRICH, "The Form of Movement in Water and Air", pp. 171-204.

Voici encore quelques articles, en italien, sur l'hydrodynamique.

A. FAVARO, "L. da V. e le scienze delle acque", dans Emporium (1919), 272-279.

F. ARREDI, "Gli studi di L. da V. sul moto delle acque", dans Annali dei Lavori Pubblici (1939), 17/4, 3-9; "Le origini dell' idrostatica", dans L'Acqua (1943), 8-18.

G. BELLINCIONI, "Leonardo e il Trattato del Moto e Misura delle Acque", dans Atti del Convegno di Studi Vinciani, Florence,

1953, pp. 323-327.

A.M. BRIZIO, "Delle Acque", dans L. da V. Saggi e Ricerche, Rome, 1954, pp. 277-289.

Enfin, il va de soi que la question du mouvement de l'eau est abordée, sous une forme plus ou moins développée, dans les ouvrages classiques de J.P. RICHTER, K. CLARK, L.H. HEYDENREICH, T. LUECKE, G. FUMAGALLI, pour ne citer que cinq noms parmi ceux des spécialistes léonardiens les plus connus.

OUR KNOWLEDGE OF THE EXTERNAL WORLD

An essay inspired by J.M.Jauch's "Are Quanta Real ?, A Galilean Dialogue" *

Jagdish Mehra

Instituts Internationaux de Physique et de Chimie (Solvay)
Université Libre de Bruxelles

In every generation, throughout history, man has wondered about the nature of the external world and asked himself : Why things are as they are ? What are the causes of natural phenomena ? What is the reality of it all ? Man has looked inside - into his thoughts, feelings, emotions, motivations and relationships. At times he has sought to harmonize the world of happenings, of phenomena, with the inner world of his mind and heart : the body of nature and the soul of man become one. This is what Kant meant when he said, "The starry heavens above, and the moral law within," - the critical reason become the moral and spiritual imperative.

Man has sought to describe the external world in the psychological and scientific idiom of his time. But the fundamental problem has always been one of understanding existence and change. Why do things exist ? How do they evolve ? Is there an object without subject ? Matter and mind, being and becoming - what is real ?

Looking at the majesty of the heavens and the marvelous spectacle of nature, man made it an article of his faith that it was a play designed for his enjoyment. Created in the image of God, man inhabited the planet which was at the center of the universe. The sun orbited around his earthly home, and he could pay homage to the Deity for this grand design. The innocence of this faith

---

* Indiana University Press, Bloomington, Indiana 1973.

Enz/Mehra (eds.), Physical Reality and Mathematical Description. 37–47.

was buttressed by the astronomy (Pythagoras, Ptolemy, Aristotle) and astrology of centuries.

The Crusades opened the routes to the East, bringing in a stream of luxuries and heresies into Europe that doomed asceticism and dogma. Parchment, that had made learning the monopoly of priests, was replaced by paper which now came cheaply from Egypt. Bold mariners, armed with compasses, ventured out on unchartered oceans and conquered man's ignorance of the earth. Patient observers, armed with telescopes, braved the confines of dogma and conquered man's ignorance of the sky. In universities, monasteries, and hidden retreats, "men ceased to dispute and began to search. Deviously, out of the effort to change baser metal into gold, alchemy was transmuted into chemistry, out of astrology men groped their way with timid boldness to astronomy, and out of the fables of speaking animals came the science of zoology."[1)] The awakening which began with Roger Bacon (d. 1294), grew with the limitless genius of Leonardo da Vinci (1452-1519), reached its fullness in the astronomy of Copernicus (1473-1543) and Galileo (1564-1642), in the researches of Gilbert (1544-1603) in magnetism and electricity, of Vesalius (1514-1564) in anatomy, and of Harvey (1578-1657) on the circulation of the blood. It was an age of adventure, enterprise, and achievement in every field. Francis Bacon had already announced that Europe has come of age.

Copernicus constructed the blueprint of man's new understanding of the universe, and decided that contrary to previous belief the earth and all other planets moved around the central sun. Man could still pay homage to the Deity for the glory of the seas and mountains and his place in the sun, but with primeval fear and pride he construed the Copernican revolution as a conspiracy against the design of God and the reflection of His image in himself.

Then a professor of mathematics in the University of Padua, Galileo Galilei, in 1609 at the age of 45, reconstructed a telescope on the basis of the news reaching him from Holland about its invention and turned it towards the heavens. He noticed the jagged mountains on the surface of the moon, noticed that planets appear as discs while fixed stars have no measurable extension, and discovered the four moons of Jupiter, the structure of the Milky Way, the phases of Venus and Mercury, the ring of Saturn, the sun spots and the rotation of the sun. In 1610, Galileo wrote his book "The Sidereal Messenger" (Sidereus Nuncius) which gave a death blow to the Aristotelian cosmology and confirmed the Copernican construction of the universe. And on 21 February 1632 was published : The Dialogue of Galileo Galilei, Lincean ... wherein, in the meetings of four days, are discussed the "Two Major Systems of the World, Ptolemaic and Copernican".

Galileo was unabashedly a Copernican. In the Dialogue, the three interlocutors, Filippo Salviati, Gianfrancesco Sagredo and Simplicio, discuss the two cosmologies, the Ptolemaic and the Copernican, and by patient argument uphold the merits of the Copernican construction. Copernicanism of the Sidereal Messenger and the Dialogue, which Galileo had confirmed on the basis of his own immortal discoveries, was his "crime" for which he was incarcerated by his fellowmen who thought that they understood the divine purpose and the scheme of the universe better.

Galileo, the man, could be sentenced and imprisoned, but his ideas and the rational soul of the man would march freely on. The Copernican system in astronomy, together with the laws of nature discerned by Johannes Kepler, Galileo and Isaac Newton, became the cornerstones of the development of man's view of the external world for the next 250 years. With the development of kinetic theory and statistical mechanics in the hands of Sadi Carnot, Robert Mayer, Kelvin, John Prescott Joule, Rudolf Clausius, Hermann Helmholtz, James Clerk Maxwell and Ludwig Boltzmann, and the laws of optical and electronagnetic phenomena in the work of Michael Faraday, Maxwell and Heinrich Hertz, the "mechanical" picture of the universe was completed by the end of the 19th century. The final honing and furnishing of this architecture of thought by Josiah Willard Gibbs and Hendrik Antoon Lorentz established the final edifice of "classical" physical science. Newton and Maxwell were the representative oracles of this worldview, the dominant notions in which were "particle", "field" and "deterministic causality".

Immanuel Kant (1724-1804) had already, in his Critique of Pure Reason (1781), established the classical notions of absolute space and time as the abstract a priori of space, time and causality which man must have in order to interpret experience and phenomena in the external world.

As soon as this marvelous architecture was complete, the new tenants found imperfections in it. Almost at the very beginning of the 20th century, Albert Einstein became the author of complaints against the classical theory and leader of the revolution that would rectify its faults. With his theories of special and general relativity and his conception of the light quantum, based upon fluctuations in kinetic theory and radiation, Einstein created a new vision of the macroscopic and microscopic universe.

The Rutherford-Bohr model of the atom, combined with classical mechanics and certain quantum conditions, and guided by Bohr's correspondence principle, continued to explain atomic phenomena for over two decades until the logical difficulties became overwhelming

The old quantum theory of Planck, Einstein, Bohr and Sommerfeld gave way in the mid-1920s to the new and consistent atomic mechanics founded on the work of Louis de Broglie, Erwin Schrödinger, Werner Heisenberg, Pascual Jordan, Max Born and Paul Dirac. Heisenberg's uncertainty principle and Bohr's principle of complementarity, together with Born's statistical interpretation of the wave function (giving rise to statistical as against deterministic causality) became the paramount philisophical basis of the understanding of microscopic phenomena, finding a unified description of nature in the so-called Copenhagen interpretation of quantum mechanics.

The essence of Bohr's principle of complementarity lies in the non-vanishing character of Planck's constant and the existence of Heisenberg's uncertainty principle : mutually exclusive aspects of nature, such as the wave and particle pictures, cannot be observed simultaneously, and the physicist must decide upon the experiment which would bring out one or the other.

The notion of statistical causality, the fundamental notion of chance in the occurrences of nature, led to serious and prolonged debates about the new worldview. Eminent physicists took opposite sides : Niels Bohr became the champion of the new quantum mechanics with its statistical interpretation, in which the role of chance and the intermingling of the subject and the object were the essence of the description of nature, while Albert Einstein adhered to the view that such a game of dice is for the Devil and the Creator has a more exact and rational design, and it is the physicist's business to find a "complete" description of physical reality.

In spite of the quantum revolution which Einstein, more than anyone else, had wrought, he expressed faith in a deterministic description of nature that the "final" theory must have - there had to be a one to one correspondence between the elements of physical reality and the mathematical formalism of the theory.

Einstein was convinced that quantum mechanics is "a brilliant shortcut which successfully avoids many of the difficulties and the hard work which the final correct field theory must solve. Atomistic phenomena must fit into a unified field theory, [because]

quantum mechanical physics avoids reality and reason."[2] Bohr was just as convinced that the notions of probability and complementarity play a decisive role in nature, that truth is statistical and quantum mechanics is the most reasonably complete description of physical reality.

At the fifth Solvay Conference in Brussels in 1927, during Arthur Compton's lecture, Paul Ehrenfest passed a note to Einstein saying, "Don't laugh ! There is a special section in the purgatory for professors of quantum theory, where they will be obliged to listen to lectures on classical physics ten hours every day." To which Einstein replied, "I laugh only at their naïveté. Who knows who would have the laugh in a few years ?"*

The Einstein-Bohr dialogue about the interpretation of quantum mechanics and the nature of "complete" physical description which began in Brussels in 1927 continued until Einstein's death in 1955. On Bohr's side were most of the principal architects of quantum mechanics and the adherents of the Copenhagener Geist der Quantentheorie. Among the major physicists who shared Einstein's concern were Planck, Schrödinger and de Broglie. Whether Schrödinger's cat was dead or alive (in an intricately designed experiment) also became a celebrated issue in determining the legitimacy of the quantum mechanical explanation.

Beginning in the early 1950s, numerous natural philosophers, most prominent among them Eugene Wigner, entered the discussion about the interpretation and epistemology of quantum mechanics and the description of physical reality. The discussion goes on, and an entire field has been growing up concerned with the mathematical and interpretative foundations of quantum mechanics that seeks to understand the problem of measurement, the role of the observer and the observed, and how completely the "reality" of nature is described by the mathematical structure. For the first time in physical theory, human consciousness has become an issue, and Wigner has even called a "friend" in aid to describe the completeness of the process of measurement.[3]

****

One of the most distinguished Brahmins of this esoteric knowledge is Josef Maria Jauch. Jauch has devoted his professional scientific career to the cleaning up of the mathematical structure of quantum mechanics, appealing to the highest standards of purity and rigor. His intellectual life has been one long act of cerebra-

* Einstein Archive, Princeton, New Jersey.

tion about the deeper meaning of quantum mechanics and the philosophical worldview which it imparts. He has created a flourishing school at Geneva devoted to clarifying all problems that arise in the mathematical formalism of quantum mechanics, and his band of disciples and collaborators range far and wide on the globe. Since 1964 he has also played an active and passionate role in discussions about the problem of measurement and the interpretation of quantum mechanics.

In Are Quanta Real ?, A Galilean Dialogue, Jauch has summed up the principal points of view about the interpretation of quantum mechanics and what the quantum principle represents concerning our knowledge of the external world.

The three interlocutors in the dialogue are again Salviati, Sagredo and Simplicio, the principals in Galileo's dialogue on the Two Major Systems of the World, meeting again since the end of the fourth day of their last dialogue in 1638.[4)] Their meeting takes place in the fall of 1970 in a villa along the shores of Lake Geneva, in the city of Calvin, far away from the other end of the lake where the tragic spirit of François Bonivard, the prisoner of Chillon, resides.

That Jauch should have chosen to bring Salviati, Sagredo and Simplicio together for a new dialogue and to become its recorder is not accidental. Apart from the fact that he is a devoted scholar of the Galilean lore and an authority on the trial of Galileo[5)], Jauch finds in the scientific premises of the quantum principle a new revolution of Copernican dimensions in man's understanding of nature. What better than that the three old Venetian friends should meet again for rational inquiry within walking distance of the Reformation Wall, away from the shadows of theological superstition and tyranny ?

Salviati declares the purpose of their new encounter. It is to discourse "as distinctly and concretely as possible on the natural reasons hitherto alleged on one side by those who maintain deterministic, materialistic philosophy, and on the other by the followers of the Copenhagen interpretation of quantum mechanics."*

Bohr denied objective reality to properties whose simultaneous presence would require mutually exclusive situations for their verification, and it has put in question the very concept of rea-

---

* All unnumbered quotations are from J.M.Jauch, "Are Quanta Real?".

lity which had become the cornerstone of the development of physics since the last time the three friends had talked : that the world has a real existence independent of our observation, that science should reveal reality of this world and determine its laws, which are absolute and immutable.

Simplicio expresses his belief that the reality of the objective world is not only common senses but can be proved logically "with the most subtle of reasons and by experimentation".

Sagredo cautions his friends not to rush to conclusions. He pretends not to have any definite views on the questions involved and invites arguments for and against the points in dispute. The discussion is joined and continues in sessions spread over four days, Sagredo preserving a subtle neutrality and trying to bring together the opposites.

Salviati gives an example of the situation which would reveal the complementarity of certain physical properties, by considering the observation of a source of light through two sheets of polaroid. As one of the sheets is rotated with respect to the first, the intensity of light goes from a maximum to a minimum and to a maximum again. This remarkable behavior is discussed from the points of view of Aristotle and Democritos, Fresnel and Maxwell, electromagnetic waves and photons. "It is as though there were a basic duality in nature which furnishes complementary description of one and the same object", says Sagredo. "How limited is man's reason before the mystery of nature", concludes Salviati and proceeds to ask more questions.

They discuss the random results of experiments, the questions of initial conditions, Brownian motion and hidden variables. Although Sagredo finds it hard to distinguish from magic a description which can relate apparently unconnected random occurrences, Simplicio finds great appeal in hidden variable theories which can reproduce all the results of quantum mechanics. He does not know, as Salviati points out, that a theory which can be made to agree with all possible observable facts is no theory at all. "A theory which can be adapted to all possible future experiments has no predictive power and is therefore entirely useless."

What is the opinion of the majority when they compare Brownian motion and the photon experiment ? Says Salviati knowingly : "We should not be led by conclusions of the majority opinions, because in science democracy is singularly ineffective. The bold and courageous ideas of the nonconformists have always represented the milestones of progress in science."

Sagredo pleads that he would feel better about the situation "if I could see a coherent physical picture which would give us a causal description" of random occurrences. Salviati points out that even in the classical realm strict causality or determinism cannot be proved in all circumstances.

Simplicio appeals to the wisdom of the fundamental truth that "all events in nature are internally connected, and that it is incorrect to consider things and events as separated from one another". In other words, says Simplicio, "There are no separate phenomena, but every phenomenon preceived individually belongs to a whole into which it is integrated." Which reminds Salviati of Timaeus where Plato considers the universe as a single living body in which every part is only a part of the whole, and the souls of men are stars which enter man when he is born and return to their places when he dies - man's fate thereby being intimately interwoven with the rest of the universe.

Bohr's principle of complementarity is discussed as "one of the greatest discoveries in the scientific history of mankind, with ramifications on many other levels of science".

Why do we have two hands and two feet ? Why do our eyes see only a tiny fraction of the spectrum ? Why is our earth finite and constructed as it is ? And why should Planck's constant have exactly the value that it has ? "That is where it all begins, that singular story of the quantum of action. It is a history of surprises, errors, confusion and daring vision, it is truly a scientific mutation". And Salviati proceeds to emphasize the basic complementarity which pervades the physical universe.

While Simplicio is enthralled by the glories of classical physics and "realistic" description, Salviati draws attention to the immensely rich world of microphysics and how all of physics could indeed be treated as being essentially quantal. This leads to a discussion of the dialectic of "reality" and the validity of physical laws. The consideration of the variation of fundamental universal constants with time is recalled, and, if it were admitted, "the spell of the sanctity of physical laws is broken, and everything might be a possible".

In the incompatibility of the uniqueness of the individual and the scientific statements one might want to make about such an individual, Sagredo finds an example of the all-pervading principle of complementarity which excludes the simultaneous applicability of concepts to the "real" objects of our world.

The central conclusion of the dialogue is that the quest for reality cannot be solved in any consistent manner without taking at the same time into consideration the principle of complementarity. Reality, therefore, is not this or that object endowed with objectively given properties, nor is it the content of our consciousness or the idealized concepts which we use for imagining the inner coherence of elementary experiences. All these things are relative, as is brought out by the striking examples treated in the dialogue.

Reality is found in the meaning, significance and relevance of a message, a perception or a concept within the context of a given situation. A variety of sense perceptions are received and registered by a process of filtering which separates the signals representing the noise from those containing the message. The important point is this : Different filters reveal different and complementary messages, all of which have equal claim to reality.

Salviati considers this the essential content of Bohr's principle of complementarity, without which quantum mechanics seems to give an incomplete description of reality. And the debate of the three friends reveals the universal significance of complementarity and its discovery in microphysics as one of the most important epistemological revolutions in the intellectual history of mankind. As Salviati says, in conclusion : "Thus does our science of microphysics lead to insights which transcend the domain from which they originated, insights which give reasons for hope of a better understanding of all our experiences, including the moral and social behavior of man."

Sagredo thanks his friends and bids farewell until a fortunate occasion may bring them together again.

****

"Are Quanta Real ?" is a remarkable dialogue. Josef M. Jauch has written a noble document which transcends the function of merely interpreting the essential philosophical and epistomological premises of the quantum principle. The device of a dialogue between the immortal Galilean interlocutors brings out the dialectic of reality in a manner that a mere philosophical essay could never do.

Such a device, in less detail and perhaps less successfully, was employed by Hermann Weyl in his dialogue between Saints Peter and Paul on the question of inertia and the cosmos[6)]. Weyl got the saints to discuss the existence of inertia in Newton's theory of gravitation and Einstein's theory of general relativity. While Weyl's saints had the blessing of a view from heaven, Jauch's three Galilean friends must come to terms here on earth with the problem of determining the reality of observation and experience and integrating this knowledge with personal destiny. Jauch has recorded how well they have acquitted themselves in their inquiry.

Well, are quanta "real" ? As Dirac remarked, "It is like asking : Is God real ?" The libraries of the world and the memory of mankind are filled with attempts to answer this question. Has it been answered once and for all ? Has the Madonna been painted, and has the Sphinx lost her smile ?

This is the deep dilemma in man's search for knowledge and truth, and Simplicio's dream on the night preceding the dialogue on the third day points it out. The mysterious boon-giver, the Green Man in his dream, exhorts Simplicio to "find the unknown road to truth." "Tell us," the Green Man declares, "what you need in your search for truth, it shall be granted. But remember, you can have only one wish, the rest is up to you." Simplicio asks for "a library containing the books with all the wisdom, all the truth, and all the beauty of all time." His wish is granted instantly. The world disappears and is replaced by the largest library, extending to infinity in all directions with stacks and stacks of books, containing every thing that ever was and will ever be written in any language of the past or future.

Simplicio asks the librarian for a book containing the correct theory of elementary particles which explains all known facts about them. Well, which correct theory ? There are 137 of them that will fit all the known facts. Simplicio falls into a deep depression, which is the universal depression of the intelligent man curious for knowledge. The sum of all the facts, of all the bits of knowledge, does not add up to the whole truth; there is a complementarity between the whole and the sum of its parts.

When the choice is given us we choose the bits. We choose the corridors of books rather than the vision of truth, the whole truth. That is the original sin, the damnation of the merely curious. For if the choice and the possibility were given to be led along the unknown road to truth, and we took it, we would be one with the whole truth and the Sphinx would throw herself into the ocean.

As it is, we trudge the corridors of knowledge and the Sphinx keeps her sullen smile.

REFERENCES

1) Will Durant, The Story of Philosophy, New York, 1953

2) In R.S. Shankland, "Conversation with Albert Einstein II", Am. Jr. of Phys. 41, 895-901 (1973)

3) For a discussion of the historical development and the current status of the interpretation of quantum mechanics, see : J. Mehra, "The Quantum Principle : Its Interpretation and Epistemology", Dialectica, 1973

4) Galileo, Two New Sciences, 1638

5) J.M. Jauch, "Le Procès de Galileo Galilei", Arch. des Sciences, Genève, 23, 543-558 (1970)

6) H. Weyl, Massenträgheit und Kosmos : Ein Dialog, Die Naturwissenschaften 12, 197-204 (1924).

GEOMETRIE UND PHYSIK

C.F. von Weizsäcker

Max-Planck-Institut
zur Erforschung der Lebensbedingungen
der wissenschaftlich-technischen Welt
Starnberg, Germany

Behauptung : Die Geometrie ist ein Teil der Physik. Frage : Was ist der Sinn dieser Behauptung ? Vermutung : Es gibt ein System geometrischer Axiome, die dasjenige korrekt beschreiben, was in der Physik der Raum genannt wird. Rückfrage : Wie entscheidet man, ob diese Beschreibung korrekt ist ? Weitere Vermutung : durch Erfahrung. Erneute Rückfrage : Kann man durch Erfahrung über geometrische Sätze entscheiden ? Gegenfrage : Wie entscheidet man denn über irgendwelche physikalische Sätze durch Erfahrung ?

Der vorliegende Aufsatz präsentiert eine erkenntnistheoretische (Abschnitt 7) und eine physikalische (Abschnitt 10) Hypothese über diesen Fragenkreis. Eine zusammenfassende Reflexion enthält Abschnitt 2.

1. Axiomatik. In der älteren Tradition der abendländischen Wissenschaft erscheinen Geometrie und Physik als streng getrennte Disziplinen. Geometrie ist in dieser Auffassung ein Teil der Mathematik, deren Erkenntnisse von der Erfahrung unabhängig sind, Physik hingegen eine Wissenschaft, deren fundamentale Sätze durch Erfahrung bewiesen werden müssen und können. Wir sind heute skeptisch genug keine dieser Meinungen für selbstverständlich zu halten. Eben dadurch sind wir dafür vorbereitet, die Entstehung dieser Meinungen als wissenschaftsgeschichtlichen Vorgang zu würdigen.

Die Babylonier besassen den Inhalt des Lehrsatzes des Pythagoras tausend Jahre vor Pythagoras. Man kann vermuten, dass das Selbstverständnis des geometrischen Wissens, das die Griechen vom

Enz/Mehra (eds.), Physical Reality and Mathematical Description. 48–90. *All Rights Reserved.*

Orient übernahmen, durch den Wortsinn des griechischen Wortes geo-metria angedeutet ist : Erdvermessung. Es scheint die Entdeckung der griechischen Mathematiker gewesen zu sein, dass die geometrischen Sachverhalte in Lehrsätzen formuliert werden können, die sich logisch aus einer kleinen Zahl von Ausgangssätzen herleiten lassen. Mit einer Vergröberung griechischer Distinktionen nennen wir heute alle diese Ausgangssätze Axiome. Wenn man von den immanenten Problemen der Logik absieht, reduziert sich die Frage nach der Wahrheit der Geometrie dann auf die Frage nach der Wahrheit (oder dem Sinn) ihrer Axiome.

Die vorherrschende Meinung der älteren neuzeitlichen Wissenschaft war, dass die geometrischen Axiome evident, also unabhängig von der Erfahrung als wahr einleuchtend sind, dass sie aber mit der Welt der Erfahrung den Zusammenhang haben, sich in ihr stets zu bewähren. Im Beispiel : Man kann aus evidenten Axiomen logisch folgern, dass die Winkelsumme im Dreieck gleich zwei Rechten ist ; und wenn man ein physisches Dreieck vermisst, so wird man (innerhalb der Fehlergrenzen der Messung) stets zwei Rechte als seine Winkelsumme finden. Diese Beschreibung fordert zu der erkenntnistheoretischen Frage heraus, wie denn diese Harmonie zwischen Evidenz und Erfahrung garantiert ist. Man darf wohl den allgemeinen Konsens der heutigen Wissenschaftstheoretiker dafür voraussetzen, dass schon diese Beschreibung selbst zu naiv ist, als dass sie eine Beantwortung der Frage zuliesse ; man muss die in ihr implizierten Auffassungen sowohl von Evidenz wie von Erfahrung zunächst auflösen.

Nur als historische Randbemerkung sei gesagt, dass die Griechen an dieser Naivität unschuldig sind. Platon wusste, dass geometrische Sätze in der Erfahrung niemals in Strenge überprüft werden können, und dass dasselbe für die Sätze der Physik gilt, d.h. er kritisierte die Naivität des Empirismus. Er wusste ferner, dass die Mathematiker vom Sinn ihrer Grundbegriffe nicht Rechenschaft geben können, sondern dass sie diesen Sinn schlicht unterstellen (ihre Grundbegriffe sind "hypotheseis" = Unter-Stellungen), d.h. er kritisierte die Naivität des Evidenzbegriffs. Platons eigener Versuch eines Aufbaus der mathematischen Physik im Timaios kann nur dann in seiner begrifflichen Struktur verstanden werden, wenn man sieht, dass in ihm aus für Platon zwingenden systematischen Gründen der Unterschied zwischen Mathematik und Physik von vorneherein gar nicht gemacht wird. Ferner hat Imre Toth[1)] sehr starke Gründe für die These vorgebracht, dass die Auffassung der Geometrie durch Aristoteles einen Wissensstand der ihm zeitgenössischen Mathematiker voraussetzt, dem die Möglichkeit einer Axiomatisierung der Geometrie ohne das "euklidische" Parallelenpostulat geläufig war. Das methodologische Niveau der griechischen Wissenschaft und Philosophie war offenbar so viel höher als das ihrer neuzeitlichen Nachfolger, dass wir bis in unsere Tage haben warten müssen, um einige der Fragen wieder zu entdecken, welche die Griechen mit ihren uns überlieferten Lehren zu lösen versucht haben. Im folgenden will ich aber die neuzeitlichen Probleme nur im neuzeitlichen Kontext behandeln.

In der Geschichte der Geometrie wurde die Naivität des Evidenzbegriffs am Beispiel des euklidischen Parallelenpostulats aufgelöst. Schon der Versuch, dieses Postulat aus anderen Axiomen zu beweisen, zeigt, dass man es faktisch nicht als evident empfand. Die Unmöglichkeit dieses Beweises wurde durch den positiven Aufbau einer nichteuklidischen Geometrie demonstriert. Dieser Nachweis verlangte aber, um streng zu sein, den Beweis der Widerspruchsfreiheit der nichteuklidischen Geometrie. Dieser wurde zuerst erbracht durch die Konstruktion euklidischer Modelle nichteuklidischer Räume (populärstes Beispiel : die Kugelfläche als Modell der "Ebene" der sphärischen Geometrie). Methodisch wichtig ist hieran u.a. die "konventionalistische" Verwendung der Vokabeln der Geometrie wie "Gerade"und "Ebene", um Gegenstände zu bezeichnen, die nach dem "evidenten" anschaulichen Verständnis der bisherigen Geometrie die Eigenschaften, "gerade" bzw. "eben" zu sein, gar nicht haben. Thematisiert wird diese Methode in der von Hilbert eingeführten und heute unter Mathematikern herrschenden Auffassung von Axiomatik. Nach ihr ist weder die Wahrheit der Axiome noch auch nur die Bedeutung der in ihnen verwendeten Worte ein Thema der Mathematik (Hilbert : Statt "Punkt", "Gerade", "Ebene" hätte ich genau so gut sagen können "Liebe", "Gesetz", "Schornsteinfeger") ; die axiomatische Mathematik in diesem Sinne des Wortes befasst sich nur mit den logischen Beziehungen zwischen formal präzisierten Sätzen.

Die mit der Gabel der Axiomatik ausgetriebenen inhaltlichen Probleme der Mathematik kehren bekanntlich durch das Fenster der Meta-Mathematik zurück (Horaz, Epistel I, 10, 24). Die Widerspruchsfreiheit der Geometrie wird durch Reduktion auf die Arithmetik bewiesen. Die Widerspruchsfreiheit der Arithmetik verlangt zu ihrem Nachweis inhaltliche metamathematische Überlegungen, welche de facto ein Stück geometrischer Anschauung von Zeichenreihen etc. enthalten dürften. Doch ist auch die Metamathematik nicht Gegenstand dieses Aufsatzes. Ich will mich im folgenden auf den Standpunkt stellen, dass uns die mathematische Analyse der Grundlagen der Geometrie eine beliebige Menge möglicher geometrischer Axiomensysteme zur Verfügung stellt. Die Frage ist dann, ob eines von ihnen, wenn ja welches, oder ob vielleicht mehrere von ihnen das zu beschreiben geeignet sind, was wir in der Physik den Raum nennen.

2. Empirismus. Gauss hat im Zug der Hannoverschen Landesvermessung das grosse Dreieck zwischen den Bergen Brocken-Inselberg-Hoher Hagen vermessen. Er wusste, dass in der nichteuklidischen Geometrie die Winkelsumme eines Dreiecks um so weiter von zwei Rechten abweicht, je grösser das Dreieck ist. Er registrierte, dass er innerhalb der Fehlergrenzen keine Abweichung fand. Er hat also die Möglichkeit in Betracht gezogen, dass in der Wirklichkeit eine nichteuklidische Geometrie gelten könnte, und dass hierüber empirisch entschieden werden könnte. Diese Denkmöglichkeit war den Mathematikern des späteren 19. Jahrhunderts geläufig. Einstein hat auf Grund theoretisch-physikalischer Überlegungen die Hypothese der Geltung einer bestimmten, nämlich der Riemannnschen Geometrie in der Wirklichkeit aufgestellt. Er hat versucht, die Konsequenzen der diese Hypothese enthaltenden Allgemeinen Relativitätstheorie (ARTh) einer empirischen Prüfung zugänglich zu machen. Seit Einstein glauben fast alle Physiker, dass über die in der Wirklichkeit geltende Geometrie eine empirische Entscheidung möglich sei. Diese These sei im folgenden als Empirismus (genauer : geometrischer Empirismus) bezeichnet.

Dieser Empirismus ist nun in seiner schlichten Form ebenso naiv wie der Glaube an die Evidenz der geometrischen Axiome. Ich werde im Abschnitt 7 eine präzisierte Fassung des Empirismus vorschlagen, in der er m.E. aufrechterhalten werden kann. Dazu ist es aber zunächst nötig, seine naive Fassung aufzulösen. Wir wollen diese als direkten geometrischen Empirismus bezeichnen, d.h. als den Glauben an die direkte empirische Entscheidbarkeit von Sätzen der physikalischen Geometrie.

Nehmen wir an, Gauss oder irgend ein heutiger Beobachter hätte empirisch in einem Lichtstrahlendreieck eine Abweichung der Winkelsumme von zwei Rechten gefunden und diese Beobachtung sei als reproduzierbar anerkannt. Spätestens seit Poincaré wird nun gegen den Empirismus so argumentiert : In diesem Falle hätte der Beobachter nicht die Gültigkeit einer nichteuklidischen Geometrie empirisch bewiesen. Wenigstens müsste er Gründe dafür angeben, dass er die physikalisch näherliegende Deutung vermeidet, die Lichtstrahlen seien keine geraden Linien. (Die übliche Ausdrucksweise für die Lichtablenkung am Sonnenrand, die als eine empirische Bestätigung der Relativitätstheorie gilt, ist "Krümmung der Lichtstrahlen im Schwerefeld"). Dieser Einwand zeigt jedenfalls, dass der Empirist ohne eine Theorie des gemessenen Vorgangs nichts beweisen kann.

Ich möchte zunächst zwei Thesen betrachten, die in der Kritik am Empirismus weitergehen und sogar behaupten, man könne durch rein erkenntnistheoretische Überlegung einsehen, dass man grundsätzlich keine empirische Entscheidung über die Geometrie treffen kann. Sie sollen hier als Hierarchismus und als Konventionalismus bezeichnet werden. Ich halte beide für falsch, glaube aber, dass vor allem der Konventionalismus eine sehr starke Position ist, ohne deren volles Verständnis unser Problem nicht gelöst werden kann.

3. Hierarchismus. Hierarchismus ist kein üblicher Terminus der Wissenschaftstheorie. Ich verwende dieses Wort, um eine Ansicht zu bezeichnen, die dort, wo man sie für wahr hält, gewöhnlich als so selbstverständlich erscheint, dass man vergisst, sie als besondere Voraussetzung auszusprechen. Es ist die Ansicht, dass es eine Hierarchie der Wissenschaften gebe, in der die jeweils niedrigeren die jeweils höheren zur methodischen Voraussetzung haben, aber nicht umgekehrt. Z.B. gilt die Logik als die hierarchisch höchste der Wissenschaften : alle Wissenschaften haben die Logik zur methodischen Voraussetzung, denn sie müssen gemäss den Regeln der Logik verfahren; die Logik aber hat keine von ihnen zur methodischen Voraussetzung, denn (so meint man) logische Wahrheiten können nicht von den Ergebnissen von Einzelwissenschaften abhängen. Die Leugnung dieses hierarchischen Verhältnisses zwischen der Logik und den anderen Wissenschaften erweckt unmittelbar den Verdacht des circulus vitiosus. Analog sieht man auch das Verhältnis zwischen der Mathematik und den empirischen Wissenschaften : zur Aufstellung und Prüfung empirischer Gesetze braucht man Mathematik und schon darum erscheint es methodisch unsauber, eine Abhängigkeit der inhaltlichen Wahrheit der Mathematik von der Erfahrung anzunehmen.

Ich möchte die Vermutung aussprechen, dass der Hierarchismus grundsätzlich und in allen Fällen falsch ist, dass vielmehr sowohl zwischen der Logik und den Wissenschaften wie zwischen der Mathematik und den empirischen Wissenschaften ein Verhältnis gegenseitiger methodischer Abhängigkeit besteht. Diese allgemeine wissenschaftstheoretische These wird, soweit hier notwendig, im Abschnitt 7 besprochen werden. Im Augenblick sei nur zweierlei hervorgehoben: die historische Herkunft des Hierarchismus und seine Bedeutung für die Geometrie.

Der Hierarchismus ist eine Folge der griechischen Entdeckung der Möglichkeit einer axiomatischen Mathematik. In einem fest vorgegebenen axiomatischen System werden die Theoreme aus den Axiomen logisch hergeleitet. Die Wahrheit der Theoreme hat also die Logik und die Wahrheit der Axiome zur (hinreichenden) Bedingung. Die Logik und die Axiome müssen als evident vorausgesetzt oder von noch höheren Voraussetzungen her begründet werden. Aristoteles hat in den Analytica Posteriora den Begriff der deduktiven Wissenschaft gemäss diesem Schema entworfen. Tatsächlich hat es aber ausser der Mathematik und der mathematischen Logik nie eine deduktive Wissenschaft gegeben. Gleichwohl hat man in der europäischen Tradition sowohl die Philosophie wie die empirischen Wissenschaften an diesem Ideal gemessen. Der Hierarchismus ist gleichsam der regulative Gebrauch dieser Idee von Wissenschaft. Dabei hat man de facto die hierarchische Überordnung der Logik und Mathematik über die empirischen Wissenschaften nicht im Sinne einer Deduzierbarkeit dieser aus jener interpretiert, sondern einer Unabhängigkeit jener von diesen. Das ist aber nicht etwa eine schwächere, sondern eine stärkere Behauptung. Ist B aus A deduzierbar, so ist eine Widerlegung von B zugleich eine Widerlegung von A. Der Hierarchismus übernimmt gerade nicht die einwandfreie logische Struktur der deduktiven Wissenschaft, sondern ihre fragwürdige Annahme evidenter Axiome.

Die Geometrie ist der Ort, an dem der Hierarchismus zuerst erschüttert worden ist. Die Geometrie galt seit den Griechen als Teil der Mathematik. Also musste sie der Physik hierarchisch übergeordnet sein. Wenn zuerst Mathematiker und dann Physiker die empirische Entscheidung über die Geltung gewisser geometrischer Axiome in der Wirklichkeit für möglich hielten, so verletzten sie diese Vorstellung. Es gab aber zwei Möglichkeiten, das Problem der empirischen Geltung der Geometrie so zu beurteilen, dass das Prinzip des Hierarchismus unangetastet blieb.

Die moderne (Hilbertsche) Auffassung der Axiomatik ist der eine Ausweg, und zwar der weichere von den beiden. Man schränkt den hierarchischen Anspruch der Mathematik in der Geometrie auf den logischen Zusammenhang zwischen Axiomen und Theoremen ein. Damit ist die eigentliche Substanz dessen, was man seit den Griechen unter Geometrie verstanden hat, nämlich der inhaltliche Sinn ihrer Begriffe und die Wahrheit ihrer Axiome, aus der Mathematik und damit aus dem Anspruch hierarchischer Überordnung ausgeschlossen. Man kann dies, im Gegensatz zum inhaltlichen Hierarchismus der älteren Auffassung als formalen Hierarchismus bezeichnen.

Nun ist aber auch der inhaltliche Hierarchismus nicht eine unsinnige Ansicht, sondern nur die dogmatische Verfestigung wirklicher Unsymmetrien zwischen den Wissenschaften ; niemand wird z.B. die Arithmetik im selben Sinne für empirisch halten wie die Botanik. Ich möchte daher zunächst auch im Fall der Geometrie die Argumentationsstrategie verfolgen, dem inhaltlichen Hierarchismus so weit wie möglich entgegenzukommen. Hier bietet sich der zweite, harte Ausweg, nämlich das strikte Festhalten an der überlieferten Geometrie. Man sagt etwa: Die euklidische Geometrie ist a priori gewiss. Wenn in einem empirisch hergestellten Dreieck die Winkelsumme nicht den aus der euklidischen Geometrie folgenden Wert hat, so ist a priori gewiss, dass die Seiten dieses Dreiecks keine Geraden sind.

Im Beispiel des Lichtstrahlendreiecks führt also gerade die Gewissheit der euklidischen Geometrie zu einer eindeutigen Folgerung für die Optik : die Lichtstrahlen, welche durch ein Schwerefeld gehen, sind keine Geraden.

Wenn der Empirist diese These widerlegen wollte, müsste er zeigen, dass es keine Interpretation der bekannten Erfahrung geben kann, die mit einer euklidischen Beschreibung der Phänomene verträglich wäre. Dieser Beweis lässt sich, wie alle Unmöglichkeitsbeweise im empirischen Bereich, voraussichtlich überhaupt nicht in Strenge führen. Der Empirist wird daher zum Gegenangriff übergehen und fragen, womit sein Gegner die Apriori-Gewissheit der euklidischen Geometrie begründen will, nachdem die logische Möglichkeit nichteuklidischer Geometrien bekannt geworden ist. In den nächsten drei Abschnitten verfolgen wir u.a. Argumentationen über dieses Problem.

4. Kants Fragestellung. Kants Auffassung der Geometrie ist insofern überholt, als sie vor der expliziten Aufstellung nichteuklidischer Geometrien entworfen wurde. Andererseits bleibt sie auch heute lehrreich, weil Kant zwar am inhaltlichen Hierarchismus streng festhielt, ihn aber nicht schlicht behauptete, sondern die Notwendigkeit betonte, ihn detailliert zu begründen. Hier sei nur soviel von seiner Auffassung skizziert als wir im folgenden brauchen werden.

Kant kannte das Ergebnis von Saccheri und Lambert, dass das Parallelenpostulat nicht aus den anderen Axiomen der Geometrie logisch hergeleitet werden kann. In diesem eingeschränkten Sinne war ihm die logische Möglichkeit einer nichteuklidischen Geometrie vertraut. Damit stand für ihn das Parallelenpostulat aber in einer Linie mit allen anderen fundamentalen Einsichten der Mathematik. Sie waren nicht aus höheren Prinzipien logisch herleitbar und gleichwohl a priori gewiss; sie waren, um seinen Ausdruck zu gebrauchen synthetische Urteile a priori. Die Grundfrage seiner Erkenntnistheorie war daher : wie sind synthetische Urteile a priori möglich ?

Diese Auffassung steht in der Mitte zwischen zwei bequemeren, aber nach Kants Überzeugung unhaltbaren Ansichten. Nach dem Logizismus sind die Urteile der Mathematik einschliesslich der Axiome a priori, aber analytisch; sie sind a priori, weil sie logisch notwendig sind. Dies kann für die Axiome der Geometrie mit Sicherheit bestritten werden; Kant bestreitet es auch für die Grundlagen der Arithmetik. Nach dem radikalen Empirismus müssten umgekehrt die Urteile der Mathematik synthetisch, aber a posteriori, also auf Erfahrung gegründet sein. Für die inhaltlich gedeuteten Axiome der Geometrie ist dies heute die herrschende Ansicht der Physiker. Für die Arithmetik aber erscheint der radikale Empirismus schwer durchführbar. Kant jedenfalls verwarf ihn für beide Zweige der ihm bekannten Mathematik mit der Begründung, dass Erfahrung die ihnen eignende Notwendigkeit und Gewissheit grundsätzlich nicht garantieren kann.

Soweit ist aber nur Kants Problem formuliert. Wie ist Mathematik möglich, wenn sie weder analytisch noch a posteriori ist ? Kants Antwort beruht auf seiner Unterscheidung von Anschauung und Denken. Anschauung ist Rezeptivität, Denken ist Spontaneität des menschlichen (d.h. endlichen) Bewusstseins. Synthetische Erkenntnis muss auf Anschauung beruhen, denn sie fügt zu den Begriffen etwas hinzu, was in ihnen nicht schon enthalten war. Erkenntnis a priori aber kann nicht auf Erfahrung beruhen; das ist ihre Definition. Synthetische Erkenntnis a priori muss also auf Anschauung beruhen, die nicht Erfahrung ist. Solche Anschauung nennt Kant reine Anschauung. Er findet sie in den Formen aller Anschauung, d.h. in der Zeit und im Raum. Mathematik beruht auf der Konstruktion der Begriffe in der reinen Anschauung, Arithmetik in der Zeit, Geometrie im Raum.

Wir können uns auf die Details dieser sehr voraussetzungsvollen Theorie hier nicht einlassen. Für die Arithmetik sei nur bemerkt, dass sie eng verwandt ist mit dem Intuitionismus Brouwers und dem Konstruktivismus, wie ihn z.B. Lorenzen vertritt. Sie ist also von aktuellem Interesse. Für die Geometrie freilich würden fast alle heutigen Mathematiker ihre Begründung auf Konstruktion in der reinen Anschauung Raum verwerfen (für Lorenzen vgl. jedoch Abschnitt 6). Gerade sie muss uns aber hier interessieren.

Kant beansprucht, mit seiner Theorie der Mathematik zugleich ein Problem zu lösen, das die Mathematiker meist nicht beschäftigt, das aber für die Erkenntnistheorie der Physik fundamental ist : das Problem der Geltung mathematischer Gesetze in der Erfahrung. Wer mit Hume erkannt hat, dass aus der empirischen Geltung von Gesetzen in der Vergangenheit ihre allgemeine Geltung und damit ihre Geltung in der Zukunft logisch schlechterdings nicht abgeleitet werden kann, der steht vor diesem Problem. Es scheint, dass die moderne empiristische Wissenschaftstheorie erst jetzt zu realisieren beginnt, dass sie dieses Problem nie gelöst hat und grundsätzlich nicht lösen kann. Kants Lösungsvorschlag ist : Die reine Anschauung (Zeit und Raum) ist zugleich die Form aller empirischen Anschauung. Deshalb müssen Sätze, die durch Konstruktion in der reinen Anschauung begründet sind, in jeder empirischen Anschauung gelten. Ich übergehe wieder die sehr schwierige Frage, was die Gleichsetzung von reiner Anschauung und Form aller Anschauung bedeuten soll und hebe nur das erkenntnistheoretische Ziel dieses Lösungsvorschlags hervor. Er ist ein Spezialfall der allgemeinen These Kants zur Lösung des Humeschen Problems. Sätze, die in jeder Erfahrung gelten sollen, können sich weder durch spezielle Erfahrung begründen lassen, noch können sie eine Begründung haben, die mit Erfahrung gar nichts zu tun hat. Sie müssen vielmehr aus den Bedingungen jeder möglichen Erfahrung folgen. Dann haben sie einen Bezug auf jede mögliche Erfahrung und man kann doch a priori wissen, dass sie nicht durch Erfahrung widerlegt werden können. Von dieser These werde ich (Abschnitt 7) die Vermutung übernehmen, dass allgemein empirisch gültige Gesetze Bedingungen aller Erfahrung formulieren, aber nicht ihre hierarchistische Verengung, dass unsere Formulierungen solcher Gesetze nicht durch Erfahrung korrigiert werden könnten.

Eine Anwendung dieser Gedanken auf die physikalische Geometrie liegt in der Messtheorie. Bohrs These, dass ein physisches Gebilde nur als Messapparat geeignet ist, wenn wir es in Raum und Zeit der Anschauung kausal beschreiben können, ist gut kantisch. Hier stellt sich aber die Frage, ob der Raum unserer Anschauung denn euklidisch ist. Die Antwort der empirischen Psychologie muss wahrscheinlich lauten, dass der Raum unseres Anschauungsvermögens und unserer Phantasie weder euklidisch noch nichteuklidisch, sondern unpräzise ist. In der anschaulichen Vorstellung können wir zwischen einem Tausendeck und einem Zehntausendeck, zwischen einer Million und einer Milliarde Kilometern, zwischen 10^{-8} und 10^{-12}cm nicht unterscheiden. Ein Kantianer würde vielleicht einwenden, dies gelte zwar für die empirische, nicht aber für die reine Anschauung. Dieser Einwand fordert aber die Frage heraus, in welchem Sinne es eine reine Anschauung gibt. In den metaphysischen Anfangsgründen der Naturwissenschaft nennt Kant den Raum eine Idee. Dort geht es um die Absolutheit des Raums, die nicht der Anschauung, sondern der Vernunft zugehört. Für unser Problem wird man sagen: die scheinbar anschauliche Evidenz der präzisierten Geometrie beruht darauf, dass wir uns erlauben, extreme Grössen ähnlich verkleinert oder vergrössert vorzustellen. Die Existenz ähnlicher Figuren ist aber bereits ein dem Parallelenpostulat äquivalentes Postulat der euklidischen Geometrie. Die reine Anschauung Kants ist also selbst schon ein Produkt des Denkens (vgl. Kritik der reinen Vernunft, 2. Auflage, S. 161 Fussnote).

Wir werden keine der speziellen Thesen Kants übernehmen, sondern nur seine Fragestellung, ob Geometrie Bedingungen der Möglichkeit von Erfahrung formuliere.

5. Konventionalismus. Ehe wir den letzten Versuch besprechen, die A priori-Gewissheit der euklidischen Geometrie zu begründen, näm-

lich den Versuch von Dingler und Lorenzen, müssen wir den von ihm methodisch vorausgesetzten Konventionalismus erörtern.

Betrachten wir als Beispiel noch einmal den fiktiven Fall, Gauss hätte im grossen optischen Dreieck eine Winkelsumme ungleich zwei Rechten gefunden. Der geometrische Empirismus in naiver Fassung hätte gefolgert, in der Natur gelte eine nichteuklidische Geometrie ; der Hierarchismus hätte gefolgert, Lichtstrahlen seien keine Geraden. Der Konventionalismus würde sagen, beide Beschreibungsweisen seien zulässig. Es sei ein Missverständnis des konventionellen Charakters unserer Sprache, eine von beiden als die empirisch richtige auszeichnen zu wollen.

Ein zweites Beispiel : Vor etwa 40 und noch einmal vor etwa 20 Jahren konnte man in Bahnhofskiosken Schriften über die sogenannte Hohlwelttheorie kaufen. Nach dieser Theorie ist die Erde eine Hohlkugel, auf deren innerer Oberfläche wir leben. Die Gestirne sind sehr kleine leuchtende Körper nahe der Mitte der Kugel. Der Eindruck eines Himmelsgewölbes entsteht nur, weil alle Lichtstrahlen nicht gerade Linien, sondern Kreise durch den Mittelpunkt der Hohlkugel sind. Die Theorie verschwand aus den Kiosken, nachdem sie irrig prophezeit hatte, ein in Russland abgeschossener Sputnik müsse nach einem Flug von höchstens 13600 Kilometern auf der anderen Seite der Erde, also z.B. in Amerika, wieder herunterfallen. Den Verfassern war wohl entgangen, dass ihre Theorie empirisch unwiderlegbar gewesen wäre, wenn sie nicht nur (wie sie es taten) die Optik, sondern auch die Kinematik materieller Körper und überhaupt die ganze physikalische Geometrie der Transformation $r \to R^2/r$ (R = Erdradius, r = laufender Radius in Polarkoordinaten um den Erdmittelpunkt) unterworfen hätten. Aber vermutlich wollten sie das auch nicht, denn ihre Theorie wäre damit zugleich empirisch unbeweisbar, nämlich von der herrschenden Theorie überhaupt empirisch ununterscheidbar geworden.

Als drittes Beispiel kann die Allgemeine Relativitätstheorie dienen. Einstein forderte die allgemeine Kovarianz der Grundgleichungen gegen beliebige topologische Koordinatentransformationen. Die beiden ersten Beispiele stützen sich auf solche Transformationen (das zweite auf eine Transformation mit einer Singularität). Man muss also fragen, wie Konventionalismus und allgemeine Kovarianz zusammenhängen.

Ein viertes Beispiel bietet die Hamilton-Jacobische Fassung der klassischen Punktmechanik an. Da die Newtonsche Bewegungsgleichung, in der die Ortskoordinaten und die Zeit als Variable vorkommen, eine Differentialgleichung zweiter Ordnung nach der Zeit ist, kann man neben den 3 n Ortskoordinaten noch 3 n unabhängige Impulskoordinaten einführen. Die dynamischen Gesetze sind dann in "kanonischer" Schreibweise invariant gegen gewisse "kanonische" Transformationen der 6 n Orts- und Impulskoordinaten untereinander. Insbesondere ist bei beliebiger Hamiltonfunktion die Transformation auf "zyklische" Variable möglich. Sie führt 3 n konstante Impulse und 3 n linear mit der Zeit wachsende Ortskoordinaten ein. D.h. sie bildet eine Bewegung mit beliebigem Wechselwirkungsgesetz auf eine reine Trägheitsbewegung ab.

Diese dem Physiker wohlbekannten Fakten müssen erkenntnistheoretisch beim ersten Blick verwirrend wirken. Was ist nun in unserer Physik Beschreibung realer Erfahrungen und was ist Konvention ?

Der heutige Physiker würde hierauf wohl antworten : Die Naturgesetze lassen gewisse Transformationsgruppen zu. Beschreibungsweisen der Natur, die bei diesen Transformationen ineinander übergehen, sind gleichberechtigt ; das ist der Wahrheitsgehalt des Konventionalismus. Eine Grösse, die von verschiedenen Standpunkten aus verschieden zu beschreiben ist, ist dann eine wohldefinierte physikalische Grösse, wenn bekannt ist, wie sie sich bei der Gruppe transformiert. Naturgesetze sind dann nicht Konventionen, wenn sie sich formal invariant bei allen Transformationen der Gruppe ("allgemein kovariant") ausdrücken lassen.

Vielleicht darf ich diese Auffassung, ehe ich sie abstrakt im Detail erörtere, durch eine anekdotische Erinnerung erläutern. Eines Sommerabends sass ich als Gast Martin Heideggers mit ihm vor seiner Hütte im Schwarzwald. Wir betrachteten den Sonnenuntergang, während sich im Osten soeben der fast volle Mond über die Tannen des Berghangs hob. Heidegger sagte zu mir : "Herr v. Weizsäcker, Sie dürfen doch eigentlich gar nicht sagen, die Sonne gehe unter. Sie müssen doch sagen, der Erdhorizont hebe sich." Ich antwortete, im Sinn der soeben geschilderten Auffassung : "Ich sage völlig unbefangen, die Sonne gehe unter. Denn ich weiss, dass ich damit dasselbe meine wie wenn ich sagte, die Erde drehe meinen Horizont über die Sonne herauf. Wenn ich auf dem Atlantik, von Amerika kommend, am Heck meines Schiffs einem Schiff nachsehe, das, von Europa kommend, uns vor kurzem passiert hat, so sage ich, das andere Schiff tauche langsam hinter dem Horizont unter, wissend, dass ein Passagier am Heck jenes Schiffs ebenso über unser Schiff reden wird, und dass wir beide rechthaben." Hatte Heidegger recht oder ich ?

Die Schärfe des Problems kommt zum Vorschein, wenn man fragt, welche Transformationen zulässig sind. Nach Felix Kleins Erlanger Programm definiert die Auswahl einer Gruppe jeweils eine Geometrie. Die Mathematiker pflegen einen Raum zunächst als eine Mannigfaltigkeit von Elementen, die "Punkte" genannt werden, aufzufassen. Man pflegt ferner vorauszusetzen, dass alle Punkte einer Geometrie gleichberechtigt sind. D.h. die definierende Gruppe muss jeden Punkt in einen beliebig gewählten Punkt desselben Raumes überführen können ; nur gewisse Relationen zwischen zwei oder mehr Punkten sollen invariant bleiben.

Für den Erkenntnistheoretiker der Physik entsteht schon hier ein Problem. Wenn alle Punkte gleichberechtigt sind, wie kann man sie dann überhaupt voneinander unterscheiden ? Wenn man sie nicht durch jeden von ihnen individuelle anhaftende Eigenschaft unterscheiden kann, wie kann man *sagen*, welcher Punkt durch eine Transformation in welchen überführt wird ? Der Mathematiker macht sich diese Sache leicht; er "denkt sich", die Punkte irgendwie unterschieden und bezeichnet.* Die Bezeichnung, der "Name" eines Punkts, ist ein Merkmal, das ihm "von aussen" angeheftet wird, das also weder eine "innere", noch eine durch die Raumstruktur ausgezeichnete Eigenschaft des Punktes, somit nicht Gegenstand der Geometrie ist. Wie aber soll der Physiker eine solche gedachte Geometrie real anwenden ? Wenn er einem Gegenstand der Erfahrung, den er als Punkt im Sinne der Geometrie auffassen will, einen Namen gibt, wie kann er feststellen, ob dieser Name an dem Punkt haftet, also ob der Punkt, den er kurz darauf mit eben diesem Namen bezeichnet, derselbe Punkt ist wie zuvor ?

Das Problem ist ein Sonderfall der Frage, wie man individuelle Gegenstände unterscheiden kann, die unter denselben Begriff fallen. Für Gegenstände des täglichen Lebens ist eine hinreichende Antwort: sie haben stets auch noch andere Eigenschaften, die durch ihren gemeinsamen Begriff nicht determiniert sind, m.a.W. sie fallen stets auch noch unter andere, und zwar verschiedene Begriffe. Hier sind zwei Katzen, aber eine weisse und eine schwarze. Hier sind zwei schwarze Katzen, aber eine grössere und eine kleinere, usw. Fingiert man jedoch zwei begrifflich ununterscheidbare Gegenstände, so besagt eine Denktradition (von der Leibniz und Kant in verschiedener Weise Gebrauch gemacht haben), sie könnten nicht zur selben Zeit am selben Ort sein, seien also wenigstens stets durch ihren Ort unterschieden. Wenn nun aber begrifflich strukturlos gedachte

---

* Alter Kalauer : Der Unterschied zwischen dem experimentellen Physiker, dem theoretischen Physiker und dem Mathematiker ist an der Aufgabe des Öffnens einer Sardinenbüchse zu erläutern. Antwort : Der Experimentator macht die Büchse einfach auf. Der theoretische Physiker gibt ein zum Öffnen der Büchse geeignetes Verfahren an. Der Mathematiker denkt sich die Büchse geöffnet.

Orte selbst, eben die Punkte, die Gegenstände sind, die man unterscheiden will, so wäre es zirkelhaft, sie durch den Ort zu unterscheiden, an dem sie sich befinden.

Die übliche Antwort ist wohl : Begrifflich gleichartige Gegenstände, im Idealfall also Punkte, kann man nur demonstrativ, durch Hinzeigen unterscheiden. Dadurch wird nunmehr das zeigende Subjekt als unerlässliche Voraussetzung des Sinns der verwendeten Begriffe in die Erkenntnistheorie der Physik eingeführt. Ich halte diesen Schritt in der Tat für fundamental und für unvermeidlich. Dies ist hier zunächst eine blosse Behauptung. Die folgenden Abschnitte sollen die semantisch konsistente Einführung der physikalischen Geometrie schrittweise diskutieren, geleitet von der progressiven Erkenntnis der konventionalistischen Freiheit.

6. Dinglers operative Begründung der euklidischen Geometrie.

Hugo Dingler[2)3)] hat in Kenntnis der Kraft der konventionalistischen Argumente Poincarés versucht, die hierarchische Überordnung genau der euklidischen Geometrie über die empirische Physik zu retten. Er behauptet nicht mehr eine ontologische Wahrheit dieser Geometrie für einen physikalischen Gegenstand "Raum", sondern ihre operative Notwendigkeit für einen methodisch eindeutigen Aufbau der Physik. In diesem Sinne meint er Kants Gedanken zu verwirklichen, Geometrie gehöre zu den Bedingungen der Möglichkeit der empirischen Physik. Lorenzen hat diese fast verschollenen Gedanken Dinglers wieder aufgenommen und zum Programm einer "Protophysik" ausgebaut. Diese Thesen Lorenzens könnten adäquat nur im Zusammenhang mit seinem Entwurf einer Protologik, also allgemein einer Wiederherstellung des Hierarchismus auf operativer oder diskursiver Grundlage erörtert werden. Das kann im gegenwärtigen Aufsatz nicht geschehen. Daher sei hier nur ein m.E. entscheidender Grundgedanke Dinglers besprochen.

Eine Geometrie im Sinne des Erlanger Programms wird definiert durch eine Gruppe von Abbildungen der zugrundegelegten Punktmannigfaltigkeit auf sich. Dabei fordert man zwar, dass jeder Punkt in jeden anderen überführt werden kann (Homogenität des Raumes). Aber eine Teilmannigfaltigkeit von Punkten (eine "Figur") soll nicht in jede gleichmächtige Teilmannigfaltigkeit überführt werden können. Figuren, die durch die Gruppe ineinander überführt werden können, heissen gleich im Sinne der betr. Geometrie. Die engsten Gruppen und damit die Einteilung in die kleinsten Klassen, die in der Geometrie betrachtet zu werden pflegen, halten eine Metrik invariant, speziell die euklidische Gruppe.

Man kann sagen, dass die euklidische Gruppe den klassischen

Schnitt zwischen Geometrie und Physik (s. Abschnitt 1) überhaupt erst definiert hat. Sie überführt "kongruente" Körper ineinander ohne Rücksicht auf ihren Ort und ihre Orientierung. Physikalisch hingegen lehrt die elementare Erfahrung den Wesensunterschied von Richtungen wie z.B. oben und unten. Die griechische Wissenschaft kannte die Denkmöglichkeit, die Auszeichnung von Orten und Richtungen auf die Relation zu bestimmten Körpern (z.B. zur Erde) oder zum Weltall zurückzuführen. Einerlei welches Modell der Physik und Kosmologie man wählte, dies modifizierte jedenfalls nicht die Geometrie, insofern diese kongruente, aber verschieden situierte und vielleicht qualitativ verschieden beschaffene Körper als äquivalent behandelt. Aus dieser Erfahrung der Existenz von Eigenschaften aller Körper, die von allen ihren übrigen empirischen Unterschieden unabhängig waren, erwuchs die Vorstellung einer hierarchisch übergeordneten Wissenschaft von Körpern überhaupt, die nunmehr auch nicht auf Erfahrung gegründet schien, eben der Geometrie.

Der Empirismus des 19. Jahrhunderts musste demgegenüber die Angabe derjenigen Erfahrung verlangen, die diese Überordnung rechtfertigte. Helmholtz fand sie in der Existenz frei verschieblicher starrer Körper. Diese Forderung reichte aber nur aus, um eine metrische Geometrie mit konstantem Krümmungsmass zu begründen. Dingler verwendete eine gruppentheoretisch weitergehende Forderung, die am bequemsten in der für ihn ohnehin zentralen operativen Fassung dargestellt wird. Wenn Glas- oder Metallschleifer eine präzise (euklidische) Ebene herstellen wollen, so schleifen sie drei Körper wechselweise aneinander ab. Zwei Körper, aneinander abgeschliffen, würden kongruente Grenzflächen konstanter Krümmung erzeugen. Wenn beide kongruent auf eine dritte Grenzfläche passen sollen, so muss die Krümmung Null sein.

Dingler argumentierte nun im wesentlichen so : Wenn wir geometrische Messungen ausführen wollen, müssen wir zunächst die geometrischen Eigenschaften der Messgeräte festlegen. Dies muss in eindeutiger und darum unstreitig wiedererkennbarer Weise geschehen und wird durch eine Reihe von real ausführbaren Operationen erreicht, von denen die soeben geschilderte Herstellung einer Ebene wohl die wichtigste ist. Diese Operationen garantieren, dass die Messinstrumente mit derjenigen Genauigkeit, mit der die eindeutigen Vorschriften bei ihrer Herstellung befolgt wurden - prinzipiell also mit beliebiger Genauigkeit - einer ganz bestimmten Geometrie genügen, und zwar der euklidischen. Folglich müssen auch alle Objekte, die man mit diesen Geräten ausmisst, kraft dieser Messungen geometrisch durch Eigenschaften charakterisiert werden, die notwendigerweise nur innerhalb dieser Geometrie scharf definiert sind und darum den Gesetzen eben dieser Geometrie genügen.

Also ist durch eine reine Operationsvorschrift a priori gewiss, dass jeder in der Erfahrung mögliche geometrische Sachverhalt der euklidischen Geometrie genügen wird. Wenn jemand nun z.B. in Lichtstrahlendreiecken Winkelsummen ungleich zwei Rechten findet, so ist a priori gewiss, dass dies nur die Deutung zulässt, dass Lichtstrahlen keine Geraden sind. Man sieht sofort, dass keines unserer vier konventionalistischen Argumente, so wie es bisher vorgebracht ist, diesem Einwand standhält. In keinem von ihnen ist erwogen, wie man es macht, geometrische Messinstrumente herzustellen (zum dritten Beispiel, der ARTh, vgl. Abschnitt 9). Das am Ende des Abschnitts 5 erörterte Problem der Bezeichnung von individuellen Punkten ergänzt Dingler somit durch eine Erwägung der operativen Kennzeichnung geometrischer Begriffe, also von Figurenklassen. Nur in Bohrs Diskussion des Messprozesses (Abschnitt 4 Ende) findet sich eine Analogie, und Bohr ist ja in der Tat zu dem Resultat gekommen, dass jede Messung klassisch (also vermutlich, obwohl Bohr m.W. darüber nichts behauptet hat, auch euklidisch) beschrieben werden muss.

Als ich Dinglers Gedanken um 1935 kennenlernte, sah ich sofort ihre Kraft und reagierte auf sie doch so, wie fast alle Physiker zuvor und danach auf sie reagiert haben : Einsteins Einführung der Riemannschen Geometrie in die Physik kann doch durch solche methodologischen Argumente nicht als falsch erwiesen werden. Also musste Dinglers Argument auch methodologisch einen Fehler enthalten. Als ich diesen Fehler (wie ich auch heute meine, zutreffend) lokalisiert hatte, schrieb ich darüber einen Aufsatz und besuchte Dingler zu einer achtstündigen, natürlich ergebnislosen Diskussion, deren Hauptargumente ich hier nach meiner (vermutlich subjektiv gefärbten) Erinnerung schildere.

Die Analogie mit Bohr gab mir einen Wink. Bohr argumentierte, wenn auch sehr viel weniger präzisiert als Dingler, ein Messinstrument müsse raumzeitlich beschreibbar sein (sonst kann man es nicht wahrnehmen) und streng kausal funktionieren (sonst kann man aus der Ablesung nicht auf das Messobjekt schliessen) ; nur in der klassischen Physik seien de facto aber Raum-Zeit-Beschreibung und Kausalforderung vereinbar, und darum müsse man ein Messinstrument klassisch beschreiben. Die oft erörterten Probleme, die dieses Argument aufwirft, bespreche ich hier nicht. Ich unterscheide nur zum Vergleich mit Dingler, zwei denkbare Interpretationen der Intention Bohrs, eine falsche und eine richtige. Bohr argumentiert n i c h t so : "Jedes Messinstrument genügt der klassischen Physik. Also definiert es alle physikalischen Eigenschaften der Messobjekte im Einklang mit der klassischen Physik. Also gilt die klassische Physik für alle messbaren Objekte." Er argumentiert vielmehr so : "Jedes Messinstrument ist nur soweit zur Messung tauglich, als man es klassisch beschreiben kann. Nun gilt aber für die Messobjekte, jedenfalls im atomaren Bereich, nicht die klassische Physik, sondern die Quantentheorie. Deshalb lassen sich diese Objekte nicht wie makroskopische Körper genähert objektivieren ; das ist die Komplementarität. Die Messgeräte also müssen makroskopische Körper sein, und wo die Näherung, in der wir sie mit den klassischen Begriffen beschreiben, zusammenbricht, sind sie eben nicht mehr als (ideale) Messinstrumente tauglich."

Analog argumentierte ich nun gegen Dingler. Dingler behauptet, sein Argument gelte a priori im Sinne Kants, d.h. unabhängig von jeder speziellen Erfahrung, abhängig nur davon, dass überhaupt Erfahrung (räumlicher Art) möglich ist. Dann müsste es aber auch gelten, wenn man die Hypothese machte, die speziellen Erfahrungen seien so, dass alle Körper und das Licht sich verhalten wie Körper und Strahlen einer bestimmten nichteuklidischen Geometrie, die nur im Kleinen (praktisch : im täglichen Erfahrungsbereich der Menschen) in hinreichender Näherung euklidisch approximiert werden kann. Unter dieser Voraussetzung müsste aber Dinglers operative Herstellung euklidischer Ebenen für hinreichend grosse Werkstücke (oder hinreichend genaue Realisierung) physisch in vorhersagbarer und reproduzierbarer Weise missraten. Ich nahm z.B. an, sie müssten sich bei den zum Anpassen und Schleifen nötigen Bewegungen so deformieren, dass sie total abgeschliffen würden, ohne je zur Deckung zu kommen. Vielleicht läge noch eher eine Annahme nahe, nach der das Schleifen zwar gelingt, aber ein auf einer so hergestellten Ebene euklidisch gezeichnetes Dreieck bei der Bewegung seine (durch Messungen im Kleinen an drei verschiedenen Orten messbare) Winkelsumme ändert. Als die Diskussion so weit fortgeschritten war, antwortete Dingler meiner Erinnerung nach nur, das von mir unterstellte Verhalten der Körper sei "eine phantastische Behauptung", für die ich die Beweislast trage. Damit glaubte (und glaube) ich die Debatte definitiv gewonnen zu haben. Denn mein Argument bedarf nicht der empirischen Richtigkeit, sondern nur der Denkbarkeit dieser Annahme über das Verhalten physischer Körper. Damit ist eine spezielle Erfahrung denkbar, die Dinglers Forderung in einer exakt angebbaren Grössenordnung unerfüllbar macht. Also kann man nicht a priori einsehen, dass Dinglers Annahme von jeder speziellen Erfahrung unabhängig sein muss, entgegen seinem Anspruch. Etwas unfreundlicher formuliert : Eine physikalische Hypothese ist nicht deshalb unmöglich, weil sie vom Standpunkt einer zum Zweck ihrer Vermeidung eingeführten Methodologie aus phantastisch erscheint.

7. Semantische Konsistenz. Wir schalten einen Abschnitt wissenschaftstheoretischer Besinnung ein und vergegenwärtigen uns zunächst den bis hierher erreichten Stand in der Frage nach den Grundlagen der physikalischen Geometrie. Wir befinden uns im Versuch einer gruppentheoretischen Präzisierung der durch den Konventionalismus erzeugten Fragestellung. Wir haben zunächst die Annahme akzeptiert, es gebe eine Mannigfaltigkeit von Punkten, genannt "der Raum", die untereinander naturgesetzlich äquivalent, aber durch Aufweisung unterscheidbar sind. Wir wissen schon, dass wir diese Mannigfaltigkeit, wenn wir die Relativitätstheorie übernehmen, durch die raumzeitliche Mannigfaltigkeit von punktuellen Ereignissen werden ersetzen müssen, die Minkowski "die Welt" nannte (Abschnitt 8). Ferner sollten wir darauf vorbereitet sein, die Annahme einer solchen Punktmannigfaltigkeit quantentheoretisch zu kritisieren und vielleicht zu begründen (Abschnitt 10). Unsere gegenwärtige Frage ist, ob und, wenn ja, wodurch eine Gruppe von Abbildungen des Raumes auf sich ausgezeichnet ist, welche eine physikalische Geometrie, also eine Äquivalenzrelation von Figuren, definieren würde.

Historisch haben wir eine solche Auszeichnung in Gestalt der euklidischen Geometrie schon vorgefunden. Wir haben die kritische Rückfrage nach der Rechtfertigung einer solchen Auszeichnung (sei es der euklidischen, sei es einer anderen Geometrie) gestellt und drei sukzessive Antworten, die wir erhielten, als unzureichend befunden : die Rechtfertigung durch Evidenz, durch direkte Erfahrung und durch hierarchistische Festlegung von Bedingungen der Möglichkeit von Erfahrung. Damit haben wir aber auch am Beispiel der Geometrie ein Stück Geschichte der Wissenschaftstheorie nachvollzogen. Wie kann man überhaupt der Wahrheit eines allgemeingültigen Satzes gewiss sein ? Die Geschichte der Reflexion auf diese Frage bietet uns, roh gesprochen, gerade die vier Antworttypen : naive Selbstverständlichkeit, Berufung auf Evidenz, Bestätigung durch Erfahrung, Reflexion auf die Bedingungen der Möglichkeit von Erfahrung. Man wird sagen dürfen, dass alle vier Verifikationsweisen, jedenfalls so, wie sie bisher präsentiert worden sind, aus allgemein einsichtigen Gründen scheitern müssen. Naive Selbstverständlichkeit einer Meinung ist überhaupt kein Argument für diese, sondern sie ist der Zustand, ehe das Bedürfnis nach Argumenten gefühlt wird. Ihre Selbstverteidigung gegen kritische Rückfragen ist die Berufung auf Evidenz. Berufung auf Evidenz ist aber dort kein Argument, wo jemand da ist, der diese Evidenz nicht in sich erlebt. Das mag daran liegen, dass er etwas nicht sieht, was ihm, wenn er es sähe, evident wäre ; es kann aber auch daran liegen, dass er ein Gegenargument oder Gegenbeispiel sieht. Der im Wort Evidenz steckende Begriff des Sehens erscheint nun der empiristischen oder sensualistischen Reflexion als eine Metapher. Sie sucht Evidenz auf sinnliche Evidenz einzuschränken und die Realwissenschaften wie die Physik auf diese zu begründen. Der von Popper wieder hervorgehobene Einwand hiergegen ist, dass die Gesetze der Physik allgemeingültig sein sollen, und dass sie als allgemeine nicht durch eine ihrem Wesen nach unvollständige Aufzählung von Fakten begründet werden können. Kants Begründung allgemeiner Gesetze als notwendige Bedingungen der Erfahrung bleibt von diesem Einwand nur dann unbetroffen, wenn die Notwendigkeit dieser Bedingungen selbst evident ist. Dass ihr diese Evidenz jedenfalls in der von Kant oder Dingler gewählten hierarchistischen Fassung fehlt, haben wir am Ende der Abschnitte 3 und 6 wohl gesehen. Wir lassen dabei, wie überhaupt in diesem Aufsatz, das Problem logischer oder struktureller Evidenz beiseite.

Wir könnten soweit der Popperschen These folgen, dass allgemeine,für die Erfahrung gültige Gesetze nicht verifiziert werden können. Nicht so klar ist, ob sie wenigstens falsifiziert werden können. Unsere Gegenbeispiele gegen Kant und Dingler standen auf einer höheren Abstraktionsstufe als derjenigen der empirischen Falsifikation. Es wurde nicht behauptet, eine Erfahrung liege vor, die mit dem euklidischen Charakter der reinen Anschauung oder der Herstellbarkeit euklidischer Ebenen unvereinbar sei, sondern nur, Hypothesen über die Anschauung oder über das Verhalten von Körpern seien möglich, welche, wenn sie wahr wären, die Realisierung der Erwartungen Kants und Poppers faktisch ausschliessen würden. Im übrigen widersteht weder der Verifikations- noch der Falsifikationsglaube der Kritik, die der Konventionalismus an den Prämissen beider übt.

Die durch Th.S. Kuhn[4)] eingeleitete neuere Entwicklung der Wissenschaftstheorie lässt diese bisher ungelösten Probleme mehr oder weniger auf sich beruhen und studiert zunächst die geschichtliche Entwicklung der Wissenschaft. Diese Wendung kann man als eine legitime Kritik der Wissenschaftstheorie an ihrem eigenen Hierarchismus verstehen. Auch die Wissenschaftstheorie ist der Wissenschaft, deren Theorie sie ist, nicht hierarchisch vorgeordnet ; sie ist durch den historischen Gang der Wissenschaft korrigierbar. Anders gesagt : Wenn Erfahrung die Grundlage unseres Wissens genannt wird, so wird man wohl erst durch Erfahrung wissen können, was Erfahrung ist ; und der hierfür relevante Erfahrungsbereich ist zunächst einmal die Geschichte der Wissenschaft. Kuhn beschreibt diese Geschichte als eine Abfolge von Paradigmen, deren jedes eine Phase normaler Wissenschaft beherrscht und in einer wissenschaftlichen Revolution durch ein neues abgelöst wird. Heisenberg[5)] hatte schon vor Kuhn denselben historischen Hergang als eine Abfolge abgeschlossener Theorien beschrieben, deren jede die früheren umfasst und sie auf einen Bereich genäherter Geltung einschränkt.

Uns hat der Unterschied zwischen Kuhn und Heisenberg zu interessieren, also der Unterschied zwischen einem zur Problemlösung brauchbaren paradigmatischen Verfahren und einer einen Erfahrungsbereich in hinreichend guter Näherung beschreibenden Theorie. Für Kuhns These, dass paradigmatische Verfahren allgemeiner verwendet werden als die zu ihrer Interpretation benützten theoretischen Regelsysteme, gibt es zahlreiche historische Beispiele. Aber sind unsere hauptsächlichen Beispiele physikalischer Geometrie, nämlich die in der klassischen Physik benutzte euklidische, die in der SRth eingeführte Minkowskische und die in der ARth verwendete Riemannsche, nicht abgeschlossene Theorien im Sinne Heisenbergs ? Sie sind wohldefinierte mathematische Theorien. Aber ihre Anwendung auf die empirische, technische Wirklichkeit setzt voraus, dass man weiss, mit welchen Phänomenen dieser Wirklichkeit man die in ihnen verwendeten mathematischen Begriffe, beginnend mit Begriffen wie Punkt, Gerade, Abstand, identifizieren soll. D.h. sie setzt gerade voraus, dass man nicht im Problem des Konventionalismus steckt.

Ich schlage nun vor, auch dieses Problem bis auf weiteres geschichtlich zu betrachten. Auch die Deutungen der mathematischen Theorien geschehen durch Paradigmen, und eben diese Paradigmen sind im allgemeinen die Rechtfertigung der Einführung der Theorien. Eine Theorie eines Phänomenbereichs geht stets von einem Vorverständnis der Phänomene aus. Wenn die Theorie volle mathematische Gestalt angenommen hat, so kann man in ihr den mathematischen Formalismus unterscheiden von der physikalischen Deutung der im Formalismus benutzten Begriffe. Die Deutung bedient sich des Vorverständnisses, z.B. : Seien $x_1y_1z_1$ und $x_2y_2z_2$ die cartesischen Koordinaten zweier Punkte an Körpern, so sei r = $\sqrt{(x_1-x_2)^2 + (y_1-y_2)^2 + (z_1-z_2)^2}$ ihr Abstand. Was Abstand ist, wissen wir alle schon.

Hier tritt nun ein zentrales wissenschaftstheoretisches Problem auf, das ich als das Problem der semantischen Konsistenz bezeichnen möchte. Das Vorverständnis liefert die Semantik, die physikalische Bedeutungserfüllung der mathematischen Theorie. Die so gedeutete Theorie aber beschreibt, wenigstens in gewissen Aspekten, eben dieselben Phänomene, in denen sich das Vorverständnis bewegt. Ist diese theoretische Beschreibung der Phänomene mit dem Vorverständnis vereinbar ? D.h. ist die Theorie, deren mathematische Konsistenz vorausgesetzt sei, in dem Sinne auch semantisch konsistent, dass der Gesamtkomplex der gedeuteten Theorie, also die mathematische Theorie zusammen mit dem Vorverständnis, widerspruchsfrei ist ? Diese Frage lässt sich oft nicht unmittelbar entscheiden, jedenfalls dann nicht, wenn das Vorverständnis selbst nicht theoretisch formuliert war. Es kann vorkommen, dass die Theorie nunmehr das Vorverständnis präzisiert. D.h. man gewöhnt sich an, das Vorverständnis im Einklang mit der Theorie auszusprechen und alle dazu nicht passenden Sprech-und Vorstellungsweisen zu verbannen. In diesem Sinne hat die euklidische Geometrie das Vorverständnis der räumlichen Anschauung präzisiert (vgl. Ende von Abschnitt 4). Es kann aber auch vorkommen, dass die Theorie das Vorverständnis explizit korrigiert. Das klassische Beispiel dafür ist Einsteins Kritik des Begriffs der Gleichzeitigkeit entfernter Ereignisse (Abschnitt 8). Jedenfalls aber wird man die semantische Konsistenz für eine abgeschlossene Theorie fordern.

Diese Forderung erweist sich jedoch als zweideutig. Sie lässt eine engere und eine weitere Fassung zu. Dies hängt damit zusammen, dass Theorien Phänomene nicht nur historisch beschreiben, sondern erklären. Die sehr komplexe wissenschaftstheoretische Debatte über den Erklärungsbegriff soll hier nicht aufgerollt werden. Der Unterschied von Beschreiben und Erklären werde vielmehr nur am Beispiel der euklidischen Geometrie erläutert. Wenn die babylonische Geometrie den pythagoräischen Lehrsatz wie ein faktisch geltendes Naturgesetz kannte, so wollen wir sagen, dass sie die Natur mit Hilfe dieses Gesetzes beschrieb. Wenn ein griechischer Mathematiker, etwa Pythagoras, diesen Lehrsatz aus wenigen Axiomen herleitete und in diesem Sinne bewies, so sagen wir, er habe die von den Babyloniern beobachteten Phänomene erklärt. Beschreiben und Erklären sind, so gefasst, Relativbegriffe. Man kann auch sagen, dass die Babylonier die Phänomene, die sie im einzelnen beschrieben, durch den empirisch fundierten, aber als allgemeingültig postulierten Lehrsatz erklärten. Andererseits kann man sagen, dass die Griechen mit den Prinzipien, aus denen heraus sie erklärten, also z.B. mit den Axiomen der euklidischen Geometrie, wieder nur gewisse sehr allgemein verbreitete Phänomene beschrieben. In dieser Verschieblichkeit der Terminologie verbergen sich die Probleme des Sinns der Begriffe "Begriff" und "Gesetz". Jedenfalls aber statuiert eine Erklärung, so wie wir hier das Wort gebrauchen, stets die Notwendigkeit eines Zusammenhangs, der ohne sie nur beschrieben werden könnte ; und Notwendigkeit ist selbst nur relativ auf andere, selbst letzten Endes entweder als evident erlebte oder nur beschriebene Zusammenhänge. Fassen wir nun das Vorverständnis einer Theorie als die Beschreibung eines Phänomenbereichs auf, so kann die Theorie einen Teil dieser Beschreibung erklären (wie der pythagoräische Beweis die allgemeine Gültigkeit des pythagoräischen Lehrsatzes erklärt), einen Teil wird sie selbst lediglich beschreiben, eventuell mit präziseren Begriffen (z.B. indem sie die Existenz der Fundamentalgebilde, wie Punkte, Geraden, Körper postuliert), einen dritten Teil wird sie als ausserhalb ihres Gegenstandsbereichs liegend auf sich beruhen lassen (z.B. den physischen Unterschied zwischen den Raumrichtungen).

Als semantisch konsistent im engen Sinne kann man nun eine Theorie bezeichnen, deren Vorverständnis, soweit es von der Theorie selbst erklärt wird, mit dieser Erklärung verträglich ist. Z.B. dürfen im Vorverständnis der euklidischen Geometrie keine Dreiecke vorkommen, für welche empirisch die Verletzung des Satzes des Pythagoras oder des Winkelsummensatzes behauptet wird. Wir wollen sagen, eine Theorie sei semantisch konsistent im engeren Sinne, wenn in ihr keine semantischen Inkonsistenzen bekannt geworden sind.

Semantisch konsistent im weiteren Sinne werden wir eine Theorie nennen wollen, in der gar keine semantischen Inkonsistenzen auftreten können. Diese Forderung lässt wieder zwei Interpretationen zu. Die erste Interpretation entspricht Heisenbergs Begriff der abgeschlossenen Theorie. Die Theorie sei als mathematisch widerspruchsfrei vorausgesetzt, und es sei verfügt, dass das Vorverständnis nur im Einklang mit der Theorie interpretiert werden darf. Dann ist die semantische Konsistenz durch Verfügung gesichert. Damit ist aber vom eigentlichen Problem der semantischen Konsistenz abgesehen. Man kann ja nicht durch Verfügung sichern, dass die Phänomene eine Beschreibung gemäss dieser Theorie überhaupt unbegrenzt zulassen. Dies verfügen zu wollen, war Dinglers Fehler. Heisenberg spricht daher konsequent vom Geltungsbereich einer abgeschlossenen Theorie. Dieser umfasst gerade die Phänomene, die sich der Verfügung fügen. An welcher Stelle der Geltungsbereich aufhört, das kann nur weitere Erfahrung lehren oder die Erklärung dieser weiteren Erfahrung in einer umfassenden abgeschlossenen Theorie. Dabei ist die nun nicht mehr mit der Theorie vereinbare Erfahrung eine Art Erweiterung des Vorverständnisses über die bisherige Theorie hinaus. Die neue Erfahrung wird im allgemeinen mit Hilfe der Begriffe der bisherigen Theorie formuliert und führt gerade dann auf Widersprüche. Erst die neue Theorie verändert mit den neuen Gesetzen, in denen die Begriffe vorkommen, den Sinn der Begriffe so, dass die Widersprüche verschwinden ("sich als scheinbar erweisen"). In diesem Sinne ist die alte Theorie Vorverständnis des neuen. Im allgemeinen wird die neue Theorie einen Teil dessen erklären, was die alte nur beschreibt. Heisenbergs abgeschlossene Theorien erweisen sich damit als semantisch konsistent im engeren Sinne, solange sie als wahr gelten, und, wenn sie überholt sind, als semantisch konsistent im weiteren Sinne nur bezüglich ihres Geltungsbereichs, ausserhalb davon als semantisch inkonsistent.

Die zweite und eigentliche Interpretation der semantischen Konsistenz im weiteren Sinne müsste eine Theorie bezeichnen, in der solche Widersprüche nicht mehr auftreten können. Dies könnte höchstens in einer letzten, endgültigen Theorie der Fall sein. Gegenüber einer historischen Abfolge von Theorien bleibt es eine regulative Idee, deren Diskussion den Rahmen dieses Aufsatzes überschreitet.

Wir können in der jetzt eingeführten Sprechweise unsere bisherigen Probleme beschreiben. Der naive Empirismus übersieht, dass jede Erfahrung ein Vorverständnis benutzt, das selbst schon Begriffe, also Theorieelemente enthält und entsprechend kritisierbar ist. Der Hierarchismus erkennt dieses Vorverständnis als solches und versucht, es ein für allemal festzulegen. Er übersieht die Möglichkeit seiner nachträglichen Korrektur. Diejenigen Züge des Vorverständnisses, deren Abänderung man für unmöglich hält, sofern überhaupt Erfahrung stattfinden soll, nennt man Bedingungen der Möglichkeit von Erfahrung. Unsere Auffassung besagt, dass man von solchen Bedingungen sinnvoll sprechen kann, dass aber ihre Formulierung, ja ihr gesamter Inhalt durch nachträgliche Prüfung der semantischen Konsistenz verändert werden kann. Der Konventionalismus schliesslich hebt das Problem aus seinem geschichtlichen Zusammenhang heraus und wäre daher vielleicht erst in einer im weiteren Sinne semantisch konsistenten endgültigen Theorie vollständig diskutierbar. Wir beschränken uns hier auf die Beispiele aus Abschnitt 5.

Das erste Beispiel (Gauss' Dreiecksmessung) formuliert nur eine Aufgabe. Es unterstellt, innerhalb des kombinierten Vorverständnisses der euklidischen Geometrie und der Theorie geradeliniger Lichtausbreitung werde ein empirischer Widerspruch konstatiert. Nach unserer jetzigen Auffassung muss erst eine Theorie gefunden werden, die diese Phänomene widerspruchsfrei beschreibt, und diese Theorie wird dann das Vorverständnis korrigieren. Ohne solche Theorie ist das Beispiel nicht diskutierbar.

Das zweite Beispiel (Hohlwelt) ist umgekehrt innerhalb einer Theorie (z.B. der klassischen Physik) formuliert. Hält man an dem Vorverständnis fest, dass ein Abstand die Anzahl von Malen misst, in der ein Einheitsmassstab angelegt werden muss, um ihn auszumessen, so ist gemäss dieser Definition die Beschreibung der Welt als Hohlwelt semantisch inkonsistent, denn sie fordert, dass sich auch die Massstäbe gemäss der Transformation $r \to R^2/r$ ändern. Im übrigen zeigt das Beispiel nur, dass in einer im engeren Sinne semantisch konsistenten Theorie Umbenennungen möglich sind, die gerade deshalb empirisch unwiderlegbar sind, weil die neuen Namen durch explizite Definition aus den alten hervorgehen.

Der Widerstand, den wir gegen solche Umbenennungen empfinden, hängt damit zusammen, dass Sprache und Vorverständnis innerhalb unserer Geschichte nie in ihrem Beitrag zu einer einzigen, wenn auch noch so umfassenden Theorie aufgegangen sind. Dies wird deutlich an Heideggers Äusserung über den Sonnenuntergang. Meine relativistische Antwort war ungenau. Die beiden Schiffe zwar sind auch mechanisch äquivalent, da ihre Bewegungen durch eine Raumspiegelung auseinander hervorgehen. Für die klassische Dynamik ist jedoch die kopernikanische Beschreibung des Sonnenuntergangs, die Heidegger mir imputierte, in der Tat ausgezeichnet, da in rotierenden Bezugssystemen "Scheinkräfte" auftreten; das Wort "Schein" bezeichnet hier das Empfinden semantischer Inkonsistenz. Freilich kann die sprachliche Freiheit in der Formulierung des Vorverständnisses so erweitert werden, dass auch die Folgen solcher Übergänge zu beschleunigten Bezugssystemen korrekt mitgedacht werden. Auf der anderen Seite umfasst die übliche Sprechweise, die Heidegger für sich in Anspruch nahm, die volle Beziehung des wahrnehmenden Subjekts zum sogenannten physischen Vorgang, also das eigentliche Phänomen. Auf diese Trennung des physikalisch objektivierten Vorgangs vom Phänomen wies Heidegger hin. Aber gerade auf diese seine Intention zielte meine Antwort. Ich war der Meinung, dass das Subjket sich anderen Subjekten gegenüber relativieren kann und soll. Deshalb sprach ich von einem Passagier auf dem anderen Schiff, der vielleicht - was ich nicht sagte - mein Freund ist. Mit veränderter Bewusstseinslage verändern sich auch die Phänomene. Doch gehört der Bewusstseinswandel durch die Wissenschaft (die Zusammengehörigkeit von "Du" und "Gestalt") zu den vielen in diesem Aufsatz nicht mehr erörterten Problemen.

Das dritte Beispiel besprechen wir im 9. Abschnitt.

Das vierte Beispiel (Hamilton - Jacobi - Theorie) ist ähnlich gebaut wie das zweite, aber physikalisch sinnvoller. Der Übergang zu den zyklischen Koordinaten ist gleichbedeutend der Lösung der Bewegungsgleichungen. Es charakterisiert jede Lösung durch ihre Integrationskonstanten. Die klassische Punktmechanik bedarf also eines Vorverständnisses, um freie von Wechselwirkungsvorgängen zu unterscheiden, und ihre Invarianz gegen kanonische Transformationen zeigt, dass sie, für sich genommen, dieses Vorverständnis selbst nicht erklären kann. Wie in den Abschnitten 9 und 10 näher besprochen, hat dieses Vorverständnis mit der Zerlegbarkeit in Einzelobjekte (hier Massenpunkte) zu tun.

Zusammenfassend kann man vielleicht sagen, dass der Konventionalismus zwei ziemlich verschiedene Problemklassen andeutet. Im Bereich des Vorverständnisses weist er auf den konventionellen oder historisch relativen Charakter der Sprache, auf die Möglichkeit verschiedener Sprachspiele hin. Dies gehört in die Sprachphilosophie und kann vielleicht überhaupt nicht mit mathematischer Strenge erörtert werden. Wenn man lange genug über solche Fragen redet und sensibel und guten Willens ist, versteht man einander am Ende ein Stück weit und kann eine Sprache entwickeln, in der dieses Verständnis wirksam wird. Im Bereich der mathematischen Theorie reduziert sich der Konventionalismus auf die Invarianz der Gesetze gegen Transformationsgruppen. Welche Gruppen das sind, müssen wir nunmehr mit Hilfe des Vorverständnisses der heutigen Physik zu formulieren suchen.

8. Spezielle Relativitätstheorie.

Das Vorverständnis der SRth enthält unter anderem die euklidische Geometrie und die Newtonsche Mechanik. Beide können hier als abgeschlossene physikalische Theorien betrachtet werden, nach deren semantischer Konsistenz gefragt werden darf.

Eine physikalische Axiomatik der euklidischen Geometrie wird zwecksmässigerweise die Helmholtz - Dinglerschen Operationen an starren Körpern zugrundelegen. Sie begründet so die 6-parametrige euklidische Gruppe der reell-dreidimensionalen Rotationen und Translationen. Die Theorie ist dann semantisch konsistent im engeren Sinne. Ihr Vorverständnis setzt jedoch die Existenz starrer Körper voraus. Dies ist erstens eine Idealisierung. Die Phänomene zeigen die Existenz starrer Körper nur genähert. Sie rechtfertigen nicht eo ipso die Fiktion beliebig genauer Annäherung an das Ideal. Man muss also auf die Möglichkeit vorbereitet sein, einen begrenzten Geltungsbereich der Theorie zu entdecken. Zweitens wird die Existenz starrer Körper von der Geometrie nur im Vorverständnis benutzt, aber nicht theoretisch erklärt. Die statistische Physik des späten 19. Jahrhunderts hat schrittweise entdeckt, dass die Anwendung der klassischen Mechanik auf das Innere der Körper auf Schwierigkeiten führt, die vermutlich prinzipiell unüberwindlich sind. Jedenfalls hat erst die Quantentheorie dieses Problem gelöst (vgl. Abschnitt 10).

Die klassische Mechanik fügt, wie man unter der gruppentheoretischen Fragestellung gegen Ende des 19. Jahrhunderts erkannte (L. Lange), eine vierparametrige Erweiterung zur euklidischen Gruppe hinzu, aus Transformationen bestehend, die die Zeit enthalten. Die einparametrige Untergruppe der Zeittranslationen, die die Homogenität der Zeit ausdrückt, hat man meist als Formulierung der Annahme, dass dieselben Naturgesetze immer gelten, leicht akzeptiert. Hingegen enthalten die "eigentlichen Galilei-Transformationen", die Inertialsysteme ineinander transformieren, das spezielle Relativitätsprinzip, das viele philosophische Diskussionen wachgerufen hat. Historisch sind diese Diskussionen in zwei Phasen abgelaufen, die man als die Phase vor Einstein und die Phase nach Einstein unterscheiden kann. Vor Einstein erschien das Relativitätsprinzip nur dann als ein allgemeines Naturprinzip begründet, wenn man die klassische Mechanik als die fundamentale Wissenschaft von der Natur ansah; man kann diese Prämisse auch als das mechanische Weltbild bezeichnen. Etwa unter dieser Prämisse wurde die Relativität der Bewegung z.B. von Leibniz (gegen Clarke, d.h. gegen Newton), von Kant (in den Metaphysischen Anfangsgründen der Naturwissenschaft) und von Mach (ebenfalls gegen Newton) behauptet und diskutiert. Die Physiker des 19. Jahrhunderts entzogen sich aber meist der Härte des Problems durch die Annahme einer speziellen im Raume ruhenden Substanz, des Lichtäthers. Deshalb hat erst der Michelson-Versuch bzw. Einsteins Deutung dieses Versuchs das philosophische Problem, nun an Hand der Lorentzgruppe, unausweichlich gemacht. Erst bei Einstein wurde das spezielle Relativitätsprinzip aus einer faktisch für gewisse Phänomene gültigen Regel zu einem unentbehrlichen Bestandteil der gewählten Beschreibung von Raum und Zeit. Einstein durfte mit Recht annehmen, damit der Intention der genannten Philosophen (vor allem von Mach und vielleicht Leibniz; Kants Ansichten über Relativität der Bewegung hat er offensichtlich nicht gekannt) erst eine präzise physikalische Gestalt gegeben zu haben.

Diese Gestalt enthält nun aber ein philosophisch beunruhigendes Problem, auf das Einstein alsbald nach der Aufstellung der SRth aufmerksam wurde. Das spezielle Relativitätsprinzip leugnet die Existenz eines absoluten Raumes, ohne doch die Annahme einer allgemeinen Relativität von Bewegungen zu rechtfertigen. Es steht als empirisch gerechtfertigte unbequeme Annahme zwischen zwei scheinbar bequemeren, aber ungerechtfertigten. Die Abgrenzung nach beiden Seiten sei getrennt diskutiert.

Die Nichtexistenz des absoluten Raumes können wir so ausdrücken: die Identität eines Raumpunktes im Lauf der Zeit lässt sich nicht in semantisch konsistenter Weise behaupten. Zeige ich zweimal nacheinander auf einen Punkt, so kann ich nicht wissen, ob ich beidemale auf denselben Punkt gezeigt habe. Ich könnte versuchen, die Identität des Punktes zu objektivieren, indem ich eine Marke, etwa eine wiedererkennbare Stelle eines Körpers (kurz ausgedrückt, einen Körper) in ihm anbringe. Aber die Gruppe, der gegenüber die Bewegungsgesetze invariant sind, transformiert eine Zustandsbeschreibung, nach welcher der Körper in dem Punkt ruht, in eine solche, in welcher der Körper mit konstanter Geschwindigkeit auf einer geraden Bahn läuft, die den Punkt nur in einem bestimmten Zeitpunkt passiert.

Diese Überlegung konnte schon an Hand der klassischen Mechanik angestellt werden. Dies tat z.B. L.Lange in der Gestalt der Einführung der zueinander äquivalenten Inertialsysteme. Einstein hat hinzugefügt, dass auch Zeitpunkte keine vom Messgerät (der realen Uhr) unabhängige Identität haben. Seine Überlegung wirkte wie das plötzliche Aufgehen eines Lichts, weil sie die erste konsequente Überprüfung dessen war, was hier semantische Konsistenz genannt wird. Es sei deshalb erlaubt, ihren methodischen Gehalt (ähnlich wie in Abschnitt 6 für Bohr) in einer falschen und einer richtigen Interpretation zu paraphrasieren. Gemeinsam ist der Ausgangspunkt: die Wellentheorie des Lichts führt zum Postulat der Konstanz der Lichtgeschwindigkeit, der Michelson-Versuch zum Relativitätspostulat auch für Licht (und damit für alle bekannten Naturphänomene); beide zusammen zur Lorentz-Invarianz der Naturgesetze. Nun geht es falsch weiter: "Also kann man absolute Gleichzeitigkeit entfernter Ereignisse nicht mit Uhren feststellen. Was man nicht feststellen kann, existiert nicht. Somit existiert die absolute Gleichzeitigkeit nicht". Richtig ist: "Der Begriff der absoluten Gleichzeitigkeit ist nicht lorentzinvariant. Wenn alle Naturgesetze lorentzinvariant sind, kann es also keine absolute Gleichzeitigkeit geben. Unser Vorverständnis ist entsprechend zu korrigieren. Nun könnte jemand einwenden, absolute Gleichzeitigkeit lasse sich doch sogar messen. Demgegenüber zeigt sich die Konsistenz der Theorie darin, dass Uhren, die lorentzinvarianten Gesetzen genügen, auch nicht fähig sind, absolute Gleichzeitigkeit zu messen." Logisch gewendet: "Was gemessen werden kann, existiert" wird als wahr vorausgesetzt. Die falsche Fassung benutzt die logisch nicht folgende Umkehrung "was nicht gemessen werden kann, existiert nicht". Die richtige Fassung bestätigt nur die korrekte Kontraposition: "was nicht existiert, kann auch nicht gemessen werden." Es sei bemerkt, dass genau dasselbe Missverständnis bei Kritikern von Heisenbergs Unbestimmtheitsrelation vorkommt.

Die Physiker, welche die konsequente Deutung der Quantentheorie aufgebaut haben, also Bohr, Heisenberg und ihre Nachfolger, haben diese Überlegungen Einsteins stets als die erste Einführung des Beobachters in eine Diskussion des Sinns physikalischer Begriffe aufgefasst. Einstein hat sich dagegen verwahrt, von seinem Standpunkt aus mit Recht, aus zwei Gründen. Erstens kann bei ihm die Messung von Längen und Zeitspannen stets als die blosse Ablesung bestimmter Zustände von Massstäben und Uhren angesehen werden, die auch dann vorliegen, wenn niemand sie beobachtet. Wie weit dieses Argument auf die quantentheoretische Messtheorie übertragbar ist, bleibe hier unerörtert; das "Paradox" von Einstein - Rosen - Podolsky soll zeigen, dass sie es nicht ist. Zweitens aber sind zwar weder Raum noch Zeit je für sich im hier definierten Sinne absolut, wohl aber das vierdimensionale Raum-Zeit-Kontinum, Minkowskis "Welt". Ein Weltpunkt, oder, wie Einstein gern sagte, ein "Ereignis" wird in der speziellen und in der allgemeinen Relativitätstheorie als objektiv identifizierbar behandelt[6]. Dies wird freilich nicht mehr im Sinne semantischer Konsistenz begründet, sondern als quasi evident vorausgesetzt. Die Mannigfaltigkeit von Raumpunkten und ebenso die Mannigfaltigkeit von Zeitpunkten eines fest gewählten Inertialsystems erscheint unter diesem Aspekt als ein konventionelles Ordnungsschema in einer nicht konventionellen Ereignismannigfaltigkeit. Die semantische Inkonsistenz dieser Annahme einer objektiven Ereignismenge wird erst in der Quantentheorie zum Thema.

Für die Begründung der physikalischen Geometrie mindestens so wichtig wie die Nichtobjektivität absoluter Geschwindigkeiten ist aber die Objektivität absoluter Beschleunigungen in der klassischen Mechanik und der SRth. Die Galilei-Transformation drückt, wie Newton klar sah und durch den Eimerversuch nachwies, nicht eine allgemeine Relativität der Bewegung aus, sondern eine Folge eines speziellen dynamischen Gesetzes, des Trägheitsgesetzes. Die Tatsache, dass die Newtonsche Bewegungsgleichung von zweiter Ordnung in der Zeitableitung ist, hat zur Folge, dass nur Geschwindigkeiten, nicht aber Beschleunigungen als relativ aufgefasst werden dürfen. Diese schlechterdings nichttrivale Tatsache wird man heute wohl am liebsten damit in Zusammenhang bringen, dass die Newtonsche Gleichung die Eulersche Gleichung eines euklidisch invarianten Variationsprinzips ist. Dieses seinerseits kann man im Sinne von Dirac[7] als Huygenssches Prinzip der Schrödingerwellen auffassen. Auch in diesem Punkte finden also die Grundfakten der klassischen Physik ihre nächste Erklärung in der Quantentheorie (vgl. dazu Abschnitt 10). Die Ungeklärtheit dieses Problems war für Einstein der Anlass zur Suche nach einer allgemeinen Relativitätstheorie. Ehe wir ihm hierin folgen, sei aber noch die Beziehung der hier besprochenen Theorien zur Erfahrung knapp methodologisch charakterisiert.

Alle diese Theorien gehen von empirisch bewährten Gesetzmässigkeiten aus, die von ihrem Vorverständnis her keineswegs selbstverständlich sind, sich aber gegen empirische Falsifikationsversuche in einer für die scientific community überzeugenden Weise als resistent erwiesen haben. Man kann diese Gesetzmässigkeiten den harten Kern der betr. Theorien nennen. Sie werden dann als schlechthin allgemeingültige Prinzipien hypothetisch postuliert. Wer dieses Postulat akzeptiert, modifiziert damit sein Vorverständnis. Vom neuen Vorverständnis aus sind die ursprünglich nichttrivialen empirischen Grundfakten notwendige, keiner weiteren Erklärung bedürftige Phänomene.

Der harte Kern des "Galileischen" Relativitätsprinzips der klassischen Mechanik ist das zunächst empirische Faktum des Trägheitsgesetzes. Postuliert man das Relativitätsprinzip als Naturgesetz, dann haben nicht Raumpunkte, wohl aber Trägheitsbahnen objektive Realität. Dann ist das Trägheitsgesetz eine selbstverständliche Konsequenz des postulierten Naturprinzips der Relativität.

Der harte Kern der SRth ist das zunächst empirische Faktum des negativen Ausfalls des Michelson-Versuchs. Fordert man Einsteins zwei Postulate als Naturgesetze, dann hat die absolute Geschwindigkeit des Michelson-Apparates keine objektive Realität. Dann ist Michelsons Ergebnis die selbstverständliche Konsequenz der postulierten Prinzipien.

9. Allgemeine Relativitätstheorie.

Der harte Kern der ARth ist das zunächst empirische Faktum der Gleichheit der schweren und trägen Masse. Man kann deshalb postulieren, dass es eine besondere Kraft "Gravitation" gar nicht gibt, sondern stattdessen eine Riemannsche Geometrie des Raum-Zeit-Kontinuums. Dann ist die Gleichheit beider Massen eine selbstverständliche Konsequenz.

Historisch fand Einstein den Weg zu diesem Ergebnis über die Suche nach einem allgemeinen Relativitätsprinzip. Dabei gingen aber einige der für seinen ursprünglichen Ansatz charakteristischen Züge verloren, und die Debatte über den Sinn des Begriffs der allgemeinen Relativität hat sich bis heute nicht ganz befriedigend aufgelöst.

Machs Motiv für die Forderung allgemeiner Relativität war die ontologische Annahme, dass nur Körper, nicht aber Raumpunkte objektive Realität haben; deshalb sollten die Kräfte zwischen Körpern nur von ihren Relativkoordinaten, nicht aber von ihrem Verhältnis zu einem "absoluten Raum" bestimmt sein. In dieser Form ist die Forderung mit der Einführung von Feldtheorien hinfällig, zumal einer Feldtheorie der Raummetrik. Einstein schreibt dem Raum nicht wie Mach keine, sondern mehr physikalische Eigenschaften zu als Newton. Ob die Forderung, die Einstein dann "Machsches Prinzip" genannt hat, physikalisch berechtigt ist, ob also das metrische Feld durch die Materieverteilung vollständig determiniert sein soll, bleibt solange fraglich, als keine physikalische Theorie der Materie vorliegt.

In der Gestalt, die die Theorie unter Einsteins Händen erreicht hat, hat jeder Weltpunkt bestimmte physikalische Eigenschaften, nämlich die Werte des metrischen Tensors und des Materietensors. Die Erfüllung der Forderung nach allgemeiner Relativität fand Einstein nun darin, dass zur Bezeichnung dieser Punkte völlig beliebige Koordinaten gewählt werden dürfen und dann die Gesetze der Theorie gegen beliebige Transformationen dieser Koordinaten invariant sein sollen (allgemeine Kovarianz).

Damit ist aber die semantische Konsistenz, die Machs und der SRth Ziel war, bis auf weiteres preisgegeben. Mach wollte die Theorie dem Vorverständnis anpassen: nur Beziehungen zwischen Körpern sollten in ihr vorkommen. Einstein hat den Feldbegriff auch für die Geometrie eingeführt und muss das Vorverständnis entsprechend modifizieren. Die ARth ist damit ein neues schönes Beispiel "antihierarchischen" Denkens. Aber ihre semantische Konsistenz würde erfordern, dass man erfährt, wie die jeweiligen Kordiantenwerte empirisch bestimmt werden, also wie sich Masstäbe und Uhren nunmehr verhalten. Gerade die Beliebigkeit der Koordinatenwahl nimmt aber auf dieses Problem gar keinen Bezug. Die Theorie braucht nicht semantisch inkonsistent zu sein, aber ihre semantische Konsistenz ist kein konstruktives Element ihres Aufbaus. In der Tat müssen diese Fragen ebenfalls dem ungelösten Problem einer Theorie der Materie zugeschoben werden.

Das hat jedoch zur Folge, dass der physikalische Sinn der allgemeinen Kovarianz undeutlich bleibt. Wie Kretschmann zuerst hervorgehoben hat, kann man durch explizite Einführung der Komponenten des metrischen Tensors in die Feldgleichungen diese stets allgemein kovariant schreiben. Demnach ist die allgemeine Kovarianz überhaupt keine Forderung an den Inhalt der Naturgesetze, sondern nur an deren Schreibweise. Einstein hielt aber daran fest, es sei eine sachhaltige Forderung, die Naturgesetze sollten in der allgemein kovarianten Schreibweise einfach sein. In der Tat sind seine Feldgleichungen (mit kosmologischem Glied) die einzigen allgemein kovarianten, die keine höheren als zweite Ableitungen der g_{ik} enthalten. Das Prinzip der Einfachheit der Naturgesetze hat eher ästhetischen als begrifflichen Charakter. Vermutlich appelliert es an eine Gestaltwahrnehmung guter Physiker für noch unverstandene, aber im Prinzip angebbare begriffliche Strukturen.

Zwei Fragen können auch heute aufgeworfen werden: Was bedeutet die Lorentz- (Poincaré-) Gruppe in der ARth ? und : Was bedeutet die Beschränkung auf stetige topologische Transformationen?

Die Poincaré-Gruppe ist die Invarianzgruppe der ARth im Kleinen, also lokal in linearer Näherung. In der klassische Mechanik entsprechen den 10 Parametern dieser (bzw. der Galilei-)

Gruppe die 10 allgemeinen Integrale der Bewegungsgleichungen. Nun hat in der Punktmechanik das n-Körper-Problem 6 n Integrale. Die offensichtlichen Symmetrien, die den 10 sog. allgemeinen Integralen entsprechen, sind durch das Vorverständnis der Punktmechanik ausgezeichnet, nach dem die 6 n verallgemeinerten Koordinaten eine Darstellung zulassen müssen als Koordinaten und Impulse von n Massenpunkten, deren jeder an den Symmetrien von Raum und Zeit gemäss der Galilei (oder Poincaré-)-Transformation teilhat. Dieses Vorverständnis kann natürlich wiederum nur weiter erklärt werden durch eine Theorie der Materie, die angibt, warum es Körper gibt, die als Massenpunkte approximiert werden können. Analog kann man für die 10 Integrale der von Einstein benutzten formalen Kontinuumsdarstellung der Materie argumentieren, z.B. durch Grenzübergang von der Punktmechanik her. Diese Betrachtungsweise entspricht nun genau der lokalen linearen Approximation in der ARth. Die nichtlinearen Glieder heben die Poincaré-Symmetrie auf und damit das Mittel der Unterscheidung der 10 allgemeinen Integrale von den übrigen. Der allgemeinen Kovarianz entsprechen dann die Bianchi-Identitäten.

Zur Topologie: in einem hierarchischen Aufbau der Geometrie nach dem Erlanger Programm beginnt man mit beliebigen Punkttransformationen, die man dann schrittweise einschränkt durch die Forderung, dass gewisse Beziehungen zwischen Punkten invariant bleiben sollen, zuerst die Topologie, dann lineare (projektive oder affine), schliesslich metrische Beziehungen. Eine umgekehrte Reihenfolge würde sich nahelegen, wenn man Beziehungen zwischen möglichst wenigen Punkten zugrundelegt. Eine metrische Beziehung (Abstand) besteht zwischen zwei Punkten, eine lineare (auf einer Geraden, Ebene, ... liegen) zwischen wenigstens drei Punkten, eine topologische (Häufungspunkt sein etc.) zwischen unendlich vielen Punkten. Nun definiert in der Tat eine Metrik auch eine Topologie. In einer semantisch konsistenten Physik erscheint es plausibel, dass räumliche und zeitliche Abstände gemessen werden können. Also sollte man das Raum - Zeit - Kontinuum wohl nicht primär als topologischen Raum von "Ereignissen" auffassen, dem dann eine Metrik aufgeprägt wird, sondern primär durch metrische Relationen bestimmen, welche bei fingierter absoluter Messgenauigkeit, auch eine Topologie festlegen. Die Metrik muss dabei nicht die pseudoeuklidische der Minkowskiwelt sein, für welche Punkte mit lichtartigem Abstand den metrischen Abstand Null haben, sondern eine durch die positiv definite Summe von räumlichen und zeitlichem Abstand definierte Metrik. Zwar ist diese Metrik selbst nicht lorentzinvariant, wohl aber die durch sie definierte Topologie. Damit hat man dann die Freiheit, die quantentheoretische Begrenzung des Raumbegriffs einzuführen.

10. Quantentheorie des Raums.

In der nichtrelativistischen Quantentheorie ist das Problem der semantischen Konsistenz der Geometrie völlig beiseitegelassen. Nach Helmholtz, Poincaré, Dingler und Einstein muss dies wie ein Rückfall in eine primitive Unreflektiertheit wirken. Das eigentlich erklärungsbedürftige Phänomen aber dürfte der ausserordentliche empirische Erfolg einer geometrisch so unreflektierten Theorie sein. Diese Behauptung sei hier zunächst erläutert.

Das Vorverständnis der euklidischen Geometrie lässt sich am besten mit der Existenz frei verschieblicher fester Körper formulieren. Die physische Existenz solcher Körper (in der Näherung, in der sie faktisch existieren) erklärt die Quantentheorie. Sie hat damit erst voll ins Bewusstsein der Physiker gehoben, eine wie schwierige, mit der klassischen Kontinuumsdynamik wohl unlösbare Aufgabe diese Erklärung ist. Unter den Philosophen war sich vielleicht Kant der Schwierigkeit dieser Aufgabe am klarsten bewusst (Metaphysische Anfangsgründe der Naturwissenschaft, 2. Hauptstück, und Opus Postumum). Die Physiker stiessen darauf in der statistischen Thermodynamik (Boltzmann). Ohne Atome, die keine inneren Freiheitsgrade mehr haben, gibt es kein thermodynamisches Gleichgewicht. Also darf man die Dynamik des Kontinuums nicht auf das Innere der Atome anwenden. Punktmechanische Atommodelle wie das von Rutherford scheitern in der klassischen Physik an der Kontinuumsdynamik der sie zusammenhaltenden Kräfte, wie Bohr erkannte. Bohr und nach ihm die Quantenmechanik erklärte die Existenz formkonstanter Atome durch einen radikalen Bruch mit der klassischen Dynamik. Vom Standpunkt des bewusstgemachten Vorverständnisses der Geometrie aus ist es dann aber zum mindesten sehr verblüffend, dass die Quantenmechanik auch auf das Innere des Atoms, bis zu beliebig kleinen Längen hinunter (empirisch gerechtfertigt jedenfalls bis zu 10^{-12}cm, also einem Zehntausendstel des Atomradius) euklidische Geometrie anwendet, obwohl es offenbar unmöglich ist, diese kleinen Räume mit starren Körpern, welche aus "Subatomen" bestehen müssten, auszumessen. Charakteristischerweise bedienen sich die quantentheoretischen Gedankenexperimente zur Ortsmessung nicht der starren Masstäbe, sondern des Mikroskops; die übliche Theorie des Mikroskops aber setzt bereits die euklidische Geometrie voraus.

Es ist kein Wunder, dass eine geometrisch so undurchdachte Theorie in Schwierigkeiten kommt, wenn sie mit der höchsten bisher erreichten Reflexionsstufe bezüglich semantischer Konsistenz der Geometrie vereinigt werden soll, nämlich der speziellen Relativitätstheorie. Formal befriedigend geglückt ist diese Vereinigung bisher nur für kräftefreie Theorien, also gerade unter Vernachlässigung des Zugs der Wirklichkeit, an dem die semantische Konsistenz (die Theorie der Messung) hängt, der Dynamik. Die Dynamik scheint eine Einschränkung der realen Anwendbarkeit der Geometrie bei sehr kleinen Längen zur Folge zu haben, für die es heute aber noch keine allgemein anerkannte Beschreibung gibt. Hingegen gibt es keine Indizien dafür, dass diese Einschränkung zugleich eine Einschränkung des Geltungsbereichs der Quantentheorie wäre. Nun bedarf ein axiomatisch konsequenter Aufbau der Quantentheorie (Jauch[8], Piron, Drieschner[9])in der Tat des Ortsbegriffs nicht. Man kann also versuchen, zunächst eine abstrakte Quantentheorie zu errichten, und dann in ihr einen Parameter oder gar Operator zu finden, der mit dem Ort identifiziert werden könnte. Diese Identifikation müsste dann sowohl den Erfolg der euklidischen Geometrie unterhalb des Bereichs erklären, in dem es feste Körper gibt, wie auch die Einschränkung für noch kleinere Längen.

Nun lässt die abstrakte Quantentheorie zunächst eine ganz ausserordentlich umfassende Transformationsgruppe zu, nämlich die unitäre Gruppe des ganzen Hilbertraums. Die Einschränkung des hierin implizierten Konventionalismus geschieht durch die Wahl der Dynamik. Die physikalische Geometrie ist auch in der Quantentheorie durch diejenige Gruppe bestimmt, welche die Dynamik invariant lässt. Realiter wird in der heutigen Quantenfeldtheorie der Aufbau freilich umgekehrt vollzogen. Man kennt vorweg die geometrisch relevante Untergruppe der dynamischen Gruppe, nämlich die Poincaré-Gruppe, und wählt die Dynamiken unter denjenigen aus, die ihr gegenüber invariant sind. Dieses Verfahren bedeutet aber, dass man über die geometrische Unreflektiertheit der unrelativistischen Quantenmechanik nicht hinausgekommen ist. Man hat lediglich die euklidisch-galileische Geometrie durch die Minkowskische ersetzt, auch für den atomaren und subatomaren Bereich, in dem Einsteins Rechtfertigung der semantischen Konsistenz dieser Theorie nicht in manifester Weise anwendbar bleibt.

Daher schlage ich ein fundamental anderes Verfahren vor. Die abstrakten Eigenschaften einer Quantentheorie der Wechselwirkung sollen vorweg studiert werden, und es soll versucht werden, aus ihnen die dynamische Gruppe und daraus die Geometrie herzuleiten.

In der abstrakten Axiomatik wird die Quantentheorie normalerweise als die Theorie der Zustände eines einzelnen Objekts, oder, wie man auch sagt, Systems formuliert. Der Anschluss dieser abstrakten Theorie an die Wirklichkeit geschieht über den Begriff der Wahrscheinlichkeit. Eine Analyse des Vorverständnisses dieses Begriffs lässt schwerlich eine andere Deutung zu als dass er eine Prognose relativer Häufigkeiten beobachtbarer Ereignisse bezeichnet.

In dieser Fassung tritt das beobachtende und prognostizierende Subjekt in die Theorie ein, zwar nicht als Gegenstand ihrer Beschreibung, aber als Voraussetzung ihres Sinns. Nun ist das Subjekt nicht nur Beobachter, sondern zugleich Teil der Welt. Wir sind, wie Bohr zu sagen pflegte, an die alte Tatsache erinnert worden, dass wir zugleich Zuschauer und Mitspieler im Schauspiel des Daseins sind. Eine semantisch konsistente Quantentheorie müsste diesen Sachverhalt beschreiben. Der erste Anlauf dazu ist die Messtheorie. Sie beschreibt die Messapparate als besondere quantentheoretische Objekte und lässt erst die Art der Kenntnisnahme von den Messapparaten durch das Subjekt undiskutiert. Sie rechtfertigt diesen Schritt durch den in hinreichender Näherung klassischen Charakter der Messapparate (Bohr), den sie dann freilich möglichst explizit nachweisen muss (Jauch[8]). Diese Seite des Problems möge hier undiskutiert bleiben. Wir studieren jetzt nur die Folgen der Tatsache, dass eine semantisch konsistente Quantentheorie notwendig eine Theorie nicht eines einzigen, sondern mehrerer wechselwirkender Objekte sein muss.

Vielleicht erscheint es sehr umständlich, die Vielheit der Objekte und ihre Wechselwirkung durch die Forderung semantischer Konsistenz zu begründen, da doch jeder Physiker weiss, dass es viele wechselwirkende Objekte gibt. Es wäre aber sonst nicht selbstverständlich, dass dieses Moment im Vorverständnis der Physiker den Übergang von der klassischen zur Quantentheorie überleben muss. Die Behandlung einer Vielheit von Objekten als ein einziges Objekt enthält ja einen besonders krassen Unterschied der Quantentheorie von der klassischen Physik und Ontologie. Auch im klassischen Denken kann man eine Vielheit von Objekten als ein Gesamtobjekt beschreiben; jeder wohldefinierte Zustand des Gesamtobjekts legt dann zugleich die Zustände seiner Teilobjekte fest. In der Quantentheorie hingegen bilden unter den Zuständen des Gesamtobjekts diejenigen, in denen die Teilobjekte wohldefinierte Zustände haben (also im eigentlichen Sinn als Teilobjekte existieren) eine Teilmenge vom Mass Null. Eben darum erlegt die semantische Konsistenz der Quantentheorie eine zusätzliche Forderung auf.

Wir versuchen diese Forderung so zu formulieren: In der Näherung, in der die Vorgänge beobachtbar sind, soll die Identität der die Beobachtung ermöglichenden Objekte über eine hinreichende Zeitspanne gewahrt bleiben. Dadurch wird die Gruppe der zulässigen Transformationen eingeschränkt. Sie müssen jedes dieser Objekte in sich transformieren. Diese Forderung lässt sich weiter spezialisieren, wenn wir vom Begriff gleichartiger Objekte Gebrauch machen. Als gleichartig sollen Objekte gelten, deren Zustandsmannigfaltigkeiten und dynamische Gesetze isomorph aufeinander abgebildet werden können. Ein isoliertes Objekt definiert eine Transformationsgruppe, die seine innere Dynamik invariant lässt. Besteht ein Gesamtobjekt aus vielen gleichartigen Objekten, so wollen wir fordern, dass seine Dynamik (die eine Wechselwirkung seiner Teilobjekte enthält) invariant sei gegen eine Transformation, in der jedes der Teilobjekte gleichzeitig durch dasselbe Element der Transformationsgruppe der isolierten Teilobjekte transformiert wird. Gewöhnlich nennt man das: die Wechselwirkung soll dieselbe Symmetriegruppe haben wie die freie Bewegung. Es sei hier nicht versucht, diese Fassung der Forderung über die vorgebrachten Plausibilitätsargumente hinaus streng zu begründen. Sie steht im Einklang mit den heutzutage üblichen Annahmen.

Die Frage ist nun, in wie kleine Teilobjekte man ein empirisch gegebenes Objekt zerlegen kann. Der Begriff des Teilchens, also auch des Elementarteilchens, setzt die Beschreibung im Ortsraum schon voraus, die wir erst begründen wollen. Auch ist die heutige Elementarteilchenphysik eine Physik der Zerlegung und Erzeugung sogenannter Elementarteilchen. Nach klassischer philosophischer Auffassung sollte ein Atom ein Gebilde sein, das unteilbar ist, weil es begrifflich keine Teile hat, nicht eines, dessen denkbare Teile empirisch bisher nicht getrennt werden können. Ausgedehnte "Atome" haben diesem Ideal nie entsprochen, auch Massenpunkte haben noch immer eine additive Eigenschaft, eben die Masse. In der Quantentheorie kann man jeden Hilbertraum multiplikativ auf zweidimensionale Teilräume zurückführen, wobei auch Dimensionszahlen, die keine reinen Zweierpotenzen sind, z.B. durch Symmetrievorschriften, erzeugbar sind. Die Observablen eines zweidimensionalen Zustandsraumes definieren einfache Alternativen. Sie bedeuten die begrifflich ärmsten möglichen empirischen Entscheidungen. Objekte, die einen solchen zweidimensionalen Zustandsraum haben, sind also in der Quantentheorie die einzigen Kandidaten für einen philosophisch strengen Begriff der Unteilbarkeit. Wir machen die Hypothese, dass alle Objekte aus solchen "Urobjekten" (kurz, um die Abstraktion kalauerhaft auszudrücken, "Uren") bestehen. Die Theorie der Ure soll anderswo im Zusammenhang dargestellt werden (vgl. [9] [10]). Hier seien nur ihre Grundzüge angedeutet.

Die Hypothese über die Symmetrie der Dynamik besagt, dass die Dynamik jedes aus Uren bestehenden Gesamtobjekts invariant sein muss gegen die Symmetriegruppe des isolierten Urs. Diese besteht (ausser der zunächst beiseitegelassenen Multiplikation des Zustandsvektors mit einem konstanten Phasenfaktor) aus der SU (2). Der Hilbertraum jedes Gesamtobjekts kann natürlich entsprechend seiner Zerlegung in Ure als Darstellungsraum der SU (2) aufgefasst werden. Nun ist es formal zweckmässig, die Vektoren eines Darstellungsraumes einer Gruppe als Funktionen in einem homogenen Raum der

Gruppe zu schreiben. Der grösste homogene Raum einer Gruppe ist die Gruppe selbst. Also wird es zweckmässig sein, die Zustandsvektoren aller Objekte als Funktionen auf der SU (2) zu schreiben. Die SU (2) ist, als Raum betrachtet, ein reeller dreidimensionaler sphärischer Raum. In diesem Raum sind die Transformationen der SU (2) selbst Cliffordsche Schiebungen eines festen Schraubungssinnes, sagen wir Rechtsschrauben. Die Dynamik aller Objekte muss gegen diese Schiebungen invariant sein. Machen wir die Zusatzhypothese, sie sei auch gegen die Linksschrauben invariant, also gegen die volle metrische Gruppe unseres sphärischen Raums; so definiert diese Invarianzgruppe gemäss Kleins Erlanger Programm eine physikalische Geometrie, d.h., sie definiert eben diesen sphärischen Raum als den Ortsraum der Physik. Ohne die Zusatzhypothese erhielte man eine Dynamik, die im selben Raum darstellbar wäre, aber unter Paritätsverletzung.

Der Vorschlag ist, diese Herleitung ernst zu nehmen, also in ihr eine Deduktion der physikalischen Geometrie aus einer semantisch konsistenten Quantentheorie einfacher empirischer Alternativen zu sehen. Zur Überprüfung muss zunächst die Beschreibung der Zeit studiert und womöglich die Dynamik festgelegt werden. Wenn dies eindeutig gelänge, müsste damit die Basis sowohl für die Theorie der Elementarteilchen wie der Kosmologie (Theorie des Weltraums) gegeben sein, und die Ausführung müsste den Vergleich mit der Erfahrung ermöglichen. Die Probleme der Geometrie werden dabei in drei sukzessiven Modellen oder Annäherungsschritten auftreten.

Betrachtet man, als einfachstes Modell, die ganze Welt als ein Objekt, das aus einer zeitlich konstanten endlichen Anzahl von Uren besteht, so ist unser sphärischer Raum der Weltraum, und das Modell entspricht dem statischen Einstein-Kosmos. Die Impulse aller Objekte (der Teilobjekte der Welt) sind gequantelt. Das isolierte Ur ist das Objekt mit minimalem Impuls. Alle Operatoren haben diskrete Spektren. Also gibt es keinen kontinuierlichen Ortsoperator. Der kontinuierliche Ort im Weltraum ist ein Parameter. Die Metrik im Weltraum ergibt sich aus der Hilbertraummetrik des Urs. Der physikalische Sinn der Hilbertraummetrik liegt im Wahrscheinlichkeitsbegriff. Als Abstand zweier Zustände des Urs kann die Wahrscheinlichkeit gelten, den einen nicht zu finden, wenn der andere vorliegt. Die Abstände werden also nicht am Einzelnur, sondern als relative Häufigkeiten an grossen Gesamtheiten von Uren gemessen. Der Ort als beobachtbare Grösse ist ein klassischer Grenzfall. Je mehr Ure in einem Teilobjekt, desto höher sein möglicher Impuls, desto genauer ist es also lokalisierbar. Die Äquivalenz aller Punkte, die bisher als unbegründetes Postulat erschien, ist nun eine Folge der Wahl eines homogenen Raums der

Gruppe als Grundlage unserer Darstellung. Sie ist in diesem Sinne konventionell, aber eben die einem Vorverständnis sich am ehesten aufdrängende (in ihm unbewusste) Konvention. Die Wechselwirkung muss zur Bildung von Teilchen führen. Sei ν die Anzahl von Uren pro Elementarteilchen, n die Anzahl der Elementarteilchen in der Welt, N die Anzahl der Ure in der Welt, so ist $N=n\cdot\nu$ Die Anzahl ν begrenzt die Lokalisierbarkeit eines Elementarteilchens. Nennt man die verbleibende Ungenauigkeit seiner Lokalisierung seinen Radius r und den Weltradius R, so wird vermutlich gelten $R=r\cdot\nu$ R und r sind die beiden "natürlichen" Längeneinheiten für kosmologische bzw. atomare Längenmessung.

Dieses Modell war nichtrelativistisch. Das ist natürlich, denn die zugrundegelegte abstrakte Quantentheorie benutzt den Raumbegriff nicht, wohl aber den Zeitbegriff. Sie benutzt einen absoluten Zeitparameter, ohne sich um die semantische Konsistenz des Zeitbegriffs, also die Zeitmessung, zu kümmern. Es gibt für sie eine absolute Gleichzeitigkeit aller Zustände eines Objekts. Will man die spezielle Relativitätstheorie berücksichtigen, so muss man schon die Axiomatik der Quantentheorie anders aufbauen. Formal ist das möglich, indem man die Anzahl der Ure zeitlich variabel oder unendlich sein lässt und die Zerlegung eines Objekts in Ure verschieden vornimmt je nach der Wahl der Zeitkoordinate, d.h. des die Zeit messenden Subjekts oder seiner Uhr. Semantisch bedeutet dies, dass die fundamentalen Alternativen nicht lorentzinvariant definiert sind. Z.B. sind zwei Spinrichtungen, die in einem Lorentzsystem entgegengesetzt sind, dies in einem anderen nicht. Die Zeitmessung kann nun genau wie die Ortsmessung als eine statistische Messung an vielen Uren definiert werden. Dieses Modell ist bisher nicht durchgeführt.

Ein drittes Modell müsste die allgemeine Relativitätstheorie einbeziehen. Im ersten Modell haben wir schon einen konstant gekrümmten Raum. Sein Krümmungsmass ist 1/R oder, wenn wir verabreden, die atomare Längeneinheit r = 1 zu setzen, ist es $1/\nu$. Mit der üblichen Schätzung $R= 10^{40}r$ ist dies gerade die Krümmung, die man für einen Einsteinkosmos annehmen müsste. D.h., <u>wenn</u> wir das Problem lösen können, die Wechselwirkung der Ure so zu bestimmen, dass sie die richtige Anzahl $\nu = 10^{40}$ von Uren pro Elementarteilchen liefert, dann werden wir automatisch für die Krümmung eines Einstein-Kosmos den richtigen Wert erhalten, wenn wir den empirischen Wert der Einsteinschen Gravitationskonstanten in Einsteins Grundgleichung einsetzen. Gäbe es nur die mittlere kosmische Raumkrümmung, so wäre dies die vollständige Theorie der Raumkrümmung, d.h., nach Einstein, der Gravitation. Nun gibt es aber vor allem das lokal variierende Gravitations-, also, nach Einstein, metrische Feld. Für dieses lässt sich nur sagen: Wenn es einen Anteil der Wechselwirkung gibt, der sich als Raumkrümmung schreiben lässt, so muss er in der Näherung, in der höchstens zweite Ableitungen der Feldgrössen vorkommen, aus Invarianzgründen der Einsteinschen Gleichung genügen. Ort und Zeit sind aber in dieser Theorie nur in einem relativ zu den Uren klassischen Grenzfall definiert. Also ist es nicht a priori klar, ob eine Quantelung der Gravitation sinnvoll ist, und ebensowenig, ob es einen Grund gibt, die höheren Ableitungen aus der Grundgleichung fortzulassen.

Wenn man überhaupt wagt, eine wegen ihrer Schwierigkeit bisher nicht durchgeführte Theorie in dieser Weise zu schildern, so ist der Sinn davon nur, durch ihre Denkmöglichkeit darauf hinzuweisen, welche Probleme notwendigerweise auftauchen werden, wenn man versucht, Quantentheorie und Relativitätstheorie semantisch konsistent zu vereinigen.

11. Erkenntnistheoretische Reflexion.

Zum Abschluss sei versucht, die etwas gewundenen Gedankengänge dieses Aufsatzes einmal in der ursprünglichen Reihenfolge knapp zu resumieren, und dann in entgegengesetzter Reihenfolge nochmals zu durchlaufen, um die Konsequenzen der bisherigen Resultate für die Ausgangsfragen zu überprüfen.

1. Axiomatik. Die von den Griechen entwickelte axiomatische Darstellung der Geometrie rechtfertigt nicht die Meinung, die Axiome seien evident und die Folgesätze müssten sich daher mit Sicherheit in der Erfahrung bewähren. In Hilberts Auffassung der Axiomatik entscheidet die Mathematik weder über die Wahrheit noch über den Sinn der Axiome. Gibt es eine Methode, zu entscheiden, ob bestimmte Axiomensysteme die räumliche Erfahrung richtig beschreiben?

2. Empirismus. Die Meinung, man könne empirisch über die Wahrheit geometrischer Axiome entscheiden, scheitert zunächst an der empirischen Unbestimmtheit der verwendeten Begriffe. Fände man in einem Lichtstrahlendreieck eine Winkelsumme ungleich zwei Rechten, so bliebe offen, ob vielleicht die Lichtstrahlen keine Geraden sind.

3. Hierarchismus. Man nimmt traditionell an, gewisse Wissenschaften seien anderen so vorgeordnet, dass Ergebnisse der letzteren die Aussagen der ersteren nicht modifizieren können; so sei die Logik allen Wissenschaften vorgeordnet, und die Mathematik, speziell auch die Geometrie, der Physik. Dies soll hier bestritten werden. Eben darum werden zunächst Verteidigungen des Hierarchismus studiert.

4. Kants Fragestellung. Kant begründet die Mathematik auf Konstruktion in der reinen Anschauung und ihre hierarchische Vorordnung vor der Physik darauf, dass sie Bedingung der Möglichkeit von Erfahrung ist, da die reine Anschauung zugleich die Form der empirischen Anschauung ist. Bohrs Auffassung der Messung übernimmt, was hiervon in der modernen Physik haltbar erscheint. Jedoch ist die empirische Anschauung weder euklidisch noch nichteuklidisch, sondern unpräzise, und die reine Anschauung ist schon ein Produkt des Denkens.

5. Konventionalismus. Mathematisch formulierte Gesetze der Physik lassen Transformationsgruppen zu. Eine hierbei nicht invariante Formulierung darf als konventionell betrachtet werden, eine invariante hingegen als echte Aussage. Welche Transformationen sind zuzulassen? Fasst man den Raum als Punktmannigfaltigkeit auf, so sind die Punkte individuell nicht verschieden, jeder sollte also in jeden transformierbar sein. Punkte sind nur durch Hinzeigen unterscheidbar. Wie garantiert man aber, dass der Punkt, auf den man zweimal gezeigt hat, derselbe Punkt war?

6. Operative Begründung der euklidischen Geometrie. Helmholtz und Dingler begründen die euklidische Geometrie auf Annahmen über die Möglichkeit bestimmter Operationen mit starren Körpern. Sie können aber die Möglichkeit nicht ausschliessen, dass diese Operationen real undurchführbar werden.

7. Semantische Konsistenz. Auch die Wissenschaftstheorie kann der speziellen Wissenschaft nicht hierarchisch vorgeordnet werden; sie bedarf der Korrektur durch die Geschichte der Wissenschaft. In der realen Geschichte geht eine Theorie stets von einem Vorverständnis aus, mit dessen Hilfe sie ihren mathematischen Formalismus auf die Wirklichkeit bezieht. In diesem Sinne ist das Vorverständnis der Theorie vorgeordnet. Aber die Theorie kann das Vorverständnis korrigieren. In diesem Sinne ist die Vorordnung nicht hierarchisch. Wenn die Theorie ihr Vorverständnis selbst widerspruchsfrei zu deuten vermag, heisse sie semantisch konsistent.

8. Spezielle Relativitätstheorie. Ihr Vorverständnis enthält die euklidische Geometrie und die Newtonsche Mechanik. Als Formulierung des Vorverständnisses für die euklidische Geometrie können die Helmholtz-Dinglerschen Forderungen an starre Körper gelten, für die Newtonsche Mechanik das Trägheitsgesetz. Die Galilei-Transformation relativiert die Identität des Raumpunkts, die Lorentz-Transformation auch die des Zeitpunkts. Einsteins Diskussion von Uhren und Masstäben ist das klassische Modell einer Herstellung semantischer Konsistenz durch Kritik des Vorverständnisses an Hand der entwickelten Theorie. Methodisch geschieht dies gemäss Ein-

steins Diktum: "Erst die Theorie entscheidet, was messbar ist". Anfangs empirische Fakten wie Trägheitsgesetz und Michelson-Versuch werden am Ende notwendige Konsequenzen der postulierten Prinzipien der Theorie.

9. Allgemeine Relativitätstheorie. Sie verknüpft die durch die empirische Gleichheit von schwerer und träger Masse angeregte Reduktion der Gravitation auf Raum - Zeit - Krümmung mit dem davon logisch unabhängigen Prinzip der allgemeinen Kovarianz. "Punkte" mit Identität sind die Ereignisse. Das Kovarianzprinzip ist nur verbunden mit der Forderung der Einfachheit physikalisch gehaltvoll. Die Forderung semantischer Konsistenz wird preisgegeben bezw. auf die zukünftige Theorie der Materie abgewälzt. Die lokale Auszeichnung der Poincaré-Gruppe lässt sich auf die in linearer Näherung mögliche Zerlegung der Materie in unabhängige Teile (z.B. Massenpunkte) zurückführen. Die Auszeichnung topologischer Transformationen dürfte auf der Raum- und Zeitmetrik beruhen.

10. Quantentheorie des Raums. Das Vorverständnis der metrischen Geometrie setzt Annahmen über formkonstante Körper voraus, die erst die Quantentheorie begründet. Die unrelativistische Quantentheorie aber setzt die euklidische Geometrie auch in Dimensionen voraus, in denen es keine festen Körper geben kann. Die Quantentheorie bedarf also einer Reflexion ihrer geometrischen Voraussetzungen. Die dazu nötige Messtheorie erfordert eine Quantentheorie vieler Objekte. Dies führt zur Forderung einer speziellen Symmetriegruppe der Dynamik eines Gesamtobjekts, die alle seine gleichartigen Teilobjekte gleichartig transformiert. Wählt man als gleichartige Teilobjekte die einfachsten möglichen, mit zweidimensionalem Zustandsraum ("Urobjekte" = "Ure"), so ist die resultierende Gruppe die SU (2). Die Zustandsvektoren jedes zusammengesetzten Objektes müssen dann einen Darstellungsraum dieser Gruppe bilden. Sie werden dazu zweckmässig als Funktionen auf die Gruppe geschrieben. Die Gruppe, als Raum betrachtet (sie ist ihr eigener grösster homogener Raum) fungiert dann als Ortsraum. Auf diese Weise lässt sich die ausgezeichnete Beschreibungsweise der Physik durch einen dreidimensionalen reellen gekrümmten Ortsraum begründen. Raum- und Zeitmetrik basieren in dieser Theorie auf der Wahrscheinlichkeitsmetrik des Zustandsraums des Urs. Orte und Zeiten sind also keine Observablen des Urs, sondern nur für grosse Anzahlen von Uren statistisch messbar; sie sind klassische Grössen.

Reflektierender Rücklauf:

10'. Quantentheorie des Raums. Wenn diese Theorie durchführbar ist, so ist der Raum kein der Physik hierarchisch vorgeordnetes Datum, sondern eine genäherte Beschreibungsweise zusammengesetzter Objekte. Der Fundamentalbegriff, der eine Bedingung jeder möglichen Erfahrung formuliert, ist in dieser Theorie der Begriff der empirisch entscheidbaren Alternative. Er setzt ein Verständnis der Zeitmodi in der Gestalt der faktischen Vergangenheit und der möglichen Zukunft voraus, aber schon keinen Begriff von Zeitpunkten. Die Theorie liefert Metrik, Topologie und Dimensionszahl des Raumes. Diese Theorie ist bisher nicht durchgeführt, aber ihre Denkmöglichkeit wirft ein verändertes Licht auf die erkenntnistheoretischen Prämissen der bisherigen Theorien.

9'. Allgemeine Relativitätstheorie. Die Bedeutung der Näherung getrennter Objekte, auf der die lokale Auszeichnung der Poincarégruppe beruht, wurde durch die Annahme der Ure aufgeklärt. Die Quantentheorie der Elementarteilchen ist die bei Einstein fehlende Theorie der Materie. Die Theorie der Ure liefert Topologie und Dimensionszahl des Weltkontinuums. Nur die lokalen Variationen der Raumkrümmung, also die Gravitation im engeren Sinne, wären dann Ausdruck desjenigen Teils der gesamten Wechselwirkung, der sich als Raumstruktur darstellen lässt.

8'. Spezielle Relativitätstheorie. Als Theorie der Poincarégruppe ist sie der ARth methodisch ebenso vorgeordnet wie die Theorie freier Objekte der Theorie der Wechselwirkung.

7'. Semantische Konsistenz. Die skizzierte Theorie wäre dem Ideal der semantischen Konsistenz näher als irgendeine bisherige, damit auch dem Ideal einer einheitlichen Physik. Soweit sie nur das Begriffsfeld voraussetzt, das zum Verständnis des Begriffs der empirisch entscheidbaren Alternative gehört, ist sie eine Art Erfüllung des Programms, allgemeine Gesetze auf Bedingungen der Möglichkeit von Erfahrung zu begründen. Sie würde damit zugleich zeigen, inwiefern sowohl alle früheren Versuche, dies zu leisten, als auch alle früheren Versuche, ohne dies auszukommen, an bestimmten Stellen scheitern mussten. Sie würde die aus empirischem Anlass gemachten, aber durch einen solchen Anlass nie in Strenge gerechtfertigten Hypothesen allgemeiner Gesetze nachträglich aus den Bedingungen der Erfahrung rechtfertigen. Selbst wenn sie durchführbar ist, stösst ihre semantische Konsistenz aber an zwei Schranken. Erstens müsste sie durch eine Theorie der Subjekte ergänzt werden, welche die Subjekte nicht durch Messapparate ersetzt, deren Ablesbarkeit schlicht vorausgesetzt wird; d.h. eine in weiterem Sinne semantisch konsistente Naturwissenschaft müsste die Biologie und Anthropologie mit enthalten. Zweitens ist die Annahme der empirischen Entscheidbarkeit irgendeiner Alternative stets nur eine Näherung. D.h. die objektivierende Naturwissenschaft ist als solche nur ein Näherungsverfahren. Es bleibt hier offen, wie weit wir jenseits dieses Verfahrens zu denken vermögen, auch wie weit es selbst Ausdruck einer geschichtlichen Entwicklungsphase des menschlichen Bewusstseins ist.

6'. Operative Begründung der euklidischen Geometrie. Durch die Quantentheorie kennen wir Recht und Grenzen der operativen Annahme über feste Körper. Es zeigt sich auch, dass die Quantentheorie der einfachen Alternative eine tiefere Begründung der fundamentalen Symmetriegruppen gibt. Die festen Körper sind selbst schon spezielle Darstellungen dieser Symmetrien.

5'. Konventionalismus. Es ist zu hoffen, dass das Problem des geometrischen Konventionalismus durch die Auszeichnung spezieller Gruppen geklärt ist. Freilich ist in diesem Aufsatz der fundamentale Schritt, die Auszeichnung der Wahrscheinlichkeitsmetrik in der Axiomatischen Quantentheorie nicht behandelt.

4'. Kants Fragestellung. Der Kern der Kantschen Theorie, die transzendentale Einheit der Apperzeption und die Selbstunterscheidung des Ich in transzendentales und empirisches Subjekt würde erst in der Theorie der Subjekte (7'.) zur Sprache kommen. Die These, Raum und messbare Zeit seien das klassische, d.h. nach Bohr beobachtbare Schema der Darstellung beliebiger zusammengesetzter Objekte, erinnert an Kants Theorie, sie seien die Formen aller Anschauung. Diese These vermeidet den unkantischen Nebenklang, die Formen der Anschauung hätten etwas mit der speziellen Beschaffenheit der Sinnlichkeit der Spezies Mensch zu tun; sie hält nur den kantischen Bezug auf bewusste endliche Subjekte überhaupt fest. Sie beansprucht auf dem für Kant unerreichbaren Weg über die Quantentheorie die mathematische Struktur des Raum-Zeit-Kontinuums (also, kantisch gesagt, den Beitrag des Verstandes zur reinen Anschauung) zu begründen.

3'. Hierarchismus. Er ist die Hoffnung, Recht und Grenzen des Hierarchismus mit Begriffen aus dem Umkreis der semantischen Konsistenz zu erläutern. Das hierarchisch Vorgeordnete war in den historischen Beispielen meist ein Element des Vorverständnisses, das zugleich als der speziellen Theorie logisch übergeordnet verstanden wurde. So sagte man, alle physischen Vorgänge seien im Raum, also gehe die Theorie des Raums der Physik voran. Das ist nach der hier vertretenen Auffassung auch in einer semantisch konsistenten Quantentheorie des Raumes in gewissem Sinne richtig. Aber dass diese Theorie des Raumes eine Quantentheorie beliebiger zusammengesetzter Objekte ist, konnte man historisch in der Phase der klassischen Physik nicht wissen. Das sachlich Übergeordnete erscheint in einer solchen früheren Phase nicht in seinem wahren, notwendigen Zusammenhang mit dem ihm Untergeordneten, und eben darum erweckt es den Anschein einer Notwendigkeit a priori, die doch nicht ausgewiesen werden kann. Eine tiefergehende Analyse solcher Zusammenhänge müsste zunächst eine Theorie der Logik und ihres Zusammenhangs mit der Erfahrung (Logik zeitlicher Aussagen) aufstellen.

2'. Empirismus. Dass man allgemeine Sätze nicht empirisch begründen kann, ist seit Popper, eigentlich seit Hume, in Wahrheit seit Platon bekannt. Andererseits ist es der historische Hergang, dass sie ständig so begründet werden; in der Darstellung der beiden Relativitätstheorien habe ich versucht, gerade dies zu pointieren. Die Erfahrung motiviert allgemeine Hypothesen, die, wenn sie wahr sind, die empirischen Fakten nachträglich als notwendig erweisen. Poppers Falsifikationsthese ist undeutlich, da Falsifikation selbst schon Theorie voraussetzt; sonst könnte man ja z.B. die euklidische Geometrie durch ein Experiment falsifizieren. Die von Kuhn beschriebene Paradigmenfolge funktioniert, wie Kuhn weiss, nur, wenn die Natur sich "gesetzmässig verhält". Der erfolgreiche Empirismus der Physiker hat die simpel realistische Struktur, dass man an wirkliche Gesetze glaubt, und diese zu erraten sucht. Die hier entworfene Theorie tendiert dahin, das Staunen über 1 000 spezielle oder 6 allgemeine Gesetze auf das Staunen über eine Grundtatsache zu reduzieren, dass nämlich Erfahrung möglich ist. Das Idealschema ist: Wenn Erfahrung möglich ist, d.h. wenn prüfbare Prognosen über präzisierte Alternativen möglich sind, dann ist ihre Theorie die semantisch konsistente Quantentheorie einschliesslich der Quantentheorie des Raums. Dann brauchen wir uns nicht zu wundern, dass die Konsequenzen dieser Theorie für spezielle Fälle empirisch in eben diesen Fällen gefunden wurden. Das philosophische Staunen träte mit dieser Feststellung nur in eine neue Phase.

1'. Axiomatik. Die Auswahl der empirisch relevanten Axiome dürfte zur Genüge besprochen sein. Der Grund der Möglichkeit einer Axiomatik, einschliesslich des geometrischen Vorverständnisses der Metamathematik, ist Gegenstand einer Theorie der Logik und Mathematik.

Anmerkungen

1) Imro Tóth, Das Parallelenproblem im Corpus Aristotelicum. Archive for the History of Exact Sciences, 3, S. 249-421, 1967.

2) Hugo Dingler, Die Methode der Physik, München 1938; Aufbau der exakten Fundamentalwissenschaft, hrsg. v. Paul Lorenzen, München 1964.

3) Nach Abschluss dieser Arbeit erhielt ich das Buch von Jörg Winkler, "Relativität und Eindeutigkeit, Hugo Dinglers Beitrag zur Begründungsproblematik", Monographien zur Philosophischen Forschung, Band 98, Meisenheim am Glan 1973.

4) Thomas S. Kuhn, "The Structure of Scientific Revolutions", Chicago 1962.

5) W. Heisenberg in Dialectica 1948, S. 331-336 und in "Einheit und Vielheit", Göttingen 1972, S. 140-144.

6) Zur Axiomatik dieser Geometrie s. J. Ehlers in "The Physicist's Conception of Nature", ed. J. Mehra, Reidel, Dordrecht, 1973, S. 70-91.

7) P.A.M. Dirac, Phys. Zeitschr. der Sowjetunion 3, Heft 1, 1933. Dazu R.P. Feynman, Rev. of Mod. Physics 20, 267, 1948. Beide abgedruckt in "Quantum Electrodynamics", ed. J. Schwinger, Dover 1958.

8) Joseph M. Jauch, "Foundations of Quantum Mechanics", Addison-Wesley 1968.

9) M. Drieschner, Dissertation Hamburg 1968. Dazu C. F. v. Weizsäcker, "Die Einheit der Natur", München 1971, II, 5 und l.c. 6) S. 635-667.

10) C. F. v. Weizsäcker, l.c. 6), S. 55-59.

QUANTUM PHYSICS AND PROCESS METAPHYSICS

David Finkelstein

Yeshiva University
New York, N.Y.

Classical quantum mechanics is a hybrid of classical concepts (space,time) and quantum concepts (states, tests). A more consistently quantum dynamics is proposed, with space, time and matter replaced by one primitive concept of process. Examples are given of relativistic propagators, mass spectrum and scattering amplitudes computed in such a quantum dynamics. Mass spectra exhibit Brillouin-like zones, as of propagation in the crystal of time. In particular, the electromagnetic and weak interactions may be mediated by the zeroth and first zone of one four-spinor process. Then the range of the weak interaction approximates the fundamental time.

****

The difficulties in theoretical physics today may be due to the uncritical mix of classical (c) space and time translation processes with quantum (q) production and detection processes. Then the history of physics has three chapters, c, cq, and q, of which two have appeared. Here are some calculations of mass spectra and propagators that might be steps towards pure q dynamics.

The primitive concepts in this q dynamics are not space, time or matter, but process. Since we already speak familiarly of transition processes, scattering processes, decay processes, we know what process means, until we are asked to define it; and making it primitive averts that question. My central constitutive principle is neither the vacuum of Newton nor the plenum of Descartes but the plexus : The world is a net of elementary q processes.

Enz/Mehra (eds.), Physical Reality and Mathematical Description. 91–99. *All Rights Reserved.*

This view that process is primary, prior to space, time and matter, goes back thousands of years outside physics. It impinged on physics not long ago in the harsh battle between Newton and Leibniz. Leibniz lost. But relativity and quanta decisively changed the direction of physics. These discoveries have so stimulated process metaphysics in art and philosophy that it would be odd not to try it in fundamental physics.[1] I am doing so.

I will call the elementary (indivisible) q process a monad after Leibniz.[2] A complex of p monads, a p-ad, is described by tensors with p indices, p-adics, in accord with the principles of quantum logic. Every process that is maximally defined has an amplitude

$$a = |E)D|$$

where $|E)$ is a tensor associated with the experimental processes of production, propagation through external environment, and detection; while $)D|$ is a tensor dual to $|E)$, fixed for the system and summing up the dynamical law. If $a = 0$, the process does not go.

One departure of q dynamics from the usual Hilbert space representation of cq mechanics: For the sake of special relativity, the group of the monad must include SL(2,C), while the group of any finite n-ary cq system, U(n,C), does not (by compactness). It is crucial for this synthesis of quantum and relativity principles that there is a significant operational non-unitary quantum mechanics.[3] The key to this extension is that we never measure expectation values but only eigenvalues, and the eigenvalue is a linear concept, not a merely unitary one. This successive extension of logic from c to cq to q is prettiest in algebraic language. c physics uses commutative * algebras. cq physics uses * algebras. q physics uses algebras.

For an n-ary system the corresponding groups are S_n (= symmetric group on n letters) $\subset$ AU(n,C) (= anti-unitary group) $\subset$ AL(n,C) (= antilinear group) ; except that the induced projective groups of the last two should be taken.

Therefore q dynamics entails an extension of relativity from geometry to logic. The fixed cq * leads to an absolute negation. In q dynamics, there are plenty of *'s, but no fixed *. At the fundamental level, which is local, the * is relative to the reference system. The corresponding logic lacks an absolute negation. I call this relativistic quantum logic in distinction to the usual unitary quantum one. It seems an inevitable part of any construction of the world from local elements. The usual * of cq physics is global.

Call a process m-ary if its tensors have indices of m values. Then relativity makes binary processes special, for the group of the binary monad alone has the Lorentz group as a normal subgroup. I also consider quaternary processes, $4 = 2 \oplus 2^H, \dots$ in order to make transient use of the Dirac algebra of γ's. An n-ary p-adic has n^p components and can be called an n^p-vector.

Chemists make static models of reality with colored polystyrene balls and wooden pegs. Now we can make dynamic models of actuality out of such parts. A dyadic process is a ball with two connectors, and the only AL(2,C) invariant dyadics are the identity $\delta = (\delta^\alpha{}_\beta)$ and its complex conjugate $\delta^{\dot\alpha}{}_{\dot\beta}$. To represent the four kinds of indices the connectors may be threaded (left or right) and polarized (male or female). There is no invariant triadic; and the enumeration of invariant tetradics, the balls with four connectors, is a familiar kind of problem.

A dynamics can be classified according to its constituent polyadics. Thus a quaternary dyadic dynamics, more briefly a 4^2 dynamics, is one whose $|D($ is a chain of dyads, pairs of quaternary processes, balls with two connectors.

Example 1. A 4^2 dynamics. Let the dynamical process D be a (linear) sequence (of an unspecified but finite number) of quaternary dyads, identity processes:

$$|D(= 1 \oplus \delta \oplus \delta^2 \oplus \dots$$

$$= \bigoplus_{n=0}^{\infty} \bigodot_{k=1}^{n} \delta(k) \quad ,$$

$$\delta = \delta^\alpha{}_\beta \, ,$$

$$\alpha, \beta = 1,2,3,4.$$

Diagrammed,

$$|D(= \succ\!\!\longrightarrow \; \dots\dots \succ\!\!\longrightarrow \, .$$

The q dynamical law takes the place of (say) Dirac's equation in cq physics.

The amplitude for propagation with momentum transfer $(\bar{p}_\mu) = \bar{p}$ is

$$a = |D(\bar{p},S|$$

where $(\bar{p},S|$ is the experimental tensor representing the production and detection with momentum transfer $\bar{p}$, and S is the further (spin) specification required to complete the description. In cq mechanics the projection for momentum transfer $\bar{p}$ is represented by the dyadic

$$P = \psi_{\bar{p}}(x)\ \psi^{*}_{\bar{p}}(x') = e^{i\bar{p}(x-x')}$$
$$= e^{i\bar{p}\ \Delta\ x}$$

I provisionally take this periodic form as characteristic of momentum control in q dynamics too but to use it I must "quantize", substitute for $\Delta\ x$ a 4-vector formed additively from the dyads in D. This gives

$$\Delta\ x^{\mu} = \tau \bigoplus_{k} \gamma^{\mu}\ (k)$$

with undetermined time τ as a universal conversion factor. This ad hoc correspondence is used for convenience in all the calculations presented here, but violates (one of the eight generators of) GL(2,C) and will be replaced soon. Thus one computes[3]

$$(E| = P \otimes S,$$
$$a = \mathrm{tr}\ AS,$$
$$A = \frac{1}{1-\exp(i\ \gamma^{\mu}p_{\mu}\tau)},$$

where S controls initial and final spins.

This propagator has simple poles at masses

$$m_n = 2\pi n\hbar/\tau c^2, \quad n = 0,1,2,\ \ldots,$$

all describing particles with spin 1/2 obeying the first-order Dirac equation and with double multiplicity for $n \neq 0$. For macroscopic purposes this theory is equivalent to a cq field $\psi(x)$ obeying the Poincaré invariant non-local field equation

$$\psi(x^{\mu} + \gamma^{\mu}\tau) = \psi(x^{\mu}).$$

The spin of this particle is the growing tip of its world process. Each mass value corresponds to a different Brillouin-like zone of propagation in the crystal of time.

Example 2. A 2^2 dynamics. Let

$$|D(= \text{seq } \delta,$$

$$\delta = (\delta^A{}_B)$$

$$A,B = 0,1$$

represent a sequence of binary dyadic identity processes. To have momentum states I add to this D its complex conjugate, representing the result as a quaternary dyadic dynamics with a projection operator Γ (on $\gamma_5 = 1$) for each dyadic :

$$|D(= \text{seq } \Gamma,$$

$$\Gamma = (1 + \gamma_5)/2.$$

(Temporary road; permanent way under construction.) Then (p,S| is as before,

$$a = \text{tr } AS,$$

$$A = \frac{\Gamma}{1-\cos(\gamma^\mu p^\mu \tau)}.$$

This is a spectrum of particles with the same masses and spin as the previous but with second-order wave equation (double poles) instead of first-order. The central zone is a two-component neutrino theory.

Example 3. A $4^2 + 4$ dynamics. This is a q dynamics modeled after a Fermi cq theory of beta decay, with dyadic propagation and one tetradic interaction, a single-scattering approximation. Let

$$|D(= \begin{matrix} I & & I \\ & X & \\ I & & I \end{matrix}$$

where each propagation I is the sequence of dyadics of the 4^2 dynamics (example 1) and X is a tetradic interaction. X can be expressed in terms of the familiar bilinear invariants of beta decay theory with as many coupling constants. Let I_1, I_2 be the input sequences of D, I_3, I_4 the output. To construct the momentum channel (E|, it is necessary to distinguish between x^μ, the external c coordinate of the central tetradic vertex relative to a macroscopic external reference system, say the concrete walls and floor of the laboratory; and $\Delta x_i (i = 1,2,3,4)$, the internal q coordinate of the 4 terminals of D relative to the central vertex, the outer or direct sum

$$\Delta x_i = \pm \bigoplus_{k \in I_i} \gamma(k),$$

over the dyads k in the corresponding sequence I_i of D, with the upper sign for outputs. I set $x_i = x + \Delta x_i$ and $(E| = (p_1 p_2 p_3 p_4 S| = = \bigodot_i e^{ip_i x_i} \otimes S$, a straightforward generalization of the experimental tensor $(\bar{p},S|$ of the previous examples. Now

$$a = |D(E\ |$$

gives the same scattering amplitude as a cq four-particle interaction X with the mass spectrum of example 1 rather than a single fixed mass. Overall energy-momentum conservation follows from the factor $e^{i(\Sigma p)x}$, a c-number.

<u>Example 4. A 2^4 dynamics</u>. Let

$$|D(= X \ldots X = \bigoplus_{n=0}^{\infty} \bigodot_{k=1}^{n} X(k)$$

where each X represents a binary tetradic $X = (\delta^A{}_B\ \delta^{\dot{C}}{}_{\dot{D}})$. This may be imbedded in a quaternary algebra as in example 2, and then

$$X = (1 - \gamma_5 \otimes \gamma_5')/2,$$

where two commuting sets of dyadics, γ^μ and $\gamma^{\mu\prime}$, are used to generate the entire space of tetradics. It is convenient to take the $\gamma^{\mu\prime}$ to be multiplication by γ^μ on the right:

$$\gamma^{\mu\prime}.a = a\gamma^\mu .$$

Then each sequence XX ... X can be thought of as made of two intertwined paths, a direct or γ path and a reverse or γ' path. For each path there is a coordinate difference,

$$\Delta x^\mu = \bigoplus \gamma^\mu \text{ and } \Delta x^{\mu\prime} = \bigoplus \gamma^{\mu\prime},$$

and a contribution to the sum over paths:

$$(\bar{p},S| = e^{ip\Delta x} \bigoplus e^{i\bar{p}\Delta x'} .$$

Now

$$a = D(\bar{p},S|$$

$$= \text{tr } AS,$$

$$A = \frac{X}{1-\cos(\gamma^{\mu}p_{\mu}\ \tau)} .$$

This spectrum of particles has the same masses as before but spin 1. The central zone is equivalent to a cq field ϕ obeying

$$\partial^2\phi = 0,$$

$$X\phi = \phi,$$

the latter constraint giving ϕ vector and axial parts only.

There is internal and external evidence that no physical particle is a linear sequence like examples 1,2,3, but the photon might be the massless transverse part of XX ... X, example 4. There are still certain differences from the usual photon theory:

The projection X of example 4 permits the field ϕ to have an axial vector part as for magnetic current. But this might even be right.

The field ϕ of example 4 is not real (Hermitian), but has a complex phase that is gauge invariant under the center of GL(2,C). I think this results from the uncritical use of the γ algebra, which is not even GL(2,C) invariant, and can readily be mended.

ϕ is not transverse. This is more serious, one of a host of integrability questions presently under study.

If XX ... X is a structural formula for the actual process we know as light, what is the physical meaning of its higher zones ? It is possible that the associated massive vector particle, the photon in its next zone of propagation, is the carrier of weak interactions. Indeed in further work the charge degree of freedom is connected with mass reflection, the $m \to -m$ transformation, the central $m = 0$ zones are automatically neutral, and the other zones charged. This puts $\tau \sim \hbar/40\text{GeV}$.

It is conceivable that the electron and muon are similarly related to the neutrino, the smaller mass spacing between these zones reflecting the greater internal complexity of the lepton process. But I cannot compute this yet.

The first three examples indicate that this discrete q language is rich enough to express the usual cq laws of propagation and interaction. In the fourth example something novel happens. In cq mechanics, duration is prior to interaction. Indeed in gauge theories duration generates interaction, in that the free Lagrangians lead to the interaction Lagrangian. In example 3 we see that in q dynamics the tetradic interaction takes time too, τ. Example 4 is the irredundant logical extreme. There interaction generates duration. In such a theory one need not assign coupling constants but may compute them. This possibility gives pure tetradic q dynamics a special interest.

Discussion

So, as the label cq suggests, there are two roads to the present field theories, from the c side and from the q side, from the classical past and from the quantum future.

The c path seems a halt creation. First (on the zeroth day?) the theorist must make a mythical c world, a Newton machine; and only then can he make the actual world, say by canonical or path integral methods.[4]

The q approach also sums over paths (see the sums in the formulae for $|D($ above) but builds up the actual process quantum jump by quantum jump, and not a mythical ancestor. The next work is the fuller determination of the dynamical law and the semantic rules. The principle conceptual problem: when the dynamics $|D($ is a more general net than the topologically trivial line of these examples, one with true loops, what is the tensor $(\bar{p},S|$ for a momentum channel ?

ACKNOWLEDGEMENT

I am indebted to Graham Frye, Gin McCollum, and Leonard Susskind for frequent discussions of these ideas; to St. Ann's Episcopal School, Brooklyn for support in this writing; and to J.M. Jauch for advice and encouragement at critical times in the work.

FOOTNOTES

1. Process quantum theory was also formulated by R.Giles, J.Math.Phys. 11, 2139 (1970); process thermodynamics, by R.Giles, Mathematical Foundations of Thermodynamics, Oxford, 1964. In a process quantum mechanics the Dirac ket vector represents a preparation process, a dual bra covector represents a detection process, and as A.Peres, Am.J.Phys. (1974, to be published), also puts it, the individual quantum system has no state.

2. But for pressing reasons Leibniz makes his monad a duration process, explicitly not a creation or destruction process, one of several concessions to Aristotle that spoil the picture. On grounds of locality, light emission-absorption, etc., others such as Dignaga, to name the most accessible of a school unknown to Leibniz I assume, factored the least duration process into elementary initial and final processes of creation and destruction, like the Dirac square root of space-time; and since duration is described by a space-time vector, it is natural to describe the elementary processes of creation and destruction by spinors. I use monad for these more elementary processes. See F.E.Manuel, Portrait of Sir Isaac Newton, Harvard, 1968 about the Newton-Leibniz controversy; T.Stcherbatsky, Buddhist Logic I, II, Dover, New York, 1930 (repr.1962) on early process philosophy; and D.Browning (ed.), Philosophers of Process, Random, New York, 1965, on modern process philosophy. I thank C.F.von Weizsäcker for discussions of these matters.

3. For further details see D.Finkelstein, G.Frye and L.Susskind, Space-time Code. V, Phys.Rev. 1974 (to be published), references there, and a paper in preparation.

4. This is possible for the Dirac equation too using c particles which not only spin but roll, 5-dimensional balls on a 4-dimensional space-time table. See A.M.Sutton, Phys.Rev. 160, 1055 (1967).

WHAT HAPPENED TO OUR ELEMENTARY PARTICLES?
(VARIATIONS ON A THEME OF JAUCH[1])

R. Hagedorn

CERN, Geneva

What are we, what is the world made of? This question has been asked ever since man started reflection about himself and his surroundings. The answers have been diverging if not contradictory:

- the really existing thing is the ALL-ONE, the variety which we we believe to see, is an illusion;

- the essence of the world lies in numbers and their ratios;

- everything is made of material atoms without quality; the richness of the material world is a consequence of the inexhaustible manifold of possible combinations;

- the world is made of ideal "geometrical atoms", not material ones; again the richness is due to the many possible combinations;

- the material world is a continuum.

Written records of such and similar thoughts are as old as any other records on philosophical reflection.

Before going further it would be good to fix our terminology; otherwise we might run into a linguistic confusion which would increase the philosophical and empirical one. We shall denote by

A-TOM: the indivisible last building block of matter in the greek sense;

Enz/Mehra (eds.), Physical Reality and Mathematical Description. 100–110.

atoms: the atoms of chemistry;

elementary particles: the last building blocks of matter as we see them today.

While it is not necessary to speak of the chemist's atoms any further (they do not deserve their name), we shall see that the remaining two notions do not mean the same thing, and that there are good reasons to believe that A-TOM's do not exist while elementary particles do.

Let us go back to the above vaguely formulated ideas about the true essence of all existing things: One would think that modern physics has discarded all other pictures in favour of material A-TOMS, changing their name into elementary particles. Indeed, since the advent of Descartes we have become used to split everything into spirit and matter, body and soul, God and world and as far as matter is concerned the choice seems clear. Do we not, after all, *see* atoms and elementary particles, do we not even know their masses, charges, spins, parities and what not? Only a very naive and ignorant person, unable to grasp the simplest experimental and theoretical arguments could deny that - or a person who is sceptical to the last consequence (for whom, if he is fair, the existence of his own nose must be as doubtful as the existence of elementary particles). So it seems.

Things, unfortunately, are not that simple. While we can, as physicists, discard the ignorant as well as the absolute sceptic, there remain serious philosophical as well as empirical difficulties. The philosophical difficulties follow the general pattern of Zeno's paradoxa or of Kant's "Antinomien":

Suppose there are A-TOMs. Then of course one can reduce the variety of the world to the variety of combinations. These A-TOMs as such, however, should then all be equal and indistinguishable from each other. Namely, if they were not, their variety should be due to different patterns made of still more elementary building blocks; finally we thus should arrive at one single sort of A-TOM. (Historically part of this sequence has been followed: molecules → atoms → electrons and nuclei → electrons, protons, neutrons).

This A-TOM is, by hypothesis, material and indivisible. Now either it has spatial extension and then we ask : what is inside? As it does not consist of more fundamental A-TOMs, it must be continuous. Or it is pointlike, but then its mass should reside in some sort of attached field as field energy - here again the continuum forces its way into the picture. Thus the assumption of purely material A-TOMs leads us to the conclusion that they consist of continuous matter. The only way out of this dilemma is to postulate the A-TOM as a logical monstrum about which further questions are not permitted. Unfortunately this does not prove that the continuum picture is true; it can be shown, by similar arguments, to be as absurd as material A-TOMs. So far the logical paradox.

However, apart from the logical difficulties there remain unanswered, in each of these pictures, physical questions of primary importance: if everything consist of material A-TOMs, where are the forces which cause movement and pattern? If everything is continuum, why does it condense into such different shapes? If it is geometry or harmony of numbers, what is the substance governed by these? If it is the ALL-ONE, what causes the illusion of variety?

These doubts do not arise from empty speculations, they find a disturbing parallel in the difficulties forced onto us by the experiment. First of all, quantum effects tell us that neither a pure particle picture nor a pure field picture can be right: matter sometimes appears in form of particles, sometimes in form of fields; neither of these (logically and experimentally) mutually exclusive pictures is "really true", although both are joined without contradiction in quantum mechanics.

Nevertheless, it *could* have turned out that there are elementary entities in the form of one single sort of field together with the associated particle (the field quantum). This would have fulfilled all basic requirements of nonrelativistic quantum theory and would have thus realized a lucky synthesis between the ALL-ONE or the continuum (the field), the material A-TOM (the field quantum) and the "geometrical atom" or the "number atom" (the field equations and their symmetries).

If quantum theory is combined with relativity, things change again and the material A-TOM in the sense of Democritos is no longer possible. This comes through the existence of anti-particles: each of the supposed A-TOMs is surrounded by pairs of A-TOMs and anti-A-TOMs in an ever fluctuating and unpredictable way; such, however, that only the whole system makes up the physical entity "particle". Whether one should call this an "elementary particle" or not is a matter of definition; it certainly is no longer an A-TOM, because just by hitting it hard enough, we may split it into parts - yet without destroying it (even the vacuum is no longer what it used to be: it is everything but empty). It is hard to display the new situation better than by the words of Werner

Heisenberg[2]:

"May I now turn to the problem of the elementary particles. I think that really the most decisive discovery in connection with the properties of the nature of elementary particles was the discovery of antimatter by Dirac. That was an entirely new feature which apparently had to do with relativity, with the replacement of the Galilei group by the Lorentz group. I believe that this discovery of particles and antiparticles by Dirac has changed our whole outlook on atomic physics completely. I do not know whether this change was realized at once at that time, probably it has been accepted only gradually; but I would like to explain why I consider it as so fundamental.

We know from quantum theory that for instance a hydrogen molecule may consist of two hydrogen atoms or of one positive hydrogen ion and one negative hydrogen ion. Generally, one can say that every state consists virtually of all possible configurations by which you can realize the same kind of symmetry. Now as soon as one knows that one can create pairs according to Dirac's theory, then one has to consider an elementary particle as a compound system; because virtually it could be this particle plus a pair or this particle plus two pairs and so on, and so all of a sudden the whole idea of an elementary particle has changed. Up to that time I think every physicist had thought of the elementary particles along the line of the philosophy of Democritus, namely by considering these elementary particles as unchangeable units which are just given in nature and are just always the same thing, they never change, they never can be transmuted into anything else. They are not dynamical systems,they just exist in themselves.

After Dirac's discovery everything looked different, because now one could ask, why should a proton be only a proton, why should a proton not sometimes be a proton plus a pair of electron and positron and so on. This new aspect of the elementary particle being a compound system has at once looked to me as a great challenge. When later I worked together with Pauli on quantum electrodynamics, I always kept this problem in my mind.

The next step in this direction was the idea of multiple production of particles. If two particles collide then pairs can be created; then there is no reason why there should only be one pair; why should there not be two pairs. If only the energy is high enough one could eventually have any number of particles created by such an event, if the coupling is strong enough. Thereby the whole problem of dividing matter had come into a different light. So far one had believed that there are just two alternatives. Either you can divide matter again and again into smaller and smaller bits or you cannot divide matter up to infinity and then you come to smallest particles. Now all of a sudden we saw a third possibility: we can divide matter again and again but we never get to smaller particles because we just create particles by energy, by kinetic energy, and since we have pair creation this can go on forever. So it was a natural but paradoxical concept to think of the elementary particle as a compound system of elementary particles. Of course then the problem arose: what kind of mathematical scheme can describe such a situation?" - So far Heisenberg.

While all this is logically possible with only one single kind of "elementary particle" or "elementary field", nature prefers multiplicity even on this level: the search for fundamental building blocks, which so promisingly reduced matter to associations of molecules, these to bound states of atoms, these further to electrons and nuclei and the nuclei to protons and neutrons, did not stop here; the result is that today we know about 200 (in words: two hundred) "elementary particles".

Does this not simply show that these particles again do not deserve the name elementary? This is more than a linguistic problem. We can, of course, make a list of particles, say neutrino, light quantum, electron, muon, pion, proton, neutron, kaon and *define* these to be elementary; this is the worst method of all because of its arbitrariness. What we would like, is a *criterion* - and here lies the difficulty. Should an elementary particle perhaps be defined as "non-composite"? Then we have difficulties with the already known fact that a particle is accompanied by a cloud of virtual particles and that only the whole thing is a physical particle; also we then should count a decaying particle not among the elementary ones: is not its decay proof of its "compositeness"? Maybe we should give the name "elementary" only to stable particles? Then the α-particle would be elementary, but the pion not. Or should we demand that an "elementary" particle be stable *and* non-composite? Then we would end up with a world containing only neutrinos, light quanta, electrons and protons with no means whatsoever to construct out of these the strange particles.

We might obtain some insight by following the historical development and consider, in each step of breaking down matter into more fundamental components, the fraction $\Delta E/M$, where ΔE is the energy needed to decompose a piece of matter M. Let us take a NaCℓ cristal:

		$\Delta E/M$
mechanical destruction:		$\sim 10^{-16}$
chemical decomposition,	$NaCl \rightarrow Na + Cl$:	$\sim 7 \cdot 10^{-10}$
decomposition of nucleus,	$Na \rightarrow 23$ nucleous:	$\sim 8 \cdot 10^{-3}$
splitting the proton: $\Delta E/M \sim 3 \cdot 10^{+1}$ insufficient!		

The double line marks a fundamental qualitative difference:

- above the line, in each splitting process, the original object disappears and is decomposed into a *definite* number of *definite* kinds of more elementary entities; the farther we proceed, the more energy per rest mass we need: each further step costs roughly six to seven orders of magnitude more; nevertheless we remain significantly below 1.

- below the double line we so far have tried under well defined laboratory conditions with splitting energies of the order of 30 x M - and failed. One might say: wait until we arrive at a factor 10^6 to 10^7 above nuclear splitting; that is, just another step of the same order as before. But there is (apart from no evidence for "proton-splitting" in cosmic ray physics where such energies are available) one other important fact: at these energies the proton *is indeed split* in the sense that a lot of particles come out of it and, at the same time, *it is not split*, because it does not disappear - we might even obtain one or more extra protons (plus an equal number of antinucleons), all exactly equal and without scratch. The proton, then, is not decomposed into more fundamental components but rather into a proton plus other particles, among them possibly protons. And the number and kinds of these other particles are not fixed.

It is this qualitative difference between the splitting reactions above and those below the double line, which suggests the following definition of an elementary particle:

> An elementary particle is a particle which cannot, in a unique and reproducible way, be decomposed into elementary particles.

This recursive definition excludes molecules, atoms and nuclei but includes, for instance, the neutrino, the electron, the proton but also unstable particles like the pion and even all the very short lived resonances of strong interactions. It emphasizes that these particles can be split and therefore must be considered as highly complicated dynamical composite systems: the A-TOMs are dead.

And yet, recent years have brought us not only the discovery of about 200 such elementary particles but also their "periodic system": a classification of them in terms of representations of the symmetry group SU(3). The regularity in atomic and nuclear states was once easily explained by the underlying structure consisting of nucleons and electrons, arranged in various patterns obeying the rules of quantum mechanics and subject to certain symmetry laws. The analogy was too tempting: should the regularity of elementary particle states not indicate that there is the next substructure waiting for discovery? From this point of view it was even fortunate that none of the known elementary particles would fit into the basic, three-dimensional representation of the symmetry group SU(3): the existence of a new triplet of "really" elementary particles, the quarks, could be postulated. Immediately the quark-model was worked out in great detail and, on the whole, with great success[3)]. That so far the mysterious quarks were never seen is no obstacle, because they might have a large mass and stick together with large binding energy. Each attempt to split, for instance, a proton into three quarks, would then involve amounts of energy so large that the one splitting-channel $p \to 3q$ is completely submerged in the many other splitting channels $p \to p$ + other particles, and thus has a unmeasurable cross section. But even if quarks did exist and would be discovered one day, they would never re-establish the old A-TOM picture, because even they would be composite, dynamical systems: relativistic quantum theory must have its way.

In one sense, however, one could consider them more fundamental than all the others: they would be the least massive material realization of the lowest dimensional representation of SU(3). Apart from that they would be only members of a large, perhaps infinite, set of elementary particles. If one realizes that, one might accept that they even do not exist as material particles and that SU(3) has no three dimensional realization; then quarks would be just a mathematical concept in which the pure idea of the "geometrical atom" resurrects.

While SU(3) symmetry and the quark concept aim at satisfying some obviously deep-rooted desire of our mind, to reduce everything to "elements", there is another, equally deep-rooted need in us to see the world as unity, as an entity in which the "elements" are no longer self-contained, isolated objects but where everything depends on everything, where the whole is more than the sum of its parts and where even the "elements" become real only through their relation to the whole. One should think that these two needs in us are contradictory and that anyone who has an interest in cleaning up his mind and arriving at a tolerable approximation to a conflict-free *Weltbild*, would finally choose either the one or the other.

I believe that this is a mistake; the two aspects are not contradictory but complementary in Bohr's sense, because none of them is complete, while each of them can rightly claim to explain the facts. Unfortunately the mistake is made by many; thus two opponent camps of physicists are the consequence. Let us call the warriors of the one camp the "quarkists" and those of the other the "bootstrappers".

What is bootstrap? Roughly, it is the philosophy of physicists, to whom the unity of the world is more obvious than its diversity and for whom the laws of nature form an interdependent and self-consistent ensemble in which nothing could be changed without changing everything. Extremists may claim that our world is, while not the best of all possible ones, indeed the *only* possible one. It is clear that this is not a scientific statement[4] and that even a moderate bootstrap philosophy can hardly be called so, because in its full generality it has no predictive power. As a philosophical attitude, however, it is very fruitful: it leads one to search for subsets of the world which are only loosely coupled to the rest such as to enable one to disregard this coupling, treat this subset as isolated and then try to take fully into account its internal relations and make the whole self-consistent. Isolating of subsets has been the conditio sine qua non of all science, but the particular emphasis of the bootstrappers lies on making it self-consistent. Applied to the problem of elementary particles, this philosophy allows us to contemplate the world of hadrons and neglect, for some time, the other interactions. The hadrons, all strongly coupled to each other, cannot easily be divided into further subsets. Nevertheless, as an illustration of a bootstrap model with predictive power, one can mention the π-ρ system[5], where the exchange of the ρ-meson provides the forces which can bind two pions into the ρ-resonance. In its predictions the model fails, as expected, because the π-ρ system is coupled to the rest of the hadron world as strongly as it is internally. But this explicit model shows the logical possibility and mathematical feasability of such an approach. It is indeed possible to build and solve a model which is not artificially restricted to a given set of hadrons and in which one begins without knowing how many hadrons there will be in the end. Its methods are borrowed from statistical thermodynamics; the interaction between the particles is introduced in two steps: first of all by allowing unlimited creation and annihilation of particles and secondly by taking the forces into account via phase-shifts and their distorting effect on phase space. All this takes place inside a hadron which thus is visualized as statistically composed of an undetermined number of other hadrons of the same structure. If one requires that this system be self-consistent, one finds that the number of elementary particles in this model is infinite and that each of them is composed of others of all types. The striking bootstrap picture emerging is then[6]:

> there exists an infinity of different elementary hadrons (with an exponentially rising mass spectrum) of which no one is in any way privileged and where each one can be considered simultaneously and without contradiction in three different ways as
> * the object of the model, namely as a closed dynamical system statistically composed of other such systems;
> * the elementary constituent of the composite system;
> * the field quantum of the forces binding the elementary constituents into the composite particle.

It is surprising that the mere requirement of self consistency yields a very detailed and definite result with great predictive power; it is even more surprising that these (numerical!) predictions are experimentally supported.

There are other models and theories based on the same philosophical grounds, though technically much more complicated and very different. The dual models[7] (duality in this context has a technical meaning and has nothing to do with a dualistic *Weltbild*) claim that the scattering amplitude can equally well be written as a sum over all t-channel exchange amplitudes or as a sum over all s-channel resonance amplitudes; the simple mathematical consequence is that there must be an infinite number of resonances and from here a general picture similar to the above bootstrap system of elementary hadrons emerges. Detailed predictions, where they overlap, agree between the two models. One can say that dual models are the present realization of a very ambitious bootstrap approach based on the hope that the concept of the S-matrix, together with the requirements of Lorentz-invariance, crossing symmetry and maximal analyticity might already be a self-contained bootstrap system with only one solution, this being the correct description of physical reality[4]. Though it will be impossible to disprove this belief, practically it has not been fulfilled; probably the task is too formidable. It has not been fulfilled in the sense that the postulates *alone*, without experimental and model input, have not led us where we now stand. If they were complete and consistent and if theoreticians were infinitely clever and productive, then they should have. What indeed was necessary was the discovery of Regge poles on the one and that of duality on the other hand; both were not derived from the basic bootstrat postulates (though perhaps one day derivable).

Thus there are successful bootstrap models of elementary particles and there are successful quark models, too.

Are these two philosophies unreconcilable or are they only the two sides of one and the same medal?

It seems they belong together and may even be synthesized in one single theory: quarkists tend more and more to consider quarks as field quanta of some basic field which governs the world of elementary particles; these, as far as they can become real, are bound states of quarks (and antiquarks) *and* of each other; the quarks themselves might not be real particles; but even if they could realize themselves as massive objects in the laboratory, they again cannot escape the rules of relativistic quantum field theory, which forces them to be composite dynamical systems. In this sense the probably most ambitious theory of elementary particles, Heisenberg's non-linear spinor theory[8)] stands on the same grounds. Finally then bootstrappers and quarkists may make peace and agree that they have everything in common, except the way of formulating the questions.

Where are we now? Material, indivisible, static A-TOMs do no longer exist; on the other hand there are material (i.e. massive) elementary particles and insofar as they are last building blocks, they are not that far from the A-TOM concept; in other respects they are very far from it and only the dialectic interplay of all the old philosophical views:

* the ALL-ONE (bootstrap)
* the harmony of numbers (SU(3) etc)
* the material A-TOMS (particle aspect)
* the continuum (the field aspect)

leads us to a still nebulous but perhaps for the first time non-contradictory (though complementary) picture of elementary particles, full of internal beauty and harmony.

Have we proved that ancient Indian and Greek philosophers gave already the correct answer to one of the most fundamental questions of natural philosophy *and* of science? Certainly not, because all of them were right in some way and none of them in a scientific sense, which means: the philosophies as such had no predictive power and therefore were not falsifiable. Is it then thus that nature has forced the present picture upon the ideal, objective and unprejudiced scientist and that the above philosophies are empty words? Again certainly not, because no science is possible without a philosophical background (cynical people may call it basic prejudices; in doing so they express already a philosophical opinion and lead themselves *ad absurdum*) which as a kind of metascience guides the scientist to the right questions and gives his work sense and direction.

REFERENCES

1) J.M. Jauch, "Are Quanta Real?". A Galilean Dialogue; Indiana University Press 1973.

2) W. Heisenberg, "From a Life of Physics". Evening Lecture at the International Centre for Theoretical Physics, Trieste 1968; published as special supplement to the International Atomic Energy Agency Bulletin, Vienna.

3) M. Gell-Mann, Phys. Letters 8, 214 (1964), G. Zweig, CERN-TH-401, Jan 17 (1964). Review in: J. Kokkedee, The Quark Model, Benjamin, New York (1969).

4) G.F. Chew, Science, 161 (1968) 762, and Physics Today 23 (1970), 23.

5) G.F. Chew and S. Mandelstam, Nuovo Cimento 19 (1961) 752. F. Zachariasen, Phys. Rev. Lett. 7 (1961) 112.

6) R. Hagedorn, Nuovo Cimento Suppl. 3 (1965) 147, and "Thermodynamics of Strong Interactions". CERN lecture notes CERN-71-12. R. Hagedorn and I. Montvay, Nucl. Phys. B59 (1973) 45.

7) See, for instance, the review article by J.H. Schwarz, Physics Reports 8 (1973) 269.

8) W. Heisenberg, "Einführung in die einheitliche Feldtheorie der Elementarteilchen". Hirzel Verlag, Stuttgart (1967).

PARTONS - ELEMENTARY CONSTITUENTS OF THE PROTON?

Sidney D. Drell

S.L.A.C, Stanford

What are we made of? Throughout recorded history this has been one of the most important and disturbing questions to challenge man's imagination.

Poets through the ages have sung of heaven and earth and the elements of which all are created. Recall the elegance and grace with which Tennyson phrased his thoughts in his poem, "Flower In The Crannied Wall":

> "Flower in the crannied wall,
> I pluck you out of the crannies,
> I hold you here, root and all, in my hand,
> Little flower - but <u>if</u> I could understand
> What you are, root and all, and all in all,
> I should know what God and man is."

Six hundred years before the birth of Christ the early Greek philosophers, speaking with more logic and less romance, gave birth to early science in their quest for the fundamental substance of nature. Thales wrote that everything was made of water; Anaximenes suggested air and Heraclitus, fire as the primordial substances. Empedocles proposed that all was made of combinations of the four fundamental substances: air, water, fire and earth.

As their ideas and reasoning became more advanced the early Greek philosophers went beyond appearances and created an invisible world of atoms, i.e. of indivisible constituents, or basic building blocks, out of which all is constructed. According to Democritus,

Enz/Mehra (eds.), Physical Reality and Mathematical Description. 111–123. *All Rights Reserved.*

who was the most important founder of the atomic theory, the atoms were simpler than all the rich, varied phenomena we see around us, but by their motion and behavior they, though not directly observable, control all we do see. In fact Democritus wrote that aside from atoms and empty space the only thing in the world was opinions!

This quest for what the Greeks called the atom - and we now call "elementary particles" - still continues, spurred enormously by the great increase in the resolving power of the experimental explorations that the giant accelerators have made possible. We now know, of course, that there are atoms and nuclei, and we are searching enthusiastically for fundamental building blocks within the proton or neutron itself on a scale of distances more than a million times smaller than the atomic scale. In little more than 60 years we have compressed the scale at this new frontier of elementary particle physics by six orders of magnitude, as illustrated in Fig. 1. We now have some exciting new clues of possible structure at this new frontier that I want to describe.

These clues, and the experiments performed at SLAC that provided them, are fundamentally the same as those that led Rutherford in 1909 to a picture of an atom that we are so familiar with today: built of a small compact nucleus around which electrons circulate in orbits. Recall that this picture of a nuclear atom was derived by scattering alpha particles from matter and observing that they were being scattered at large angles and even into the backward hemisphere far more often than one would have predicted on the basis of the then current ideas of atomic structure. Prior to Rutherford's work the electric charge in atoms was believed to be diffusely distributed and hence should not have exhibited the concentrated electric fields needed to produce such large deflections of the electrically charged alpha particles. The pattern of scattering told him that the original picture of an atom as a homogeneous body, or plum pudding as described by J.J. Thomson who discovered the electron, was wrong. In contrast the atom was found to have a nucleus that was very small and hard, containing all the mass, about which the electrons that carried electricity circulated like planets around the sun. Fig. 2 illustrates the difference between scattering patterns from a nuclear atom and a distributed charge.

On the next scale of probing within the atomic nucleus, structure was once again found. The nucleus itself was revealed to be neither homogeneous nor elementary, but built of individual protons and neutrons that are bound to one another by a still imperfectly understood strong nuclear force. This discovery was made originally by study of the debris emerging from a nucleus when it is given a hard smash by a projectile such as an alpha particle emitted by a

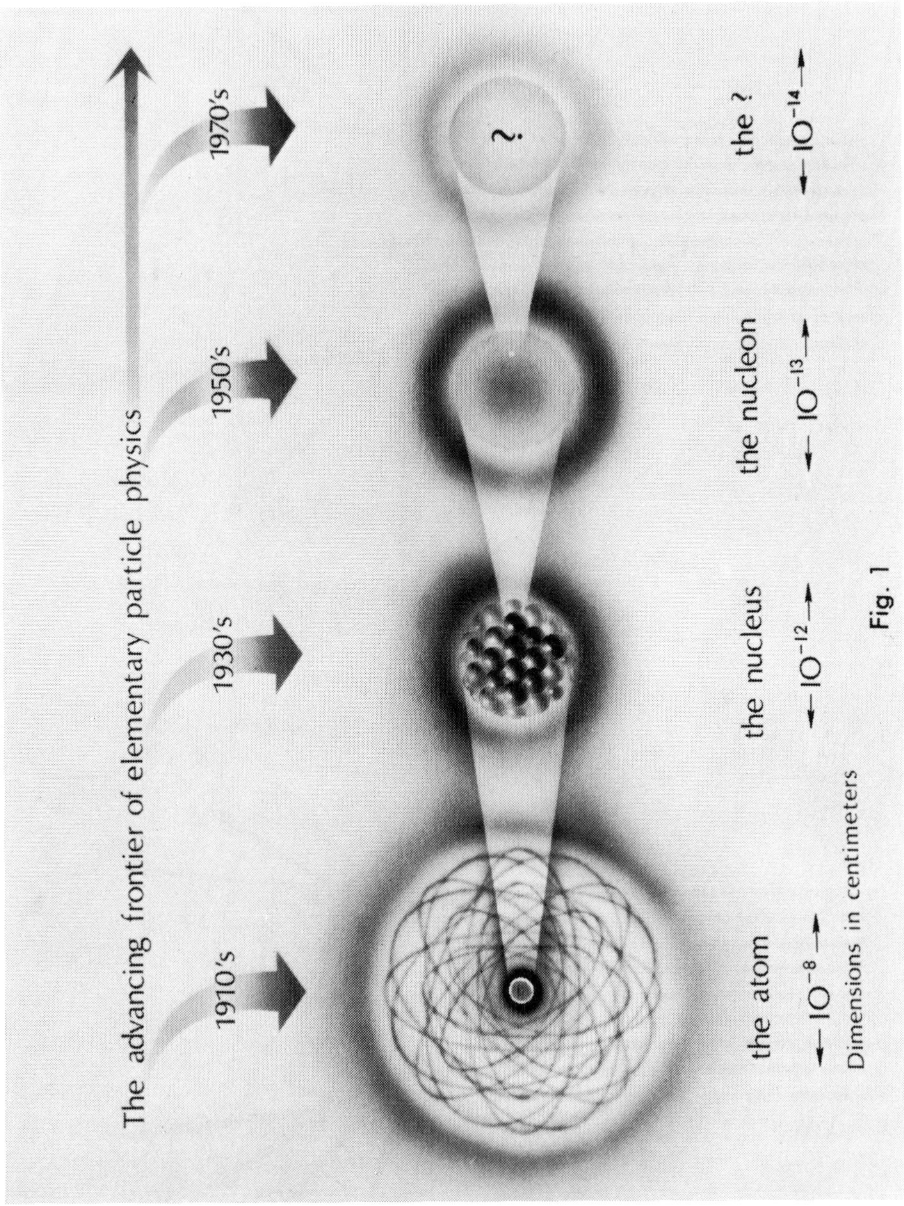
The advancing frontier of elementary particle physics
1910's
1930's
1950's
1970's
?
the atom
the nucleus
the nucleon
the ?
10^{-8}
10^{-12}
10^{-13}
10^{-14}
Dimensions in centimeters

Fig. 1

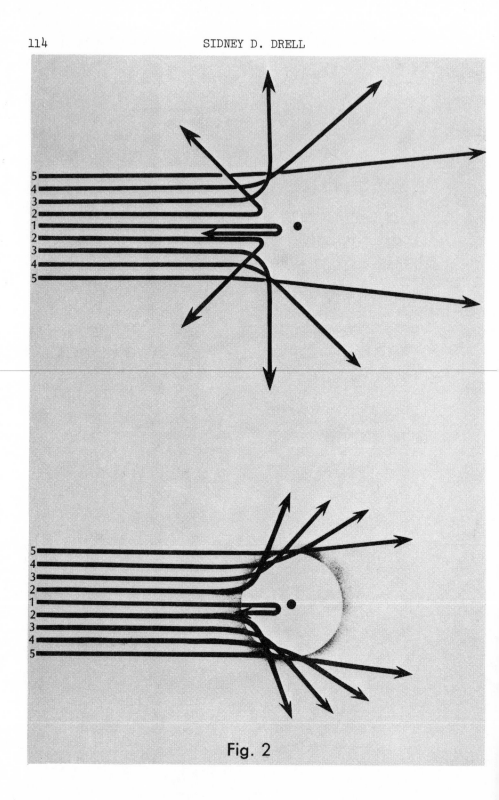

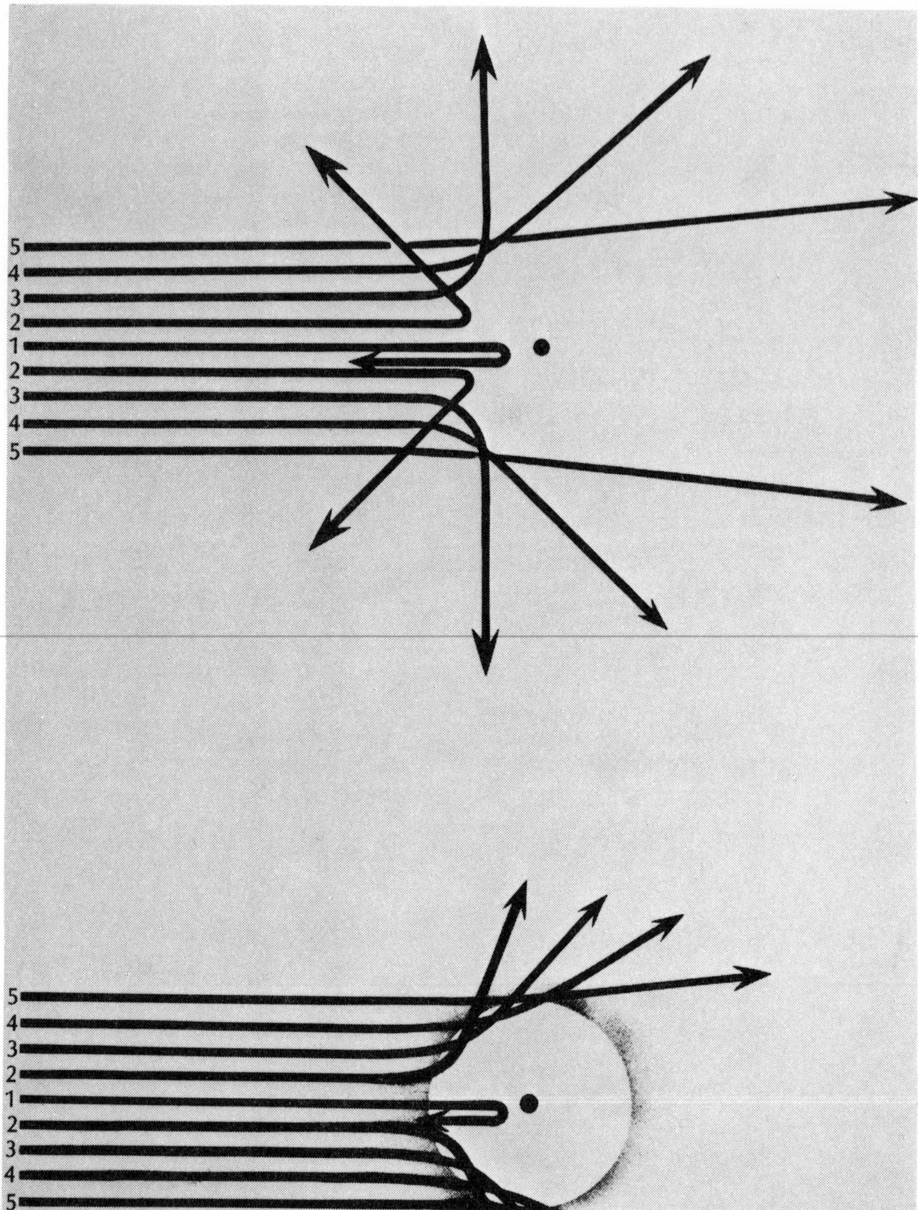

Fig. 2

naturally radioactive source.

At the frontier of exploration of the 1930's, on a scale of distances measured in units of 10^{-13} cm., or ten-millionths of an angstrom, the world view was simple. This view was that all matter, living and dead, was built of combinations of three fundamental constituents - electrons, protons and neutrons.

Though appealing in its simplicity this was indeed a short-lived view and was soon buried in its own debris. We have known now for more than twenty-five years that when target protons are hit very hard - as was done in collisions of particle beams incident on hydrogen targets at the large high energy accelerators - that many fragments are produced from them in inelastic collisions; i.e., a lot of debris emerges when a target proton is smashed by a high energy incident beam. That the proton and neutron are not the indivisible objects or basic building blocks of matter sought by the ancient Greeks quickly became apparent when physicists probed with higher energies on a scale of length that is reduced to fractions of 10^{-13} cm., or fractions of a Fermi. Typically the proton changes its internal state of motion when hit by a high energy projectile such as another proton, but these internal excitations are very short-lived and the proton returns to its normal state in less than 10^{-22} seconds with the balance of energy carried off by mesons, or other short-lived, unstable particles of intermediate mass between that of an electron and a proton. Indeed, some of the fragments emerging from the proton have heavier masses than even the proton itself, since the more energy the greater the mass that can be created by the famous Einstein relation $E = MC^2$.

To find new evidence that the proton is composite and has structure is not surprising to anyone at the present time. The great excitement generated by the recent SLAC inelastic electron scattering experiments is that they reveal for the first time a scattering pattern indicating the presence of strong concentrated electromagnetic fields within a proton.

This scattering pattern is the same as if the proton itself contains point-like electrically charged constituents. More suggestively, the proton seems to have seeds and is like raspberry jam, rather than jelly; in analogy to the original Rutherford observations the proton scatters like a jam rather than a smooth jelly, or Thomson's atomic "plum pudding."

What, then, is inside the proton? What are those seeds in the raspberry jam? Feynman has called them, with characteristic directness, the "partons", or parts of proton. Do they really exist, and are they the unknown elementary particles? Why have they not been already seen in the debris of high energy proton scattering?

At this time we do not know what is going on inside the proton. Part of the problem of identifying these elementary constituents, or partons, in the proton is tied up in the question of what holds them together. What are the forces in nature that glue together the elementary parts of the proton? We know only that these forces are very strong - literally hundreds of millions of times stronger than the electrical forces that bind the electrons in their orbits about the nucleus, and at least hundreds of times greater than the nuclear energies binding the protons and neutrons into a nucleus.

This very strong nature of the forces creates the following problem as illustrated in Fig. 3. When we look into an atom we "see" the individual electrons because they are almost free, and the fact they are joined together in a community - the atom - has very little effect on their individual character or appearance. The electrons and the nucleus of the atom are quite clearly identified as such when we ionize, or break apart an atom. In a system such as an atom with weak forces so that the binding energies are typically less than ten parts per million of the electron rest energy we can identify the constituent electrons and nuclei either by a study of the debris or of the inelastic scattering patterns.

Even though the nuclear binding forces are very much stronger than the electrical ones in an atom this same identification remains clear. Although the constituent nucleons in a nucleus are bound very strongly, their binding energies of typically 10 MeV are no more than 1% of the nucleon's rest energies and protons and neutrons inside a nucleus are still qualitatively, if not quantiatively, much the same as if they were free and separated from one another - that is they are similar as nuclear constituents or nuclear debris. However a difference is already beginning to be apparent between the debris and the constituents.

For an example of the effect of strong forces consider the nucleus of a deuterium atom which is composed of a proton and a neutron. Deuterium is a stable substance in nature - that is, if undisturbed by outside forces, it lives forever. If we were to take an instantaneous snapshot of whatever it is that is whirling around inside of the deuterium nucleus we would see two stable constituents - one proton and one neutron. However, if we separate the neutron from the proton we soon learn that, isolated from the nuclear force, a neutron β-decays in roughly fifteen minutes. This example serves to show how strong nuclear forces completely change the cha-

ATOM: $\dfrac{\text{BINDING}}{\mathcal{M}_e C^2} \sim 10^{-5}$

"SEE" ELECTRON

NUCLEUS: $\dfrac{\text{BINDING}}{M_P C^2} \sim 10^{-2}$

≈ "SEE" NUCLEON
BUT ENVIRONMENT HAS
EFFECT ON LIFETIME,
MAGNETIC MOMENT

PROTON: $\dfrac{\text{BINDING}}{\mathcal{M} C^2} \gtrsim 1$

PROBE OF CONSTITUENTS
& DEBRIS SEARCH MAY
BE VERY DIFFERENT

2462A1

Fig. 3

racter of a neutron. It is stable inside the nucleus of a deuterium atom with a packing fraction of 0.1%, whereas it dies off when it is freed and emerges as debris.

When we move up several more orders of magnitude in strength of the binding energy to the study of the structure of the proton itself, this problem is very much more difficult. Now we are faced with a circumstance in which the binding energy of whatever has been observed to emerge in the debris of the proton exceeds its rest energy. The relation between this debris and the fundamental constituents that may reveal themselves to the instantaneous snapshots of very inelastic electron scattering is one of particle physics' greatest mysteries and most profound challenges. We have here a problem due to the very strong binding that did not confront Rutherford - nor the nuclear physicists.

Several times now I have alluded to the notion of taking instantaneous snapshots by studying the very inelastic collisions of high energy electrons. Electrons have very important and special value as projectiles, or cue balls, for studying the structure of the proton. We have learned from independent experiments done in the past few years at many places that electrons have no detailed structure. Electrons and positrons clashing into one another in colliding rings have been very important in these studies. We know now that electrons are simple and well understood particles. The existing theory of electrons and electromagnetism successfully meets all experimental challenges extending over the entire range of 24 decades in distance from many earth radii, and our understanding of the earth's magnetic field, down to the smallest distances of 10^{-14} cm. which are probed by the very precise measurements of the electron's gyromagnetic ratio, the Lamb Shift contributions to the hydrogen atom's fine structure, and the hyperfine structure of the hydrogen atom. Indeed we think of the electron as one of the probable elementary particles in fact. Therefore, if we scatter electrons from protons, whatever complexities are observed can be blamed on the structure of the proton. This picture is more complicated if, say, two protons collide because we understand neither the projectile nor the target to start with and so we generally have a lot more trouble deciphering the results. A second charm of electrons is that a proton is diffuse, but the electron probes an individual point and so gives a picture of what is happening at one point. In this sense electrons and protons are complementary probes - electrons for taking structure pictures with a known electromagnetic interaction, and protons for probing the nucleon forces and producing much debris.

The reason for emphasizing the very inelastic scattering of

the electrons is the following: As illustrated in Fig. 4 elastic electron scattering refers to an event in which the target particle does not change its internal state but recoils as if it were a rigid billiard ball. Such a scattering measures only the average charge distribution of the target, and this average will be smooth whether the proton does or does not have point-like constituents. In order to determine the details we need to take an instantaneous snapshot and by the uncertainty principle, $\Delta E \Delta t \sim h$, this in turn means a large energy transfert ΔE to the target, i.e. a very inelastic collision if the duration of the interaction, Δt, is to be short. In this case the target is excited to more energetic states or simply disintegrates as illustrated as a result of the energy transferred to it.

The clue that these are point-like constituents or partons within the proton is shown in Figs. 5 and 6 and lies in the appearance of a large continuum tail for very inelastic scattering.

In the case of scattering of electrons from a nucleus this continuum arises from the electrons bouncing off of the individual protons. Each of the protons scatters as a hard billiard ball and each scatters independently when the nucleus is given a hard kick with a large transfer of momentum. The total area under the curve tells us that there are Z such protons, and the location of the peak occurs where we expect it to on simple kinematical grounds for elastic scattering from one of the protons inside the nucleus. The disappearing structure in Fig. 5 as we go from weak to hard kicks, or from small to large momentum transfers to the nucleus, indicates that the nucleus is unlikely to remain bound together either in its normal ground state or in one of its many known excited nuclear states after it receives a hard blow. However the persistence of the continuum tail is evidence of Coulomb scattering from the individual, rock-hard, point-like protons within the nucleus. The scattering is inelastic because there is a transfer of energy breaking apart the nucleus - this is very much like a break shot in pocket billiards. The cue ball bounces elastically from the first ball it encounters but the latter in turn recoils into its neighbors and several fly off in various directions.

When we increase the electron energy by several orders of magnitude to the 20 BeV available at SLAC and scatter from a proton target we again see the large continuum as shown in Fig. 6. The proton for such hard interactions is no longer indivisible. It too can be rent asunder. Once again we find that the scattering pattern for disintegrating the proton depends on the scattering angle of the electron and on the energy transferred from the electron in just precisely that way that we would predict if the

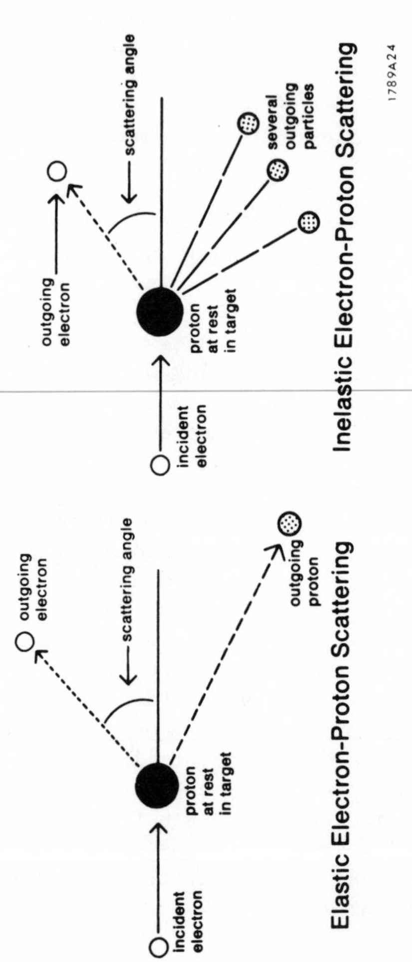

incident electron
proton at rest in target
outgoing electron
scattering angle
outgoing proton
Elastic Electron-Proton Scattering
incident electron
proton at rest in target
outgoing electron
scattering angle
several outgoing particles
Inelastic Electron-Proton Scattering
1789A24

Fig. 4

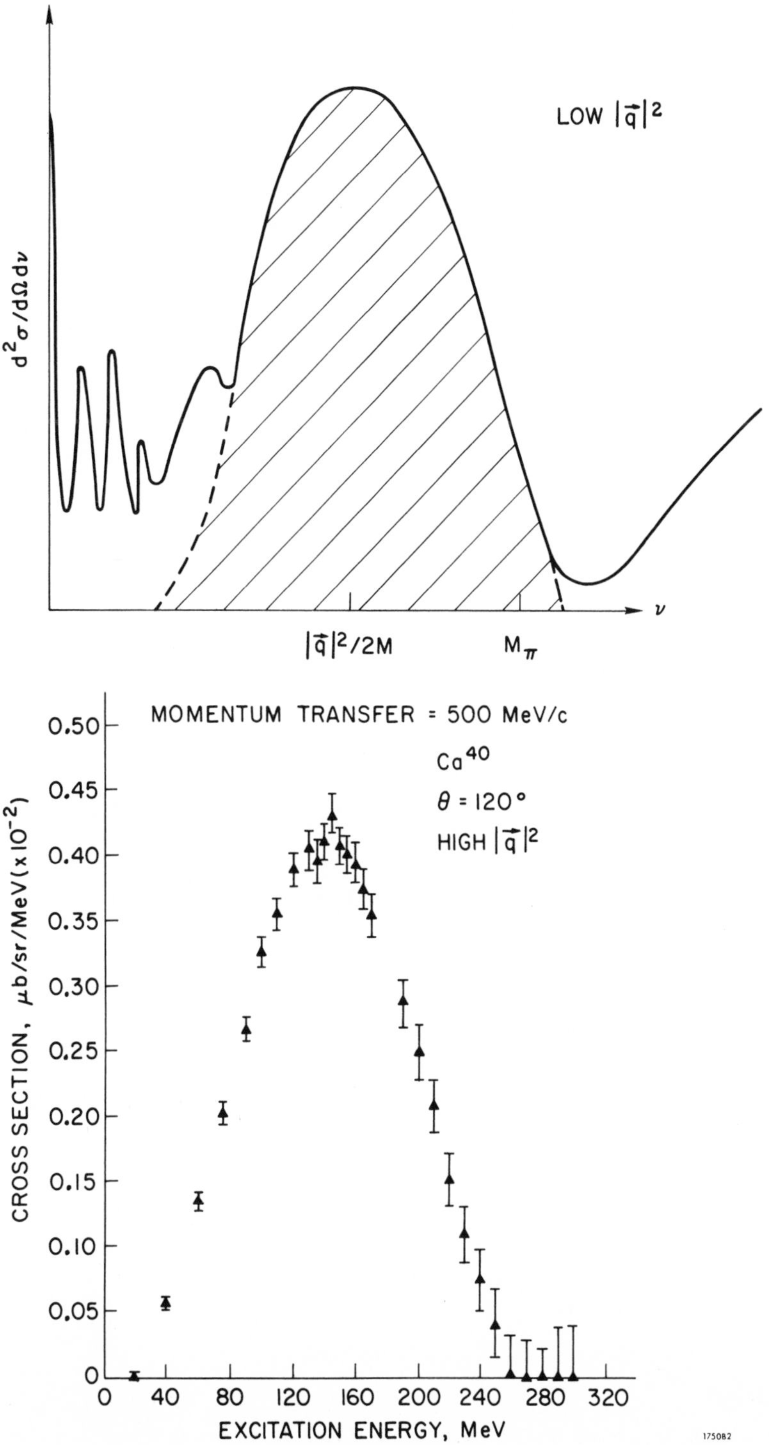
LOW |q|²
d²σ/dΩdν
ν
|q|²/2M
Mπ
MOMENTUM TRANSFER = 500 MeV/c
Ca40
θ = 120°
HIGH |q|²
CROSS SECTION, μb/sr/MeV(x10⁻²)
0.50
0.45
0.40
0.35
0.30
0.25
0.20
0.15
0.10
0.05
0
0
40
80
120
160
200
240
280
320
EXCITATION ENERGY, MeV
175082

Fig. 5

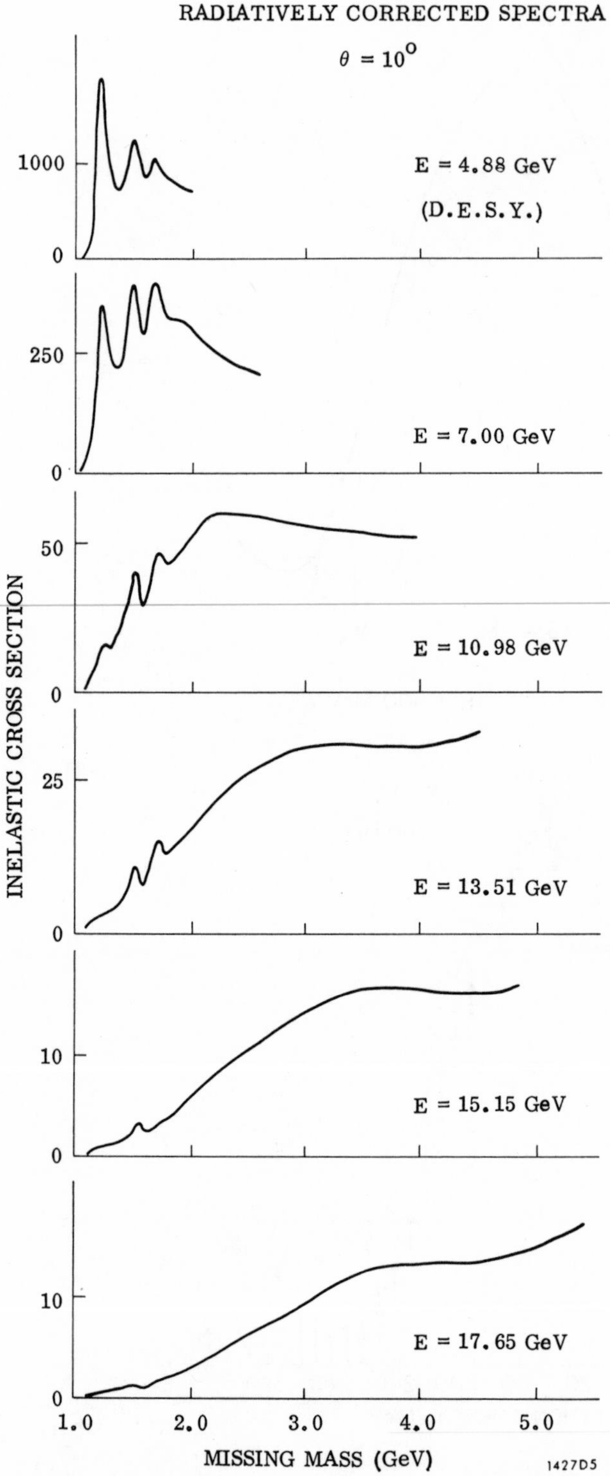
RADIATIVELY CORRECTED SPECTRA
$\theta = 10^{\circ}$
INELASTIC CROSS SECTION
1000
0
E = 4.88 GeV
(D.E.S.Y.)
250
0
E = 7.00 GeV
50
0
E = 10.98 GeV
25
0
E = 13.51 GeV
10
0
E = 15.15 GeV
10
0
E = 17.65 GeV
1.0
2.0
3.0
4.0
5.0
MISSING MASS (GeV)
1427D5

Fig. 6

electron were being elastically scattered from hard billiard ball-like chunks or partons inside the proton. This is the very same pattern that led Rutherford to the nuclear atom.

We have yet to see the tail of the continuum come back down again and therefore cannot say how many of these objects there are. We cannot identify their mass since there is no sharp peak. And what is more, as I have already described, we have no clues of the proton's elementary constituents by searching the debris because of the very strong binding forces, so that is our problem. Perhaps there really are elementary constituents - or Feynman's "partons" - within the proton. We are not yet sure.

Physicists have been searching now for some years for the so-called "quarks", a family of three "things" that are mathematically attractive candidates for the basic building blocks or fundamental constituents of nature. But so far the quarks have avoided detection. They are supposed to have fractional electric charges - either 1/3 or 2/3 as large as the electron's charge. Do the quarks exist? Are they the same as the partons from which the electrons at SLAC seem to be bouncing? One important new result of the past year has been the discovery that the proton and neutron have different inelastic scattering patterns. They both scatter as if composed of elementary partons but their distributions of partons differ. This new clue reminds us of the familiar result that there are also different scattering patterns from atoms with different configurations of valence electrons. More complete and detailed studies of the difference of the neutron and proton patterns will be valuable additions to this new parton spectroscopy for solving the mysteries of the proton's structure and for establishing the properties of its constituent parts. This is analogous to the earlier atomic and nuclear spectroscopies that provided the basic information on which were built our present understandings of atomic and nuclear structure.

So that is where we are now on the new frontier. Our quest has advanced the frontier to a scale six orders of magnitude smaller than atomic dimensions over the past six decades; by its nature it is a never ending quest, for today's candidates for elementary particles may become tomorrow's composite structures. But remember how far we still have to go - we don't yet know the nuclear forces. I hope that a decade hence we shall have pushed ahead yet another decade on the elementary particle physics frontier with the aid of the great new accelerators and clashing rings operating (400 BeV at NAL, ISR at CERN, and SPEAR at SLAC) and under study and design. We may then gain understanding of the very strong nuclear forces and through them of the cataclysmic origin of our universe.

IS THE ZERO-POINT ENERGY REAL ?

Charles P.Enz

Département de Physique Théorique
Université de Genève
1211 Genève 4

The differences between the points of view of Planck and Nernst are discussed by making the distinction between the zero-point energy of material oscillators and of the radiation field, as emphasized by Pauli. The strongest confirmation of Planck's value $\frac{1}{2}h\nu$ is found in the non-existence of a permanent magnetic moment of the electron gas. Attention is called to the compatibility of this value with the freedom to order the factors in the free Hamiltonian at will, as emphasized by Enz and Thellung. Finally, an unpublished letter by Otto Stern revives the episode of his discussions with Pauli on the subject.

***

When Jauch invited Salviati, Sagredo and Simplicio to the shores of Lake of Geneva for a continuation of their Galilean dialogue on the problem of the quanta[1] he could have let the discussion drift into the question of the zero-point energy. There were of course reasons for Jauch's choice, for, historically the problem of the zero-point energy, although intimately related to that of the quanta[2], was soon overshadowed by the more urgent questions of the foundations of quantum mechanics[3].

Recently, however, the problem of the zero-point energy (ZPE) has surfaced again, and this in a rather unexpected spot : classical mechanics of interacting particles[4]. In fact, it had been noticed for some time in computer calculations that in such systems there often exists a threshold energy E^c below which the motion is ordered while above E^c it is stochastic (see Ref.4 for a review).

Enz/Mehra (eds.), Physical Reality and Mathematical Description. 124–132. *All Rights Reserved.*

In this context it is interesting to note that in 1916, that is five years after Planck introduced the notion of zero-point energy[5], Nernst declared ZPE to be ordered energy at zero temperature[6].

The surprizing new fact noted by Galgani and Scotti[7] is that the time-averaged energy of a chain of particles with nearest neighbour Lenard-Jones interactions exhibits a Planck-like distribution. And, even more surprizingly, the action constant entering this distribution is of the order of magnitude of Planck's constant h while, as shown by Cercignani, Galgani and Scotti[8], the threshold energy for stochasticity, E^c, turns out to be of the order of magnitude of Planck's ZPE.

This revival of interest in the ZPE has led me to reconsider the view expressed earlier[9] that one could do without ZPE in all calculations of observable effects. The motivation for these lines, however, is not to revoke the mentioned view but rather to invite to further discussion.

Planck's introduction of the ZPE by his second quantum hypothesis of 1911[5] was actually already a move back in direction of classical physics, in that it confined the quantum discontinuity to the process of emission of radiation by an oscillator[2]. In contrast to the much bolder first hypothesis of 1900[10] in which both emission and absorption were discontinuous, the second hypothesis led to a non-vanishing ground state energy of the oscillator. The reason was that with continuous absorption the energy of an oscillator in the nth cell of phase space with volume h was an average of the n quanta $\varepsilon_n = n\,h\nu$ over the cell[11],

$$\bar{\varepsilon}_n = \int_n^{n+1} h\nu x W_n(x)\, dx$$

with

$$\int_n^{n+1} W_n(x)\, dx = 1.$$

Since in the interior of the cell there is no emission, Planck concluded that for the stationary state the probability W_n was constant and hence $W_n = 1$. Thus[11]

$$\bar{\varepsilon}_n = h\nu(x^2/2)\Big|_n^{n+1} = (n+\tfrac{1}{2})h\nu$$

and the ZPE $\frac{1}{2}h\nu$ is determined by the averaging process !

After this first backward step by Planck's second hypothesis it was clear that people would try to do away completely with discontinuities which, as Poincaré emphasized[12], "challenged the general applicability of our foremost logical tool, the differentia equation" (in the words of Nernst, Ref.6, p.84). The first attempt in this direction probably was the paper of 1913 by Einstein and Stern[13] "in which it is to be shown how, on the basis of the zero-point energy, Planck's radiation formula can be derived in a relaxed, although not quite rigorous, fashion and without any assumption of discontinuities whatsoever" (Ref.13, p.556). For this derivation Einstein and Stern were forced to assume a ZPE of $h\nu$. This value which is twice that of Planck was also postulated by Nernst[6] and seems to be favoured by Cercignani, Galgani and Scotti[8,4].

But Nernst differed from Planck in a more fundamental respect[6]. While Planck associated the ZPE to material oscillators Nernst gave it a universal meaning by associating it with the surroundings, the aether. By attributing ZPE to the radiation field of the universe Nernst ventured into bold but fascinating cosmological speculation : the universe could be saved from the awe inspiring heat death by the mechanism that "from time to time, although exceedingly seldom as can easily be estimated, an atom of a (probably heavy) element pops out, or that from time to time Helium atoms polymerize into such an element" (Ref.6, p.85). Thus Nernst already had the idea of nuclear synthesis which in his view was possible because of the immense energy stored in the aether in the form of ZPE. (With a cutoff frequency of the order of 10^{20}sec^{-1} Nernst calculated an energy density of the order of 200 g/cm^3).

This energy exchange of material systems with the aether led Nernst to the assumption of only statistical energy conservation in atomic systems, an idea which later was taken up by Bohr for atomic transition processes[14] and to avoid the neutrino hypothesis[15] (see also Ref.16, p.785). Nernst[6] also discussed other consequences such as the radiation pressure and the gravitational effect of the ZPE of the aether. I will come back to these questions later. For the moment it suffices to note that Planck's value $\frac{1}{2}h\nu$ of the ZPE has been confirmed both by quantum mechanics and by a number of experiments (see the examples given in Refs.2 and 9).

It seems however, that the strongest confirmation has been overlooked in the literature[2,9]. Indeed, one year after the contribution to the memorial issue for Pauli[9] I noted[17] that a deviation from Planck's value of the ZPE would give rise to a permanent magnetic moment of the electron gas and to a modified value of the Landau diamagnetism. This seems to me of sufficient importance to be discussed in more detail here.

The free energy per unit volume of the electron gas in a magnetic field H can be written as[17)]

$$F = n\,\zeta + \int_0^\infty \phi_\gamma(E)\,\frac{df_o}{dE}\,dE$$

where n is the density of electrons, ζ the Fermi energy and $f_o = (1+\exp\left[(E-\zeta)/k_BT\right])^{-1}$ the Fermi distribution function in the absence of the field. The function

$$\phi_\gamma(E) = \frac{2}{3}\,\beta(\hbar\omega_c)^{5/2} \sum_{0\leq N<(\varepsilon-\gamma)} (\varepsilon-N-\gamma)^{3/2}$$

is written here for an arbitrary value $\gamma\hbar\omega_c$ of the ZPE of the cyclotron frequency $\omega_c = eH/(mc)$; $\varepsilon = E/(\hbar\omega_c)$ where $\hbar = h/(2\pi)$ and $\beta = (2\pi^2\hbar^3)^{-1}(2m)^{3/2}$. The evaluation of $\phi_\gamma(E)$ was given for $\gamma=\frac{1}{2}$ in Eq.(17) of Ref.17 which, apart from oscillatory terms, reads

$$\phi_{\frac{1}{2}}(E) = \frac{2}{3}\,\beta\{\frac{2}{5}\,E^{5/2} - \frac{1}{16}(\hbar\omega_c)^2\,E^{\frac{1}{2}}\}$$

With the identity

$$\phi_\gamma(E) = \phi_{\frac{1}{2}}\,(E+(\tfrac{1}{2}-\gamma)\hbar\omega_c)$$

one then finds for arbitrary γ, by developing in powers of $\hbar\omega_c$,

$$\phi_\gamma(E) = \phi_{\frac{1}{2}}(E)+(1-2\gamma)Z_o(E)\mu_BH-\{1-3(1-2\gamma)^2\}\frac{1}{6}D_o(E)\mu_B^2H^2+\ldots$$

where $Z_o(E) = (2/3)\beta E^{3/2}$ is the number of states with energy smaller than E and $D_o(E) = dZ_o/dE$ the density of states in the absence of the field; $\mu_B = e\hbar/(2mc)$ is the Bohr magneton. With the last expression of $\phi_\gamma(E)$ the permanent magnetic moment at T=0 and H=0 becomes

$$M_o = -\left(\frac{\partial F}{\partial H}\right)_{\substack{T=0\\H=0}} = (1-2\gamma)\mu_B\,\frac{k_F^3}{3\pi^2}\,,$$

where $\hbar k_F = (2m\zeta)^{\frac{1}{2}}$ is the Fermi momentum, and the susceptibility takes the form

$$\chi = -\left(\frac{\partial^2F}{\partial H^2}\right)_{H=0} = \{1-3(1-2\gamma)^2\}\,\chi_{Landau}$$

where

$$\chi_{Landau} = \frac{1}{3}\mu_B^2 \int_0^\infty D_o(E)\frac{df_o}{dE}\,dE$$

is the Landau diamagnetism[17]. Thus for $\gamma \neq \frac{1}{2}$ one has $M_o \neq 0$ which contradicts numerous established facts and $\chi \neq \chi_{Landau}$.

Thus Planck's ZPE of excitations in solids seems to have a well established reality and reconsideration of the view of Ref.9 is timely. So let me rewrite the Hamiltonian of these excitations,

$$H = \sum_i H_i = \sum_i \frac{m_i}{2}\left(\frac{p_i^2}{m_i^2} + \omega_i^2 q_i^2\right)$$

and its Wick-ordered form[9]

$$:H: = \sum_i :H_i: = \sum_i \frac{m_i}{2}\left(\frac{p_i}{m_i} + i\omega_i q_i\right)\left(\frac{p_i}{m_i} - i\omega_i q_i\right)$$

$$= H - \sum_i \hbar\frac{\omega_i}{2}$$

Since the quantization rules do not fix the order of the operators p_i and q_i, H and :H: must be equivalent, together with any hermitean linear combination

$$\sum_i \{(1+2\gamma_i)H_i - 2\gamma_i :H_i:\} = H + \sum_i \gamma_i\hbar\omega_i$$

Hence any value $\gamma_i\hbar\omega_i$ of the ZPE is permissible. How does this agree with the established fact that $\gamma_i = \frac{1}{2}$?

The point made in Ref.9 is that these Hamiltonians differ only in an additive constant and that any measurable quantity is of the form (k_B is Boltzmann's constant)

$$\langle A\rangle = \mathrm{Tr}(Ae^{-H/k_BT}) \,/\, \mathrm{Tr}(e^{-H/k_BT})$$

which is evidently invariant under a transformation $H \to H+\text{const}$. Therefore I think today that the significance of Ref.9 lies not so much in the fact that any calculation can be made by renormalizing the ZPE away but in the proof that Planck's ZPE is not in contradiction with the freedom to order non-commuting factors in the free Hamiltonian at will. It is only the interaction that fixes the value of the ZPE, as exemplified above by the electron gas in a magnetic field.

While Nernst certainly does not seem to have the approval of modern physics for his (or Einstein and Stern's) value $h\nu$ of the ZPE[6,13] his view about the transformability between the ZPE of the aether and heat energy, surprizingly, contains a grain of truth. Indeed, the specific heat of an excitation of the type mentioned above,

$$c_v(T;\nu) = \frac{R}{k_B}\frac{\partial}{\partial T}\frac{h\nu}{e^{h\nu/k_BT}-1},$$

(R is the gas constant), has the property that

$$\lim_{T\to\infty}\int_0^T \left[c_v(\infty;\nu) - c_v(T';\nu)\right] k_B\, dT' = R\,\tfrac{1}{2}h\nu$$

This means that for fixed ν the area between the curves $c_v(T;\nu)$ and $c_v(\infty;\nu)$ equals Planck's ZPE per mole. This equality remains true for an arbitrary spectrum $g(\nu)$ of excitations such that

$$c_v(T) = \int g(\nu)c_v(T;\nu)\, d\nu$$

is the specific heat of any insulating solid (see problem 20 and the Fig. on p.136 of Ref.18). In Nernst's words this means "that with increasing temperature zero-point energy transforms into heat motion and that for this reason the truly absorbed energy $[c_v(T)]$ drags behind that required by the mechanistic theory of heat $[c_v(\infty)]$". (Ref.6, p.90. The insertions in square bracket are my interpretation.).

This is certainly a surprizing relationship, but I very much doubt that it is strong enough to save Nernst's cosmological views. In fact, field theory has done away with vacuum ZPE expediently "by subtraction", i.e., Wick-ordering, and thus got rid of its most trivial divergence problem[16]. The most vigorous opponent of these views probably was Pauli who in his famous "Wellenmechanik" of 1933[19] writes on p.250 :

"At this point it should be noted immediately that it is more consistent here, in contrast to the material oscillator, not to introduce a zero-point energy of $\frac{1}{2}h\nu_r$ per degree of freedom. For, on the one hand, the latter would give rise to an infinitely large energy per unit volume due to the infinite number of degrees of freedom, on the other hand, it would be principally unobservable since nor can it be emitted, absorbed or scattered and hence, cannot be contained within walls and, as is evident from experience, neither does it produce any gravitational field".

This statement by Pauli refutes point by point but without mentioning the name of Nernst, all the properties that Nernst attributed to the ZPE of the radiation field (the aether). But it also makes a clear distinction between the ZPE of material oscillators and that of the radiation field. This distinction already emphasized by Pauli in this famous paper of 1927 on the paramagnetism[20] (see also Ref.16, pp.776 and 783) leaves no material property whatsoever to the ground state of the radiation field, the vacuum. This conclusion is compelling since, as Pauli tells us, any of the properties assumed by Nernst leads to a contradiction.

Pauli had been introduced to the problem of the ZPE by Otto Stern (see Ref.9) who in 1913 made the first correct application of Planck's ZPE[2] in his calculation of the vapour (or sublimation) pressure of solids[21]. Let me, therefore, close with the following recollections by Otto Stern[22]

"Pauli and I have continually discussed the question of the zero-point energy in Hamburg in the early twenties. We always went together from the Institute to the [?] house for lunch, and thereby there were two problems which we talked over time and again."

"Pauli was convinced already at that time of the spherical symmetry of the H-atom against which I bristled violently (you should remember that this happened before Heisenberg and electron angular momentum !)"

"I for my part always tried to convert Pauli to the zero-point energy against which he had the gravest hesitations."

"My main argument was that I had calculated the vapor pressure differences of the Neon isotopes 20 and 22 which Aston had tried in vain to separate by distillation. If one calculates without zero-point energy there results such a large difference that the separation should have been quite easy. The argument seemed (and seems) to me so strong because one does not assume anything else than Planck's formula and the fact that the isotopes are distinguished only by the atomic weight. Myself I have not undertaken any vapor pressure measurements because Aston's experiments seemed to me quite decisive. I have not published this matter at the time which now appears to me very incorrect."

"As to the gravitation of the zero-point energy Pauli has communicated this to me essentially in the form you mention, only much later, at one of my regular visits in Zurich after the war, about 1950 or later. He mentioned, however, that it concerned a calculation which he had done much earlier. When, he did not mention or I have forgotten it. Sorry."

"So, this is all I presume."[22)]

I gratefully acknowledge stimulating discussions with L.Galgani, A.Scotti, A.Rose and J.Mehra.

REFERENCES

1) J.M.Jauch, "Are Quanta Real ?". A Galilean Dialogue (Indiana University Press, Bloomington, USA, 1973).

2) See, e.g., J.Mehra and H.Rechenberg, "Planck's Half-Quanta : History of the Concept of the Zero-Point Energy in the General Development of Quantum Theory", C.R.Acad.Royale Belge (to be published).

3) J.M.Jauch, Foundations of Quantum Mechanics (Addison-Wesley, Reading, USA, 1968).

4) L.Galgani and A.Scotti, Rivista del Nuovo Cimento 2, 189 (1972).

5) M.Planck, Verh.Deutsch.Phys.Ges. 13, 138 (1911).

6) W.Nernst, Verh.Deutsch.Phys.Ges. 18, 83 (1916).

7) L.Galgani and A.Scotti, Phys.Rev.Letters 28, 1173 (1972).

8) C.Cercignani, L.Galgani and A.Scotti, Physics Letters 38A, 403 (1972).

9) C.P.Enz and A.Thellung, Helv.Phys.Acta 33, 839 (1960).

10) M.Planck, Verh.Deutsch.Phys.Ges. 2, 237 (1900).

11) See, M.Planck, Theorie der Wärmestrahlung, 4.Auflage (Barth, Leipzig, 1921), §§169, 170.

12) H.Poincaré, J.de Physique (5) 2, 5 (1912).

13) A.Einstein and O.Stern, Ann.Phys. (Leipzig) 40, 551 (1913).

14) N.Bohr, H.A.Kramers and J.C.Slater, Z.Phys. 24, 69 (1924).

15) N.Bohr, J.Chem.Soc. (London), p.349 (February, 1932), on p.383.

16) C.P.Enz, in The Physicist's Conception of Nature, edited by J.Mehra (Reidel, Dordrecht,Holland, 1973), p.766.

17) C.P.Enz, in Proc.Internat.School of Physics "Enrico Fermi", Course XXII (Academic Press, New York, 1963), p.458.

18) R.Kubo, Statistical Mechanics (North-Holland, Amsterdam, 1971).

19) W.Pauli, "Die allgemeinen Prinzipien der Wellenmechanik", in Handbuch der Physik, edited by H.Geiger and K.Scheel, 2nd edition (Springer, Berlin, 1933), Vol.24, Part 1, p.83.

20) W.Pauli, Z.Phys. 41, 81 (1927).

21) O.Stern, Physik.Zeitschr.14, 629 (1913).

22) Letter from Otto Stern to C.P.Enz, dated Berkeley, 21 January 1960 (in German handwriting, unpublished).

REFLECTIONS ON "FUNDAMENTALITY AND COMPLEXITY"

Max Dresden

Institute for Theoretical Physics
State University of New York at Stony Brook
Stony Brook, New York

It seems fitting that the impetus for the sketchy considerations presented here came from many luncheon conversations in Iowa City, where both Joseph Jauch and the author spent many pleasant and productive years.

****

1. The notion of fundamentality in physics

a) Prologue in Iowa. General comments

There is an enormous and obvious variety in the activities of physicists, this variety refers to the systems studied, the methods used, the types of questions asked. The systems vary from atoms to solids, from electrons to galaxies, from quarks to quasars. Because of the tremendous breadth of the field it is reasonable that physicists would be concerned about what areas, what type of investigations, what methods and especially what laws are the most basic, the most fundamental. There is obviously no general consensus on this point, although most physicists would agree that among the sciences physics is surely the most fundamental discipline. But this unanimity disappears rapidly when different areas within physics are considered.

As an example[1)] one could compare the "fundamental nature" of the information obtained from experiments with space probes with

1) The title of this section is not unrelated to the example picked to illustrate the point at issue.

Enz/Mehra (eds.), Physical Reality and Mathematical Description. 133–166.

that obtained from high energy deep inelastic electron scattering of SLAC, or proton-proton scattering experiments at CERN. It might be argued that even though the information obtained from the space probes is important and interesting; the basic laws governing the phenomena are the known laws of quantum electrodynamics. (In this context classical mechanics, classical electrodynamics and non-relativistic quantum theory are all considered as appropriate limiting versions of quantum electrodynamics). The relevant interactions and the necessary concepts[2)] are known, so that the intricate phenomena of space physics can be considered as unusual combinations of known and understood parts of physics. By contrast there is no a priori theoretical structure in which the high energy experiments fit, there are marked uncertainties about the interactions. The experimental data must provide the needed concepts and the laws which govern the phenomena. To the extent that the investigation of a new domain, with the likelihood (or perhaps the certainty) of finding altogether new underlying dynamical laws is "more fundamental" than the investigation of new classes of phenomena based on already known dynamical laws, high energy physics is "more fundamental" than space physics. It is clear that what has been presented here is not really a definition, of "fundamentality" but rather a particular view of what that concept might be. The mere knowledge (or belief) that a bewildering set of phenomena in space physics must in some sense be describable in terms of known dynamical laws is of no particular help in either organizing the data, in constructing the necessary concepts for such an organization, or the prediction of future phenomena. It is more a general dictum whose main function is to serve as a basis for optimism for further studies. This therefore raises the question of a "less partisan" more appropriate definition of fundamentality, which could sensibly be applied to a variety of areas in physics. It is of course not at all obvious, that such a definition would have any relevance or importance as far as physics is concerned. It is the purpose of this paper to show that, in spite of its semi-philosophical and somewhat vague beginnings, the study of a more precise definition of "fundamentality" leads to very interesting physical questions, and interesting physical insights. The questions are not altogether of the traditional type within physics, they deal primarily with the logical deductive and conceptual interrelations between various levels[3)] of description.

---

2) At least most of the necessary concepts are known.

3) The word "level" is defined below, often "scale" is used instead of level.

The results appear to indicate that there is a surprising (although not complete) degree of autonomy, which these various levels possess. By the description "in a certain level"is meant a description which employs certain concepts, laws which relate these concepts to another, however, the description itself is closed, no concepts, laws other than those enumerated enter the description. An example of a "level" of description would be the "classical hydrodynamical level". The physical quantities involved here are the density, local velocity, local pressure and local temperature. The relevant laws are those of classical continuum mechanics, classical thermodynamics. No other laws enter. Another level is the "kinetic or molecular level", where the physical entities are the molecular parameters, their interactions, the operating laws are those of classical particle mechanics. It is pretty clear that to the extent that the descriptions overlap the "molecular level" is indeed the more fundamental[4)]. The relationship between these two levels constitutes the much-studied subject of classical statistical mechanics. But even in this case, which is the simplest known relationship between different scales, the distinction between the deductive logical and conceptual features of the scales can already be discerned. For example the force on a body moving in a fluid under the influence of given external forces can be calculated in the hydrodynamical description, from an appropriate solution of the hydrodynamical equations. In principle this same force can be computed within the kinetic level, from an appropriate solution of the Boltzmann equation. (Actually the solution of the Boltzmann equation would also allow the calculation of the fluctuations of this force; this is impossible within the hydrodynamical level. This is one reason why the kinetic level is more fundamental). Disregarding the rather formidable technical difficulties of the force-calculation on the kinetic level, the results obtained form the hydrodynamical level calculation, can be independently, straightforwardly and unambiguously obtained from the more basic kinetic level. This is an instance of a deductive or logical relationship between these levels. However, in the hydrodynamical description other concepts and features appear, such as the Reynolds number, turbulent motion. A complete understanding of the relationships between the two levels would also explain the kinetic basis for these features.

---

4) A more formal definition of "fundamentality" will be given in section 3 . Since this definition agrees with the intuitive expectations, it is legitimate to use the intuite notion at this point to describe the relationship between the two levels.

This is an instance of the conceptual relationship between the two scales. This is generally a much more subtle matter than the logical relationship between the scales. It is probably possible[5)] to derive the phenomenology of turbulence from an examination of the mathematical stability characteristics of the solutions of the Boltzmann equation. It is, however, very doubtful that without prior experimental knowledge or prior mathematical experience on the hydrodynamical scale, the appropriate concepts would have been introduced purely on the basis of a mathematical examination of the kinetic level. What has been described here for the kinetic and hydrodynamical levels, seems to be a fairly general situation : even though one level may appear more basic than another, it often happens that the less basic level employs descriptive concepts, and intuitive notions which are not only extraordinarily useful, but often essential, and which yet do not appear in a particular natural or obvious manner in the deeper level. This paper attempts to make these pretty vague general statements more precise and concrete. In this process the original question about just how to define fundamentality, will also be answered, although this is less important and less interesting than the discussion to which it gave rise. By contrast the appearance of new concepts on different levels gives an interesting insight in the relationships between various parts of physics. But probably more important than these insights are the questions which come up when physical principles are applied to chemical or biological systems. Surely the mere statement that physical principles govern the behavior of proteins, macromolecules or genetic material does not give a great deal of insight in what is actually happening. These areas all have their own phenomenology based on partially empirical, partially descriptive concepts. The basic problem is how the notions of physics in these complex systems are transcribed to the characteristic chemical or biological concepts. It is clear from experience that these latter concepts only come into play if the underlying system is sufficiently complex. Thus the necessity of introducing new concepts is in some way related to the degree of complexity of the system. The analysis of the relationships between scales, therefore leads in a natural manner to

---

5) Even this statement is mainly an expression of optimism. The mathematical control over the solution of the non-linear Boltzmann equation is very poor. Whether the stability analysis of the linearized Boltzmann equation (which can be carried out) is indeed sufficient to describe the hydrodynamical level to sufficient accuracy is an open question.

the investigations of the "complexity" of systems. But natural or not, the ideas expressed here, suffer from an extraordinary degree of vagueness. For that reason the last section of this paper (section 4) attempts to give a somewhat more formal description of the "complexity" notion. The remainder of this section, section one, contains further comments on the traditional understanding of the ideas of scales and fundamentality. The extraordinary difference between principle and practice will be described and analyzed in sections two and three. The "autonomy" of scales introduced by practical procedures and their "inprinciple" significance for the idea of fundamentality is discussed in sections two and three. The resulting view is not nearly as definite and clear cut as the earlier discussion might indicate.

b) The traditional view of fundamentality

In many ways the discussion of fundamentality in section a, may seem unnecessarily complicated and obtuse. A straightforward definition could be phrased this way "A field A is more fundamental than B, if A explains and describes everything that B does and more." There are a variety of ways in which the "more" could be interpreted, it could mean "with greater accuracy" or "more different phenomena" but the definition as given makes obvious sense, and would appear to express all that is needed. It is important to observe that, however interpreted, the definition of "fundamental", just given uses the idea of "explanation" in an essential fashion.

The notion of "explanation" itself has been extensively investigated in the philosophical literature, but only a very naive and pragmatic approach is necessary for the purpose of this discussion. In the current practive in physics the explanation and description of the behavior of a physical system is carried in terms of the known properties of the constituents of the system. Their properties, together with the interactions between the constituents and the dynamical laws governing the behavior of the constituents are ingredients used in the explanation of the physical behavior. Stated succinctly, for the present purposes an explanation of the behavior of a system consists of "the reduction of the properties of the system to those of the triple : constituents, interactions, dynamics". Consequently (by the definition just given) the study of the interactions, characteristics, and dynamics, of universal constituents is the most fundamental activity in physics. Speaking somewhat pedantically, it is the only "genuinely fundamental" activity. Without being phrased in such a ponderous fashion, the ideas expressed here describe the generally adopted attitude in physics toward the idea of "fundamentality". For example investigations of electron-electron interactions are more fundamental than the study of the collision cross section of apples; Maxwell's equations are more fundamental, than the equation of state of CO_2. Truly fundamental studies (by the same definition) deal with the fundamental interactions (weak, strong, electromagnetic, gravitational) of fundamental particles and the dynamical laws which they obey (the description of quantum electrodynamics, S matrix theory, classical mechanics general relativity, etc). The remainder of physics, whether it is the character of solids, the structure of molecules, the flow of classical or quantum fluids, the structure of stars, or nuclear structure becomes a gigantic addition process based on these three "fundamental" ingredients[6)]. For example, all the characteristics of a solid should follow completely from the electromagnetic interactions between electrons and nuclei and the laws of quantum mechanics. This is an extension of a famous phrase of Dirac "Quantum mechanics explains most of physics and all of chemistry" (Reference 1). There are various extensions of this dictum; chemistry explains all of biology, biology explains all of medicine and psychology. Whatever the status of these last claims may be; it is clear that in physics the prescription given allows a distinction between "more" and "less" fundamental at least for those parts of physics which are connected. For example van der Waals forces are electrical in origin, but they are considered as the combined results of the more basic Coulomb forces. But as long as gravitational and nuclear forces are viewed as distinct and unrelated, no assessment about their relative fundamentality can be made.

It is not essential for the use of this version of the fundamentality notion that the constituents referred to actually exist as free, independent physical entities. If a satisfactory theory of hadrons could be constructed based on a quark picture, the study of the quark properties, quark dynamics and quark interactions would become fundamental; more so in fact than the investigation of the properties of individual hadrons. This fundamental significance is only indirectly tied to the existence and occurrence of free quarks. These brief remarks are probably sufficient to indicate the manner in which most physicists approach "fundamentality" questions. It does not usually play a significant role in the investigations. This is in part because the manner in which a study is conceived and carried out actually defines the level (or scale)

---

6) This particular viewpoint is sometimes called "the fundamentalist" approach.

of its discussion and there is no need to continually restate that level. For phenomena in the electron theory in solids, matters of nuclear structure are generally (not invariably) of no special concern. The nuclear parameters may enter the description, but do not play a significant dynamical or conceptual role. For purposes of nuclear structure, protons and neutrons are the constituents; their eventual make up in terms of quarks or partons, (although important for high energy or hadron physics,) is less important for nuclear physics. The question of "ultimate" fundamentality is more a logical and philosophical than a physical question; and although particular approaches toward that question undoubtedly influence theoretical schemes the success and acceptance of a theoretical structure is more determined by its organizing and predictive power, than by its philosophical basis.

c) Contractions and Scales

The description of a physical system in terms of the known characteristics of the constituents is of course the domain of statistical mechanics. If indeed the "fundamentalist approach" is followed, everything in physics except the search for fundamental laws, between fundamental objects becomes a part of statistical physics. The notion of "scales" and "levels" which are so important for the present discussion are really derived from statistical mechanics. To see how this comes about it is best to refer back to a formulation of statistical mechanics often stressed by Uhlenbeck (Reference 2). The ideas pertinent to the discussion in this paper are summarized by the following.

I. The starting point in this framework is an underlying microscopic dynamics described in terms of dynamical variables $x_1 \ldots x_N$. This corresponds in the formulation of the previous section to the given constituents, given dynamics, given interactions. It is important to note[7], that the variables $x_1 \ldots x_N$ need not be classical coordinates or momenta, they could be wave functions, field operators, S matrix elements. In short the $x_1 \ldots x_N$, stand for any set of variables, which give a complete specification of the microscopic state.

II. The next point (in many ways the most important for this discussion) is the introduction of macroscopic variables $y_1 \ldots y_s$. Written somewhat symbolically

[7] In the formulation given in Reference 2, the $x_1 \ldots x_N$ variables are assumed to be the classical phase space variables.

$$y_i = f_i(x_1 \ldots x_N) \qquad i = 1 \ldots s \tag{1}$$

It is important that S<<N. In the case that the x_i are numbers, the f_i are functions, the y_i again are numbers. In the more general situation, where the x_i might themselves be functions or operators, the f_i are functionals, or projection type operators, respectively. The essential point to stress is that

$$x \rightarrow y \tag{2a}$$

$$y \not\rightarrow x \tag{2b}$$

This states formally that a knowledge of x implies a knowledge of y, but the converse is not true, the "y" specification does not yield the x specification. Furthermore, it is implied that the y description is a good deal less detailed than the x description. If the x and y are real variables, this means that the number of y variables s, is much less than the number of x variables N. Such a description in terms of many fewer variables is called a contracted description in Reference 2, it is a "cruder" description than the x description.

III. A macroscopic or contracted description[8] is called a macroscopic level, or a description on a definite macroscopic scale, if the description is both closed and causal. The closure refers to the circumstance that the y description involves just the y variables and no reference to the x variables need ever be made. This is a somewhat more precise version of the "autonomy" of the scale referred to in section 1.a. The causality of the macroscopic level, (sometimes referred to as the "macroscopic causality"), asserts that the specifications of the initial data on the y level (which is therefore an incomplete specification on the x level) is sufficient to determine the y variables at a latter time. It is clear that if a description possesses the properties of causality and closure it has indeed achieved "total autonomy".

IV. A macroscopic variable y(t) must be a secular variable. This means that there exists time intervals during which the y variables change very little in comparison to the change in the x variables. There must exist a time τ_0 so that for all $\tau < \tau_0$

$$\left|\frac{y(t+\tau) - y(t)}{x(t+\tau) - x(t)}\right| \ll 1 \qquad \text{for } \underline{\text{all}}\ t \tag{3}$$

---

8) It is at this point that the discussion here begins to deviate from that given in References 2 and 3.

Roughly speaking the macroscopic variables $y(t)$ must be slowly varying on the time scale of the microscopic variables.

It is no doubt evident from what has been said that there can be a sequence of successive contracted descriptions. Each one describes the system in a less detailed manner, in terms of fewer dynamical variables. The remarks just made about "macroscopic variables" suffice[9] to discuss the connection between this version of statistical problems and the concerns of this paper; scales and fundamentality. This connection is most conveniently discussed by considering a number of specific questions. 1. If a microscopic description (x description) is given, how does one find a contracted (y) description ? 2. How is it possible that a contracted description which by its very construction gives an incomplete characterization of the physical system, can be closed and causal ? This raises the question of the existence of autonomous scales.

In the traditional form of classical statistical mechanics the question of the characterization of the macroscopic variables in terms of the phase space variables really does not come up. This is precisely because these macroscopic variables are already known from previous experimental or descriptive information. The information from thermodynamics, hydrodynamics, elasticity guarantees the existence of the macroscopic parameters, and the problem is finding the appropriate combinations of microscopic variables to produce the macroscopic parameters. For example if the $x(t)$ are the phase space variables, and if a macroscopic thermodynamic description is desired, (its existence is of course known), it follows that the appropriate y variables must have the property that $\lim_{t\to\infty} y_i(t)$ exists. This is in contrast to $\lim_{t\to\infty} x_i(t)$ which for classical mechanical systems need not exist. The necessary existence of $\lim_{t\to\infty} y_i(t)$ is an expression of the macroscopic approach to thermal equilibrium. The macroscopic parameters and the macroscopic laws are already known, statistical mechanics becomes mainly an a posteriori theory relating two distinct known levels of description, without a great deal of predictive power. However, in more general situations the definition of appropriate y variables is

---

9) Further details about the "structure of statistical mechanics" provided by these properties of the macroscopic variables can be found in Reference 2 and especially Reference 3.

not at all straightforward and usually experimental or phenomenological information is needed to arrive at a suitable contracted description. In that sense, there is no a priori knowledge of the macroscopic variables, nor an exhaustive enumeration of the possible macroscopic variables. If it were possible to construct on a priori - theoretical grounds a set of macroscopic variables, yielding a closed description, then the theory would thereby achieve great predictive power.

From the fact that the contracted description is incomplete relative to the y description, it follows that in some sense any contracted description must contain stochastic elements. For to complete the description certain probability assumptions must be made. The result is a description which employs probability functions of the y variables. In general this means that $\overline{y^2} \neq (\overline{y})^2$, there are fluctuations. The average is usually a time average. In classical statistical mechanics where the y(t) are the macroscopic thermodynamic parameters, the fluctuations are small[10]:

$$| \overline{y^2} - (\overline{y})^2 |^{\frac{1}{2}} \ll \overline{y} \tag{4}$$

Also in classical statistical mechanics, the time averages can via ergodic theorems be related to geometrical averages over appropriate ensembles. In general, all that can be said is that the y description is a stochastic description. The meaning of closure of the description can easiest be explained in connection with classical statistical mechanics. Let $x_i(t)$ $i = 1...N$ be the phase space variables, $\rho(x,t)$ the phase space density, which satisfies the Liouville equation. Let further the macroscopic variables be

$$y_i = f_i(x_1 ... x_n) \qquad i = 1...s \tag{5}$$

Consider now the (ensemble) averages

$$\langle y_i(t) \rangle = \int f_i \, \rho(x_1 ... x_N, t) \, dx \tag{5a}$$

Because of the Liouville equations these averages will satisfy equations of motion. These will relate $\frac{d}{dt}\langle y_i(t) \rangle$ to $\langle y_i \rangle$, to $\langle y_i^2 \rangle$, $\langle y^n \rangle$ and possibly other quantities. In the case that

10) This is imposed as a condition on the macroscopic variables by Uhlenbeck in References 2 and 3. It is justified for those variables by the requirement of a unique equilibrium state.

$\frac{d}{dt} \langle y_i(t) \rangle$ can be written as

$$\frac{d}{dt} \langle y_i(t) \rangle = F(\langle y_i \rangle) \quad (5b)$$

where F denotes some functional dependence, of the single independent variable $\langle y_i \rangle$, the y description is closed. It is clear that only if the fluctuations in y are neglected this can be the case. Thus the closed equations of a contracted level are just the equations for the average values of the fluctuating variables. In these equations the terms depending on the fluctuations have been neglected. The possibility of closed contracted equations is therefore directly related to the legitimacy of an approximation procedure which omits fluctuations. One could improve the description on a given level, by for example including in the closed equation (5b) for $\langle y \rangle$ an external fluctuating term, which simulates the overall effect of the more fundamental level, as a fluctuating background. This is reminiscent of the way in which a fluctuating force $\mathfrak{F}$ in the Langevin equation represents the external fluctuating environment[11]. The resulting equation is of the form

$$\frac{d}{dt} \langle y_i(t) \rangle = F\langle y_i \rangle + \mathfrak{F}(t) \quad (5c)$$

Comments, the resulting picture

I. The picture emerging from these general considerations, shows that from a basic and fundamental dynamics a number of contracted descriptions may be obtained. Experiments or phenomenology show the existence of these contracted descriptions which allow an adequate, more limited description, in terms of fewer variables and new concepts. These contracted descriptions are closed, if it is possible to neglect the fluctuations. This brings in the important matter of the accuracy of the description. Obviously any set of experimental data is of limited accuracy. The best way of phrasing the conditions of the closure or autonomy of the level is that the average fluctuations in the measured variables, of that level shall be less than the experimental accuracy. If this is the case than the contracted description would appear closed. It is worth pointing out, that the closure criteria as given de-

11) One of the few places where this is actually carried out is in the book of Landau and Lifshitz "Fluid Mechanics", Chapter 17.

pends explicitly on the variable under consideration. In this sense the closure of a level does not just depend on the required accuracy of description, but also on the type of experimentations to which the system is subjected.

II. It might well appear that for a known dynamics on the x level, it ought to be possible to enumerate the possible types of contracted closed descriptions. Stated differently the basic dynamics should provide an exhaustive classification of all the possible macroscopic levels. As noted in remark 1, such an enumeration would depend both on the precision required and the class of experiments envisaged. In practice of course the procedure is inverted, the macroscopic level is determined from macroscopic experimentations. But even if it seems very difficult (if not impossible) to say much in an a priori manner about the variety of contracted levels, it is pretty evident that there will not exist just one class of macroscopic variables leading to a closed description of a prescribed accuracy. The very fact that there exist fields which are based on physics, (chemistry is an outstanding example) which have their own at least approximately closed description, already indicates that there must be appropriate macroscopic variables. The same ought to be true for biology, but there the task of finding the relevant macroscopic variables is even more difficult. But even in the much simpler physical situations there is no general way of obtaining the macroscopic variables and the corresponding levels. In those circumstances where the macroscopic formulation is well known as in the case of classical electrodynamics, extensive and far from trivial considerations[12)] are still necessary to obtain the macroscopic electrodynamic equations from those of the classical electron theory (Reference 4). From the difficulties encountered there it would appear doubtful that in circumstances where the appropriate macroscopic formulation is unclear or unknown an analysis of the basic dynamics would suggest the appropriate macroscopic variables. This would require for example, that from a quark dynamics, one would be able to construct the relevant macroscopic variables, which would correctly predict, new hadronic properties.

III. It may appear strange that the existence of possible macroscopic variables, with an accompanying variety of macroscopic levels, which is such a central point in the present considerations,

---

12) The subtlety of the derivations may be inferred from the lengthy history of this subject (References 4 and 5), and from the painstaking and detailed character of the arguments.

is only infrequently stressed in the usual statistical treatments. The reason for this is that in a good deal of statistical mechanics, equilibrium situations are studied. For such situations one can prove that the actual choice of the macroscopic variables doesn't matter,[13] so the whole question disappears. (But already for situations near equilibrium, the choice of macroscopic variables is more important and far away from equilibrium the general situation pertains[14]).

2. The Autonomy of the scale : practice or principle ?

a) Examples, more, examples

To relieve the rather general discussion of the previous section, let us examine the procedures actually employed for the calculation and prediction of physical properties.

a. Consider as a first simple example the motion, particularly the possible wave motion, of fluids in a channel of depth h, length l, width b under the influence of gravity (g). It seems quite obvious that the starting point for the investigation would be the classical hydrodynamical equations. The description takes place on the hydrodynamical level, with parameters such as viscosity, heat conduction, diffusion coefficients introduced empirically. It is possible to analyse this same problem in a molecular (one body) level, starting from the Boltzmann equation[15] However, it would appear very difficult to get any physical information about the behavior of waves in channels, starting from the molecular many body level, by an analysis of the Liouville equation.

Actually the observed qualitative differences in behavior can not be obtained that directly from just the hydrodynamical level. If the channels are deep and wide (b and h large)[16] the wave ve-

---

13) This is in part because the thermodynamic laws are independent of the particular system.

14) The statistical mechanics for systems far away from equilibrium is the least well understood part of the subject, and it is there that new features may appear.

15) It is clear that this level is considerably more difficult than the hydrodynamical, for both the existence and numerical value of the viscosity, heat conduction terms, must now be derived, together with the motion of the fluid.

16) What "large" and "small" means in this context will be discussed below.

locity v depends on the wave length as $v = \sqrt{\lambda g}$. If the channel is shallow and wide (l large, h small) v is roughly independent of λ. In both these circumstances, the force of gravity plays a dominant role. If the channel is very narrow (b, h small) capillary action becomes important and the physical behavior is now primarily determined by the surface tension S. In that case the wave velocity $v \sim \frac{S}{\rho\lambda}$, ($\rho$ is the density), so the dependence of the wave velocity v on the wave length is quite different. This is only one instance of the very different qualitative features of waves in capillaries, which are mainly determined by the surface tension, and waves in wide channels, which are mainly determined by the action of gravity. Strictly speaking both influences : surface tension and gravity, are of course always there, but depending on the numerical values of the parameters, h, b, l, one or the other dominates[17]) giving rise to qualitative differences in the physical behavior. This information can be incorporated in the hydrodynamical level of description; but it doesn't really follow as a logical consequence of that description. An anlytic investigation of the solutions of the hydrodynamical equations, without surface tension terms, as functions of the parameters b, h, and l will not yield the phenomenon of surface tension. On the other hand this phenomenon can be incorporated in terms of the concepts on the hydrodynamical level by introducing either appropriate boundary condtions or additional forces. When this is done the level of the description does not really change but it now includes both gravity and surface tension effects. However, because of the qualitative differences just mentioned the solutions of the equations on this level, can be anticipated to show a non-analytic character (or at least a rapidly changing character) as functions of the numerical parameters, b, h, l.

It is unfortunately not possible to give explicit solutions of the hydrodynamical equations, with external forces, viscosity, heat conduction, and surface tension effects, included for general geometries. Consequently the change from a regime dominated by gravitational effects to a regime dominated by surface tension effects (with the accompanying qualitative changes) can not be exhibited as a mathematical consequence from the hydrodynamical equations. Instead in practice a decision is made as to what regime will be investigated; solutions are constructed, specifically adapted to those special circumstances. Different regimes - then yield different descriptions and once this is done, there is no further possibility to relate these regimes to each other, since the

17) There is obviously an intermediate region as well.

solutions found are intrinsically tied to their respective regimes. This is the pragmatic reason for the "autonomy of the scales", the only way solutions can be constructed in practice is to exploit the simplifications introduced by the special circumstances of a regime. Furthermore these distinct scales are distinguished in practice by different numerical ranges of the relevant parameters. So even within the hydrodynamical level, there exist quite distinct methods of treatment.

It is a natural question to inquire, whether and how, such features as the relative importance of gravitational and surface tension effects, in fluid flow through channels emerge from a "more fundamental" level, of which the simplest would be a molecular level. Even the most casual reflection shows that this is a highly non-trivial matter. It means that the Boltzmann equation has to be written down, for a finite geometry (this geometry defines the "channel", through which the fluid flow takes place); molecular forces between the walls and the fluid have to be introduced, appropriate boundary conditions must be formulated. It is not clear that all this can be done; but the resulting scheme is undoubtedly of great mathematical complexity, surely more difficult to handle than the usual Boltzmann equation[18]. Thus the likehood that the solutions can be obtained in sufficient detail, so that their dependence on the channel dimensions can be investigated is very small indeed. As a further indication of the different character of the molecular level and the hydrodynamical level, consider the same fluid flow problem again, but suppose it is found that the elastic properties of the channel (the capillary) play an important role. (This might occur for example, in physiological problems.) This would not be so difficult to incorporate in the hydrodynamical description, but it would complicate the molecular description quite appreciably. In this case one could not just describe the channel by an additional geometrical boundary condition for the Boltzmann equation of the fluid. Since the capillary participates in the dynamics, a coupled set of Boltzmann equations would be needed. It seems clear that to obtain a preliminary physical insight in the new phenomena, caused by the elastic properties of the capillary, the hydrodynamical, (thermodynamical) level is much to be preferred. This discussion suggests that even this very simple classical problem of the fluid flow through a channel in practice already is treated on at least these autonomous levels:

I. The Navier Stokes equations, a description valid for large channels, surface tension effects are omitted. This is the le-

18) Precise mathematical proofs about the existence and uniqueness of the solution of the ordinary Boltzmann equation are known to be difficult. See Reference 6.

vel usually employed by hydraulic engineers[19].

II. Flow equations for capillaries, the main force is the surface tension; other forces, friction, gravity, are occasionally included. This is the level usually employed by physical chemists.

III. General description of flow in physiological systems, surface tension, osmotic pressures, diffusion processes, elasticity of the channels are introduced phenomenologically in the flow equations. This level is employed by mathematical physiologists.

IV. Although the three previous levels could be based on a molecular level, very little use is made of this possible reduction. The three macroscopic levels are distinguished by the numerical values of macroscopic parameters.

b) The main moral to be learned from the example just discussed is that although the basic laws do no change, they are in different physical circumstances combined in "different proportion", so that the qualitative and quantitative features change dramatically. The "different circumstances" usually refer to different ranges of values of parameters, so that as a consequence of <u>quantitative</u> changes in physical variables, <u>qualitative</u> changes of behavior result. As another illustration of this phenomenon, consider an object, mass m, velocity v, charge Q, spin J, wave length λ under the influence of an external force F, which enters a medium. <u>All</u> the necessary properties of the medium are assumed given, its density, charge density, refractive index, atomistic constitution. The question now is the description of the physical effects after the object enters the medium, such as the motion of - and forces on - the object, the behavior of the medium, the radiation emitted, particles produced. Questions : 1. What is the "appropriate level" of description of this set of phenomena ? 2. What are the appropriate macroscopic variables ?

But it is immediately clear, that without further <u>numerical</u> specification of the physical circumstances, nothing can be said at all. As phrased the question of the appropriate level of the description is a little silly, since widely different physical situations, requiring quite different concepts and methods, are all included in the specifications as given. It is perhaps illumi-

---

19) There is a whole branch of engineering devoted to the study of flow through porous media, its level is a combination of I, II, III, emphatically <u>not</u> IV.

nating to give some examples which indicate just how different these levels can be.

1. classical hydrodynamic level - The medium is a fluid, described on a classical, continuum level, the particle is an oil drop (mass $\sim$ 0.1 gram, velocity $\sim$ 1 cm/sec), the force is gravity.

2. classical Brownian motion level - The medium is a fluid described classically, however, the particle is so light (mass $\sim 10^{-5}$ grams) that Brownian motion must be taken into account.

3. non-relativistic quantum mechanical level - The medium considered is a metal, an electron of energy 20eV is injected. The discussion now is intrinsically quantum mechanical, presupposing knowledge of the band structure of the solid.

4. non-relativistic nuclear level - A nucleus can be considered as a medium,the particle introduced in it might be a neutron (energy 0.1eV). The information given describes a (slow) neutron. The resulting phenomenon is a nuclear reaction and the appropriate description would be in terms of non-relativistic quantum mechanics and a preconstructed nuclear model.

5. relativistic level - The phenomena accompanying the scattering of a $\sim$ 5BeV electron, hitting (as medium) a proton, must be described in a frame-work which allows the creation and annihilation of various types of particles. It must further incorporate the principles of special relativity, and quantum theory. No totally satisfactory formalism exists; it is not clear whether additional concepts or new objects such as partons are necessary on this level.

6. ultrarelativistic level - If the particle injected is a 50BeV pion, and the medium is a proton, another level of description appears necessary, although there is as yet no definite formulation of this level, it appears that apart from relativistic concepts, some type of fragmentation notion (Reference 7) is an essential element in the description. It is important to stress that this notion seems peculiar to this level of description; it does not occur in the other levels mentioned.

It is clear that the list of systems meeting the general conditions can be extended indefinitely. The main point illustrated by these examples is the recognition that in different size or scale ranges, different physical factors out of the total govern the phenomena, so that these different regimes exhibit qualitatively and quantitatively different types of behavior. The actual physical description is matched to the respective scales, leading to different equations and different concepts, for these scales. This general state of affairs is given the name "autonomy of the scale". (It is worth noting that usually the concepts, particular to a scale, are of empirical origin .)

c) The examples just presented do demonstrate that as a matter of practice, physics is carried out in terms of a number of distinct autonomous levels. It might be argued that this is not really a matter of principle at all, but merely a combination of common sense and intellectual economy. Common sense for example strongly suggests that for the investigation of molecular properties it is manifestly silly to consider anything except a point nucleus (which of course can have a spin), but the form factors are of no importance. They provide a uniform background, which does not affect and is unaffected by experiments on the molecular level, and can therefore safely be omitted. Furthermore it is presumably perfectly possible to write down all the interactions, in some way, for example, in a Hamiltonian. In practice most of these interactions contribute very little, so it indeed is a matter of economy not to include those terms, and those interactions which will be neglected at an early stage.

To see whether these remarks which certainly appear very reasonable and suggestive are completely to the point, consider the following situation. Imagine that an extensive series of experiments is carried out on Hydrogen. Specifically the compressibility is measured over enormous (all!) ranges of pressure and temperatures. Imagine further that all these observations are extremely accurate.

The precise question now is on what level are these observations to be explained. According to the admonitions of the previous paragraph it should be more or less evident what factors should be taken into account, which ones can safely be omitted, this then defines the level. Further, the needed starting equations can at least in principle be written down. To see whether and how these suggestions can be implemented in this case it is necessary to decide on the theoretical frame-work and the pertinent interactions. There are a surprising number of possibilities:

(1) Classical kinetic theory of hard spheres of prescribed diameter.

(2) The one particle Schrödinger equation with $V = \frac{1}{r}$.

(3) The many-body Schrödinger equation with Coulomb interactions.

(4) The many-body Schrödinger equation with Coulomb, retarded, and magnetic interactions.

(5) The one (or many)-particle Pauli equation including spin-spin interactions.

(6) The one-body Schrödinger equation with an effective mass.

(7) The Bethe-Salpeter equation.

(8) A relativistic field theory (assuming point protons).

(9) A relativistic field theory, including a description of the proton form factor.

(10) The one particle Dirac equation with $V = \frac{1}{r}$.

Apart from the possibilities enumerated here, one can still ask whether within the frame-work adopted, it is sufficient to include just electromagnetic interactions. In other words, should the frame-work adopted include weak, strong, gravitational interactions in addition to the electromagnetic interactions. If this is so, still other formalisms must be considered.

Faced with these many alternatives, let us now reconsider the previous admonitions. These assert that :

1. In a given situation the <u>level</u> of description is fixed by common sense, physical considerations.

2. It is in principle possible to write down a formalism which includes <u>all</u> the important interactions. It is then only necessary to specialize this general formalism to the case under study, omitting those terms which give a negligeable contribution.

Let us apply these suggestions to the explanations of the compressibility data of Hydrogen. It is best to consider suggestion 2 first. This calls for a choice of the fundamental formalism[20] from (1) to (10). It is clear that the "most fundamental"[21]

20) For argument's sake assume that the listed frameworks (1)-(10) above are indeed the only possible one's.

21) "Most fundamental" is used here in the sense, that the other descriptions are in principle limiting cases of the fundamental description.

approach would be to use (8) in which the system of protons and electrons is described by a relativistic field theory with point protons. In this description the electrons must be treated fully relativistically, with the protons participating in the dynamics but treated non-relativistically. A field theory could probably be set up for such a situation but it is not particularly simple, and it is not a completely rigorous description.

If the extended protons, the electrons and all interactions must be treated fully relativistically, no usable formalism is available. Thus the suggestion that the description should start from a framework including "all interaction" has more of the character of an exhortation, than a usable procedure. The difficulty with the suggestion is that it is much too vague and general. That of course is the trouble with the whole discussion given so far. Mere examination of the above list of possibilities is no guide at all toward deciding on the appropriate formalism. What is necessary is a statement about the needed accuracy of the description and above all a numerical specification of the pressure, density, temperature of the Hydrogen. For unless such numerical information is provided, the system might be Hydrogen gas (at ordinary room pressure and temperature), solid Hydrogen (at high pressure, and low temperature) a neutron star (at very high pressure, high temperatures). The knowledge that the system is composed of electrons and protons, that the interactions are electromagnetic, does not suffice to determine the appropriate level. Numerical information is essential to allow a reasonable physical decision, as to what factors should be included, which ones to omit. This choice depends crucially on the scale (in fact it defines the scale), which in turn depends on the external parameters. For example, to explain the compressibility data, in a temperature range from 50°K - 330°K, for pressure ranges from 0.1 to 1 atmosphere; a description in terms of a classical gas of constituent molecules[22] with intermolecular forces derived from level (2) or (5), is probably quite good. An appropriate Hamiltonian can be set up and the Schrödinger equation can be discussed in a systematic manner, which allows a calculation of the partition function. The compressibility data in the temperature range 0 - 5°K, pressure range 10^3 - 10^5 atmospheres can not be obtained in this manner at all. The investigation here would be carried out using solid state concepts, describing the system in terms of electrons moving in a crystal lattice. The starting point would probably be taken as (6). These two situations, which only differ in the nu-

[22] The possibility of a description in terms of H_2 molecules was not even included as a possible level of description.

merical values of the temperature and pressure, provide an interesting illustration of the independence of the respective levels, and their relationship to a more basic common level. There would appear little doubt that in a formal sense both the description in terms of gas of weakly interacting molecules of H_2 and that in terms of a crystalline solid can be traced back to a Hamiltonian containing just electrons and protons, Coulomb forces and spin interactions (level 5). However, the physical description and the physical concepts used are so different that it seems wholly justified to refer to these descriptions as distinct autonomous levels. Although these frameworks have a common origin, it is clear that the actual reduction of these formalisms to that original Hamiltonian is extremely difficult. It would involve the derivation of molecule formation in one case; the derivation of the existence of a crystalline solid in the other. These concepts are extensively used on their respective levels, but neither occurs as such in a Hamiltonian which contains just electrons, protons and electromagnetic interactions (levels 3, 4 or 5). It is further worth observing that under extreme conditions none of the frameworks listed gives an adequate description of the compressibility of the Hydrogen system. If the densities of the system are in the range 10^{13} - 10^{16} grams/cm^3 (as is the case in neutron stars) the inverse β-decay process $p+e^- \rightarrow n+\nu$, begins to play an increasingly important role. This is a weak interaction; its appearance could in no way be inferred or derived from the class of frameworks employing just electromagnetic interactions. Under similar extreme circumstances, general relativistic features and strong interactions also begin to play a role. In a strict sense a completely general discussion of the compressibility of proton electron systems, (Hydrogen) would require a formalism including all these interactions. It is therefore - without further numerical information, not possible to single out the main interactions. But as the present example again illustrates once the numerical specification is given, a satisfactory framework for that level can be obtained. Such a framework will employ the concepts, useful to that level and even though different levels may in a formal sense have a common "pedigree", they function independently. This example shows further that this is not just a matter of convenience and practice; but that the formal structure of a general framework must be supplemented by numerical information, the accuracy required and estimates and bounds on the neglected terms. This is necessary before such formalisms can be used to describe realistic physical situations. The supplementary information is precisely what defines and specifies the level.

d) Scales and the propagation of errors

There is another factor which plays an important role in producing and maintaining the autonomy of the physical description of a level. That is the unknown sometimes accumulative, sometimes cancelling effect of the omitted terms. In the previous sections the somewhat teneous character of the initial framework was stressed, expressed for example by the doubts about the adequacy of the initial Hamiltonians. But even when this framework is fixed, it is still necessary to make many approximations before the results can eventually be compared with experiment. The point stressed here is that in practically all circumstances one does not have any effective control over the accuracy of the approximations involved. In a typical approximation procedure certain terms are omitted because they are small, other larger terms (or otherwise more important terms) are kept. Even if such a distinction can be made, it is still not always true that the accumulative effect of the omitted terms can be neglected. Occasionally it happens that an apparently unimportant term, gives rise to a qualitative alteration in the behavior of the system. As systems get more complex, as for example in N body problems with N large (or $N \to \infty$) this happens more often[23]. Conversely it also happens that the combined effect of major terms in a Hamiltonian yields an irrelevant background. Thus an examination of what terms are large and small is not invariably a reliable guide in assessing the importance of those terms. Since this situation is not always fully appreciated, the following examples, might be helpful:

I. It is usually assumed that it is legitimate to neglect special relativity in atomic and chemical calculations. This however, may lead to errors of as much as 100 eV (out of 1500 eV) in the calculation of ionization energies of such elements as Fe.

II. Calculations neglecting retarded interactions of electrons can lead to serious errors in absorption spectroscopy in the 1μ range (Reference 8).

III. Surprisingly enough quantum electrodynamics leads to

---

[23] For complex, large systems this possibility becomes increasingly difficult to exclude. The simplest case of such an accumulative effect would be a perturbation series which does not converge. Perturbation theory for large systems ($N \to \infty$) is known to be tricky. If a number of distinct perturbations are involved, the correct discussion becomes very involved.

observable effects in the van der Waals interactions between _neutral_ atoms (Reference 9).

IV. Coulomb effects which appear as large terms in the typical solid state Hamiltonian, have a relatively minor effect on the qualitative aspects of conductivity.

The unknown propagation of errors (enhancing some terms and suppressing others) further changes the qualitative features and so hampers the reduction of the properties of one level to a more basic level. Because of this added difficult in relating the description to a deeper level the "autonomy" of each scale is further increased. In complicated systems and especially in biology, where competition plays a crucial role, a numerically very slight difference might confer a special stability or a competitive advantage, to a system, so that these numerically small effects are amplified to an enormous extent. In those circumstances the small size of a term or an effect is not indicative of its ultimate significance, and the distinction between various levels becomes especially pronounced (Reference 10)[24].

It should be stressed that this effect of the propagation of errors, does _not_ influence _different_ observables in the identical manner. It may very well be that certain "correctly chosen" observables behave in a controlled systematic manner in a perturbation series, while other entities constructed in the same scale, fluctuate or diverge, under an identical perturbation development. This situation is very reminiscent of the choice of the _macroscopic_ variables in terms of the microscopic variables discussed in section 1.c . Instead of insisting that the macroscopic variables must be normal variables (they must not fluctuate too much), one would insist here that the correct "next level macroscopic" observables exhibit a certain stability under successive refinements. These remarks provide some mathematical insight in the way these autonomous levels came about. There are two _extreme_ types of behavior, a system of many degrees of freedom can exhibit relative to perturbations, alterations or errors (in initial data for example). The equations can be stable, so that no major qualitative changes result from these alterations. Or the equations (their solutions) can depend sensitively on those changes yielding a practically chaotic situation. Most complicated physical

---

24) This point has been stressed particularly in Reference 10. As can be seen from the approach presented here this reference has had a great influence on the ideas expressed in this paper.

systems possess both trends. (This interesting and important observation was first explicitely made by Grad, Reference 11, for classical gases). Thus certain variables or variable combinations, behave in a very unstable manner under additional refinements. These variables would become increasingly difficult to follow. This would correspond to the situations where the errors accumulate. Other variable combinations, would exhibit a smooth behavior, they would correspond to the correct macroscopic variables for the new (contracted) level. These comments still do not identify the macroscopic variables, nor do they yield a procedure for the construction of a contracted level, they nevertheless do indicate, what type of behavior one should look for.

e) The solid state Hamiltonian as an example

The matters discussed so far, autonomous scales, the propagation of errors, can all be illustrated by examining the various approaches to solid state physics. Consider a solid as composed of nuclei and electrons, assume as interactions, electromagnetic and spin interactions. The level of the description is non-relativistic quantum mechanical. In spite of the concerns voiced in the last section relativistic and retardation effects will be omitted. These specifications provide a well defined starting point. The appropriate Hamiltonian is

$$H=\sum_{\alpha}\frac{P_{\alpha}^{2}}{2M_{\alpha}}+\sum_{i}\frac{P_{i}^{2}}{2m}+\sum_{ij}\frac{e^{2}}{r_{ij}}+\sum_{\alpha,\beta}\frac{Z_{\alpha}Z_{\beta}e^{2}}{R_{\alpha\beta}}-\sum_{i,\alpha}\frac{Z_{\alpha}e^{2}}{|R_{\alpha}-x_{i}|} \tag{6}$$

$$+ V_{S,S}(e) + V_{S,S}(N) + V_{S,S}(e,N)$$

$$+ V_{S,O}(e) + V_{S,O}(N) + V_{S,O}(e,N)$$

$$+ V_{O,O}(e) + V_{O,O}(N) + V_{O,O}(e,N)$$

$$+ H_{ext}(e) + H_{ext}(N)$$

The first two terms in (6) are the kinetic energy of the nuclei (mass M_{α}) and electrons (mass m). The next three terms are the electron-electron, nucleus-nucleus and electron-nucleus electrostatic energies. The $V_{S,S}$ terms are the electron-electron, nucleus-nucleus, nucleus-electron spin-spin interactions. The $V_{S,O}$ are the corresponding spin-orbit interactions, $V_{O,O}$ the orbit-orbit interactions. Finally $H_{ext}(e)$ and $H_{ext}(N)$ are the interactions of the electron and nuclear systems with prescribed external electromagnetic fields. This Hamiltonian (6) defines "the fundamental level" or basic scale. However, as formulated (6) contains not only solid state physics, but also atomic physics, chemisty and many other systems. Further (6) contains all kinds of interactions, which are not traditionally included, a particular choice of interaction is usually made, adapted to the system considered and the experimental conditions imposed. If the compressibility of Hydrogen, discussed before also falls within the scope of the Hamiltonian (6), if it were found experimentally that the compressibility of ortho and para Hydrogen is radically different, the spin terms would assume special importance but barring such information, they would be omitted.

None of the characteristic solid state physics concepts such as a lattice structure, energy bands, free electrons, phonons, electron-phonon interactions are as yet included in the Hamiltonian (6). These concepts can only emerge after numerical specifications, with appropriate approximations, and simplifications have been introduced. For example a solid state physicist will borrow information from atomic physicists, the Hamiltonian (6) is not used to simultaneously determine the spectrum of Fe, and its ferromagnetic transition temperature. Instead information from atomic physics is introduced into the Hamiltonian (6). (As written (6) does not even contain the information that a solid is made up of atoms !) In so doing a new scale is defined and it is this continual process that makes a sensible description of solids on a number of scales possible (Reference 12)[25]. As new information and new specifications are introduced the description proceeds from scale to scale. In actual practice the process is again inverted : the calculation of the cohesive energy of solids, starts from a very different approximation to (6), (it would be better to say a different version of (6)), than does the calculation of the magneto-conductivity. The descriptions of paramagnetic relaxation phenomena and the Ising model of ferromagnetism, both have their conceptual origin in (6), but the very different factors believed to be important in the respective cases, have been isolated, and incorporated in the models constructed to describe them. This again is an example of the factual autonomy of the two scales. The procedure followed is rarely an analysis of the basic Hamiltonian (6) but rather particular factors or features are singled out as especially important. Then a simplified model is constructed embodying those features, omitting all others. This is analyzed for

[25] More details of this process are described in Reference 12.

further physical consequences. The level in which the analysis is carried out, really defines the scale - and the autonomy of the scales and the lack of any simple relationship between the scales, follows directly from the manner in which they were obtained. This means that every level of description has to develop its own concepts and methods, for organizing and structuring its information. In this process reference to underlying, deeper levels, to universal results of those levels might be useful and important. But as the previous discussion shows the search for regularities and concepts, especially adapted to the level under study can not exclusively (or primarily) be based on a deductive analysis of the deeper levels. Thus the independence of the scales is every bit as important as their interrelations.

If this discussion were limited to pure physics, the ideas expressed here might or might not be of interest. Inasmuch as they deal with the structure of physics (as contrasted to physics itself) they could be legitimately ignored. However, for the application of physical principles to other fields, such as chemistry and biology, the ideas of the autonomy of the scales, the qualitative change in behavior from scale to scale, becomes of paramount importance.

3. Scales and Complexity

a) The chemical scale

The existence and importance of distinct scales becomes especially important in chemistry. It might be argued (and the Dirac dictum formalizes this view) that chemistry, the prediction of chemical properties, becomes effectively an exercise in finding many electron wave functions. It was already noted in connection with the Hamiltonian (6) that strictly interpreted, all questions of molecular structure can be discussed on this level. Through the use of computers great progress has been made in these "abinitio" chemical calculations, starting from a Hamiltonian of type (6) (with many of the interactions written there omitted). On this "fundamental" scale the only difference between large and small molecules lies in the duration of the calculations. There also exists a very powerful chemical scale of analysis, which employs a mixture of intuitive chemical concepts such as velency, bonding, chemical forces, together with numerical information about many-electron wave functions. The difference between the two methods can be clearly seen in the way in which they would answer the question : why is the binding energy of HF larger than that of FF ? In the fundamental approach the Schrödinger equation would be solved (numerically) for the two cases. Whatever information is available about wavefunctions, trial functions, would be employed, in the calculation but the answer would be of the type : the calculation yields a larger value of the binding energy of HF than for FF. A purely intuitive approach would describe the HF molecule in terms of strong bonds due to the sharing of electrons, while the FF bonds are weak, due to an effective repulsion between the electrons. The basic (fundamental) method has a well defined procedure but many of the intuitive, geometrical notions so useful in chemical descriptions are lost. Furthermore calculational methods which yield excellent values for the energy often give much poorer results for other quantities such as dipole moments or transition probabilities. The chemical scale gives an intuitively appealing description of the molecular structure, using such concepts as bond strength, orientation of bonds and provides a useful pictorialization of the system. However, there is not really a strict definition of these objects and most important the precise limits of their validity is unknown. The general discussion about the relationship between scales given in the previous section would lead one to anticipate just this situation.

Crucial for the following analysis is the observation (Reference 10), that there is in fact not one chemical scale, but as the size of the molecules increases, new phenomena and new types of behaviors can occur, which are not just scaled up versions of the smaller molecules. There are consequently a number of distinct chemical scales. Their existence can not be inferred from the basic level sketched above; the descriptive chemical level (especially through its emphasis on geometrical patterns) suggests that there might well be such different levels, but the direct experimental evidence is of course most convincing. This is not the place to attempt an exhaustive enumeration of the size-dependent features of molecular properties (see again Reference 10) : but to indicate the basic ideas some selected examples will be presented.

There are a number of properties which molecules in the 5-50 atom range possess which have no counterpart in diatomic and most triatomic molecules.

I. They exhibit an angular dependence of bonds.

II. Quite a few of the larger molecules (> 5 atoms) show anisotropic absorption spectra.

III. Beginning with 4 non-planar, non-identical atoms they exhibit sterochemical properties - manifested experimentally by optical rotation and magnetic anisotropy.

These properties are all more or less obvious from the geometrical structure of molecules. Recall that this pictorial description uses the concepts of the chemical level in an essential way. There are other new molecular properties, on this scale.

IV. The possibility exists that because of the involved rigid body mechanics, an "internal conversion" type process of excitation energy can take place. This experimentally leads to the possibility of rather long lived (triplet) excited states.

V. In the C_{10}-C_{14} chains among others, strong electronic absorption processes can take place.

Properties IV and V are not such obvious geometrical properties as I, II, III. They (IV and IV) result from the greater variety in dynamical behavior these larger molecules can exhibit. A similar (selected) listing of properties of molecules in the 50-500 shows that yet another number of new features make their appearance.

I. Some of these features arise from the connectedness properties of the long chain molecule, for example the possibility of internal rotations around certain bonds can cause substantial changes in the dimensions of the molecule. This is a much more pronounced effect in longer than in short molecules. This contractability would produce a variable viscosity.

II. Such long chains may have a selective response or enzyme type behavior i.e., a particular input can be transformed to an amplified output, after which the system returns to its original state. In order to perform this function with any degree of reliability the molecule can not be too simple; this possibility appears to occur for the first time for molecules in the 300 + atom range.

III. The molecules in this range also exhibit certain types of long range order, which shows up as the exact or partial repetition of specific sequences. The form of the long range order is related to the state of contraction of the system. Because of this the state of the long range order is solvent dependent. The existence of long range order can only occur for molecules of sufficiently large size (300).

IV. As is now well known, a long chain molecule can carry information by the sequential arrangements of the side groups. A 50 atom chain carrying 3 or 4 amino acid groups, might be one of about 10^4 possible compounds of the same size. A particular

500 atom chain would be one of 10^{40} possible compounds. Thus a long chain is capable of dramatically increasing the information it carries.

There is a celebrated theorem of v. Neumann (Reference 13) which asserts that in an appropriate environment of component parts "an automaton" or "sufficiently complex" operating machine with a sufficiently long (but finite) set of instructions can duplicate itself together with the set of instructions. The theorem is of course an existence theorem, it neither specifies the size of the system, nor the mechanism of duplication. However, applied to a chain molecule, in an environment of amino acids, the theorem appears to state that self-replication is a possible mode of behavior for sufficiently complex chain molecules beyond some critical size.

These selected properties (should be sufficient to illustrate the qualitative changes in behavior of molecules with changes in size). The fact that self-replication,because possible once molecules exceed a certain size and complexity, shows just how dramatic these changes of behavior can be. In the terminology of this paper the level or scale of description is altered radically with the size and complexity of the molecule. The conclusion to be drawn is that the behavior of large aggregates of elementary atoms should not just be understood in terms of a simple extrapolation of the properties of the system with just a few atoms. There seem to be many levels of complexity, and at each such level entirely new properties begin to appear. The further development of these ideas requires first of all a more detailed examination of just what the idea of "complexity" entails. Once this is clarified one might hope to gain some understanding of where and how these new levels of description appear.

b) Comments on complexity[26]

The previous considerations suggest very strongly that there might be good reasons to study the "complexity" of a system. Although intuitively it seems easy enough to distinguish a quite complex from a less complex system, it is not at all trivial to go beyond a vague intuitive characterization. The previous comments about molecular properties indicate that the type of beha-

[26]The comments made in this section are intended as exploratory. Some aspects of the complexity idea are discussed, but no precise definition can be obtained without further detailed study. Hopefully the suggestions given here are interesting in spite of their admittedly tentative character.

vior is tied to the size, structure and complexity of the system. In particular changes in the possible qualitative behaviors were in a general way associated with the changes in the complexity of the system. Thus it would appear worthwhile to explore in first instance the appropriate definition of the complexity of a molecule. Some reflection will immediately suggest two separate (but not unrelated) aspects of complexity. One is a static or structural complexity. It, in some way, must be a measure of the degree and type of geometrical complexity of the molecules, it is determined by the spatial organization of the molecule. By contrast the dynamical complexity should indicate something about the number and types of interaction which the molecule exhibits in response to changes in its environment. Because of its geometric character the static complexity, can be expected to be closely related to structural molecular formulae. Structural formulae in turn, have just the character of mathematical graphs. In the graph-theoretical literature two quite distinct definitions of the complexity of a graph are given. It is reasonable to examine these definitions and see whether they express the physical idea of complexity.

Let G be a graph with n vertices and m edges. Call the set of vertices V and suppose that $A_1 \ldots A_h$ is a partition of V, so that

$$\sum_{i=1}^{h} A_i = V \tag{7}$$

Let $|A_i|$ be the number of vertices in the set A_i. Then in certain studies the complexity j of the graph G relative to the partition A is defined as (Reference 14)

$$j_A = -\sum_{i=1}^{h} p_i \log p_i \qquad \text{with} \qquad p_i = \frac{1}{n}|A_i| \tag{8}$$

The logs are taken to the base 2.

Although this definition is frequently used, especially in connection with operational analysis, it is not a very obvious or intuitive notion. The definition is obviously patterned after the entropy (or information content) but it does not give any special insight in the structure of a graph or of the molecules represented by the graph. As given, the definition depends not only on G but also on the particular partition A of the graph. A definition more in keeping with the usual entropy definition would be

$$j = \text{Min } j_A \tag{9}$$

The complexity would be the minimum over all partitions. But both (8) and (9) have the feature that they depend on just the vertex set. This for many purposes would seem inappropriate[27].

There exists an older definition of complexity. Let G be a connected graph; B is the graph matrix of G.

B is defined by

$$\begin{aligned} b_{ii} &= \text{the degree of vertex } i \\ b_{ij} &= (-1) \text{ the number of edges incident on both } i \text{ and } j. \end{aligned} \tag{10}$$

Then the complexicity C of G is the determinant of any principal minor of B (Reference 5). This definition <u>does</u> depend on both the vertex and edge structure. It is, however, very inconvenient to use, because of the great difficulty of calculating C, for all but the simplest graphs. Another difficulty is that this definition does not seem to tie in closely with the geometrical character of the molecule. In many ways an immediately appealing definition for the static complexity would be to just count the average number of independent paths between vertices. If v and w are vertices, and N(v,w) is the number of <u>paths</u> (independent) connecting v and w, then the complexity for a simple graph would be

$$j(G) = \binom{n}{2}^{-1} \sum_{v,w} N(v,w) \tag{11}$$

If the graph is not simple (11) would generalize to

$$j(G) = \binom{n}{2}^{-1} \sum_{v,w} N(v,w) + \frac{1}{n}\sum_{v} N(v,v) \tag{12}$$

j(G) is somewhat simpler to calculate; it has the additional advantage that molecules which intuitively appear as more complicated, do possess a larger numerical value of j(G) as given by (12). This indicates the general direction one might pursue in order to formalize the notion of <u>static</u> complexity. For a serious

[27] Use of j or j_A would imply that all Feynman diagrams of a given order have the identical complexity index.

study, certain graph theoretic notions will need generalization, the fact that several types of atoms occur in a molecule means that several types (colors!) of vertices have to be introduced. The characteristic three dimensional features of the molecules have to be incorporated in the graph theoretic language.

The dynamic complexity must provide an indication of the manifold dynamical behaviors which a molecule can exhibit. It is therefore, necessary to consider a molecule as a dynamical system. The simplest approach would be to first study reasonable definitions of complexity for classical dynamical systems[28]. Since one clearly has a qualitative type of criteria in mind, one might consider the number of possible topological types of trajectories in the phase space as an appropriate indication of the complexity of a system. The use of this definition would require an exhaustive enumeration of the trajectory types, not an inconsiderable mathematical task. In many ways it appears physically more sensible to relate the complexity notion, to the number and type of responses a system can exhibit. It is probably best to combine these ideas and define the complexity of a classical dynamical system as

$$C \equiv \Delta\tau \equiv \tau_2 - \tau_1 \qquad (13)$$

Here τ_1 is the number of topological types of trajectories of the system, τ_2 is the number of topological types of trajectories if the system is perturbed by a specified class of external influences. (13) expresses the intuitive idea, that if a system "can not do much to alter its qualitative structure" it indeed must be considered as simple. This formula has the further advantage, that systems which are structurally stable (Reference 16), so that $\Delta\tau = 0$, are considered the simplest, which is a reasonable view. A further advantage of phrasing the complexity notion in terms of the class of responses, is that this method can more easily be carried over to quantum theory, where a well developed response formalism already exists.

These brief and tentative comments may suffice to indicate that "complexity" is a worthwhile notion to investigate. The discussion can be quite concrete and specific. Hopefully suitable definitions will in time be found, by the examination of concrete

28) It should be clear that as a first attempt to define complexity, one would try to distinguish systems which have the same number of degrees of freedom.

examples. In that way these ideas may be helpful in organizing and relating the many distinct facets of physical descriptions.

Acknowledgement

It is a particular pleasure to thank Professor C.N. Yang for many hours of interesting and thought-provoking discussions about these and related topics.

References

1. P.A.M. Dirac, *Principles of Quantum Mechanics*, Oxford University Press, 1938.

2. G.E. Uhlenbeck, in *Fundamental Problems in Statistical Mechanics*, edited by E.G.D. Cohen, North-Holland, John-Wiley & Sons, 1968.

3. G.E. Uhlenbeck, in Proc. 1966, Midwest Conf. on Theoret. Phys., Indiana University.

4. Suttorp, Thesis Amsterdam, 1968.

5. S.R. de Groot, *Studies in Statistical Mechanics*, vol. IV, edited by deBoer and Uhlenbeck, Interscience, 1969.

6. S. Harris, *Boltzmann equation*. Rinehart-Winston, 1971.

7. C.N. Yang, in *High Energy Collisions*, Gorden & Breach, N.Y., 1971.

8. J.R. Platt, Bull. Lowell Obs. 4, 278, (1960).

9. J. Sucher, Phys. Rev. D6, 1798, (1972).

10. J.R. Platt, Th. Biology, 1, 342, (1961).

11. H. Grad, Delaware Seminar in the Foundations of Physics. Springer, New York, 1967.

12. M. Dresden, Rev. Mod. Phys. 33, 265, (1961).

13. J.V. Neumann, in *Cerebral Mechanisms and Behavior*, John Wiley, New York, 1951.

14. G. Karreman, Bull. Math. Biophys. 17, 279, (1955).
A. Moshowitz, Bull. Math. Biophys. 30, 175, (1968).

15. Essam and Fisher, Rev. Mod. Phys. 42, 271, (1970).

16. M. Dresden, Fundamental Interactions in Physics, Plenum Press, New York, (1972).

Part II

MATHEMATICAL PHYSICS

WEIGHTS ON SPACES

E.R.Gerelle, R.J.Greechie* and F.R.Miller**

Kansas State University
University of Geneva

INTRODUCTION

Boolean σ-algebras and probability measures arise in the study of the logic of propositions associated with a single experiment, as was pointed out by Kolmogorov [15]. Recently [4, 16, 17] a program has been initiated to generate models for the logic of propositions associated with multiple experiments. These investigations and others on the quantum logic approach to quantum mechanics (e.g., [10, 12, 14]) have motivated us to study generalized probability measures, called weights on a space $(X,\mathcal{E})$.

Intuitively $\mathcal{E}$ represents a collection of experiments. Each experiment E in $\mathcal{E}$ is identified with a set of possible outcomes which determine the experiment, usually the elementary outcomes. $X=\bigcup\mathcal{E}$ is the set of all outcomes under consideration. Two outcomes x, $y \in X$ are said to be <u>orthogonal</u>, denoted $x \perp y$, if there is some experiment E in $\mathcal{E}$ such that x and y are mutually exclusive outcomes of E. (An expanded account of these heuristics may be found in [4].) When $\mathcal{E}$ is the set of all cliques in the graph $(X, \perp)$, $(X, \mathcal{E})$ is called an orthogonality space. We are most interested in

* This research was partially supported by the National Science Foundation Grant No. PO-11241 and by the Battelle Memorial Institute.

** This research was partially supported by the Swiss National Science Foundation.

Enz/Mehra (eds.), Physical Reality and Mathematical Description. 169–192.

orthogonality spaces which give rise to orthomodular posets $\mathcal{L}(X,\mathcal{E})$ (as defined in Section 1) for a logic of propositions; these are the Dacey spaces [2]. The combinatorial properties of the collection $\mathcal{E}$ influence the structure of $\mathcal{L}(X,\mathcal{E})$.

In Section (1) we give two characterizations of when a space is an orthogonality space and establish a correspondence between weights on an orthogonality space $(X,\mathcal{E})$ and states on the associated logic. It is this correspondence that allows us to restrict our attention to the study of weights on spaces.

In Section (2) we see that there are point closed spaces having a full set of weights for which the projection postulate of quantum mechanics fails. These spaces have the flavor of a theory of measurement for which the postulate does not apply. Such a situation could occur for other than academic reasons, for example when the "filtering-device" needed is socially, politically, economically or morally unacceptable or unfeasible.

In Section (3) we study the relationship between properties of a space $(X,\mathcal{E})$ and the geometry of the weight space Ω. Under topological hypotheses we obtain a bijection between the elements of certain spaces and the facets of the weight space. We obtain the result that an ortho-bijection between orthogonality spaces is equivalent to an affine structure map between the weight spaces for a nontrivial class of spaces. Section (4) continues this investigation for finite spaces.

Finally we consider symmetry groups of the space, its dual space and weight space showing, in certain cases, that these are isomorphic to the symmetry group of the logic of propositions.

This work shows a relationship between several fields. It is predominantly combinatorial in nature in that it is related to incidence structures, convex set theory and graph theory. Indeed the "stochastic functions on hypergraphs" of [1] are essentially the same as our weights on spaces. However our motivation was originally derived from the foundations of quantum mechanics [14, 20], in particular quantum logic [10, 12] and empirical logic [4, 16, 17]. It is precisely this willingness to depart from the well-honed paths of knowledge and the restrictive paradigms of accepted theory while adhering to the canons of mathematical rigor which is so much in the spirit of him to whom this paper is dedicated.

1. SPACES AND WEIGHTS

By a space we mean a pair $(X,\mathcal{E})$ where X is a nonempty set and $\mathcal{E}$ is a family of nonempty subsets of X. By a weight on a space $(X,\mathcal{E})$ we mean a mapping $\omega : X \to [0,1]$ such that $\sum_{a\in A} \omega(a) = 1$ for all $A \in \mathcal{E}$.

An orthogonality graph is a pair $(X,\perp)$ where $\perp$ is a nonempty symmetric irreflexive relation on the non empty set X. Given an orthogonality graph $(X,\perp)$ we obtain a space $(X,\mathcal{E})$ by taking $\mathcal{E}_\perp = \{E|E\subseteq X,\ E \text{ is a maximal } \perp\text{-set}\}$. (Recall that $M\subseteq X$ is a $\perp$-set if $x,y\in M$ and $x\neq y$ imply $x\perp y$). Conversely, given a space $(X,\mathcal{E})$ we obtain an orthogonality space $(X,\perp_{\mathcal{E}})$ by defining $x\perp_{\mathcal{E}} y$ to mean that $x \neq y$ and there is a set $A\in\mathcal{E}$ such that $x \in A$ and $y\in A$. We now characterize when $\mathcal{E}$ is the family of maximal orthogonal sets of an orthogonality space.

Let $(X,\mathcal{E})$ be any space. For each $x\in X$ we define

$$\mathcal{E}_x = \{A|A\in\mathcal{E} \text{ and } x \in A\} \text{ and } \mathcal{E}(x) = \bigcup\{A|A \in \mathcal{E}_x\}.$$

Lemma 1.1. Let $(X,\perp)$ be an orthogonality graph and $\mathcal{E} = \mathcal{E}_\perp$. Then $\mathcal{E}$ satisfies

(1) If $E\in\mathcal{E}$ and $E\notin\mathcal{E}_x$ then $E\nsubseteq\mathcal{E}(x)$.

(2) If $A\subseteq X$ and $A\subseteq\bigcap\{\mathcal{E}(y)|y \in A\}$ then there exists $E\in\mathcal{E}$ such that $A\subseteq E$.

Proof. Ad(1). Assume $E\in\mathcal{E}$ and $E\notin\mathcal{E}_x$. Now E is a maximal $\perp$-set and $x\notin E$. Thus there is a $y\in E$ such that $x\perp y$ fails. $y\notin\mathcal{E}(x)$.

Ad(2). Assume $A\subseteq\bigcap\{\mathcal{E}(y)|y \in A\}$. Let $x\in A$, $y\in A$ and $x \neq y$. Then $x\in\mathcal{E}(y)$ so that $x\perp y$. Hence A is a $\perp$-set and the result follows.

Theorem 1.2. Let $(X,\mathcal{Q})$ be a space such that $X=\bigcup\{A|A\in\mathcal{Q}\}$, and $\mathcal{E}$ be the family of all maximal $\perp_{\mathcal{Q}}$-sets. Consider the following conditions.

(1) $A\in\mathcal{Q}$ and $A\notin\mathcal{Q}_x$ implies $A\nsubseteq\mathcal{Q}(x)$.

(2) $M\subseteq X$ and $M\subseteq\bigcap\{\mathcal{Q}(x)|x\in M\}$ implies that $M\subseteq A$ for some $A\in\mathcal{Q}$. Then, (1) implies that $\mathcal{Q}\subseteq\mathcal{E}$ and (2) implies that $\mathcal{E}\subseteq\mathcal{Q}$. Thus $\mathcal{Q}=\mathcal{E}$ if and only if (1) and (2) hold.

Proof. Ad(1). Assume (1) holds. Let $A\in\mathcal{Q}$. Then A is a $\perp_{\mathcal{Q}}$-set.

Suppose $y \notin A$ and $A \cup \{y\}$ is a $\perp_{\mathcal{A}}$-set. For each $a \in A$ there is an $A_a \in \mathcal{A}$ such that $\{a,y\} \subseteq A_a$. Hence $A \subseteq \mathcal{A}(y)$ but $A \notin \mathcal{A}_y$, a contradiction. Thus A is maximal $\perp_{\mathcal{A}}$-set and $A \in \mathcal{E}$.

Ad(2). Assume (2) holds. Let $E \in \mathcal{E}$. Then E is a $\perp_{\mathcal{A}}$-set so that $E \subseteq \bigcap \{\mathcal{A}(y) \mid y \in E\}$. But this means that there is an $A \in \mathcal{A}$ such that $E \subseteq A$. Since A is a $\perp_{\mathcal{A}}$-set and $E \in \mathcal{E}$ we have $E = A \in \mathcal{A}$. Finally, we have proved that (1) and (2) imply $\mathcal{A} = \mathcal{E}$. If $\mathcal{A} = \mathcal{E}$ the previous lemma shows that (1) and (2) hold.

<u>Corollary 1.3.</u> Let $(X, \mathcal{A})$ be a space and $\mathcal{E}$ be the family of all maximal $\perp_{\mathcal{A}}$-sets. Then $\mathcal{A} = \mathcal{E}$ if and only if the following conditions are all satisfied.

(1) $\bigcup \{A \mid A \in \mathcal{A}\} = X$.

(2) If $M, N \in \mathcal{A}$ and $M \neq N$ then $M \not\subseteq N$.

(3) If $M \subseteq X$ and for all $x,y \in M$ there is an $N_{x,y}$ in $\mathcal{A}$ with $\{x,y\} \subseteq N_{x,y}$, then there exists $N \in \mathcal{A}$ with $M \subseteq N$.

We will use the name <u>orthogonality space</u> to refer to a space $(X, \mathcal{E})$ such that $X = \bigcup \mathcal{E}$, and (1) and (2) of Theorem 1.2. hold. (Note that this definition is equivalent to the usual one given earlier.) In this situation we will freely use the symbol $\perp$ to refer to $\perp_{\mathcal{E}}$.

Let $(X, \mathcal{E})$ be an orthogonality space and $\mathcal{O}(X, \perp)$ denote the set of all $\perp$-sets. For each $M \subseteq X$ define $M^{\perp} = \{x \in X \mid x \perp m$ for all $m \in M\}$ and $M^{\perp\perp} = (M^{\perp})^{\perp}$. By the <u>quasilogic</u> of $(X, \mathcal{E})$ (as of $(X, \perp)$) we mean the set $\mathcal{L} = \{D^{\perp\perp} \mid D \in \mathcal{O}(X, \perp)\}$ partially ordered by set theoretic inclusion [2]. If ambiguity threatens we sometimes write $\mathcal{L}(X)$, $\mathcal{L}(X, \mathcal{E})$, or $\mathcal{L}(X, \perp)$ for $\mathcal{L}$. A quasilogic $\mathcal{L}$ is said to be a <u>logic</u> if $M \in \mathcal{L}$ implies $M^{\perp} \in \mathcal{L}$. An orthogonality space $(X, \mathcal{E})$ such that $\mathcal{L}(X, \mathcal{E})$ is a logic is called an <u>orthocomplemented</u> space. Note that if M and N are elements of a logic $\mathcal{L}$ and $M \subseteq N^{\perp}$ then the join of M and N exists in $\mathcal{L}$ and $M \vee N = (D_1 \cup D_2)^{\perp\perp}$ where $D_1^{\perp\perp} = M$ and $D_2^{\perp\perp} = N$. An orthogonality space $(X, \mathcal{E})$ is said to be <u>point closed</u> if $\{x\}^{\perp\perp} = \{x\}$ for all $x \in X$.

An orthogonality space $(X, \mathcal{E})$ is called a <u>Dacey space</u> if, for all $x,y \in X$, $x^{\perp} \cup y^{\perp} \supseteq A$ for some $A \in \mathcal{E}$ implies $x \perp y$. It can be shown that each Dacey space $(X, \mathcal{E})$ is an orthocomplemented space and that $\mathcal{L}(X)$ is an orthomodular poset [2]. We denote the cardinality of a set M by $|M|$.

<u>Proposition 1.4.</u> Suppose that $(X, \mathcal{E})$ is an orthogonality space such that $|A| \geqslant 3$ for each $A \in \mathcal{E}$ and $|(A_1 \cap A_2)| \leqslant 1$ for all $A_1 \neq A_2 \in \mathcal{E}$. Then $(X, \mathcal{E})$ is a Dacey space. Such a space is also

point closed.

Proof. Assume that $x,y \in X$, $A \in \mathcal{E}$ and $x^{\perp} \cup y^{\perp} \supseteq A$. Then $x \neq y$. If $x \in A$ then $x \in y^{\perp}$ so that $x \perp y$. Similarly, $y \in A$ implies $x \perp y$. Suppose $x \perp y$ fails. Then $x \notin A$ and $y \notin A$. Now $|A| \geqslant 3$ so that

$$|A \cap x^{\perp}| = |A \cap \mathcal{E}(x)| \text{ or } |A \cap y^{\perp}| = |A \cap \mathcal{E}(y)| \text{ is at least 2.}$$

We may assume $|A \cap \mathcal{E}(x)| \geqslant 2$. Choose z_1, z_2 in $A \cap \mathcal{E}(x)$. Now $z_1 \perp z_2$ so that $\{x,z_1,z_2\}$ is a $\perp$-set. Choose $A_1 \in \mathcal{E}$ such that $A_1 \supseteq \{x,z_1,z_2\}$. Since $x \notin A$ we have $A_1 \neq A$. But $|A \cap A_1| \geqslant 2$, a contradiction. The last sentence follows from the previous observation.

By a state on a logic $\mathcal{L} = \mathcal{L}(X,\mathcal{E})$ we mean a mapping $\alpha : X \to [0,1]$ such that $\alpha(X) = 1$ and if $M \subseteq N^{\perp}$ then $\alpha(M \vee N) = \alpha(M) + \alpha(N)$ for any $M,N \in \mathcal{L}$. A set $\mathcal{S}$ of states on $\mathcal{L}$ is said to be full if, for any $M,N \in \mathcal{L}$, $M \subseteq N$ if and only if $\alpha(M) \leqslant \alpha(N)$ for all $\alpha \in \mathcal{S}$. A set $\mathcal{W}$ of weights on an orthogonality space $(X,\mathcal{E})$ is said to be full if, for any $x,y \in X$, $x \perp y$ if and only if $\omega(x) + \omega(y) \leqslant 1$ for all $\omega \in \mathcal{W}$. A pair $(\mathcal{L},\mathcal{S})$ where $\mathcal{S}$ is a full set of states on $\mathcal{L}$ is called a quantum logic [10].

Remark 1.5. Every weight on a Dacey space $(X,\mathcal{E})$ gives rise to a state $\overline{\omega}$ on $\mathcal{L}(X)$ by defining, for $M \in \mathcal{L}(X)$, $\overline{\omega}(M) = \sum_{d \in D} \omega(d)$ where $D \in \mathcal{O}(X,\perp)$ and $D^{\perp\perp} = M$.

Theorem 1.6. Let $(X,\mathcal{E})$ be a Dacey space and $\mathcal{L} = \mathcal{L}(X,\mathcal{E})$. Then the mapping $\Phi : \Omega_X \to \mathcal{S}_0$ is an injection where Ω_X is the set of all weights on $(X,\mathcal{E})$, $\mathcal{S}_0$ is the set of all states on $\mathcal{L}$, and $\Phi(\omega) = \overline{\omega}$ as defined in Remark 1.5. Now let $\mathcal{S}$ be any subset of $\Phi(\Omega)$. Then the following are equivalent :

(1) $\mathcal{S}$ is full.

(2) $M, N \in \mathcal{L}$ and $\alpha(M) + \alpha(N) \leqslant 1$ for all $\alpha \in \mathcal{S}$ imply $M \subseteq N^{\perp}$.

(3) $\mathcal{W} = \Phi^{-1}(\mathcal{S})$ is full.

Furthermore if X is finite Φ is a bijection.

Proof. See [4] or [8].

We have seen that the condition for an arbitrary space $(X,\mathcal{E})$ to be orthogonality space is a relationship between $\mathcal{E}$ and $\perp_{\mathcal{E}}$; namely, that $\mathcal{E}$ is the collection of maximal $\perp_{\mathcal{E}}$ sets. If one wishes to obtain the classical measure algebras as logics of spaces, and hence represent measures as weights on a space, this relationship must be generalized. Such a generalization has been given in [4].

We say that $(X,\mathcal{E})$ is a generalized sample space if conditions (1) and (2) of corollary 1.3. are satisfied together with

(3') If $A,B \in \mathcal{E}$, $C \subseteq A \cup B$, and $x \perp_{\mathcal{E}} y$ holds for all x,y in C with $x \neq y$ then there exists $D \in \mathcal{E}$ so that $C \subseteq D$.

Remark 1.7.

(1) Every orthogonality space is a generalized sample space.

(2) Every finite generalized sample space is an orthogonality space.

The quasilogic of a generalized sample space $(X,\mathcal{E})$ is given by $\mathcal{L}(X,\mathcal{E}) = \{D^{\perp\perp} \mid \exists\, A \in \mathcal{E} \ni D \subseteq A\}$. $(X,\mathcal{E})$ is said to be a Dacey space if $A \in \mathcal{E}$ and $A \subseteq x^{\perp} \cup y^{\perp}$ implies $x \perp y$. Then theorem 1.6. holds for generalized sample spaces. For more details see [4]. The important fact is that the (regular) states on the logic of a generalized sample space are included in our investigations of weights on spaces.

2. SPECTRUM OF AN ELEMENT

Let $(X,\mathcal{E})$ be any space and Ω be the set of weights. If $\omega_1, \ldots, \omega_n \in \Omega$ and $\lambda_1, \ldots, \lambda_n$ are positive real numbers so that $\Sigma\lambda_i = 1$ then $\omega(x) = \sum_{i=1}^{n} \lambda_i\omega_i(x)$ defines an element of Ω so that Ω is a convex set. Let

$$\underline{\underline{A}}(X,\mathcal{E}) = \{f : X \to R \mid \sum_{x \in E} |f(x)| < \infty \text{ for all } E \in \mathcal{E}\} \subseteq R^X$$

with the product topology. Then $\Omega \subseteq \underline{\underline{A}}(X,\mathcal{E})$ is a convex set in the locally convex topological vector space $\underline{\underline{A}}(X,\mathcal{E})$. If we define $S: \underline{\underline{A}}(X,\mathcal{E}) \to R^{\mathcal{E}}$ by $S(f)(E) = \sum_{x \in E} f(x)$ then $\Omega = S^{-1}(u) \cap K$ where $u \in R^{\mathcal{E}}$ by $u(E) = 1$ for all E and $K = \{f \in \underline{\underline{A}}(X,\mathcal{E}) \mid f(x) \geq 0 \ \forall x \in X\}$

For each $x \in X$ we define the spectrum of x, spec(x) as follows : spec(x) = closure $\{\omega(x) \mid \omega \in \Omega\}$; spec$(x)$ is a closed bounded convex subset of R, thus a closed interval. Define numbers m_x, M_x by spec$(x) = [m_x, M_x]$ if spec$(x) \neq \emptyset$, and $m_x = 1$, $M_x = 0$ if spec$(x) = \emptyset$. The consideration of spaces in which spec$(x) \neq [0,1]$ is important because of the possibility of developing a theory of measurement which utilizes a full set of weights rather than a strong set (cf. 3.1.) of weights. In such a theory the well-known projection postulate of Quantum Mechanics may not be valid. Such a theory would not have at its disposal a class of filters which produce with certainty the properties of interest. Such spaces may suggest an extension of the domain of theoretical physics and appear necessary in the behavioral sciences where subjective judgements restrict the purely academic possibilities in experimentation.

For $Y \subseteq X$ we define the induced space $(Y, \mathcal{E}_Y)$ by $\mathcal{E}_Y = \{Y \cap E \mid E \in \mathcal{E} \text{ and } Y \cap E \neq \emptyset\}$.

Proposition 2.1. Let $(X, \mathcal{E})$ be any finite space.

(1) $M_x < 1$ if and only if the space $(X \setminus \mathcal{E}(x), \mathcal{E}_{X \setminus \mathcal{E}(x)})$ admits no weights.

(2) $m_x > 0$ if and only if the space $(X \setminus \{x\}, \mathcal{E}_{X \setminus \{x\}})$ admits no weights.

Proof. We leave the straightforward proof to the reader.

We can now construct orthogonality spaces exhibiting various properties of spec(x). Let $(X_I, \mathcal{E}_I)$ be given by Figure I.

Figure I.

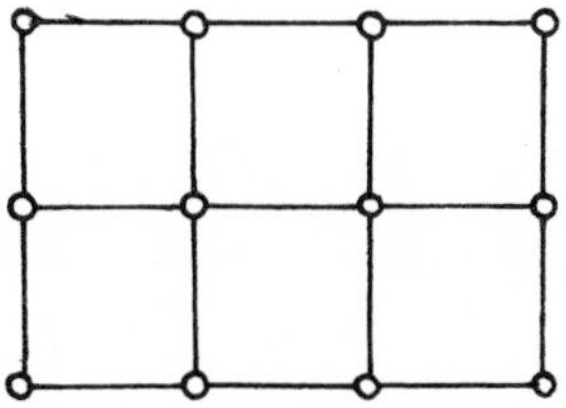

where X_I consists of the twelve points and $\mathcal{E}_I$ consists of the sets of points on maximal line segments, cf. [7,13]. By Proposition 2.1. $(X_I, \mathcal{E}_I)$ is a point closed Dacey-space, so that $(X_I, \mathcal{E}_I)$ is a logic. However, there are no weights on $(X_I, \mathcal{E}_I)$. Suppose that μ is a weight on $(X_I, \mathcal{E}_I)$. Let E_i, $1 \leqslant i \leqslant 3$ be the three horizontal members of $\mathcal{E}_I$ and E_i, $4 \leqslant i \leqslant 7$ be the four vertical members of $\mathcal{E}_I$. Then

$$3 = \sum_{i=1}^{3} \sum_{x \in E_i} \mu(x) = \sum_{x \in X_I} \mu(x) = \sum_{i=4}^{7} \sum_{x \in E_i} \mu(x) = 4, \text{ a contra-}$$

diction.

We now exhibit, in Figure II, a point closed Dacey-space $(X_{II}, \mathcal{E}_{II})$ admitting an element z such that $\text{spec}(z) = [0, j/k]$

where j and k are arbitrary pre-assigned integers with $1 \leqslant j < k$.

Figure II.

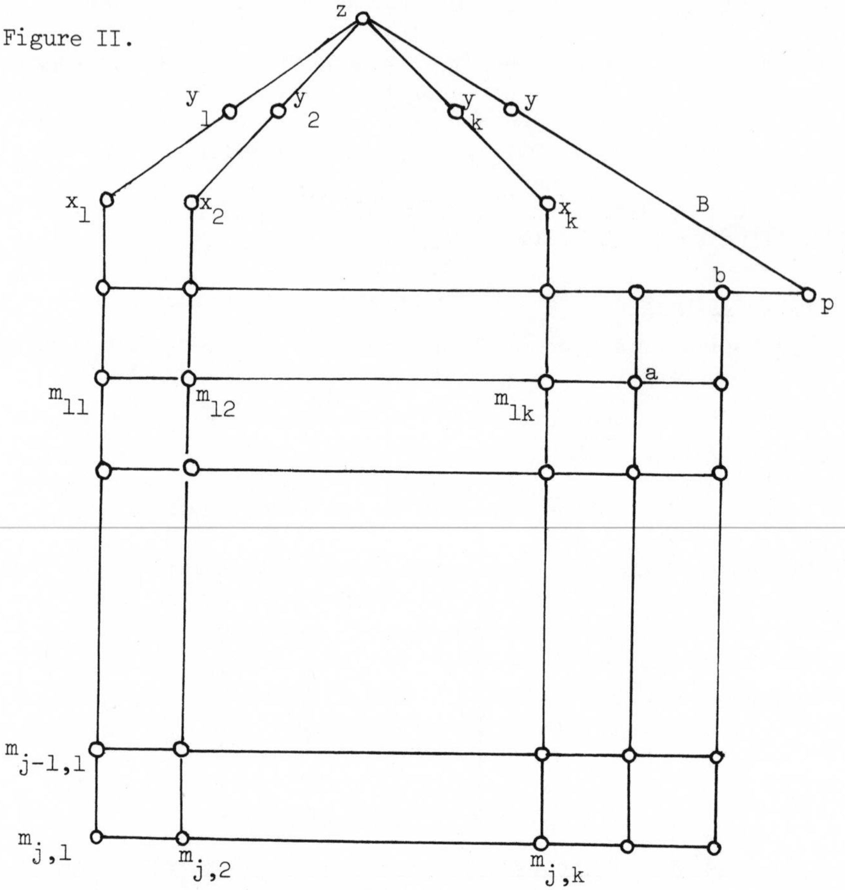

Let ω be a weight on $(X_{II}, \mathcal{E}_{II})$. Summing the images under ω of the elements of X_{II} in two rays we obtain

$$j + 2 + k -(k-1)\omega(z) + \omega(y) = k + 2 + \omega(z) + \omega(y) + \omega(p) + \sum_{i=1}^{k} \omega(y_i)$$

so that $j = k\omega(z) + \omega(p) + \sum_{i=1}^{k} \omega(y_i)$. Thus $\omega(z) \leqslant j/k$ and $\mathrm{spec}(z) \subseteq [0, j/k]$.

To see that $\mathrm{spec}(z) = [0, j/k]$ we need only exhibit two weights ω_1 and ω_2 such that $\omega_1(z) = 0$ and $\omega_2(z) = j/k$. For $x \in X_{II}$ and $h, i = 1, \ldots, k$ and $g = 1, \ldots, j$ define ω_1 and ω_2 as follows :

$$\omega_1(x) = \begin{cases} 1 - j/k & \text{if } x = x_i \\ j/k & \text{if } x = y_i \\ 1 & \text{if } x \in \{a,b,y\} \\ 1/k & \text{if } x = m_{g,h} \\ 0 & \text{otherwise} \end{cases}$$

$$\omega_2(x) = \begin{cases} j/k & \text{if } x = z \\ 1 - j/k & \text{if } x = y \\ 0 & \text{if } x = y_i \\ \omega_1(x) & \text{otherwise} \end{cases}$$

Proposition 2.2. Let $(X_{II}, \mathcal{E}_{II})$ be the point closed Dacey-space given in Figure II. Then $\Omega_{X_{II}}$ is full if and only if $1/2 < j/k$.

We omit the easy but lengthy proof. The interested reader will note that the line B and the elements p and y are essential only if $k - j = 1$. If $j < k - 1$ then this line and these two points may be deleted from $(X_{II}, \mathcal{E}_{II})$. The result on fullness and the restriction on spec(z) remain valid.

We have observed, via Figure II, that it is possible for a point closed Dacey-space X to admit an element x such that $M_x<1$ and Ω_x be full. From X we may construct a Dacey-space Y such that Ω_Y is full and $m_y > 0$ for some $y \in Y$. Simply let $Y = \mathcal{L}(X)\setminus\{\emptyset\}$ with the induced orthogonality relation and let $y = x^{\perp}$. It is somewhat more difficult to obtain a point closed Dacey-space admitting a full set of weights and an element y with $m_y > 0$. In fact we find the existence of such a space rather surprising. One such space is given in Figure III.

Figure III

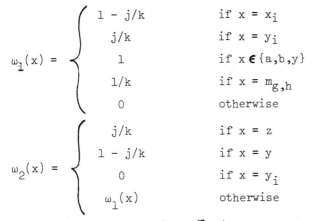

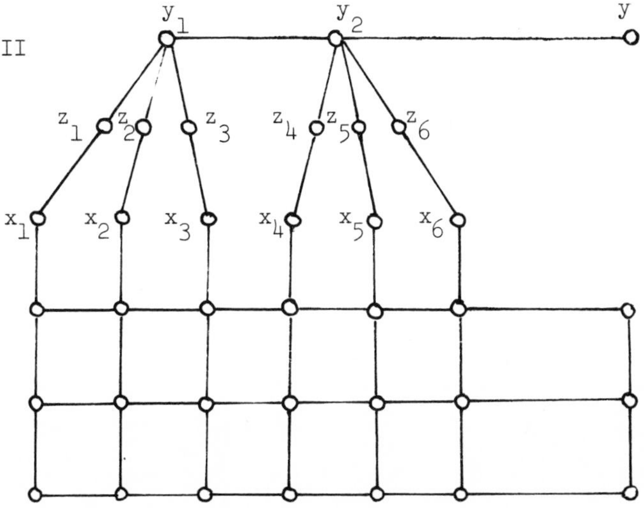

We outline the argument that $m_y > 0$. Let ω be any weight on the space $(X_{III}, \mathcal{E}_{III})$ of Figure III,

$$\sigma = \sum_{x \in X_{III}} \omega(x), \quad \tau = \sum_{i=1}^{6} \omega(x_i), \quad \zeta = \sum_{i=1}^{6} \omega(z_i).$$ Then counting σ

three different ways we get

(1) $\sigma = 4 + \tau + \zeta$

(2) $\sigma = 8 + \zeta$

(3) $\sigma = 7 + 3\omega(y)$

and hence

(4) $\tau = 4$ and $1 + \zeta = 3\omega(y)$.

Therefore $\omega(y) \geqslant 1/3$. Keeping (4) in mind it is easy to list enough states to show that ΩX_{III} is full and $\mathrm{spec}(y) = [1/3, 1]$.

3. TOPOLOGICAL-GEOMETRIC PROPERTIES OF SPACES

Let $(X, \mathcal{E})$ be any space and Ω be the set of weights on $(X, \mathcal{E})$. In section 2 we obtained $\Omega \subseteq \underline{\underline{A}}(X, \mathcal{E})$ making Ω a convex set in a topological vector space. In this section we will define two topologies for X, which are nontrivial even in the finite case, and we will see that the sets spec(x) defined in section 2 relate X to the geometry of Ω .

For each $\omega \in \Omega$ let $E(\omega) = \{x \in X \mid \omega(x) > m_x\}$. Recall $\mathrm{spec}(x) = [m_x, M_x]$. Assume that $X = \bigcup_{\omega \in \Omega} E(\omega), \Omega \neq \emptyset$ and let $\tau_{\mathcal{E}}$ be the topology on X which is generated by the sets $E(\omega), \omega \in \Omega$, as a sub-base.

<u>Remark 3.1.</u> (1) If $m_x = 0$ for all $x \in X$ then the closed sets of $\tau_{\mathcal{E}}$ are generated (as a sub-base) by the sets $Z_\omega = \{x \mid \omega(x) = 0\}$, $\omega \in \Omega$.

(2) $(X, \tau_{\mathcal{E}})$ is a T1-topological space if and only if $(\forall x \neq y)(\exists\, \omega \in \Omega)(\omega(x) = m_x$ and $\omega(y) > m_y)$.

(3) Let $(X, \tau_{\mathcal{E}})$ be a T1-topological space. If $(X, \mathcal{E})$ is a Dacey space then $(X, \mathcal{E})$ is point closed. If X is finite then $m_x = 0$ for each x in X.

<u>Definition 3.2.</u> Let $F \subseteq \Omega$. We say that F is a <u>face</u> of Ω if

(1) $\emptyset \subsetneq F \subsetneq \Omega$ and F is convex.

(2) $\omega_1 \in \Omega$, $\omega_2 \in \Omega$, $(\omega_1,\omega_2) \cap F \neq \emptyset \implies [\omega_1,\omega_2] \subseteq F$.

A face maximal under set theoretic inclusion is called a <u>facet</u>.

For each $x \in X$ let $S(x) = \{\omega \in \Omega | \omega(x) = m_x\}$. Now we have

<u>Lemma 3.3.</u> Let $(X,\mathcal{E})$ be any space and assume that $(X,\tau_{\mathcal{E}})$ is a T1-topological space. Then, for each $x \in X$, $S(x)$ is a face of Ω.

<u>Proof.</u> Let $x \in X$. Since $\tau_{\mathcal{E}}$ is a topology there exists $\omega \in \Omega$ so that $x \in E(\omega)$. Thus $\omega \in S(x)$ and we have $S(x) \subsetneq \Omega$. Since $\tau_{\mathcal{E}}$ is T1, if we choose $y \neq x$, there is $\omega \in \Omega$ so that $y \in E(\omega)$ and $x \notin E(\omega)$. But then $\omega \in S(x)$ so that $S(x) \neq \emptyset$. Now suppose $\omega_1 \neq \omega_2 \in \Omega$, $0<\lambda<1$, and $\lambda\omega_1 + (1-\lambda)\omega_2 \in S(x)$. Then $\lambda\omega_1(x) + (1-\lambda)\omega_2(x) = m_x$. Since $\omega_1(x) > m_x$ and $\omega_2(x) > m_x$ we must have $\omega_1(x) = \omega_2(x) = m_x$. Thus $[\omega_1,\omega_2] \subseteq S(x)$.

We define a second topology $S_{\mathcal{E}}$ on X as follows : $S_{\mathcal{E}}$ is generated (as a sub-base) by the sets $E(\omega,n) = \{x | \omega(x)>1/n\}$ for $\omega \in \Omega$ and n is a positive integer. Clearly if $\tau_{\mathcal{E}}$ is a topology on X then so is $S_{\mathcal{E}}$. Other interesting topologies can be defined on X using $\mathcal{E}$ including the one used in [5]. The properties and relationships between these topologies give a class of combinatorial problems for infinite spaces.

<u>Lemma 3.4.</u> Let $(X,\mathcal{E})$ be any space and $\omega \in \Omega$ such that $\omega(x) \geqslant m > 0$ for some m. If $\omega_1 \neq \omega$ then there exists ω_2 so that $\omega \in (\omega_1,\omega_2)$. Thus if F is a face of Ω we must have $\omega \notin F$.

<u>Proof.</u> Let $\mu(t) = \omega + t(\omega_1-\omega)$. Then $\mu(t) \in \Omega$ for $-m \leqslant t \leqslant 1$.

<u>Theorem 3.5.</u> Let $(X,\mathcal{E})$ be a space so that $(X,\tau_{\mathcal{E}})$ is a T1-topological space and $S_{\mathcal{E}}$ is a compact topology . Then $m_x=0$ for each x in X and the mapping $x \to S(x)$ is a bijection between the points of X and the facets of Ω.

<u>Proof.</u> Suppose $F \subseteq \Omega$ is a face and $\nexists$ x in X so that $F \subseteq S_0(x) = \{\omega \in \Omega | \omega(x)=0\}$. Then, for each $x \in X$, we can choose $\omega_x \in F$ so that $\omega_x(x) > 0$. Also choose an integer n_x so that $\omega_x(x) > \frac{1}{n_x}$.

Now $X = \bigcup_{x \in X} E(\omega_x, \frac{1}{n_x})$ and using $S_{\mathcal{E}}$ -compactness we can choose $x_1,\dots,x_n$ so that $X = \bigcup_{i=1}^{n} E(\omega_{x_i}, \frac{1}{n_{x_i}})$. Let $\omega = \sum_{i=1}^{n} \frac{1}{n} \omega_{x_i}$.

then for each $x \in X$ we have $\omega(x) \geq \frac{1}{n} \min(\frac{1}{n_{x_i}}) > 0$ and $\omega \in F$.

But this contradicts lemma 3.4. Thus there is an x so that $F \subseteq S_o(x)$. Now if $y \in X$, by Lemma 3.3, $S(y)$ is a face. Thus there exists an x so that $S(y) \subseteq S_o(x)$; $m_x=0$; and the T1-property of $\tau_{\mathcal{E}}$ gives x=y. Hence $m_y=0$. Since we have shown that any face of Ω is contained in an $S(x)$ the T1-property of $\tau_{\mathcal{E}}$ gives the bijection.

Definition 3.6. Let $(X,\mathcal{E})$ be any space and Ω the set of weights on $(X,\mathcal{E})$. Define $S_1(x) = \{\omega \in \Omega | \omega(x) = 1\}$

$$S_0(x) = \{\omega \in \Omega | \omega(x) = 0\}.$$

Definition 3.7. The set of weights Ω is strong on an orthogonality space $(X,\mathcal{E})$ when $x \perp y \Longleftrightarrow S_1(x) \subseteq S_o(y)$. If Ω is strong then it is also full and hence $(X,\mathcal{E})$ is a Dacey space. For the rest of this section $|X|>1$.

Lemma 3.8. If $(X,\mathcal{E})$ is a point closed orthogonality space with Ω strong on $(X,\mathcal{E})$ then

(1) $S_1(x) \neq \emptyset$ and $S_o(x) \neq \emptyset$.

(2) $(X,\tau_{\mathcal{E}})$ is a T1-topological space.

Proof. Both parts are straightforward and the proofs are left to the reader.

Corollary 3.9. Let $(X,\mathcal{E})$ be a point closed orthogonality space. If Ω is strong and $S_{\mathcal{E}}$ is a compact topology, then the mapping $x \to S_0(x)$ is a bijection between the points of X and the facets of Ω.

Proof. The proof is given by theorem 3.5 and the previous lemma.

Corollary 3.10. Let $(X,\mathcal{E})$ be a finite Dacey space with weights Ω. If $(X,\tau_{\mathcal{E}})$ is a T1-topological space then $\Omega(X,\mathcal{E})$ is not a planar pentagon.

Proof. The proof is by inspection and theorem 3.5. In particular the result holds if Ω is a strong set of weights and thus answers a question posed by F.Shultz [19].

Definition 3.11. A space $(X,\mathcal{E})$ is (homogeneous) of dimension n if $|E| = n$ for all $E \in \mathcal{E}$. For a pair of weights ω and ν in Ω and $0 \leq \lambda \leq 1$ define $\langle\omega,\nu,\lambda\rangle = (1-\lambda)\omega + \lambda\nu$. If $(X,\mathcal{E})$ is of dimension n the weight e, given by $e(x) = \frac{1}{n}$ for all x, is called the unit of Ω and the pair (Ω,e) is called the weight structure for $(X,\mathcal{E})$. If Ω_1 and Ω_2 are two convex sets, an affine map f from Ω_1 to Ω_2 is a function for which $f(\lambda\nu + (1 - \lambda)\omega) = \lambda f(\nu) + (1 - \lambda)f(\omega)$ [11].

If $(X_1,\mathcal{E}_1)$ and $(X_2,\mathcal{E}_2)$ are two orthogonality spaces and $f : X_1 \to X_2$ is a bijection satisfying $f(x)\perp_2 f(y) \Longleftrightarrow x\perp_1 y$, we say f is an ortho-bijection and we write $(X_1,\perp_1) \simeq (X_2,\perp_2)$. If $f : (\Omega_1,e_1) \to (\Omega_2,e_2)$ is an affine map and $f(e_1) = e_2$, then f is called a structure map. If there is a bijective structure map from (Ω_1,e_1) onto (Ω_2,e_2) we write $(\Omega_1,e_1) \simeq (\Omega_2,e_2)$.

Lemma 3.12. Let (Ω,e) be the weight structure for the homogeneous orthogonality space $(X,\mathcal{E})$ of dimension n. If $x \in E$ and $S_o(x) \neq \emptyset$ then

$$S_1(x) = \{\nu \in \Omega \mid \langle\omega,\nu,\tfrac{1}{n}\rangle = e,\ \omega \in S_o(x)\}$$

and if $M_x = 1$,

$$\frac{1}{n} = \inf_{\nu,\omega} \{\lambda \mid \langle\omega,\nu,\lambda\rangle = e,\ \omega \in S_o(x),\ \nu \in \Omega\}.$$

Proof. The proofs are straightforward and left to the reader.

Lemma 3.13. Let $(X_i,\mathcal{E}_i)_{i=1,2}$ be homogeneous point closed orthogonality spaces of dimension n_i with weight structures $(\Omega_i,e_i)_{i=1,2}$ and Ω_i strong on $(X_i,\mathcal{E}_i)$ with $S_{\mathcal{E}_i}$ compact. A bijective structure map $f : (\Omega_1,e_1) \to (\Omega_2,e_2)$ induces a natural ortho-bijection $\bar{f} : X_1 \to X_2$.

Proof. By corollary 3.9. there is a bijection ψ_j between the points of X_j and the facets of Ω_j, where $\psi_j(x) = S_o(x) \subseteq \Omega_j$. It is easy to show that f induces a bijection $\tilde{f}$ between the facets of Ω_1 and those of Ω_2. Hence $\bar{f} = \psi_2^{-1}\,\tilde{f}\,\psi_1 : X_1 \to X_2$ is a bijection and $f(S_o(x)) = S_o(\bar{f}(x))$.

It remains to show that

$$x \perp_1 y \text{ if and only if } \bar{f}(x) \perp_2 \bar{f}(y).$$

We now show that the dimensions of X_1 and X_2 are equal.

By lemma 3.12 if $x \in X_j$ $(j=1,2)$ then

$$\frac{1}{n_j} = \inf_{\omega,\nu} \{\lambda \mid \langle\omega,\nu,\lambda\rangle = e_j,\ \omega \in S_o(x), \nu \in \Omega\}$$

Let $x \in X_1$ then there exist $\omega \in S_o(x)$ and $\nu \in S_1(x)$ such that $<\omega, \nu, \frac{1}{n_1}> = e_1$. Since f is a structure map $<f(\omega), f(\nu), \frac{1}{n_1}> = e_2$; but $f(\omega) \in S_o(\bar{f}(x))$ so $n_2 \geq n_1$. By symmetry $n_1 = n_2$.
Let $n_1 = n_2 = n$ and so for $x \in X_j$ $(j = 1,2)$,

$$S_1(x) = \{\nu \in \Omega_j \mid <\omega, \nu, \tfrac{1}{n}> = e_j, \ \omega \in S_o(x)\} .$$

Since f is a bijective structure map, for $x \in X_1$, $f(S_1(x)) = S_1(\bar{f}(x))$.

Let $\{x,y\} \subseteq X_1$. Since Ω_1 and Ω_2 are strong

$$x \perp_1 y \leftrightarrow S_1(x) \subseteq S_o(y) \leftrightarrow f(S_1(x)) \subseteq f(S_o(y)) \leftrightarrow$$

$$S_1(\bar{f}(x)) \subseteq S_o(\bar{f}(y)) \leftrightarrow \bar{f}(x) \perp_2 \bar{f}(y).$$

Theorem 3.14. Let $(X_i, \mathcal{E}_i)_{i=1,2}$ and $(\Omega_i, e_i)_{i=1,2}$ satisfy the hypothesis of the previous lemma, then $(X_1, \perp_1) \simeq (X_2, \perp_2) \leftrightarrow (\Omega_1, e_1) \simeq (\Omega_2, e_2)$.

Proof. The forward implication is left to the reader and applying the previous lemma completes the proof.

4. EXISTENCE OF WEIGHTS

Throughout this section $(X, \mathcal{E})$ will be assumed finite. A weight ω on $(X, \mathcal{E})$ is said to be dispersion free or deterministic if $\omega(X) \subseteq \{0,1\}$.

Theorem 4.1. The set of dispersion free weights on a space $(X, \mathcal{E})$ is in one-to-one correspondence with the set of nonempty subsets M of X having the following properties.

(1) If $x, y \in M$ and $x \neq y$ then $\mathcal{E}_x \cap \mathcal{E}_y = \emptyset$.

(2) $\mathcal{E} = \bigcup \{\mathcal{E}_x \mid x \in M\}$.

Proof. Let M be a set and $\psi_M : X \to \{0,1\}$ be the characteristic function of M. We claim that ψ_M is a weight. Let $E \in \mathcal{E}$. By (1) and (2) there is a unique $x \in M$ such that $E \in \mathcal{E}_x$. Let $y \in E$, $y \neq x$. Since $\mathcal{E}_x \cap \mathcal{E}_y \neq \emptyset$ we must have $y \in M$ by (1). Thus $\sum_{y \in E} \psi_M(y) = 1$.

Conversely, if ω is a dispersion free weight on X. let $M = \omega^{-1}(1)$.

Corollary 4.2. If $(X, \mathcal{E})$ is a space such that $|\mathcal{E}|$ is odd, and $|\mathcal{E}_x|$ is even for each $x \in X$ then there are no dispersion free weights on $(X,\mathcal{E})$.

Let $(X,\mathcal{E})$ be an orthogonality space with the relation $\perp = \perp_{\mathcal{E}}$. We say that $x \perp' y$ if and only if $x \neq y$ and not $x \perp y$. $(X,\mathcal{E})$ is said to be an F-space if $E \cap F \neq \emptyset$ for every maximal $\perp'$-set F and every $E \in \mathcal{E}$. The above theorem shows that in an F-space every maximal $\perp'$-set supports a dispersion free state and thus the set of dispersion free states is full. The converse, however is not true. An example of a point closed Dacey-space which has a full set of dispersion free states but is not an F-space is given by Figure IV.

Fig.IV

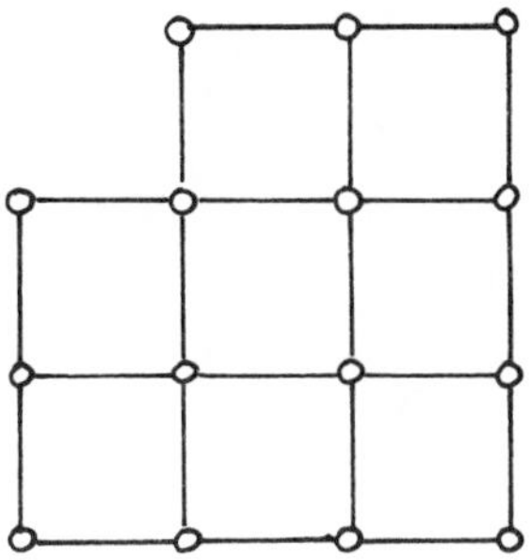

Suppose that $(X,\mathcal{E})$ is any space. The weights on $(X,\mathcal{E})$ are determined by a system of linear equations and this fact can be used to derive properties of the set of weight functions. We now give a realization of this structure which does not depend upon any labeling of the points of X.

For any set A, $F(A)$ will denote the free real vector space over A, and $F(A)^*$ its dual vector space. For each $a \in A$ define $\delta_a \in F(A)^*$ by $\delta_a(b) = 1$ if $a = b$, $= 0$ if $a \neq b$, and linear extension. If A is finite, $F(A)^* \simeq F(\{\delta_a | a \in A\})$. Let $(X,\mathcal{E})$ be a (finite) space. The linear transformation $T : F(\mathcal{E}) \to F(X)$ defined by $T(E) = \sum_{x \in E} x$ and linear extension, is called the linear realization of $(X,\mathcal{E})$. The transpose $T^* : F(X)^* \to F(\mathcal{E})^*$ is given by $T^*(\delta_x) = \sum_{E \in \mathcal{E}_x} \delta_E$, or in general,

$$(I) \quad T^*(\mu) = T^*\left(\sum_{x \in X} \mu(x)\delta_x\right) = \sum_{x \in X} \mu(x)\left(\sum_{E \in \mathcal{E}_x} \delta_E\right)$$

$$= \sum_{E \in \mathcal{E}}\left(\sum_{x \in E} \mu(x)\right)\delta_E.$$

Since X is finite the vector space $\underline{\underline{A}}(X,\mathcal{E})$, defined in section 2, is $R^X \simeq F(X)^*$. With the identification it is easy to see that $S=T^*$ where $S : \underline{\underline{A}}(X,\mathcal{E}) \to R^{\mathcal{E}}$ was defined in section 2.

We will thus consider $\Omega_X \subseteq F(X)^*$. Since $F(X)^*$ is finite dimensional each element of Ω_X has a representation as a convex combination of the extreme points of Ω_X. Also, if μ_1 and μ_2 are in Ω_X then $\mu_2 - \mu_1 \in \ker T^*$ and since $\mu_2 = \mu_1 + (\mu_2 - \mu_1)$ we see that $\Omega_X = (\mu_1 + \ker T^*) \cap C$ for any $\mu_1 \in \Omega_X$. By a pure weight on $(X, \mathcal{E})$ we mean an extreme point of Ω_X and by a perturbation on $(X, \mathcal{E})$ we mean an element of $\ker T^*$.

Let Y be any subset of X. Recall that the space $(Y, \mathcal{E}_Y)$ is defined by $\mathcal{E}_Y = \{E \cap Y \mid E \in \mathcal{E} \text{ and } E \cap Y \neq \emptyset\}$. Note that this need not be an orthogonality space. We say that Y is supporting if $Y \cap E \neq \emptyset$ for each $E \in \mathcal{E}$. For each $\mu \in \Omega_X$ let $Z_\mu = \{x \mid x \in X \text{ and } \mu(x) = \emptyset\}$.

Theorem 4.4

(1) If $\mu \in \Omega_X$ then $X \setminus Z_\mu$ is supporting, and if $(X, \mathcal{E})$ is an orthogonality space then $(X \setminus Z_\mu, \mathcal{E}_{X \setminus Z_\mu})$ is an orthogonality space.

(2) If $X \setminus Y$ is supporting then the set of weights on $(X, \mathcal{E})$ which vanish on Y is in one-to-one correspondence with the weights on $(X \setminus Y, \mathcal{E}_{X \setminus Y})$.

(3) $Y = Z_\mu$ for some $\mu \in \Omega_X$ if and only if $X \setminus Y$ is supporting and $(X \setminus Y, \mathcal{E}_{X \setminus Y})$ admits a positive weight.

Proof. Ad(1). Let $\mu \in \Omega_X$ and $E \in \mathcal{E}$. Since $\sum_{x \in E} \mu(x) = 1$ there is an $x \in E$ such that $\mu(x) > 0$. Hence $E \cap (X \setminus Z_\mu) \neq \emptyset$. Suppose that $(X, \mathcal{E})$ is an orthogonality space. We will show that $\mathcal{E}_{X \setminus Z_\mu}$ satisfies the conditions of corollary 1.3. Clearly properties 1) and 3) are inherited from $\mathcal{E}$. Now suppose that $E_1 \cap (X \setminus Z_\mu) \subseteq E_2 \cap (X \setminus Z_\mu)$ and there is a point x in $E_2 \cap (X \setminus Z_\mu) \setminus E_1 \cap (X \setminus Z_\mu)$. We have

$$\sum_{y \in E_2} \mu(y) = \sum_{y \in E_2 \cap (X \setminus Z_\mu)} \mu(y) = \sum_{y \in E_1 \cap (X \setminus Z_\mu)} \mu(y) + \mu(x) = 1 + \mu(x) > 1, \text{ since}$$

$x \notin Z_\mu$. But this is a contradiction.

Ad(2). Let $\mu \in \Omega_X$ and $Y \subseteq Z_\mu$. For each $E \in \mathcal{E}$ $\sum_{y \in E \cap (X \setminus Y)} \mu(y) = \sum_{y \in E} \mu(y) = 1$ so that $\mu | X \setminus Y$ is a weight on $(X \setminus Y, \mathcal{E}_{X \setminus Y})$. Now suppose that $\mu : X \setminus Y \to [0,1]$ is a weight on $(X \setminus Y, \mathcal{E}_{X \setminus Y})$. Extend μ by 0 to all of X. Since $X \setminus Y$ is supporting we get a weight on $(X, \mathcal{E})$. Ad(3). (3) follows easily from (2).

We have seen that each weight on a space $(X, \mathcal{E})$ can be written as a convex combination of pure weights. Thus the existence of weights on $(X, \mathcal{E})$ is equivalent to the existence of pure weights.

Lemma 4.5. A weight μ on $(X, \mathcal{E})$ is pure if and only if $Z_\mu \not\subseteq Z_{\mu_1}$

for all weights $\mu_1 \neq \mu$.

Proof. Assume there is a weight $\mu_1 \neq \mu$ for which $Z_\mu \subseteq Z_{\mu_1}$. Let $\nu = \mu - \mu_1$. Then ν is a perturbation. That is, $\sum_{x \in E} \nu(x) = 0$ for all $E \in \mathcal{E}$.

Recalling that X is finite, let $\delta = \min\{\mu(x) | \mu(x) \neq 0\}$. Suppose that $\lambda \in [-\delta, \delta]$ and $\mu_\lambda = \mu + \lambda\nu$. Clearly, $\sum_{x \in E} \mu_\lambda(x) = 1$ for all $E \in \mathcal{E}$. Let $x \in X$. If $\mu(x) = 0$ we have $\mu_1(x) = 0$ so that $\nu(x) = 0$ and $\mu_\lambda(x) = 0$. Suppose $\mu(x) \neq 0$. Then $|\lambda((\mu - \mu_1)(x))| = |\lambda| |\nu(x)| < \delta \cdot 1 = \delta \leq \mu(x)$. Thus μ_λ is a weight for each $\lambda \in [-\delta, \delta]$ and, since $\mu_0 = \mu$, μ is not a pure weight. This establishes the necessity of the condition. Conversely, assume $\mu = \lambda\mu_1 + (1 - \lambda)\mu_2$ where $\mu_1 \neq \mu_2$ are weights and $0 < \lambda < 1$. Then $Z_\mu = Z_{\mu_1} \cap Z_{\mu_2} \subseteq Z_{\mu_1}$ and $\mu_1 \neq \mu$.

Proposition 4.6. A weight μ on $(X, \mathcal{E})$ is pure if and only if the space $(X \setminus Z_\mu, \mathcal{E}_{X \setminus Z_\mu})$ has exactly one weight and that weight is positive (i.e., never vanishes).

Proof. By Theorem 4.4 the set of weights μ_1 on $(X, \mathcal{E})$ such that $Z_\mu \subseteq Z_{\mu_1}$ is in one-to-one correspondence with the weights on the space $(X \setminus Z_\mu, \mathcal{E}_{X \setminus Z_\mu})$. By Lemma 4.5 μ is pure if and only if $\mu_1 = \mu$ is the only such weight. Of course $\mu | X \setminus Z_\mu$ never vanishes.

Corollary 4.6. (1) Every dispersion free weight is pure. (2) The set of extreme points is finite. (3) A pure weight ω assumes only rational values.

Proof. (1) and (2) are immediate from proposition 4.6 and lemma 4.5 respectively. Ad(3). Let $X = Y - Z$ where $Z = \{x | \omega(x) = 0\}$ and $F = \{E \cap Y | E \in \mathcal{E}\}$. By 4.4 and proposition 4.6 $\mu = \omega | Y$ is the only weight on (Y, F). Thus the values of μ, listed as a vector, is the unique solution of a system of linear equations with integer coefficients. Such a vector has rational components.

The above results show the importance of orthogonality spaces which admit only one weight. An interesting question is the existence of orthogonality spaces which are also Dacey spaces and have only one weight. A construction of such a space is given in [8]. See [19] for the case when the associated logic is a lattice.

Let $(X, \mathcal{E})$ be any space and $Y \subseteq X$. We say that Y is connected if for any two distinct points x and y in Y there are sequences of points, $x_0, x_1, x_2, \ldots, x_n$ and of blocks $E_1, E_2, \ldots, E_n$ such that $x_0 = x$, $x_n = y$ and $\{x_{i-1}, x_i\} \subseteq E_i$, $1 \leq i \leq n$. A maximal connected is said to be a <u>connected component</u>.

Theorem 4.7. Let μ be a weight on $(X,\mathcal{E})$. Then $X\setminus Z_\mu$ has a decomposition $X\setminus Z_\mu = F\cup C_1\cup C_2\cup\ldots\cup C_n$ which is unique relative to the following conditions, (1), (2), and (3).

(1) $\mathcal{E}_F = \{\{x\}\mid x\in F\}$ so that F represents the dispersion free part of μ.

(2) Each E ε intersects exactly one of the sets in $\{C_i\mid 1\leq i\leq n\}\cup\{\{x\}\mid x\in F\}$.

(3) Each C_i is connected and $E\in\mathcal{E}$, $E\cap C_i\neq\emptyset$ implies $|E\cap C_i|\geq 2$. That is, each element $\mathcal{E}_{C_i}$ has cardinality at least 2.

Moreover,

(4) If $(X,\mathcal{E})$ is an orthogonality space then each $(C_i,\mathcal{E}_{C_i})$ is an orthogonality space.

(5) μ is pure if and only if each $(C_i,\mathcal{E}_{C_i})$ admits exactly one weight and that weight is positive.

Proof. The C_i's are the connected components of $(X\setminus Z_\mu,\mathcal{E}_{X\setminus Z_\mu})$ which are not isolated points and $F = X\setminus(Z_\mu\cup\bigcup_{i=1}^{n} C_i)$. (1) and (2) are clear.

Suppose that $E\cap C_i = \{x\}$. Since $\{x\}$ is not isolated (otherwise it would be in F) there is an $E_1\in\mathcal{E}$ such that $E_1\cap C_i\supset\{x,y\}$ for some $y\neq x$. Now $E\cap C_i = \{x\}$ and by (2) we must have $\mu(x)=1$. But $\mu(y) > 0$ so that we get $\mu(x) + \mu(y) > 1$, $\{x,y\}\subseteq E_1$, a contradiction. Thus we get (3). (4) follows from the fact that $\mu|_{C_i}$ is a positive state on $(C_i,\mathcal{E}_{C_i})$ and (5) follows from Proposition 3.6.

We can easily give an example of a finite point closed Dacey-space which has a pure weight that is not dispersion free, cf.Figure V.

Fig.V

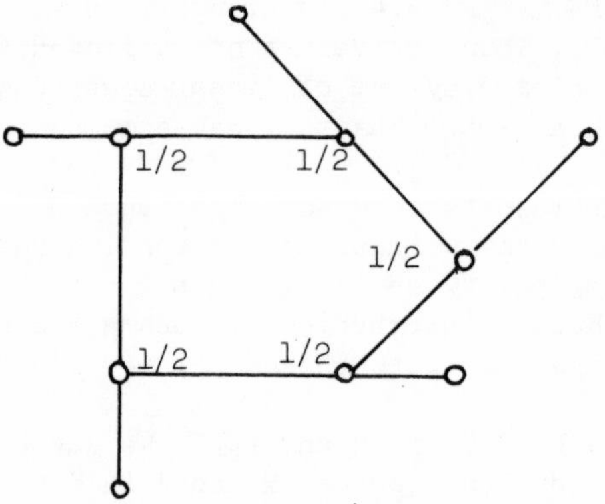

Note that the results of Theorem 4.1 concerning the existence of dispersion free weights are a special case of Theorem 4.7 corresponding to the situation $X\setminus Z_\mu = F$. Thus the existence of weights on $(X,\mathcal{E})$ depends upon the existence of a partition of $\mathcal{E}$ determined by a set $F \subseteq X$ as is described in Theorem 4.1 or upon the existence of orthogonality spaces $(C, \mathcal{E}_C)$ in $(X,\mathcal{E})$ which admit exactly one weight and that weight is positive.

By a <u>generalized weight</u> we mean any element of $\Omega_g = T^{*-1}(\sum_{E\in\mathcal{E}} \delta_E)$.

Recall that a <u>perturbation</u> is any element of Ker T*. It is clear that μ_1 and μ_2 in Ω_g implies $\mu_1 - \mu_2$ Ker T* and if μ is any element in Ω_g then $\Omega_g = \mu + \text{Ker } T^*$.

We now consider the relationship between generalized weights weights, and perturbations.

<u>Proposition 4.8.</u> Let $(X,\mathcal{E})$ be any finite space, $\Omega \subseteq F(X)^*$ be the weight space of $(X,\mathcal{E})$, and $\Omega_{\mathbf{P}} = \{P_1, P_2, \ldots, P_n\}$ be the set of pure weights. Let $R = \{\nu \in F(X)^* \mid \text{There is a number } r > 0 \text{ and weights } \mu_1, \mu_2 \text{ such that } r\nu = \mu_1 - \mu_2\}$. Let d be the number of vectors in a maximal linearly independent subset of $\{P_2 - P_1, P_3 - P_1, \ldots P_n - P_1\}$. Then

1) $R \subseteq \text{Ker } T^*$ and R is a vector subspace.

2) dimension R = d.

3) If $(X,\mathcal{E})$ possesses a positive weight then R = Ker T*.

<u>Proof.</u> (We assume $\Omega \neq \emptyset$).

1) Choose $\mu \in \Omega$. Then $1\cdot 0 = \mu - \mu$ so that $0 \in R$. Suppose $\nu \in R$ and s is any number. If s = 0 then $s\cdot\nu = 0 \in R$. Suppose $s \neq 0$. Choose $r > 0$ and weights μ_1, μ_2 so that $r\nu = \mu_1 - \mu_2$. If $s > 0$ we get $(\frac{r}{s})(s\nu) = \mu_1 - \mu_2$. If $s < 0$ we get $(\frac{r}{-s})(s\nu) = \mu_2 - \mu_1$. Thus $s\nu \in R$. Suppose $0 \neq \nu_1 \in R$ and $0 \neq \nu_2 \in R$. Choose $r_1 > 0, r_2 > 0, \mu_1, \mu_2, \delta_1, \delta_2$ so that $r_1\nu_1 = \mu_1 - \mu_2$ and $r_2\nu_2 = \delta_1 - \delta_2$.

Let $s = \dfrac{r_1 r_2}{r_1 + r_2} > 0$. Then $0 < \dfrac{s}{r_1} = \dfrac{r_2}{r_1 + r_2} < 1$, $0 < \dfrac{s}{r_2} = \dfrac{r_1}{r_1 + r_2} < 1$,

and $\dfrac{s}{r_1} + \dfrac{s}{r_2} = 1$. Now

$$s(\nu_1 + \nu_2) = s\left\{\frac{1}{r_1}(\mu_1 - \mu_2) + \frac{1}{r_2}(\delta_1 - \delta_2)\right\}$$

$$= (\frac{s}{r_1}\mu_1 + \frac{s}{r_2}\delta_1) - (\frac{s}{r_1}\mu_2 + \frac{s}{r_2}\delta_2).$$

This proves that $\nu_1 + \nu_2 \in R$ so that R is a vector subspace.

2) Clearly $\{P_2 - P_1, P_3 - P_1, \ldots, P_n - P_1\} \subseteq R$. Let $\nu \in R$. Write $r\nu = \mu_1 - \mu_2$ where $r > 0$ and μ_1, μ_2 are weights. We can write $\mu_1 = \sum_{i=1}^{n} \alpha_i P_i$ and $\mu_2 = \sum_{i=1}^{n} \beta_i P_i$, where $\sum_1^n \alpha_i = 1 = \sum_1^n \beta_i$.

Thus $r\nu = \sum_1^n (\alpha_i - \beta_i) P_i = ((1-\sum_2 \alpha_i) - (1 - \sum_2 \beta_i)) P_1 + \sum_2 (\alpha_i - \beta_i) P_i$

$$r\nu = (\sum_2^n (\beta_i - \alpha_i)) P_1 + \sum_2^n (\alpha_i - \beta_i) P_i = \sum_2^n (\alpha_i - \beta_i)(P_i - P_1).$$

Thus $\nu = \sum_2^n (\frac{\alpha_i - \beta_i}{r}) (P_i - P_1)$ and we have shown that

$\{P_i - P_1 \mid 2 \leqslant i \leqslant n\}$ generates R. The result now follows.

3) Suppose there is a weight μ such that $\mu(x) > 0$ for each $x \in X$. Let $\nu \in \text{Ker } T^*$. Let $m = \min\{\mu(x) \mid x \in X\}$. Choose $r > 0$ so that $|r\nu(x)| < m$ for all $x \in X$. Let $\mu_1 = \mu + r\nu$. Then $\mu_1 \in \Omega$ and since $r\nu = \mu_1 - \mu$ we have that $\nu \in R$. Thus $\text{Ker } T^* = R$.

Let $(X, \mathcal{E})$ be any finite space and let $\Omega_p \subseteq \Omega$ be the set of pure weights. We known by Corollary 4.6 that Ω_p is a finite set. Let $\Omega_p = \{\mu_0, \mu_1, \mu_2, \ldots, \mu_n\}$. It is a well known fact that Ω is a simplex if and only if $\{\mu_k - \mu_0 \mid 1 \leqslant k \leqslant n\}$ is a linearly independent subset of $F(X)^*$. $(X, \mathcal{E})$ is said to be a <u>classical space</u> if $\mathcal{E} = \{X\}$. In this case Ω is a simplex and $|\Omega_p| = |X|$. We now obtain a converse to this result.

Let $(X, \mathcal{E})$ be any finite space. Let $E \subseteq F(X)^*$ be a complement of $\text{Ker } T^*$. That is $F(X)^* = \text{Ker } T^* \oplus E$. Then $F(X) = F(X)^{**} = (\text{Ker } T^*)^* \oplus E^*$ so that $(\text{Ker } T^*)^\perp = E^*$. Now $\text{Ker } T^* = (\text{im } T)^\perp$ so that we get $\text{im } T = (\text{im } T)^{\perp\perp} = (\text{Ker } T^*)^\perp \Rightarrow E^*$. Now $|X| = \dim \text{Ker } T^* + \dim E = \dim \text{Ker } T^* + \dim E^*$ so that $|X| = \dim \text{Ker } T^* + \dim \text{im } T$.

<u>Theorem 4.9.</u> Let $(X, \mathcal{E})$ be any finite space which has a positive weight. Then Ω is a simplex if and only if $|X| + 1 = \dim (\text{im } T) + |\Omega_p|$.

<u>Proof.</u> Suppose $|X| + 1 = \dim (\text{im } T) + |\Omega_p|$. Then $|\Omega_p| - 1 = \dim \text{Ker } T^*$ and by Proposition 4.8 we can conclude that Ω is a

simplex.

Suppose Ω is a simplex. Then dim Ker T* = $|\Omega_p| - 1$ and the sult follows from $|X|$ = dim Ker T* + dim (im T).

<u>Corollary. 4.10.</u> Suppose $(X,\mathcal{E})$ is a finite space, Ω is a simplex, and there is a positive weight. If $|\Omega_p| \geq |X|$ then $(X,\mathcal{E})$ is a classical space.

<u>Corollary 4.11.</u> Let $(X,\mathcal{E})$ be a finite space such that $x \neq y$ implies the existence of weight ω so that $\omega(x) = m_x$ and $\omega(y) > m_y$. If Ω is a simplex then $(X,\mathcal{E})$ is a classical space.

<u>Proof.</u> By theorem 3.5 there is a bijection between the points of X and the facets of Ω. Since Ω is a simplex the number of facets of Ω is the same as the number of pure weights. Thus $|\Omega_p|=|X|$ and the result follows from corollary 4.10.

Recall that $\Omega(X,\mathcal{E})$ satisfies the projection postulate in case for all $x \in X$ there exists $\omega \in \Omega$ with $\omega(x) = 1$.

<u>Remark 4.12.</u> Let $(X,\mathcal{E})$ be a finite point closed space such that Ω is full and satisfies the projection postulate. If Ω is a simplex then $(X,\mathcal{E})$ is a classical space.

<u>Proof.</u> We may assume $|X| > 1$. For $\omega \in \Omega_p$ let F_ω denote the facet of Ω opposite ω. As in the first part of the proof of Theorem 3.5 there exists $x_\omega \in X$ such that $F_\omega \subseteq S_o(x_\omega)$ and, since $M_{x_\omega} > 0$, $F_\omega = S_o(x_\omega)$. The mapping $\phi : \omega \to x_\omega$ is an injection of Ω_p into X. Let $\omega_1,\ldots,\omega_n$ be the pure weights. Then $\omega_i(x_{\omega_j}) = \delta_{ij}$, the Kronecker delta. Fullness implies that image $(\phi) = E$ is an (in fact, maximal) orthogonal set. If $y \in X \setminus E$ then there exists $\omega \in \Omega_p$ with $\omega(y) = 1$ and there exists $e \in E$ with $\omega(e) = 1$. It follows that $y^\perp \subseteq e^\perp$ and hence $\{e\} = e^{\perp\perp} \subseteq y^{\perp\perp} = \{y\}$, i.e. e=y a contradiction. Thus X = E and $\mathcal{E} = \{E\}$.

5. <u>SYMMETRIES AND DUALITY</u>

Let $(X,\mathcal{E})$ be any space with $X = \bigcup\mathcal{E}$. We define the <u>dual space</u> $(X^*,\mathcal{E}^*)$ by $X^* = \mathcal{E}$ and $\mathcal{E}^* = \{\mathcal{E}_x |\ x \in X\}$. We say that $(X,\mathcal{E})$ is <u>separating</u> if $\mathcal{E}_x \not\subseteq \mathcal{E}_y$ or $\mathcal{E}_x = \mathcal{E}_y$ whenever $x \neq y$. We say that $(X,\mathcal{E})$ is <u>distinguishing</u> if $\mathcal{E}_x \neq \mathcal{E}_y$ whenever $x \neq y$. Note that the logic of a separating and distinguishing Dacey-space need not be hyper-irreducible [6].

Proposition 5.1.

1) If $(X,\mathcal{E})$ is any space then $(X^*,\mathcal{E}^*)$ is a distinguishing space.

2) If $E\in\mathcal{E}$, $F\in\mathcal{E}$, $E \neq F$ implies $E \not\subseteq F$ then $(X^*,\mathcal{E}^*)$ is also a separ ting space.

3) $(X,\mathcal{E})$ is distinguishing if and only if $(X^{**},\mathcal{E}^{**}) \simeq (X,\mathcal{E})$.

4) Suppose that X is finite and $\Omega_g \neq \emptyset$. Then $\sum_{x\in X} \mu_1(x) = \sum_{x\in X} \mu_2(x$ for all μ_1, μ_2 in Ω_g if and only if there exists a generalized weight on $(X^*,\mathcal{E}^*)$.

Proof. For the proof of 5.1 and more results on dual spaces sec [9].

Let $(X,\mathcal{E})$ and $(Y,\mathcal{F})$ be spaces, then every bijection $\phi : X \to$ induces a bijection $\tilde{\phi} : 2^X \to 2^Y$ defined by $\tilde{\phi}(E) = \phi(E)$ for all $E\in$ If ϕ is such that $\tilde{\phi}(\mathcal{E}) = \mathcal{F}$, then we say that ϕ is an isomorphism from $(X,\mathcal{E})$ to $(Y,\mathcal{F})$ and write $(X,\mathcal{E}) \simeq (Y,\mathcal{F})$. If $(X,\mathcal{E})$ and $(Y,\mathcal{F})$ are orthogonality spaces then the isomorphisms from $(X,\mathcal{E})$ to $(Y,\mathcal{F}$ are precisely the ortho-bijections (c.f. 3.11) of the induced graphs $(X,\perp_{\mathcal{E}})$ and $(Y,\perp_{\mathcal{F}})$.

An isomorphism from $(X,\mathcal{E})$ to itself is called an automorphis the set of all automorphisms on $(X,\mathcal{E})$ forms a group $\mathrm{Aut}(X,\mathcal{E})$. One can show [9] that if $(X,\mathcal{E})$ is a distinguishing space then $\mathrm{Aut}(X,\mathcal{E}$ is isomorphic to $\mathrm{Aut}(X^*,\mathcal{E}^*)$. Gerald Schrag [18] has combined this observation and some of our earlier results with some known resul of graph theory to prove the following theorem.

Theorem (Schrag) : Every finite group is the automorphism group of some orthomodular lattice.

If $(X,\mathcal{E})$ is a homogeneous space of dimension n then the set of all bijective structure maps of (Ω,e) onto itself is a group. Denote it by $\mathrm{Aut}(\Omega,e)$. We conclude with the following theorem.

Theorem 5.2. For a homogeneous point closed $S_{\mathcal{E}}$ compact spac of dimension n with a strong set of weights Ω,

$$\mathrm{Aut}(\Omega,e) \simeq \mathrm{Aut}(X,\mathcal{E}) \simeq \mathrm{Aut}(X^*,\mathcal{E}^*).$$

Proof. Define ϕ: $\mathrm{Aut}(X,\mathcal{E}) \to \mathrm{Aut}(\Omega,e)$ by the following : for $g\in \mathrm{Aut}(X,\mathcal{E})$, $[\phi(g)]\ \omega(x) = \omega(g^{-1}(x))$. It is straight forward to see that ϕ is an injective homomorphism. That it is surjective fo lows from the proof of 3.13.

BIBLIOGRAPHY

1. Berge, C., Graphs and Hypergraphs, North Holland Publishing Co., Amsterdam-London, 1973.

2. Dacey, J.C., "Orthomodular Spaces and additive measurement", Caribbean Journal of Sciences and Mathematics, Vol. I, No II, July, 1969.

3. Dembowski, P., Finite Geometries, Springer-Verlag, New York, 1968.

4. Foulis, D.J. and Randall, C.H., "Operational Statistics I Basic Concepts", J.Math Physics, Vol. 13, No. 11, November 1972.

5. Gerelle, E.R., "A Preliminary Report on the Affine Geometry of Quantum Logics", Kansas State University, Technical Report No. 34, July 1973.

6. Greechie, R.J., "Hyper-irreducibility in an orthomodular lattice", J.Nat.Sc. and Math Vol. VIII, No. 1, April 1968.

7. Greechie, R.J., "Orthomodular lattices admitting no states", Journal of Combinatorial Theory, Vol. X, No. 2, March 1971.

8. Greechie, R.J. and Miller, F.R., "On Structures Related to States on an Empirical Logic I. Weights on Finite Spaces", K.S.U. Technical Report No. 14, April 1970.

9. Greechie, R.J. and Miller, F.R., "On Structures Related to States on an Empirical Logic II. Weights and Duality on Finite Spaces", K.S.U. Technical Report No. 26, August 1972.

10. Greechie, R.J. and Gudder, S.P., "Quantum Logics" in Contemporary Research in The Foundation and Philosophy of Quantum Theory, D.Reidel Publ. Co., Dordrecht-Holland, Boston-U.S.A., (C.A.Hooker ed.), pp. 143-173.

11. Grünbaum, B., Convex Polytopes, Interscience, New York-London, 1967.

12. Gudder, S.P., "Hilbert Space, independence, and generalized probability", J.Mathematical Analysis and Applications, Vol. 20, No. 1, Oct. 1967.

13. Holland, S.S., Jr., "The current interest in orthomodular lattices", Trends in Lattice Theory, (J.C.Abbott, ed.), Van Nostrand, Princeton, 1970.

14. Jauch, J.M., Foundations of Quantum Mechanics, Addison Wesley, Reading, London, 1968.

15. Kolmogorov, A.N., Foundations of the Theory of Probability, 2nd edition, Chelsea, New York, 1956.

16. Randall, C.H., "A mathematical foundation for empirical science with special reference to quantum theory. Part I. A calculus of experimental propositions", Knolls Atomic Power Lab. Report, KAPL-3147, June 1966.

17. Randall, C.H. and Foulis, D.J., "States and the free orthogonality monoid", Math Systems Theory 6, No. 3 (1972) 268-276.

18. Schrag, G., "Combinatories and Graph Techniques in Orthomodular Theory", Ph.D.Dissertation, Kansas State Univ., 1971.

19. Shultz, F., "A representation of the state spaces of orthomodular lattices", (to appear).

20. Varadarajan, V.S., "Probability in physics and a theorem on simultaneous observability", Comm. Pure and Applied Math., Vol. 15 (1962), 189-217.

A SECOND LOOK AT THE ESSENTIAL SELFADJOINTNESS OF THE SCHRÖDINGER OPERATORS (*)

Tosio Kato

Department of Mathematics, University of California, Berkeley

1. INTRODUCTION

There is an extensive literature on the essential selfadjointness of the Schrödinger operators. Basically, the question has been whether or not the minimal operator constructed from a given, formally selfadjoint differential operator L has a unique selfadjoint extension. For simplicity consider the case

$$L = -\Delta + q(x) \quad \text{on } R^m , \tag{1}$$

where q is a real-valued measurable function. The minimal operator T_{min} for L is by definition the symmetric operator in $\underline{H} = L^2(R^m)$, given by $T_{min}u = Lu$, with domain $\underline{D}(T_{min}) = C_0^\infty(R^m)$. The question is

(Q1) Has T_{min} a unique selfadjoint extension?

For T_{min} to exist, however, L must map $C_0^\infty(R^m)$ into $\underline{H}$. Hence it is necessary that

$$q \in L^2_{loc}(R^m) . \tag{2}$$

(2) is in general not sufficient for (Q1) to have a positive answer, but some recent results show that it is in fact sufficient if q is bounded from below (see [1,2]).

(*) This work was partly supported by NSF Grant GP37780X.

Enz/Mehra (eds.), Physical Reality and Mathematical Description. 193–201. *All Rights Reserved.*

It does not seem to the author, however, that (2) is a natural restriction. One may admit a more general q by changing one's viewpoint and considering the maximal operator T_{max} rather than the minimal operator T_{min}. T_{max} is by definition the operator in $\underline{H}$, given by $T_{max}u = Lu$, with $\underline{D}(T_{max})$ consisting of all $u \in \underline{H}$ such that $qu \in L^1_{loc}(R^m)$ and $Lu \in \underline{H}$, where $Lu = -\Delta u + qu$ is taken in the distribution sense. Here we require $qu \in L^1_{loc}(R^m)$ explicitly because otherwise the meaning of qu as a distribution is not clear. (A function like qu may be a distribution without belonging to $L^1_{loc}(R^m)$. But since there is no general criterion for it, we prefer to require the stated condition.) Now the question is

(Q2) Is T_{max} selfadjoint?

Several remarks are in order.

(a) An affirmative answer to (Q2) is as satisfactory as one for (Q1), since it means that practically L determines a unique selfadjoint operator in $\underline{H}$. Furthermore, T_{max} may well be defined when T_{min} does not exist. Although it is not at all obvious that T_{max} can have a dense domain when (2) is not satisfied, the theorem given below shows in fact that (Q2) can have a positive answer for q not satisfying (2); for example, this is the case if

$$\text{const} \le q \in L^1_{loc}(R^m) \tag{3}$$

(b) If we do assume (2), on the other hand, then it is easy to see that

$$T_{max} = T^*_{min}. \tag{4}$$

In this case (Q1) and (Q2) are equivalent. Thus (Q2) is a natural generalization of the problem (Q1).

(c) In (Q1), we defined T_{min} with $\underline{D}(T_{min}) = C_0^\infty(R^m)$. But this choice of $C_0^\infty(R^m)$ is more or less arbitrary. If q has a strong singularity at a single point, say x = 0, we could choose $\underline{D}(T_{min}) = C_0^\infty(R^m-\{0\})$. More generally, we could choose an open set $\Omega \subset R^m$ such that $R^m-\Omega$ has Lebesgue measure zero, and set $\underline{D}(T_{min}) = C_0^\infty(\Omega)$ so that T_{min} is still a symmetric operator in $\underline{H}$. In this case (2) can be weakened to

$$q \in L^2_{loc}(\Omega). \tag{2'}$$

In this way there arise many different problems of type (Q1). (Actually the case $\Omega = R^m-\{0\}$ has been studied extensively). Now the same situation exists also for (Q2). Here the question is what we mean by "distribution". In the definition of T_{max} given above, we tacitly mean "distribution in R^m". But we could as well choose "distribution in Ω", and this would give rise to different problems of type (Q2). In this paper, however, we shall consider only distributions in R^m.

In what follows we shall prove a theorem that answers (Q2) in the affirmative. The proof of the theorem is based on two ideas; one is the theory of quadratic forms, and the other is a distributional inequality proved in [2].(*) For convenience we state here the latter in a specialized form we need.

LEMMA A. If u and Δu are in $L^1_{loc}(R^m)$, then

$$\Delta|u| \geq \operatorname{Re}\left[(\operatorname{sign} \bar{u})\Delta u\right]$$

in the distribution sense.

2. THE MAIN THEOREM

In what follows we assume that

$$q = q_1 + q_2 , \tag{5}$$

where

(A) $\quad q_1 \in L^1_{loc}(R^m), \quad q_1(x) \geq - q^*(|x|),$

with $q^*(r)$ positive, monotone increasing for $0 < r < \infty$, and $q^*(r) = O(r^2)$ as $r \to \infty$.

For q_2 we assume

(B1) $\quad \int_{|x|<r} |q_2(x)|^2 dx \leq K^2 r^{2s} , \quad 1 \leq r < \infty ,$

where K and s are some constants, and

(B2) $\int_{|y|<r} |q_2(x-y)|\,|y|^{2-m} dy \to 0$ as $r \to 0$, uniformly in $x \in R^m$;

here $|y|^{2-m}$ should be replaced by $-\log|y|$ if $m = 2$ and by 1 if $m = 1$.

If $m \geq 5$, (B1-2) may be replaced by

(*) The theory of forms is useful in constructing selfadjoint restrictions of T_{max}, but it is not powerful enough to show that T_{max} itself is selfadjoint (cf. [3]). Some other means is necessary for our purpose.

(B') $\quad q_2 \in L^{m/2}(R^m)$.

THEOREM. Assume (A), (B1), and (B2). Then T_{max} is self-adjoint. If $m \geq 5$, (B1-2) may be replaced by (B').

REMARK. (a) The assumptions of the theorem are almost identical with those made in [2], with the only significant difference that $q_1 \in L^1_{loc}$ replaces $q_1 \in L^2_{loc}$ of [2]. We have also improved $q^*(r) = o(r^2)$ of [2] into $O(r^2)$, but this is a minor point; actually $O(r^2)$ should have been sufficient in [2].

(b) (B1) involves quadratic integrals and is unsatisfactory from the new point of view, but it seems necessary to assume it in the proof given below.

(c) There would be no difficulty in replacing L by a more general second-order differential operator of elliptic type, in particular one involving a magnetic potential.

The proof of the theorem is somewhat complicated; it runs roughly parallel to the one given in [2] for the problem (Q1). We first consider the case when q_1 is bounded from below, so that we may take $q^*(r) = \text{const}$.

LEMMA 1. If $q^*(r) = q^* = \text{const}$, then T_{max} is selfadjoint with $\underline{D}(T_{max}) \subset H^1(R^m) \cap \underline{D}((q_1+q^*)^{1/2})$, where H^1 is the Sobolev space of order 1 and $D((q_1+q^*)^{1/2})$ is the domain of the selfadjoint operator $(q_1+q^*)^{1/2}$ in $\underline{H}$.

Proof. Define the quadratic form

$$h = h_0 + h' \quad , \quad h' = h_1' + h_2' \quad , \tag{6}$$

$$h_0[u] = \|\text{grad } u\|^2 \quad , \quad h_j'[u] = \int_{R^m} q_j(x)|u|^2 dx \quad , \quad j = 1,2,$$

with $\underline{D}(h) = \underline{D}(h_0) \cap \underline{D}(h_1') = H^1(R^m) \cap \underline{D}((q_1+q^*)^{1/2})$. h is densely defined, bounded from below, and closed, since h_2' is relatively bounded with respect to h_0 with an arbitrarily small relative bound. (The proof is given in 2(*); here we use (B2) or (B') but not (B1)). We note also that

$$h[u] \geq (3/4)\|u\|_1^2 - (q^*+1+M)\|u\|^2 \quad , \quad u \in \underline{D}(h) \quad , \tag{7}$$

where $\|u\|_1 = (h_0\, u + \|u\|^2)^{1/2}$ is the usual Sobolev norm in $H^1(R^m)$ and M is a constant depending on q_2 (see (2.4) of [2])(**).

(*), (**) The form h was not used in [2], but essentially the same calculation can be used here.

Let H be the selfadjoint operator associated with h. We shall first show that

$$H \subset T_{max} . \tag{8}$$

To this end let $u \in \underline{D}(H)$. Then we have for each $v \in C_0^\infty(R^m)$

$$(Hu,v) = h[u,v] = -(u,\Delta v) + \int qu\bar{v}\,dx,$$

and hence

$$-(u,\Delta v) = \int (Hu - qu)\bar{v}\,dx \quad \text{for all } v \in C_0^\infty(R^m). \tag{9}$$

Here $qu \in L^1_{loc}(R^m)$, since $2|qu| \le |q| + |q||u|^2$ where $|q| \in L^1_{loc}$ by hypothesis and $|q||u|^2 \in L^1$ because $u \in \underline{D}(H) \subset \underline{D}(h) \subset \underline{D}(q)$. Since $Hu \in \underline{H} = L^2$, (9) implies that $-\Delta u = Hu - qu \in L^1_{loc}$ exists as a distribution. In other words, $Lu = -\Delta u + qu = Hu \in \underline{H}$. Hence $u \in \underline{D}(T_{max})$ by definition, with $T_{max}u = Hu$. This proves (8).

Next we show that actually we have equality in (8), so that the assertion of the lemma is true. To this end, it suffices to show that $\underline{D}(T_{max}) \subset \underline{D}(H)$. Suppose $u \in \underline{D}(T_{max})$ and set $v = (H+z)^{-1}(T_{max}+z)u$, where z is a real number sufficiently large that $(H+z)^{-1}$ exists on $\underline{H}$ (recall that H is bounded from below). Then $(H+z)v = (T_{max}+z)u$ and hence by (8), $(L+z)w = (T_{max}+z)w = 0$ for $w = u-v$. Hence $\Delta w = (q+z)w$, where $qw \in L^1_{loc}$ because $w \in \underline{D}(T_{max})$. Application of Lemma A now yields

$$\Delta|w| \ge (q+z)|w| \ge (-q^*+q_2+z)|w|.$$

Writing $c^2 = z-q^*$, we obtain

$$(c^2-\Delta)|w| \le -q_2|w| \le |q_2||w|.$$

Then an argument used in [2] shows that $|w| = 0$, $w = 0$, $u = v \in \underline{D}(H)$, hence $\underline{D}(T_{max}) \subset \underline{D}(H)$ as required, provided c^2 is sufficiently large (in relation to a number depending only on q_2), which can be achieved by choosing z sufficiently large in the beginning. |||

For later use we deduce the following estimate.

LEMMA 2. Let $q^* = \text{const}$. If $u \in \underline{H}$, $qu \in L_{loc}(R^m)$, and $Lu \in H^{-1}(R^m)$, then $u \in H^1(R^m) \cap \underline{D}((q_1+q^*)^{1/2})$, with

$$\|u\|_1 \le 2^{1/2}\left[(q^*+1+M)^{1/2}\|u\| + \|Lu\|_{-1}\right], \tag{10}$$

where $H^{-1}(R^m)$ is the Sobolev space of order -1 and $\|\ \|_{-1}$ is the associated norm dual to $\| \|_1$.

Proof. Let $\underline{K}$ be the Hilbert space $\underline{D}(h)$ with the norm $(h[u]+c^2\|u\|^2)^{1/2} \geq \|u\|$, where $c^2 \geq q^*+2+M$. Then we have the continuous inclusions

$$\underline{D}(H) \subset \underline{K} \subset H^1(R^m) \subset \underline{H} \subset H^{-1}(R^m) \subset \underline{K}^*.$$

As is well known, the map H ($= T_{max}$) on $\underline{D}(H)$ to $\underline{H}$ can be extended to a continuous map H' on $\underline{K}$ to $\underline{K}^*$. Actually H' is a restriction of L; more precisely, $w \in \underline{K}$ implies that $qw \in L^1_{loc}(R^m)$ and $Lw = -\Delta w + qw = H'w$, as is easily verified by noting that $qw \in L^1_{loc}$ as remarked above.

It is also well known that $c^2+ H'$ maps $\underline{K}$ onto $\underline{K}^*$ bicontinuously.

Suppose now that u satisfies the conditions stated in Lemma 2. Then $(c^2+L)u \in H^{-1}(R^m) \subset \underline{K}^*$. Hence there is $w \in \underline{K}$ such that $(c^2+L)w = (c^2+H')w = (c^2+L)u$, hence $Lv = -c^2v$ for $v = u-w$. But $v \in \underline{H}$ and $qv \in L^1_{loc}(R^m)$, since $qu \in L^1_{loc}$ by hypothesis and $qw \in L^1_{loc}$ by $w \in \underline{K} = \underline{D}(h)$ (see the remark above). Hence $v \in \underline{D}(T_{max})$ with $T_{max}v = -c^2v$. This implies $v = 0$ since $T_{max}+c^2 \geq 1$. Thus $u = w \in \underline{K} = \underline{D}(h) = H^1(R^m) \cap \underline{D}((q_1+q^*)^{1/2})$.

To prove (10), we first note that $(c^2+H')^{-1}$ maps $\underline{K}^*$ onto $\underline{K}$ continuously. A fortiori, it maps $H^{-1}(R^m)$ into $H^1(R^m)$ continuously.

Suppose $w \in \underline{D}(H)$. Then $h[w] = (Hw,w) \leq \|Hw\|_{-1} \|w\|_1 \leq \|Hw\|_{-1}^2 + (1/4)\|w\|_1^2$. Combined with (7), this leads to (10) with u replaced by w and Lu by Hw (see a corresponding calculation in [2]). If we write $f = (c^2+H)w$, $w = (c^2+H)^{-1}f$, $Hw = f-c^2(c^2+H)^{-1}f$, we obtain

$$\|(c^2+H)^{-1}f\|_1 \leq 2^{1/2}\left[(q^*+1+M)^{1/2}\|(c^2+H)^{-1}f\| + \|f-c^2(c^2+H)^{-1}f\|_{-1}\right], \tag{11}$$

where $f \in \underline{H}$ is arbitrary because $(c^2+H)\underline{D}(H) = \underline{H}$. Since $(c^2+H)^{-1}$ has an extension $(c^2+H')^{-1}$ that maps $H^{-1}(R^m)$ into $H^1(R^m)$ continuously as noted above, (11) is extended to all $f \in H^{-1}(R^m)$ if H is replaced by H'. If we write $(c^2+H')^{-1}f = u$ or $f = (c^2+H')u = (c^2+L)u$, (11) returns to the form (10) in which u may be any element of $\underline{K}$ such that $(c^2+L)u \in H^{-1}(R^m)$ or, equivalently, $Lu \in H^{-1}(R^m)$. Since u in Lemma 2 satisfies these conditions as shown above, we have proved Lemma 2. |||

3. PROOF OF THEOREM (continued)

We now consider the general case in which q^* need not be constant.

LEMMA 3. Let $u \in \underline{D}(T_{max})$. Then $u \in H^1_{loc}(R^m)$, and we have for $0 < r < R$

$$\|u\|_{H^1(B_r)} \leq 2^{1/2}(\|T_{max}u\| + q^*(R)^{1/2}\|u\|) + C_{R-r}\|u\| , \tag{12}$$

where B_r denotes the ball in R^m with center 0 and radius r, and C_{R-r} is a constant depending only on m, R-r, and q_2. Furthermore, $\phi u \in \underline{D}(T_{max})$ for any $\phi \in C_0^\infty(R^m)$.

Proof. The same as in [2] (see Propositions 5 and 7)(*).|||

In what follows we fix a real-valued $\phi \in C_0^\infty(R^m)$ such that supp $\phi \subset B_2$ and $\phi(x) = 1$ for $x \in B_1$, and define $\phi_n(x) = \phi(x/n)$, $n = 1,2,\ldots$. Then it is obvious that $\phi_n w \to w$ in $\underline{H}$ for any $w \in H$.

LEMMA 4. Let $u \in \underline{D}(T_{max})$. Then $T_{max}(\phi_n u) \to T_{max}u$ weakly in $\underline{H}$.

Proof. Note that $\phi_n u \in \underline{D}(T_{max})$ by Lemma 3. The proof of Lemma 4 is essentially the same as in [2, Proposition 9]. A slight modification is required, since we have replaced the assumption $q^*(r) = o(r^2)$ of [2] by $O(r^2)$. (This is the reason why we assert only weak convergence now.) Reviewing the proof of Proposition 9, we see that it suffices to show that $(\partial_k\phi_n)(\partial_k u) \to 0$ weakly in $\underline{H}$ as $n \to \infty$. But the proof of Proposition 9, with $o(r^2)$ replaced by $O(r^2)$, shows that the sequence in question is bounded as $n \to \infty$. To prove its weak convergence to 0, therefore, it suffices to show that $((\partial_k\phi_n)(\partial_k u),w) \to 0$ for $w \in C_0^\infty(R^m)$. But this can be done easily by integration by parts. |||

In what follows we also use "truncations" of T_{max}. Let $L^{(n)}$ be the formal operator $-\Delta + q_1^{(n)} + q_2$, where

$$q_1^{(n)}(x) = \begin{cases} q_1(x) & \text{for } x \in B_{2n}, \\ q_1(x) + q^*(|x|) - q^*(2n) & \text{for } x \notin B_{2n}. \end{cases} \tag{13}$$

Since $q_1^{(n)}$ is bounded from below and is in $L^1_{loc}(R^m)$, the associated maximal operator $T^{(n)}_{max}$ is selfadjoint by Lemma 1.

(*) The proof of these propositions is valid due to Lemma 2.

LEMMA 5. $\underline{D}(T_{max}^{(n)}) = \underline{D}'$ is independent of n. If either $u \in \underline{D}(T_{max})$ or $u \in \underline{D}'$, then $\phi u \in \underline{D}(T_{max}) \cap \underline{D}'$ for each $\phi \in C_0^\infty(R^m)$. Moreover, $T_{max}^{(n)}(\phi u) = T_{max}(\phi u)$ if supp $\phi \subset B_{2n}$.

Proof. It is easy to see that $q_1^{(n)} - q_1^{(k)}$ is a bounded function. Hence $T_{max}^{(n)}$ and $T_{max}^{(k)}$ have the same domain.

We have already proved that (multiplication by) ϕ maps $\underline{D}(T_{max})$ into itself (Lemma 3). Hence ϕ maps $\underline{D}'$ into $\underline{D}'$. It is easy to see, on the other hand, that if supp $u \subset B_{2n}$, then $u \in \underline{D}(T_{max})$ and $u \in \underline{D}(T_{max}^{(n)})$ are equivalent, with $T_{max}u = T_{max}^{(n)}u$. This proves the lemma. |||

LEMMA 6. $(T_{max}u,v) = (u,T_{max}v)$ for u, $v \in \underline{D}(T_{max})$.

Proof. For each n we have $(T_{max}^{(n)}(\phi_n u),\phi_n v) = (\phi_n u,T_{max}^{(n)}(\phi_n v))$ because $T_{max}^{(n)}$ is selfadjoint and $\phi_n u$, $\phi_n v$ are in $\underline{D}'$ by Lemma 5. The desired result follows on letting $n \to \infty$, by virtue of Lemmas 4, 5 and the strong convergence $\phi_n u \to u$ etc. |||

LEMMA 7. T_{max} is closed.

Proof. Suppose $u_n \in \underline{D}(T_{max})$, $u_n \to u \in \underline{H}$ and $T_{max}u_n \to f \in \underline{H}$. We have to show that $u \in \underline{D}(T_{max})$ and $T_{max}u = f$.

Since $\{u_n\}$ and $\{T_{max}u_n\}$ are Cauchy sequences in $\underline{H}$, (12) shows that $\{u_n\}$ is Cauchy in $H^1(B_r)$ for any $r > 0$. Thus $\{u_n\}$ converges in $H^1_{loc}(R^m)$, and the limit must coincide with u so that $u \in H^1_{loc}(R^m)$.

A simple computation gives

$$T_{max}^{(k)}(\phi_k u_n) = L(\phi_k u_n) = \phi_k Lu_n - 2\ \text{grad}\ \phi_k \cdot \text{grad}\ u_n - u_n \Delta\phi_k. \tag{14}$$

Let $n \to \infty$ with a fixed k. Then $Lu_n = T_{max}u_n \to f$. Since $u_n \to u$ and grad $u_n \to$ grad u in $L^2(B_{2k})$, the right member of (14) converges in $\underline{H}$ to the right member of (15) below. Since $\phi_k u_n \to \phi_k u$ and since $T_{max}^{(k)}$ is closed, we must have

$$T_{max}^{(k)}(\phi_k u) = \phi_k f - 2\ \text{grad}\ \phi_k \cdot \text{grad}\ u - u\Delta\phi_k \quad . \tag{15}$$

If we evaluate (15) at $x \in B_k$ and note that $\phi_k(x) = 1$, $q_1^{(k)}(x) = q_1(x)$ there, we obtain

$$Lu(x) = f(x) \quad \text{for} \quad x \in B_k,$$

where $\phi_k qu \in L^1(R^m)$ because $\phi_k u \in \underline{D}'$. Since k is arbitrary, it follows that $Lu = f$ with $qu \in L^1_{loc}(R^m)$. Since $f \in \underline{H}$, we conclude that $u \in \underline{D}(T_{max})$ with $T_{max}u = f$. |||

LEMMA 8. $\underline{R}(T_{max} \pm i) = \underline{H}$. ($\underline{R}$ denotes the range.)

Proof. Let $f \in \underline{H}$. Since $T^{(n)}_{max}$ is selfadjoint, there is $u_n \in \underline{D}'$ such that

$$(T^{(n)}_{max} - i)u_n = f \;, \quad \|u_n\| \le \|f\| \;, \quad \|T^{(n)}_{max}u_n\| \le \|f\| \;. \tag{16}$$

Set $k = n$ in the identity (14) and note that $T^{(n)}_{max}(\phi_n u_n) = T_{max}(\phi_n u_n)$ by Lemma 5. Since

$$\phi_n L u_n = \phi_n L^{(n)} u_n = \phi_n T^{(n)}_{max} u_n = \phi_n (iu_n + f) \;,$$

we obtain

$$\phi_n f - 2 \operatorname{grad} \phi_n \cdot \operatorname{grad} u_n - u_n \Delta\phi_n = (T_{max} - i)(\phi_n u_n) \in \underline{R}(T_{max} - i) \;. \tag{17}$$

Let $n \to \infty$ in (17). Then $\phi_n f \to f$ and $u_n \Delta\phi_n \to 0$ in $\underline{H}$, since $\|u_n\| \le \|f\|$ and $\sup_x |\Delta\phi_n(x)| = O(n^{-2})$. Furthermore, $\operatorname{grad} \phi_n \cdot \operatorname{grad} u_n \to 0$ <u>weakly</u> in $\underline{H}$. The proof is similar to the one noted in the proof of Lemma 4, being based on the estimate $\|u_n\|_{H^1(B_{2n})} = O(n)$ that follows from (12) and (16). It follows that f is the weak limit of a sequence of elements of $\underline{R}(T_{max} - i)$. But the latter is closed as a result of Lemmas 6 and 7, and hence weakly closed. Consequently $f \in \underline{R}(T_{max} - i)$. Since $f \in \underline{H}$ was arbitrary, this proves $\underline{R}(T_{max} - i) = \underline{H}$. Similarly for $\underline{R}(T_{max} + i)$. |||

Completion of the proof of the theorem. As is easily seen, Lemmas 6 and 8 imply that T_{max} is selfadjoint. (It is easy to show <u>directly</u> that $\underline{D}(T_{max})$ is dense in $\underline{H}$, but this is not necessary).

REFERENCES

[1] B. Simon, Essential self-adjointness of Schrödinger operators with positive potentials, Math. Ann. 201 (1973), 211-220.

[2] T. Kato, Schrödinger operators with singular potentials, Israel J. Math. 13 (1972), 135-148.

[3] M. Schechter, Hamiltonians for singular potentials, Indiana Univ. Math. J. 22 (1972), 483-503.

SOME ABSOLUTELY CONTINUOUS OPERATORS

P.A.Rejto*

School of Mathematics
University of Minnesota
Minneapolis, Minn. 55455
USA

1. INTRODUCTION

The property of a self-adjoint operator, or part of it, being absolutely continuous has a remarkable consequence for the unitary semigroup that it generates. This consequence is remarkable inasmuch as for Schrödinger operators it admits a direct physical interpretation in terms of the axioms of quantum mechanics. For a specific description of this interpretation we refer to the recent lectures of Amrein [27].

Professor Jauch has always emphasized that physical intuition and mathematical rigour complement each other. In a paper with Amrein and Georgescu [38] they mention that in one of the methods used in scattering theory: "... one introduces a subset $\mathfrak{X} \subset \mathcal{H}$ of the Hilbert space considered itself a Banach space with its own independent norm, and one interprets some of the operators of the theory in this space. In the end result, ..., this space disappears. This fact and the fact that the choice of $\mathfrak{X}$ is to some extent arbitrary may make this theory, in the view of the physicist, appear somewhat artificial."

The aim of the present note is to give a partial answer to this criticism. Specifically in Section 2, with the aid of such a Banach space we formulate an abstract criterion for absolute conti-

* Supported by NSF Grant GP 28933.

Enz/Mehra (eds.), Physical Reality and Mathematical Description. 202–225.

nuity. Then in Section 3 we illustrate that the introduction of such a norm can be motivated by the JWKB-approximation method [29] [36].

Finally let us emphasize that several different authors have established absolute continuity for several different classes of self-adjoint operators. In particular the following authors have done this; Agmon [15], Aguilar-Combes [17], Balslev-Combes [18], Faddeev [39], Georgescu [38], Howland [8], Ikebe-Saito [19], Kato [11], Kato-Kuroda [13], Kuroda [21], Lax-Phillips [9], Lavine [23], Neumark [32], Rejto [16], Titchmarsh [1], Walter [20] and Weidmann [7].

2. AN ABSTRACT CRITERION FOR ABSOLUTE CONTINUITY

Let A be a given self-adjoint operator acting on a given abstract Hilbert space $\mathfrak{H}$. We start with a lemma which gives a simple sufficient condition for a part of A to be absolutely continuous. To formulate it we need some notation. To a given interval of reals, ϑ, and angle α, we assign two open regions of the complex plane by setting

$$(2.1)\qquad \mathcal{R}_{\pm}(\vartheta) = \{\mu\colon \operatorname{Re}\mu \in \vartheta^0 \ ,\ 0 < \arg\mu < \alpha\}$$

where ϑ^0 denotes the interior of the interval ϑ. As usual, we denote by $R(\mu)$ the resolvent of A at the point μ, that is,

$$(2.2)\qquad R(\mu) = (\mu I - A)^{-1} \quad .$$

LEMMA 2.1. *Let ϑ be a bounded interval and let $\mathfrak{S}$ be a dense subset of $\mathfrak{H}$. Suppose that $\mathfrak{S}$ is such that for each fixed pair of vectors (f,g) in $\mathfrak{S} \times \mathfrak{S}$*

$$(2.3)\qquad \sup_{\mu \in \mathcal{R}_+(\vartheta)} |(R(\mu)f,g) - (R(\bar{\mu})f,g)| < \infty$$

Then $A(\vartheta)$, the part of A over the interval ϑ is absolutely continuous.

It was observed elsewhere [6.a] that this lemma is an elementary consequence of the resolvent loop-integral formula. We stated if for completeness only.

For a class of Schrödinger operators it is possible to factorize the resolvent in a manner which allows one to establish the rather general assumptions of Lemma 2.1. To describe such factorizations we make a digression on forms. Accordingly let $\mathfrak{B}$ be an abstract Banach space and $[F]$ a functional on $\mathfrak{B} \times \mathfrak{B}$ which is linear in the first argument and conjugate linear in the second argument, in short a sesquilinear form. In analogy to the notion of the norm of an operator we define the norm of the form $[F]$ by

$$\|F\| = \sup_{f\neq 0, g\neq 0} \frac{|[F](f,g)|}{\|f\|_{\mathfrak{B}} \cdot \|f\|_{\mathfrak{B}}} .$$

Next let A be a bounded operator on $\mathfrak{B}$. We define the product $[F]A$ to be the form determined by

$$[F]A(f,g) = [F](Af,g) \;,\; (f,g) \in \mathfrak{B} \times \mathfrak{B} .$$

Then clearly

$$\|[F]A\| \leq \|[F]\| \cdot \|A\| .$$

So far the Banach space $\mathfrak{B}$ was independent of our Hilbert space $\mathfrak{H}$. Now we impose our first requirement, namely, that both $\mathfrak{B}$ and $\mathfrak{H}$ can be embedded in a metric space $\mathfrak{M}$ in such a manner that $\mathfrak{B} \cap \mathfrak{H}$ is dense in each of the spaces $\mathfrak{B}$ and $\mathfrak{H}$. In view of this fact each operator A on $\mathfrak{B} \cap \mathfrak{H}$ defines a form on $\mathfrak{B} \cap \mathfrak{H} \times \mathfrak{B} \cap \mathfrak{H}$; namely the form defined by

$$[A]_{\mathfrak{B}}(f,g) = [A]_{\mathfrak{H}}(f,g) = (Af,g) .$$

If this form is bounded with reference to the $\mathfrak{B}$-norm, we denote its closure by the same symbol $[A]_{\mathfrak{B}}$ and say that the operator A determines a bounded form on $\mathfrak{B}$. Note that the boundedness of this form does not imply and is not implied by the property that A is in $\mathfrak{B}(\mathfrak{H})$, the class of everywhere defined bounded operators on $\mathfrak{H}$. Similarly, if the operator A maps $\mathfrak{B} \cap \mathfrak{H}$ into itself and it is bounded with reference to the $\mathfrak{B}$-norm, we denote its closure by $A_{\mathfrak{B}}$ and say that A determines an operator in $\mathfrak{B}(\mathfrak{H})$.

After this digression on forms we formulate the previously mentioned condition for a given factorization of the resolvent.

CONDITION G(ϑ). <u>For each</u> μ <u>in</u> $\mathfrak{R}_{\pm}(\vartheta)$ <u>the given factorization of the resolvent</u>

$$R(\mu) = S_0(\mu)\,.\,Q(\mu)\ ,\ S_0(\mu),\ Q(\mu) \in \mathfrak{S}(\mathfrak{H}) \tag{2.4}$$

is such that the first factor satisfies Condition I. ϑ *and the second factor satisfies Condition* II. ϑ *that follow*.

CONDITION I. ϑ. *For each* μ *in* $\mathfrak{R}\pm(\vartheta)$ *the operator* $S_0(\mu)$ *on* $\mathfrak{H}$ *determines a bounded form on* $\mathcal{B} \times \mathcal{B}$ *for which*

$$\sup_{\mu \in \mathfrak{R}\pm\vartheta} \left\| \left[R_0(\mu)\right]_{\mathcal{B}} \right\| < \infty \tag{2.5}$$

CONDITION II. ϑ. *For each* μ *in* $\mathfrak{R}\pm(\vartheta)$ *the operator* $Q(\mu)$ *on* $\mathfrak{H}$ *determines an operator in* $\mathfrak{S}(\mathfrak{H})$ *for which*

$$\sup_{\mu \in \mathfrak{R}\pm(\vartheta)} \left\| Q_{\mathcal{B}}(\mu) \right\| < \infty\ . \tag{2.6}$$

Remembering the way the product of a form and an operator was defined we see that if the factorization (2.1) satisfies Condition G(ϑ) then the resolvent also satisfies the condition of Lemma 2.1 provided that the set $\mathcal{B} \cap \mathfrak{H}$ is dense in $\mathfrak{H}$ with reference to the $\mathfrak{H}$ -norm. This leads to the lemma that follows.

LEMMA 2.2. *Let* ϑ *be a compact interval*. *Suppose that the resolvent of* A *admits a factorization satisfying* Condition G(ϑ) *and that the set* $\mathcal{B} \cap \mathfrak{H}$ *is dense in* $\mathfrak{H}$. *Then* A(ϑ), *the part of* A *over the interval* ϑ , *is absolutely continuous*.

The assumption of this lemma, that is, Condition G(ϑ), is still too general. In the following we introduce more special ones which imply it.

CONDITION $G_1(\vartheta)$. α. *The operator* $S_0(\mu)$ *of factorization* (2.4) *satisfies Condition* I. ϑ. β. *To each* ω *in* ϑ *there is a self-adjoint operator* $A_0(\omega)$ *such that* $\mathfrak{D}(A_0(\omega)) \supset \mathfrak{D}$ (A) *and the operator*

$$R_0(\mu) = (\mu I - A_0(\mathrm{Re}\ \mu))^{-1} \tag{2.7}$$

satisfies Condition I. ϑ. γ. *The limits of the forms* $\left[R_0(\mu)\right]_{\mathcal{B}}$ *and* $\left[S_0(\mu)\right]_{\mathcal{B}}$ *are equal*. *Specifically for* ω *in* ϑ

$$\lim_{\mu=\omega} \left[R_0(\mu)\right]_{\mathcal{B}} = \lim_{\mu=\omega} \left[S_0(\mu)\right]_{\mathcal{B}}\ , \tag{2.8}$$

and this is uniform in μ in $\mathcal{R}_{\pm}(\vartheta)$.

CONDITION $G_2(\vartheta)$. For each μ in $\mathcal{R}_{\pm}(\vartheta)$ the operator $Q(\mu)$ admits an inverse in $\mathfrak{B}(\mathfrak{H})$. This inverse determines an operator in $\mathfrak{B}(\mathcal{B})$.

CONDITION $G_3(\vartheta)$. The family of operators on $\mathfrak{H}$,

$$(2.9) \qquad T(\mu) = (A - A_0(\operatorname{Re}\mu))R_0(\mu)$$

satisfies Condition II.ϑ. Furthermore for each point ω of ϑ,

$$(2.10) \qquad \lim_{\mu=\omega} \|(I - T(\mu))_{\mathcal{B}} - Q^{-1}_{\mathcal{B}}(\mu)\| = 0 ,$$

uniformly in μ in $\mathcal{R}_{\pm}(\vartheta)$ and the limit operator is continuous.

CONDITION $A_1(\vartheta)$. For each ω in ϑ each of the two limit operators $(I - T^{\pm}(\omega))_{\mathcal{B}}$ admit inverses in $\mathfrak{B}(\mathcal{B})$.

Incidentally note that Conditions $G_{1,2,3}(\vartheta)$ are satisfied for the case of a perturbation, $(A_0, A-A_0)$, which is gentle over the interval ϑ . In this case we may take $A_0(\omega) = A_0$ and $S_0(\mu) = R_0(\mu)$ = unperturbed resolvent.

THEOREM 2.1. Let ϑ be a compact interval. Suppose that the resolvent of A admits a factorization which satisfies Conditions $G_{1,2,3}(\vartheta)$ and Condition $A_1(\vartheta)$ with reference to the space $\mathcal{B}$ and that $\mathcal{B} \cap \mathfrak{H}$ is dense in $\mathfrak{H}$. Then $A(\vartheta)$, the part of the operator A over the interval ϑ , is absolutely continuous.

To establish Theorem 2.1 we maintain that the assumptions of this theorem imply the assumptions of Lemma 2.2. First we claim that the first factor in (2.4) satisfies Condition I.ϑ. For, according to Condition $G_1(\vartheta)$ the operator $R_0(\mu)$ of definition (2.7) determine bounded forms on $\mathcal{B} \times \mathcal{B}$ for which the uniform boundedness relation (2.5) holds. This together with the existence of the limit (2.8) shows that the limit forms $[R_0^{\pm}(\omega)]_{\mathcal{B}}$ are also uniformly bounded. Since this limit is uniform in ω in ϑ the uniform boundedness relation (2.5) also holds for the forms $[S_0(\mu)]_{\mathcal{B}}$. That is to say the first factor in (2.4) satisfies Condition I.ϑ as we have claimed. Second we claim that the inverse of the second factor in (2.4) satisfies Condition II.ϑ. That is to say,

$$(2.11) \qquad \sup_{\mu \in \mathcal{R}_{\pm}(\vartheta)} \|Q^{-1}_{\mathcal{B}}(\mu)\| < \infty \quad .$$

For, according to Condition $G_2(\vartheta)$ for each μ in $\mathfrak{R}_{\pm}(\vartheta)$ this norm is bounded. To establish that this bound is uniform we need Condition $G_3(\vartheta)$. According to this condition the operator $T(\mu)$ of definition (2.9) determines a bounded operator on $\mathfrak{B}$ for which the uniform boundedness relation (2.6) holds. This together with the existence of the limit (2.10) shows that the limit operators $(I-T^{\pm}(\omega))_{\mathfrak{B}}$ are also uniformly bounded. This, in turn, shows the validity of relation (2.11) if we remember that both limits in (2.10) are uniform. Finally we claim that the second factor itself in (2.4) satisfies Condition II.ϑ. That is to say

$$(2.12) \qquad \sup_{\mu\in\mathfrak{R}_{\pm}(\vartheta)} \| Q_{\mathfrak{B}}(\mu_{\pm})\| < \infty \quad .$$

For, according to Condition $A_1(\vartheta)$ the limit operators on the left of relation (2.10) do admit inverses in $\mathfrak{L}(\mathfrak{B})$. As is well known [34] [35] the resolvent set of an operator in $\mathfrak{L}(\mathfrak{B})$ is open. Hence for μ close enough to ϑ the operator $Q^{-1}(\mu)$ also admits an inverse in $\mathfrak{L}(\mathfrak{B})$. Since the limit relation in (2.10) holds uniformly for both sides the sets of such μ will contain regions of the form $\mathfrak{R}_{\pm}(\vartheta)$, which correspond to an angle α possibly smaller than the one in definition (2.1). This shows that for each μ in $\mathfrak{R}_{\pm}(\vartheta)$ the second factor in (2.4) determines operators in $\mathfrak{L}(\mathfrak{B})$. To show that these operators are uniformly bounded we need an elementary fact. Namely, that if an operator valued function defined on a compact subset of $\mathfrak{R}$ is norm-continuous and pointwise invertible then these inverses define another norm-continuous operator valued function. Condition $G_3(\vartheta)$ ensures that the functions $(I-T^{\pm}(\omega))_{\mathfrak{B}}$ depend continuously on ω in ϑ and Condition $A_1(\vartheta)$ ensures that each of these two functions is pointwise invertible. Hence the operator valued function $(I-T^{\pm}(\omega)_{\mathfrak{B}})^{-1}$ is norm continuous and

$$(2.13) \qquad \sup_{\omega\in\vartheta} \|(I-T^{\pm}(\omega))_{\mathfrak{B}}^{-1}\| < \infty \quad .$$

Combining this relation with relation (2.10) we arrive at the validity of relation (2.12).

Thus we have shown that the assumptions of Theorem 2.1 do imply the assumptions of Lemma 2.1. Therefore we can conclude from this lemma the validity of this theorem.

In applying the abstract Theorem 2.1 to Schrödinger operators the difficult assumption to verify is Condition $A_1(\vartheta)$. Note that this is a local condition and we call the point ω exceptional if either of the two operators $(I-T^{\pm}(\omega))_{\mathfrak{B}}$ is not one to one. The basic lemma that follows describes a useful property of exceptional points.

It is an abstract version of a result of Povzner [3] and Ikebe [4]. Various versions of this result have been formulated by various authors [8] [13] [15] for various classes of operators. The present version is similar to the one of Lavine [26] and to the one formulated elsewhere [6.b].

LEMMA 2.3. _Suppose that the point_ ω _in_ ϑ _and the vector_ h _in_ $\mathcal{B}$ _are such that_

$$(2.14)_\pm \qquad (I-T^+(\omega))_{\mathcal{B}} h = 0 \quad \text{or} \quad (I-T^-(\omega))_{\mathcal{B}} h = 0 \ .$$

Then

$$(2.15) \qquad [S_0^+(\omega)]_{\mathcal{B}}(h,h) - [S_0^-(\omega)]_{\mathcal{B}}(h,h) = 0 \ .$$

To establish conclusion (2.15) first make the additional assumption

$$(2.16) \qquad h \in \mathcal{B} \cap \mathcal{H} \ .$$

For brevity also assume that relation $(2.14)_+$ holds. Then clearly

$$[R_0^+(\omega)] \ (h,(I-T^+(\omega))_{\mathcal{B}} h) = 0 \ .$$

and

$$[R_0^-(\omega)] \ ((I-T^+(\omega)_{\mathcal{B}} h,h) = 0 \ .$$

These two relations yield

$$(2.17) \qquad [R_0^+(\omega)]_{\mathcal{B}}(h,h) = [R_0^+(\omega)]_{\mathcal{B}}(h,T^+(\omega)\ h)$$

and

$$(2.18) \qquad [R_0^-(\omega)]_{\mathcal{B}}(h,h) = [R_0^-(\omega)]_{\mathcal{B}}(T^+(\omega)\ h,h) \ .$$

Remembering definition (2.9) of Condition $G_3(\vartheta)$ we see that

$$(2.19) \qquad [R_0^+(\omega)]_{\mathcal{B}}(h,T^+(\omega)_{\mathcal{B}} h) = \lim_{\varepsilon=+0}(R_0(\omega+i\varepsilon)h,\ (A-A_0(\omega))R_0(\omega+i\varepsilon)h).$$

Similarly, we see that

$$(2.20) \qquad [R_0^-(\omega)]_{\mathcal{B}}(T^+(\omega)_{\mathcal{B}} h,h) = \lim_{\varepsilon=+0}(R_0(\omega-i\varepsilon)(A-A_0(\omega))R_0(\omega+i\varepsilon)h,h).$$

According to definition (2.7) the operator $R_0(\omega-i\varepsilon)$ is the resolvent of a self-adjoint operator. Hence the adjoint is given by

$$R_0^*(\omega-i\varepsilon) = R_0(\omega+i\varepsilon) \quad .$$

Remembering that according to Condition $G_1(\vartheta)$ the operator $A-A_0(\omega)$ is self-adjoint this yields

$$(2.21)\quad \begin{aligned}&(R_0(\omega-i\varepsilon)(A-A_0(\omega))R_0(\omega+i\varepsilon)h,h) = \\ &= (R_0(\omega+i\varepsilon)h \ , \ (A-A_0(\omega))R_0(\omega+i\varepsilon)h) \quad .\end{aligned}$$

Combining relations (2.19), (2.20) and (2.21) we obtain

$$\left[R_0^+(\omega)\right]_{\mathcal{B}}(h,T^+(\omega)_{\mathcal{B}}h) = \left[R_0^-(\omega)\right]_{\mathcal{B}}(T^+(\omega)_{\mathcal{B}}h,h) \quad .$$

Combining this relation, in turn, with relations (2.17) and (2.18) we obtain

$$\left[R_0^+(\omega)\right]_{\mathcal{B}}(h,h) - \left[R_0^-(\omega)\right]_{\mathcal{B}}(h,h) = 0$$

Finally combining this with relation (2.8) of Condition $G_1(\vartheta)$ we arrive at the validity of conclusion (2.15). Note that this conclusion was derived under the additional assumption (2.16). To remove this additional assumption we need a limiting argument, similar to the one used elsewhere [6]. For brevity at present we omit the details and consider the proof of Lemma 2.3 complete.

3. AN APPLICATION

In this section we illustrate how to derive a result of Titchmarsh [1] - Neumark [32] - Walter [20] from the abstract theorem of Section 2. To describe their result, first we state a condition in a form which is, essentially, due to Walter [20].

CONDITION S(T-N-W). The potential p is real valued and twice continuously differentiable and is such that

$$(3.1)\qquad \lim_{\xi=\infty} p(\xi) = -\infty$$

$$(3.2)\qquad \int^{\infty} \frac{d\xi}{|p^{1/2}(\xi)|} = \infty$$

$$(3.3)\qquad \sup_{\mu\in\mathcal{R}_{\pm}(\vartheta)} \int_0^\infty \left(\frac{1}{4}\left|\frac{p''(\xi)}{(\mu-p(\xi))^{3/2}}\right|+\frac{5}{16}\left|\frac{p'(\xi)^2}{(\mu-p(\xi))^{5/2}}\right|\right) d\xi < \infty .$$

This condition is more special than the one of Walter [20] inasmuch as he allows an arbitrary limit in assumption (3.1). At the same time he emphasized that the interesting case is the one in which assumption (3.1) does hold. Walter also gave a different formulation of assumption (3.3). Specifically, all that he required was that for some λ in the interval $(p(\infty),\infty)$,

$$(3.4)\qquad \int^\infty \left[\frac{1}{4}\frac{p''(\xi)}{(\lambda-p(\xi))^{3/2}}+\frac{5}{16}\frac{p'(\xi)^2}{(\lambda-p(\xi))^{5/3}}\right] d\xi < \infty .$$

It is an interesting fact discovered by him that assumption (3.4) together with a lemma of Atkinson-Coppel [33] does imply assumption (3.3) for any bounded subinterval of $(p(\infty),\infty)$.

Next assume that the potential p satisfies this condition and with the aid of it define $\mathcal{D}(L(p))$ to be the set of those functions f in $\mathfrak{h} = \mathcal{L}_2(\mathcal{R}^+)$ which have absolutely continuous first derivatives and for which

$$(3.5)\qquad f(0) = 0 \ \underline{\text{and}}\ \ -f'' + pf\in\mathcal{L}_2(\mathcal{R}^+)$$

Then define the operator L(p) mapping $\mathcal{D}(L(p))$ into $\mathcal{L}_2(\mathcal{R}^+)$ by the equation

$$(3.6)\qquad L(p)f = -f'' + pf$$

For convenience we also assume that p is such that

$$(3.7)\qquad \lim_{\xi=\infty} \int_\xi^{\xi+1} \left(\frac{p'(\eta)}{p(\eta)}\right)^4 d\eta = \int_\xi^{\xi+1} \left(\frac{p''(\eta)}{p(\eta)}\right)^2 d\eta = 0$$

Let D denote differentiation and let M(p) denote the operator of multiplication by the potential p. In contrast to the case of short range potentials the operator M(p) need not be $-D^2$ compact. Accordingly the domains of L(p) and $-D^2$ need not be equal, that is

$$(3.8)\qquad \mathcal{D}(L(p) \neq \mathcal{D}(L(0)) = \mathcal{D}(-D^2) .$$

Nevertheless, symbolically we write

(3.9) $$L(p) = -D^2 + M(p) \quad .$$

THEOREM 3.1 : *Suppose that the potential* p *satisfies Condition* S(T-N-W) *and define the operator* L(p) *by equations* (3.5) *and* (3.9). *Then the part of the operator* L(p) over ϑ *is absolutely continuous, that is,*

(3.10) $$L(p)(\vartheta) = L(p)(\vartheta)_{ac}$$

To derive Theorem 3.1 from the abstract Theorem 2.1 we first introduce an auxiliary potential. This is done in the lemma that follows.

LEMMA 3.1 *Suppose that the potential* p *satisfies Condition* S(T-N-W). *Then there is a potential* p_0 *which also satisfies* Condition S(T-N-W) *and is such that*

(3.11) $$\sup_{\mu\in\mathfrak{R}_\pm(\vartheta)} \sup_{\xi} \left|\frac{1}{\mu-p_0(\xi)}\right| < \infty$$

Furthermore $p-p_0$ *is bounded and*

(3.12) $$p-p_0\in\mathcal{L}_1(\mathfrak{R}^+)\cap\mathcal{L}_2(\mathfrak{R}^+).$$

To establish this lemma recall assumption (3.1). It clearly implies the existence of a positive number ξ_0 such that

$$\sup_{\mu\in\mathfrak{R}_\pm(\vartheta)} \sup_{\xi>\xi_0} \left|\frac{1}{\mu-p(\xi)}\right| < \infty \quad .$$

Now let $\ell(\xi)$ be the Lagrange interpolating polynomial corresponding to the points $(\xi_0, 2\xi_0)$ and weights (2,2). That is to say $\ell(\xi)$ is a polynomial degree 5 such that

$$\ell(\xi_0) = p'(\xi_0) = \ell''(\xi_0) = 0$$

and

$$\ell(2\xi_0) = 1 \ , \quad \ell'(2\xi_0) = \ell''(2\xi_0) = 0$$

Then setting

$$(3.13)\qquad p_o(\xi) = \begin{cases} 0 & 0\leqslant\xi\leqslant\xi_o \\ \ell(\xi)p(\xi) & \xi_o\leqslant\xi\leqslant 2\xi_o \\ p(\xi) & \xi\geqslant 2\xi_o \end{cases}$$

it is almost evident that this auxiliary potential p_o satisfies the conclusions of Lemma 3.1.

Second with the aid of this auxiliary potential p_o and the JWKB-approximation method we construct a factorization of the resolvent of the original operator $L(p)$. This will be done in the lemma that follows. To formulate it we need some notations. Let $\sqrt{z}$ denote the branch of the square root function defined by the property

$$(3.14)\qquad \mathrm{Re}\sqrt{z} > 0 \quad \text{for} \quad z \notin [0,\infty) \ .$$

With the aid of this branch define four more functions by setting

$$(3.15)^{\pm}\qquad w_o^{\pm}(\mu)(\xi) = \pm\sqrt{p_o(\xi)-\mu} - \frac{1}{4}\,\frac{p_o{}'(\xi)}{p_o(\xi)-\mu}$$

and

$$(3.16)^{\pm}\qquad y^{\pm}(\mu)(\xi) = \exp\left[\int_0^{\infty} w_o^{\pm}(\sigma)d\sigma\right] \ .$$

Then elementary algebra shows that for the ξ-derivatives

$$(3.17)^{\pm}\qquad y^{\pm}(\mu)'' = \left[w_o^{\pm}(\mu)' + w_o^{\pm}(\mu)^2\right] y^{\pm}(\mu) \ .$$

Further algebra shows that

$$(3.18)\qquad w_o^{\pm}(\mu)'(\xi)+w_o^{\pm}(\mu)^2(\xi)=p_o(\xi)-\mu+\frac{5}{16}\left(\frac{p_o'(\xi)}{p_o(\xi)-\mu}\right)^2-\frac{1}{4}\,\frac{p_o''(\xi)}{p_o(\xi)-\mu} \ .$$

Note that the right member is independent of the square root function used to define $w_o^{\pm}(\mu)$. This fact allows us to define

$$(3.19)\qquad q_o(\mu) = p_o-\mu-w_o^{\pm}(\mu)'-w_o^{\pm}(\mu)^2 \ .$$

LEMMA 3.2. <u>Let</u> p_o <u>be the potential of</u> Lemma 3.1 <u>and for each non-real complex number</u> μ <u>define the potential</u> $q_o(\mu)$ <u>by equation</u> (3.19). <u>Then the inverses in the following definitions do exist,</u>

(3.20) $$S_o(\mu) = (\mu' - L(p_o - q_o(\mu)))^{-1} \in \mathfrak{B}(\mathfrak{h}) \quad ,$$

and

(3.21) $$Q(\mu) = \left[I - M(q_o(\mu) + p - p_o) S_o(\mu)\right]^{-1} \in \mathfrak{B}(\mathfrak{h}) \quad .$$

Furthermore they define a factorization of the resolvent of the operator $L(p)$ of definition (3.9). Specifically

(3.22) $$R(\mu) = S_o(\mu) . Q(\mu) \quad .$$

To establish conclusion (3.20) first we introduce two more functions by setting

$(3.23)_o$ $$y_o(\mu) = y^+(\mu) - y^-(\mu) \quad ,$$

$(3.23)_\infty$ $$y_\infty(\mu) = y^-(\mu) \quad .$$

Then with the aid of these functions define a kernel by setting

(3.24) $$S_o(\mu)(\xi,\eta) = \frac{1}{W(y_o(\mu), y_\infty(\mu))} \cdot \begin{cases} y_o(\mu)(\eta) y_\infty(\mu)(\xi), & \text{for } \eta < \xi \\ y_o(\mu)(\xi) y_\infty(\mu)(\eta), & \text{for } \eta > \xi \end{cases}$$

Here the denominator of the first factor is the Wronskian of the functions $y_o(\mu)$ and $y_\infty(\mu)$. We claim that the corresponding integraloperator is inverse to the operator $\mu I - L(p_o - q_o(\mu))$; more specifically

(3.25) $$S_o(\mu)(\mu I - L(p_o - q_o(\mu))) = I \ \underline{\text{on}} \ \mathfrak{D}(L(p)) \cap \dot{\mathfrak{C}}_\infty(\mathfrak{R}^+)$$

and

(3.26) $$(\mu I - L(p_o - q_o(\mu))) S_o(\mu) = I \ \underline{\text{on}} \ \dot{\mathfrak{C}}_\infty(\mathfrak{R}^+) \quad .$$

Here $\dot{\mathfrak{C}}_\infty(\mathfrak{R}^+)$ denotes the class of infinitely differentiable functions with bounded support in $\mathfrak{R}^+$. To establish these two relations recall definitions $(3.16)^\pm$, $(3.22)_{o,\infty}$, (3.19) and relations $(3.17)^\pm$, (3.18). They show that each of the functions $y_{o,\infty}(\mu)$ satisfies the differential equation

(3.27) $$(\mu - p_o + q_o(\mu)) y(\mu) + y''(\mu) = 0 \quad .$$

Relation (3.27), in turn, together with a well known elementary argument [30] shows the validity of relations (3.25) and (3.26). Next we claim that the range of this operator is dense; more specifically

$$(3.28)\qquad \overline{(\mu I-L(p_o-q_o(\mu)))\mathcal{D}(L(p))\cap\dot{\mathcal{C}}_\infty(\mathcal{R}^+)} = \mathfrak{h} .$$

It is a general operator theoretic fact that the ortho-complement of the range equals the nullspace of the adjoint [35]. Hence it suffices to show that for each vector f^* in the domain of the adjoint

$$(3.29)\qquad (\mu I-L(p_o-q_o(\mu)))^*f^* = 0 \ \underline{\text{implies}}\ \ f^*=0 .$$

An application of the mollifying technique [31.b] shows that such a vector f^* corresponds to a continuous function satisfying the boundary condition $f^*(0)=0$. Since this application is nearly identical to the case of a real potential we do not carry it out. At the same time it follows that f^* is twice continuously differentiable since p_o and $q_o(\mu)$ are. Furthermore f^* is a pointwise solution of the differential equation (3.27) where μ is replaced by its complex conjugate. The technical lemma of the Appendix implies that up to a constant multiple this equation has exactly one solution which is square integrable near infinity. In fact this solution is given by $y^-(\overline{\mu})$. Since according to definition $(3.16)^-$ this function does not vanish at zero it follows that $f^*=0$. This establishes the validity of relation (3.29) and hence of (3.28).

To complete the proof of conclusion (3.20) we show that the operator defined by the kernel (3.24) is bounded. In fact we show that its closure, that we denote by the same symbol, is such that

$$(3.30)\qquad S_o(\mu)\in\mathfrak{B}(\mathfrak{h}) .$$

To establish this relation define the positive function $t(\mu)$ by

$$(3.31)\qquad t(\mu)(\eta) = |(p_o(\eta)-\mu)^{-1/2}| .$$

Then formula (3.24) together with conclusion (3.11) of Lemma 3.1 and the technical lemma of the Appendix, show that

$$\sup_{\xi\in\mathcal{R}^+}\frac{1}{t(\mu)(\xi)}\int_{\mathcal{R}^+}|S_o(\mu)(\xi,\eta)|t(\mu)(\eta)\,d\eta<\infty ,$$

and

$$\sup_{\eta\in\mathcal{R}^+}\frac{1}{t(\mu)(\eta)}\int_{\mathcal{R}^+}|S_o(\mu)(\xi,\eta)|t(\mu)(\xi)\,d\xi<\infty .$$

This, in turn, according to a result of Schur-Holmgren-Carleman implies the validity of relation (3.30). Finally combining relations (3.25), (3.26), (3.29) and (3.30) we arrive at the validity of conclusion (3.20).

To establish conclusions (3.21) and (3.22) we need an abstract operator-theoretic fact which was used implicitly by Kato [34] Schechter [37] and elsewhere [28]. Namely that if P is A-compact then for each μ in the intersection of the resolvent sets of A and A+P we have

$$(3.32)\qquad [I-P(\mu I-A)^{-1}]^{-1} \in \mathcal{B}(\mathcal{H})$$

and

$$(3.33)\qquad (\mu I-A-P)^{-1} = (\mu I-A)^{-1}[I-P(\mu I-A)^{-1}]^{-1}.$$

The simplifying assumption (3.7) together with definition (3.19) and conclusion (3.12) of Lemma 3.1 shows that

$$\lim_{\xi=\infty} \int_{\xi}^{\xi+1} |q_o(\mu)(\eta)+p(\eta)-p_o(\eta)|^2 d\eta = 0$$

The technical lemma of the Appendix together with formula (3.24) and conclusion (3.11) of Lemma 3.1 implies that

$$\sup_{\xi\in\mathcal{R}^+} \int |S_o(\mu)(\xi,\eta)|^2 d\eta . \exp[+2v(\mu)(\xi)] <\infty .$$

These two relations, in turn, according to the books of Kato [34] and Schechter [37] imply that

$$(3.34)\qquad M(q_o(\mu)+p-p_o)S_o(\mu) \quad \underline{\text{is compact}}.$$

Since the operator L(p) is self-adjoint its spectrum is real and relation (3.34) allows us to apply the abstract relation (3.32) to the operators

$$A = L(q_o(\mu)-p_o), \quad P = -M(q_o(\mu)+p-p_o) \quad .$$

This establishes the validity of conclusion (3.20) of Lemma 3.2. Then application of the abstract formula (3.33) to these operators yields the validity of conclusion (3.22). This completes the proof of Lemma 3.2.

Finally we illustrate without proof that the factorization of Lemma 3.2 satisfies the rather stringent conditions of the abstract Theorem 2.1. To construct such a space $\mathcal{G}$ first with the aid of the potential p_o of Lemma 3.1 and the JWKB-approximation method define a function n by setting

$$n(\xi) = |p(\xi) - p_o(\xi) + \sup_{\mu \in \mathfrak{R}^+(\vartheta)} \left| \frac{p_o'(\xi)}{\mu - p(\xi)} \right|^2 \tag{3.35}$$

$$+ \sup_{\mu \in \mathfrak{R}^+(\vartheta)} \left| \frac{p''(\xi)}{\mu - p(\xi)} \right| + \exp(-\xi) \quad .$$

Then define the space $\mathfrak{B}$ to consist of those functions f in $\mathfrak{h} = \mathfrak{L}_2(\mathfrak{R}^+)$ for which the norm,

$$\|f\|_{\mathfrak{B}} = \|(\tfrac{1}{n})^{\frac{1}{2}} f\|_{\mathfrak{h}} \quad ,$$

is finite. This norm is similar to the smoothness norm of Kato and Kuroda [11] [13] inasmuch as it defines a Hilbert space and, in turn, it is defined with the aid of a factorization.

The key fact in establishing the assumptions of the abstract Theorem 2.1 with reference to this norm is that the kernel of the operator $S_o(\mu)$ of Lemma 3.2 remains bounded in the sense of the lemma that follows. Roughly speaking it says that for this kernel the same estimates hold as for the kernel of the operator $(\mu I - D^2)^{-1}$.

LEMMA 3.3. Let p_o be the potential of Lemma 3.1 and let the function $q_o(\mu)$ be defined by equation (3.19). Then the kernel of the operator $S_o(\mu)$ of Lemma 3.2 is bounded in the sense that

$$\sup_{\mu \in \mathfrak{R}^{\pm}(\vartheta)} \quad , \sup_{(\xi,\eta) \in \mathfrak{R}^+ \times \mathfrak{R}^+} |S_o(\mu)(\xi,\eta)| < \infty \quad . \tag{3.36}$$

To establish conclusion (3.36) first we maintain that the Wronskian in formula (3.24) is given by

$$W(y_o(\mu), y_\infty(\mu)) = 2\sqrt{p_o(0) - \mu} \quad . \tag{3.37}$$

For, by definition

$$W(y_o(\mu), y_\infty(\mu))(\xi) = y_o(\mu)(\xi) y_\infty(\mu)'(\xi) - y_o(\mu)'(\xi) y_\infty(\mu)(\xi) \quad .$$

Since equation (3.27) does not contain a first order therm this Wronskian is independent of ξ [30]. Hence we may and shall evaluate it at $\xi = \infty$. Definition $(3.23)_\infty$ shows that

$$y_\infty(\mu)' = w^-(\mu)y^-(\mu) \quad .$$

Definition $(3.23)_\infty$ together with relation (A.21) of the Appendix shows that

$$y_o(\mu)'(\xi) \sim w^+(\mu)'(\xi)y^+(\mu)(\xi)$$

$$\underline{\text{near}} \quad \xi \sim \infty \ .$$

Combining these three relations we obtain

$$W(y_o(\mu), y_\infty(\mu))(\xi) \sim \left[w_o^-(\mu)(\xi) - w_o^+(\mu)(\xi)\right] y^+(\mu)(\xi)y^-(\mu)(\xi)$$

$$\underline{\text{near}} \quad \xi \sim \infty \ .$$

According to definitions $(3.15)^\pm$

$$w_o^-(\mu)(\xi) - w_o^+(\mu)(\xi) = -2\sqrt{p_o(\xi)-\mu} \quad ,$$

and remembering definitions $(3.16)^\pm$ we see that

$$y^+(\mu)(\xi)y^-(\mu)(\xi) = \frac{1}{\sqrt{p_o(\xi)-\mu}} \sqrt{p_o(0)-\mu} \quad .$$

These three relations together establish the validity of relation (3.37).

To establish conclusion (3.36) next we note that definitions $(3.23)_{o,\infty}$ and $(3.16)^\pm$ imply

$$|y_o(\mu)(\eta)y_\infty(\mu)(\xi)| \leqslant 2|y^+(\mu)(\eta)y^-(\mu)(\xi)| \ .$$

Another application of definitions $(3.16)^\pm$ together with definitions $(3.15)^\pm$ shows that for $\eta<\xi$,

$$|y^+(\mu)(\eta)y^-(\mu)(\xi)| \leqslant |(p_o(0)-\mu)^{1/2}| \ .$$

(3.38)

$$\cdot \ \frac{1}{p_o(\eta)-\mu}^{1/4} \ \frac{1}{p_o(\xi)-\mu}^{1/4} \quad .$$

According to conclusion (3.11) of Lemma 3.1 the supremum of the second factor is finite. Inserting this fact and relations (3.37) and (3.38) in formula (3.24) we arrive at the validity of conclusion (3.36).

This completes the proof of Lemma 3.3.

APPENDIX : A TECHNICAL LEMMA ON POTENTIALS

In this appendix we formulate a lemma on potentials satisfying the first two assumptions of Condition S(T-N-W). These assumptions are stated in the condition that follows:

CONDITION A. *The potential* p *is real valued, continuous, and such that*

$$\int^{\infty} \frac{d\xi}{|p^{1/2}(\xi)|} = \infty \tag{A.1}$$

and

$$\lim_{\xi=\infty} p(\xi) = -\infty \quad . \tag{A.2}$$

In the lemma that follows we use the same branch of the square root function as in Section 3. Namely the branch defined by the property,

$$\mathrm{Re}\sqrt{z} > 0 \quad \text{for} \quad z \notin (-\infty,0] \quad . \tag{A.3}$$

LEMMA. *Suppose that the potential* p *satisfies* Condition A *and for each non-real complex number* μ *define the real valued function* $v(\mu)$ *by*

$$v(\mu)(\eta) = \mathrm{Re} \int_0^{\eta} \sqrt{p(\sigma)-\mu}\, d\sigma \quad . \tag{A.4}$$

Then for each positive number ξ

$$\int_{\xi}^{\infty} \exp\left[-2v(\mu)(\eta)\right] \left| \sqrt{\frac{1}{p(\eta)-\mu}} \right| d\eta \leq \frac{1}{|\mathrm{Im}\mu|} \exp\left[-2v(\mu)(\xi)\right] \tag{A.5}$$

and

$$\int_{\xi}^{\infty} \exp\left[+2v(\mu)(\eta)\right] \left| \sqrt{\frac{1}{p(\eta)-\mu}} \right| d\eta = \infty \quad . \tag{A.6}$$

To establish conclusion (A.5) we need a lower estimate for $\mathrm{Re}\sqrt{z}$. This is formulated in the proposition that follows.

PROPOSITION. *Let* $\sqrt{z}$ *be defined by relation* (A.3). *Then for each non-real complex number* z

(A.7) $$\frac{|\mathrm{Im} z|}{2} \left| \sqrt{\frac{1}{2}} \right| < \mathrm{Re}\, \sqrt{z} \quad .$$

To establish conclusion (A.6) set

(A.8) $$z = \rho \exp(i\alpha) \quad .$$

Then definition (A.3) shows that

(A.9) $$\mathrm{Re}\, \sqrt{z} = \sqrt{\rho}\ \left|\cos \frac{\alpha}{2}\right| \ .$$

As is well known

$$\left|\cos \frac{\alpha}{2}\right| = \frac{1}{\sqrt{2}} \sqrt{(1+\cos\, \alpha)} \quad .$$

This yields

$$\left|\cos \frac{\alpha}{2}\right| \geqslant \frac{1}{\sqrt{2}} \sqrt{1- \sqrt{1-\sin^2 \alpha}} \quad .$$

It is an elementary consequence of the mean value theorem that

$$\sqrt{1- \sqrt{1-\sin^2\alpha}} \geqslant \frac{1}{\sqrt{2}} |\sin\, \alpha| \quad .$$

Inserting this estimate in the previous one we obtain

(A.10) $$\left|\cos \frac{\alpha}{2}\right| \geqslant \frac{1}{2}|\sin\, \alpha| \quad .$$

Definition (A.7) shows that

(A.11) $$\sin\, \alpha = \frac{\mathrm{Im} z}{\rho} \quad .$$

Inserting relations (A.10) and (A.11), in turn, in relation (A.9) we arrive at the validity of conclusion (A.7).

Having established the Proposition we return to the proof of conclusion (A.5) of the Lemma. Definition (A.4) shows that

(A.12) $$\frac{dv(\mu)(\eta)}{d\eta} = \mathrm{Re}\, \sqrt{p(\eta)-\mu} \quad .$$

Since according to Condition A the potential p is real, application of the Proposition to the complex number

$$z = p(\eta)-\mu \ ,$$

yields

$$\text{(A.13)} \qquad \frac{|\mathrm{Im}\mu|}{2}\left|\sqrt{\frac{1}{p(\eta)-\mu}}\right| \leq \mathrm{Re}\,\sqrt{p(\eta)-\mu} \quad .$$

Combining relations (A.12) and (A.13) we obtain

$$\text{(A.14)} \qquad \left|\sqrt{\frac{1}{p(\eta)-\mu}}\right| \leq \frac{2}{|\mathrm{Im}\mu|}\,\frac{dv(\mu)(\eta)}{d\eta} \ .$$

From this, in turn, we obtain

$$\text{(A.15)} \int_{\xi}^{\infty} \exp\left[-2v(\mu)(\eta)\right]\left|\sqrt{\frac{1}{p(\eta)-\mu}}\right| d\eta \leq \frac{2}{|\mathrm{Im}\mu|}\int_{\xi}^{\infty} \exp\left[-2v(\mu)(\eta)\right]\frac{dv(\mu)(\eta)}{d\eta}\,d\eta .$$

Clearly the integral on the right equals

$$\frac{\exp\left[-2v(\mu)(\xi)\right] - \exp\left[-2v(\mu)(\infty)\right]}{2} \ .$$

Inserting this fact in estimate (A.15) we arrive at the validity of conclusion (A.5). Note that in the proof of this conclusion we did not use assumptions (A.1) and (A.2) of Condition A.

To establish conclusion (A.6) recall assumption (A.2). This assumption shows that setting

$$\text{(A.16)} \qquad p(\eta)-\mu = \rho(\eta)\exp(i\alpha(\eta))$$

we have

$$\text{(A.17)} \qquad \lim_{\eta=\infty} \alpha(\eta) = \pi \ .$$

This, in turn, shows that for the complex number

$$\text{(A.18)} \qquad z(\eta) = p(\eta)-\mu$$

the lower estimate of the Proposition is actually an asymptotic formula. More specifically, relation (A.17) together with definition (A.18) shows that

$$\text{(A.19)} \qquad \lim_{\eta=\infty} \ \mathrm{Re}\,\sqrt{z(\eta)} \cdot \frac{2}{|\mathrm{Im}z(\eta)|}\left|\sqrt{\frac{1}{z(\eta)}}\right| = 1 \ .$$

This follows by a repetition of the arguments used to establish conclusion (A.7) of the Proposition. For brevity we omit this repetition and consider the proof of relation (A.19) complete. Combining relations (A.12) and (A.19) we obtain that for large enough η

$$\sqrt{\frac{1}{p(\eta)-\mu}} > \frac{1}{|\mathrm{Im}\mu|} \frac{dv(\mu)(\eta)}{d\eta} .$$

Hence for large enough ξ

$$(A.20) \int_{\xi}^{\infty} \exp\left[+2v(\mu)(\eta)\right] \left|\sqrt{\frac{1}{p(\eta)-\mu}}\right| d\eta \geqslant \frac{1}{|\mathrm{Im}\mu|} \int_{\xi}^{\infty} \exp\left[+2v(\mu)(\eta)\right] \frac{dv(\mu)(\eta)}{d\eta} d\eta .$$

Clearly the integral on the right equals

$$\frac{\exp\left[+2v(\mu)(\infty)-2v(\mu)(\xi)\right]}{2} .$$

We claim that assumptions (A.1) and (A.2) imply that

$$(A.21) \qquad \lim_{\eta=\infty} v(\mu)(\eta) = \infty .$$

For, they clearly imply that

$$\int_0^{\infty} \left|\sqrt{\frac{1}{p(\eta)-\mu}}\right| d\eta = \infty .$$

Inserting relation (A.19) and definitions (A.4) and (A.18) in this relation we arrive at the validity of relation (A.21). Inserting relation (A.21), in turn, in estimate (A.20) we arrive at the validity of conclusion (A.6). This completes the proof of the Lemma.

REFERENCES

1. Titchmarsh, E.C., Eigenfunction Expansions Associated with Second-Order Differential Equations. Oxford, Clarendon Press, 1948. See Section V.

2. Kodaira, Kunihiko, The eigenvalue problem for ordinary differential equations of the second order and Heisenberg's theory of S-matrices. Amer. J. Math. LXXI (1949) 921-945.

3. Povzner, A. Ya., On the expansion of an arbitrary function in terms of the eigenfunctions of the operator $-\Delta u+cu$. (in Russian Math. Sbornik 32 (1953), 109-156.

4. Ikebe, T., Eigenfunction expansion associated with the Schrödinger operators and their application to scattering theory. Arch. Rat. Mech. Anal. 5(1960), 1-34.

5. Eidus, D.M., The Principle of Limiting Absorption, Amer. Math. Soc. Transl. 47(1965) 157-192.

6. Rejto, P.A., On partly gentle perturbations, I.J. Math. Anal. Appl. 17, 453-462 (1967).
   a. property [*] and the proof of Lemma 2.1.
   b. Theorem 4.2.

7. Weidmann, Joachim, Zur Spektraltheorie von Sturm-Liouville-Operatoren. Math. Zeitschr. Vol. 98, pp. 268-302 (1967). See Satz 5.1 and Korollar 5.2.

8. Howland, J.S., A perturbation theoretic approach to eigenfuntion expansions. J. Math. Anal. Appl. 20(1967) 145-187.

9. Lax, Peter D., and Phillips, Ralph, S., Scattering Theory, Academic Press, (1967).

10. Goldstein, Charles, Eigenfunction expansions associated with the Laplacian for certain domains with infinite boundaries. I.II and III. Trans. Amer. Math. Soc. 135(1969) 1-31, 33-50 and 143 (1969) 283-301.

11. Kato, T., Some results on potential scattering. Proc. Intern. Conf. Functional Anal. and Related Topics, Tokyo 1969, Univ. Tokyo Press 1970, 206-215.

12. Jäger, Willi, Ein gewöhnlicher Differentialoperator zweiter Ordnung für Funktionen mit Werten in einem Hilbertraum. Math. Z. 113 (1970) 68-98.

13. Kato, T., and Kuroda, S.T., On the abstract theory of scattering. Rocky Mountain J. Math. 1, 127-171, (1971).

14. Amrein, W.O., Georgescu V. and Jauch, J.M., Stationary state scattering theory, Helv. Phys. Acta 44(1971) 407-434.

15. Agmon, S., Spectral theory of self-adjoint elliptic operators in $L_2(R^n)$. Abstracts of the 1971 Oberwolfach symposium on the Mathematical Theory of Scattering.

16. Rejto, P.A., Some potential perturbations of the Laplacian. Helv. Phys. Acta. 44(1971), 708-736.

17. Aguilar, J. and Combes, J.M. A class of analytic perturbations for Schrödinger Hamiltonians: I. The one body problem. Comm. Math. Phys. 22(1971), 269-279.

18. Balslev, E. and Combes, J.M., Spectral properties of many-body Schrödinger operations with dilatation-analytic interactions. Comm. Math. Phys. 22(1971), 280-294.

19. Ikebe, Teruo and Saito, Yoshimi, Limiting absorption method and absolute continuity for the Schrödinger operator. J. Math. Kyoto Univ. 12(1972) 513-542.

20. Walter, Johann, Absolute continuity of the essential spectrum of $-\frac{d^2}{dt^2} + q(t)$ without monotony of q. Math. Z. 129, (1972), 83-94.

21. Kuroda, S.T., Scattering theory for differential operators, I, operator theory, J. Msth. Soc. Japan 25 (1973) 75-104, part II, preprint.

22. Rejto, P.A., On a theorem of Titchmarsh-Weidmann concerning absolutely continuous operators. Res. Rep. Univ. Minn. 1974.

23. Lavine, Richard, Absolute continuity of the positive spectrum for Schrödinger operators with long range potentials. J. Funct. Anal. 12. (1973) 30-54.
a. proof of Theorem 4.

24. Prugovečki, E. and Zorbas, J., Modified Lippmann-Schwinger equations for two-body scattering with long range interactions J. Math. Phys. 14(1973) 1398-1409.

25. Combes, J.M., Dilation analytic perturbations. In Proceedings of the 1973 Nato Summer School on Scattering Theory (to appear

26. Lavine, Richard, Commutator methods in scattering theory. In Proceedings of the 1973 Nato Summer School on Scattering Theory (to appear).

27. Amrein, W.O. Some fundamental questions in quantum mechanics. In Proceedings of the 1973 Nato Summer School on Scattering Theory (to appear).

28. Gustafson, K.E. and Rejto, P.A., Some essentially self-adjoint Dirac operators with spherically symmetric potentials. Israel J. Math. 14. (1973) 63-75. See Theorem 3.1.

29. Kemble, Edwin, C., The Fundamental Principles of Quantum Mechanics With Elementary Applications. 1937 Reprinted by Dover Publications, N.Y.
   a. Equations (21.11) and (21.12)
   b. Appendix D.

30. Coddington, Ear, A., and Levinson, Norman, Theory of Ordinary Differential Equations. McGraw-Hill (1955). See Section III.8.

31. Friedrichs, K.O., Spectral Theory of Operators in Hilbert Space, Courant Institute Lecture Notes, (1959).
   a. Section 20
   b. Section 35.

32. Neumark, M.A., Lineare Differentialoperatoren. Berlin: Akademie-Verlag (1960). See Section 22.

33. Coppel, W.A., Stability and asymptotic behavior of differential equations. Heath & Company, 1965. See Lemma 6.

34. Kato, T., Perturbation Theory for Linear Operators. Springer-Verlag (1966).

35. Dunford, Nelson and Schwartz, Jacob, T., Linear Operators, Parts I and II. Wiley (Interscience) New York, 1963 and 1967.

36. Newton, R.G., Scattering Theory of Waves and Particles, McGraw

Hill, (1966).

37. Schechter, Martin, Spectra of partial differential operators, American Elsevier Publishing Co., New York 1971.

38. Georgescu, V., Ph.D. Thesis, Univ. Geneva, to appear.

39. Faddeev, L.D., Mathematical Aspects of the Three-Body Problem in the Quantum Theory of Scattering. Israel Program for Scientific Translations, Jerusalem (1965).

40. Van Winter, C., Complex dynamical variables for multi-particle systems with analytic interactions. Preprint, Department of Mathematics and Physics, University of Kentucky.

A REMARK ON THE KOCHEN-SPECKER THEOREM

A. Lenard

Department of Mathematics
Indiana University
Bloomington, Indiana

A famous theorem of Andrew Gleason[1] states that every finite measure on the lattice of closed subspaces of a Hilbert space H of dimension ≥ 3 is of the form $\mu(K) = \mathrm{Tr}(WP_K)$, where W is a non-negative operator of trace class, uniquely determined by the measure, and P_K is the projection operator whose range is the subspace K of H. By a finite measure one means in this context a function μ: $L(H) \to \mathbb{R}_+$ ($\mathbb{R}_+$ the non-negative real numbers, L(H) the lattice of closed subspaces of H) with the property that $\Sigma\mu(K_n) = \mu(\oplus K_n)$ for every finite or countably infinite system $\{K_n\}$ of mutually orthogonal closed subspaces of H. If $\dim(K) = 1$, and $f \in K$ is a unit vector, then the theorem yields $\mu(K) = (f,Wf)$. This shows that if μ is restricted to one-dimensional subspaces only, then its range is a connected set of real numbers. Indeed, if f and g are two independent unit vectors, and $\alpha = (f,Wf)$, $\beta = (g,Wg) > \alpha$ say, then for any γ in the interval (α,β) one can find a suitable unit vector h in the span of f and g such that $\gamma = (h,Wh)$. The only way the range of μ(restricted to one-dimensional subspaces) can be a finite set is when $\dim(H) < \infty$ and W is a multiple of the identity operator. In particular, no measure has the range $\{0,1\}$.

This latter fact is sometimes expressed by saying that quantum mechanics admits no "hidden variable interpretation". Let us recall briefly what justifies this statement. Laying aside many features of traditional quantum theory (e.g. wave-functions, probability interpretation, etc.) we assume only that there is a one-to-one correspondence between all binary observables pertaining to the system under study, and the set of all projection operators

Enz/Mehra (eds.), Physical Reality and Mathematical Description. 226–233. *All Rights Reserved.*

acting in some Hilbert space. A binary observable is one that has only two possible "values" which, without restriction of generality, may be taken as 0 and 1. This one-to-one correspondence is subject to the consistency condition that if the value 1 for an observable O_1 implies the value 1 for an observable O_2 then the corresponding projections P_1 and P_2 satisfy $\mathrm{Ran}(P_1) \subseteq \mathrm{Ran}(P_2)$ (also written symbolically $P_1 \leq P_2$). Also if O_1 has value 1 precisely when O_2 has value 0, the corresponding projections are related by $P_1 = 1-P_2$ (also written $P_1 = P_2^{\perp}$). Imagine a physicist who wishes to believe that every binary observable must <u>have</u> a well determined value 1 or 0 for a given sample system (unknown though this value may be for a variety of practical reasons). He would claim that at least a prediction for these values is possible, though he would perhaps admit to the knowledge of no experimental arrangement, or preparation procedure, that would make such a prediction certain of fulfillment for any observable chosen by an experimenter. Mathematically, such a "prediction" is a function μ assaigning a value 1 or 0 to each projection P, in such a way that $P_1 \leq P_2$ implies $\mu(P_1) \leq \mu(P_2)$ and $P_1 = P_2^{\perp}$ implies $\mu(P_1) = 1- \mu(P_2)$. It is easy to see that such a function is a measure on L(H) if we identify the projections with their ranges. But, it follows from Gleason's Theorem, that no such measure exists! Thus the would-be hidden variable theorist is frustrated by not even being able to <u>specify a single one consistent set of values</u> for all observables, as much as he is willing to renounce the practical testability of such predictions!

A significant strengthening of this "impossibility theorem" was discovered by Kochen and Specker [2)]. They constructed a <u>finite</u> set of projections in a 3-dimensional real Hilbert space such that already on this finite set no {0,1}-valued measure was possible. This is shown by constructing a finite set $F = \{p_j\}$ of points on the sphere $S^2 = \{x \in \mathbb{R}^3 \mid \; ||x|| = 1\}$ such that if this set is partitioned into two subsets in any manner (label the parts as the set of 0-points and the set of 1-points), then there is at least one orthogonal triplet of points having not exactly one 1-point. The construction of F, as well as the proof that it has the above property is straightforward, though seemingly unmotivated. The present remarks are offered in the hope of throwing further light on this construction.

Let us assume that there exists a partitioning of S^2 into two subsets, say F_o and F_1 (the sets of 0-points and 1-points), such that among the three points of every orthogonal triple there is precisely one 1-point or, as we say, every orthogonal triple is of the type (1,0,0). How is it shown, in the simplest manner, that this hypothesis leads to a contradiction? Let a be a 1-point.

Then the great circle γ_a of points orthogonal to a consists of 0-points alone. Let a and b be two 1-points, at spherical distances $<\pi/2$. Then any points $x \in \gamma_a$ and $y \in \gamma_b$ are 0-points. Choose x orthogonal to y and let z be orthogonal to both; then z is a 1-point. As x travels on γ_a and y on γ_b the point z describes a continuous curve on S^2 which passes through a and b. We conclude then that any two 1-points, of distance $<\pi/2$, are connected by a continuous arc of 1-points (1-arc).

Still assuming our hypothesis, let us now analyse the nature of γ_a when a is a 0-point. Let b and c be two 1-points on γ_a whose distance is $<\pi/2$. It is not hard to see that all points on the arc of γ_a between b and c are also 1-points. Indeed, assume not, and let d be a 0-point on γ_a lying between b and c. Then the point $e \in \gamma_a$ orthogonal to d is a 1-point and therefore γ_e a 0-circle which separates b and c. But this is impossible because then there can be no continuous 1-arc between them. If we call a the North Pole and γ_a the Equator, and adopt a suitable spherical coordinate system, we see then that all points of the Equator whose longitude is in $[0,\pi/2) \cup [\pi,3\pi/2)$ are 1-points, and the rest 0-points. But then <u>all</u> points of S^2 of longitude in $[\pi/2,\pi) \cup [3\pi/2,2\pi)$ are necessarily 0-points, since they lie on great circles of constant longitude orthogonal to the 1-points of the Equator.

Let x and y be two 0-points orthogonal to each other, of longitude in $[\pi/2,\pi)$, and such that the third point of the orthogonal triple (x,y,z) has latitude $\pi/4$ north of the Equator. (Such triples exist.) Let x',y',z' be the points obtained by reflection of x,y,z in the equatorial plane. Then z and z' are orthogonal, but both are 1-points. This contradicts the hypothesis, and thereby proves the impossibility of the postulated partitioning $S^2 = F_0 \cup F_1$.

The essential feature of this reasoning is the last step: The existence of the configuration of six points (x,y,z,x',y',z') between which there are seven orthogonality relations and such that x,y,x',y' are forced to be 0-points. And that comes about because x and x' lies on a great circle orthogonal to some 1-point u, say, y,y' are orthogonal to a 1-point v, say, u and v lying on a 1-segment of the Equator. These eight points may be visualised as the vertices of a graph Γ_1 ("orthogonality graph") on which two points are joined by an edge if and only if they are orthogonal on S^2 (see Fig. 1). This graph clearly exhibits the impossibility of the postulated partition, provided it is placed on the sphere so that u and v are 1-points.

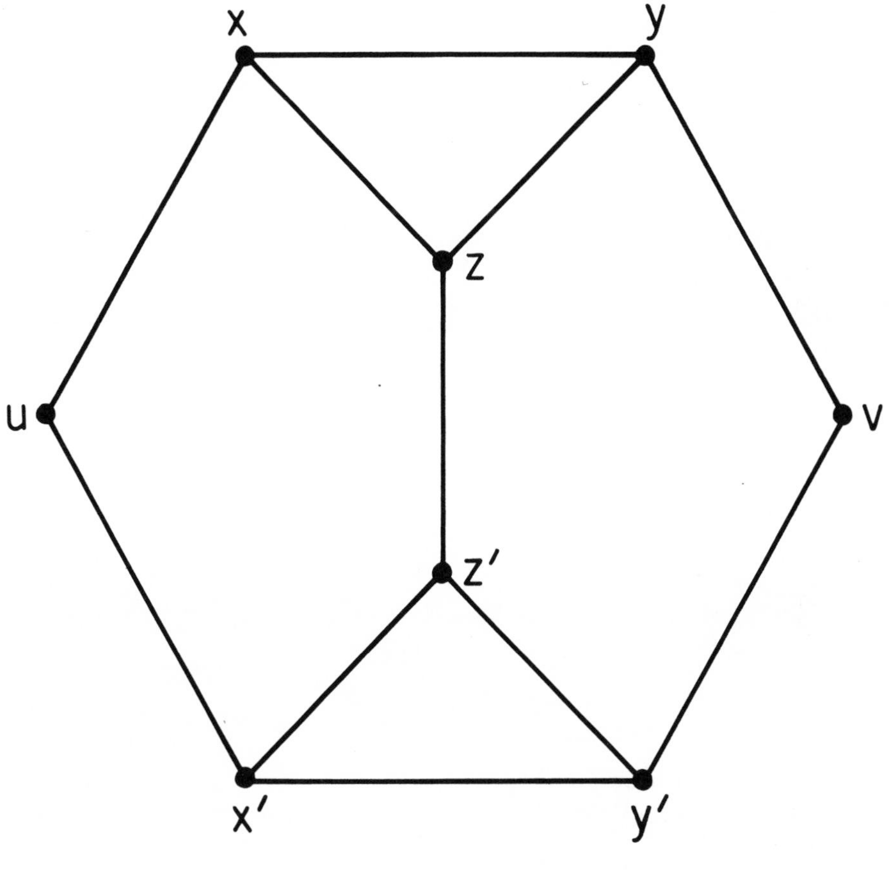

FIG. 1.

Can we always locate two 1-points on S^2 so that they can become the points u,v respectively of a graph Γ_1? To investigate this, let us assign coordinates, in a suitable cartesian system, to the points of Γ_1 as given in the following table.

Point	Rectangular coordinates		
x	γ	0	$-\alpha$
y	γ'	0	$-\alpha'$
z	0	1	0
x'	0	γ	$-\beta$
y'	0	γ'	$-\beta'$
z'	1	0	0
u	α	β	γ
v	α'	β'	γ'

Here $\alpha^2 + \beta^2 + \gamma^2 = \alpha'^2 + \beta'^2 + \gamma'^2 = 1$, and for the sake of ease of writing not all points are "normalised" to lie on the unit sphere. All orthogonality relations of Γ_1 are automatically satisfied except for $x \perp y$ and $x' \perp y'$. These yield the conditions $\alpha\alpha' + \gamma\gamma' = \beta\beta' + \gamma\gamma' = 0$. The cosine of the angle (spherical distance) between u and v is $\cos\theta = \alpha\alpha' + \beta\beta' + \gamma\gamma' = -\gamma\gamma'$. Thus $\alpha' = (-\cos\theta)/\alpha$, $\beta' = (-\cos\theta)/\beta$, $\gamma' = (-\cos\theta)/\gamma$ and so

$$\frac{1}{\alpha^2} + \frac{1}{\beta^2} + \frac{1}{\gamma^2} = \frac{1}{\cos^2\theta}$$

The left side of this equation achieves a maximum of 9 when all three coordinates are equal, and this shows that

$$|\cos\theta| \leq \frac{1}{3}$$

must hold. Conversely, when the distance of two points on S^2 satisfies this inequality they are the points u and v of an orthogonality graph Γ_1.

Now let γ_e be a great circle and suppose a point x_o on it is known to be a 1-point. Can we then choose it together with some other points of γ_e, finite in number, so that some two of them will be at a distance θ with $|\cos\theta| \leq 1/3$ and at the same time these two will be 1-points? The answer is yes. Let x_o have polar coordinate $\phi = 0$, say, and let x_n be the points with polar coordinate $\phi = n\pi/10$ ($n = -4,-3, \ldots, 3, 4$). The arrangements of the 9 points on the circle is shown on Fig. 2, and their orthogonality graph is shown on Fig. 3, where the dotted lines connect those pairs of points whose distance is $\theta = 2\pi/5$ (note $\cos(2\pi/5) = \frac{1}{4}(\sqrt{5}-1) < 1/3$). It is evident that at least one pair of 1-points

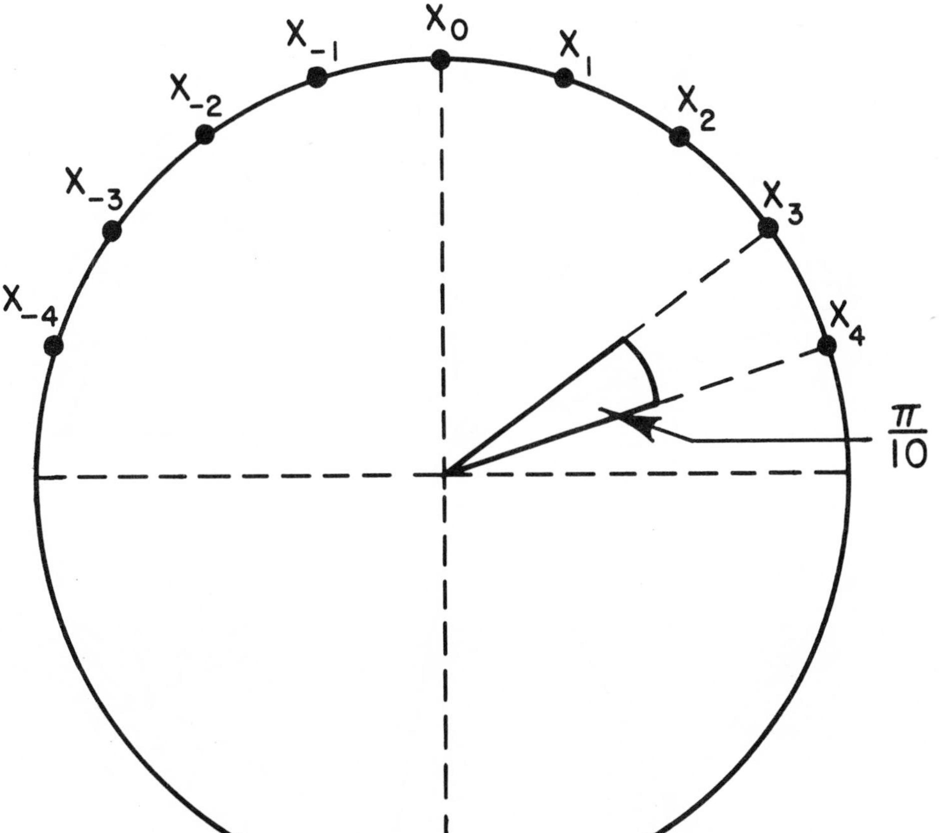

FIG. 2.

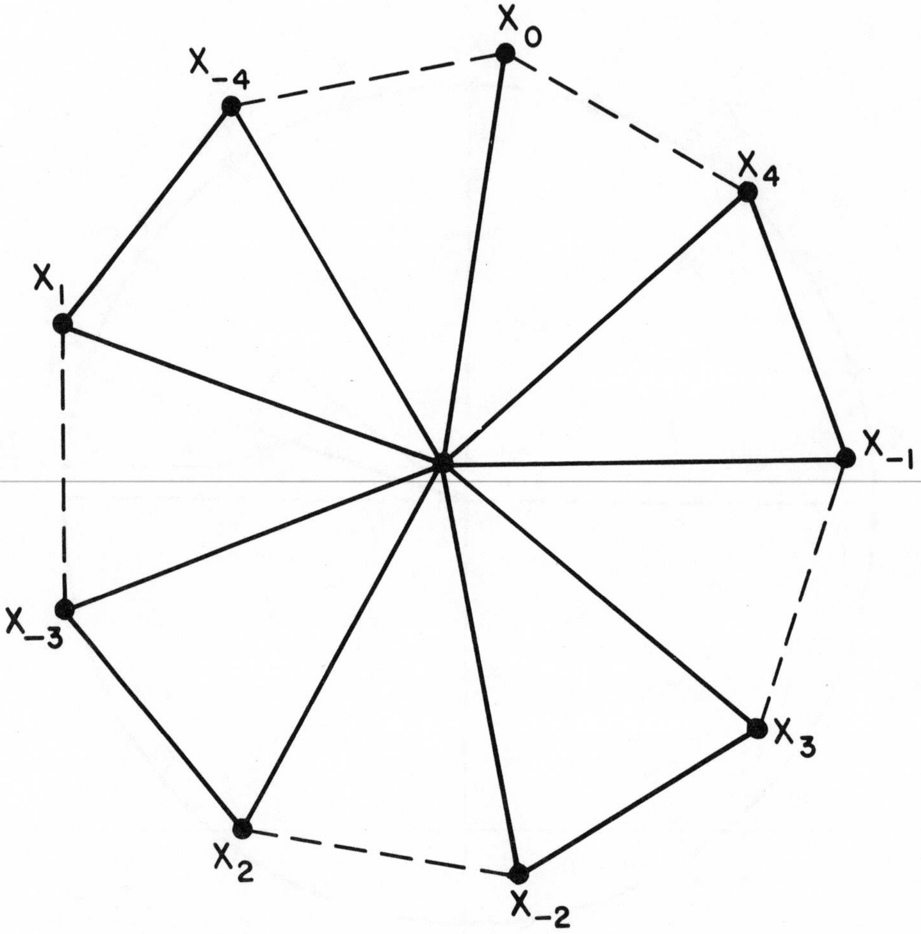

FIG. 3.

of the latter kind exists, or else our hypothesis is violated, i.e., either there is the pair (e,x_o) of 1-points, or there is an orthogonal triple (e,x_n,x_{n-5}) of 0-points.

Let us attach a copy of Γ_1 to each of the five pairs (x_n,x_{n-4}) so that they become the (u,v) pair of the respective Γ_1. The resulting orthogonality graph Γ_2 has this property: If e is a 0-point and x_o a 1-point, Γ_2 always contains either an orthogonal triple of 0-points or else an orthogonal pair of 1-points. Finally, choose any orthogonal triple (a,b,c) on S^2. Attach three copies of Γ_2 in such a manner that the pairs (a,b), (b,c) and (c,a) become respectively the pairs of the three Γ_2 copies called (e,x_o) before. Suppose now that the points of Γ_3 (117 in number) are partitioned in any manner into 0- and 1-points. Then it follows from the construction that Γ_3 contains either an orthogonal triple of the type (0,0,0) or an orthogonal pair of the type (1,1). The graph Γ_3 is precisely the one constructed by Kochen and Specker[2]. The preceeding considerations show how one is quite naturally led to find it in the process of proving the impossibility of a $\{0,1\}$ valued measure on the lattice L(H) when H is the 3-dimensional real Hilbert space $\mathbb{R}^3$.

REFERENCES

1) A. Gleason, J. Rat. Mech. Anal. 6, 885 (1957).

2) S. Kochen and Specker, J. Math. and Mech., 17, 59 (1967).

DIE HEISENBERG-WEYLSCHEN VERTAUSCHUNGSRELATIONEN : ZUM BEWEIS DES VON NEUMANNSCHEN SATZES.

Res Jost

E T H, Zürich

"Einem geschenkten Gaul schaut man nicht ins Maul". Eingedenk dieses tröstlichen Sprichwortes widme ich diese minuskule Note Josef Jauch zum 60. Geburtstag. Der Inhalt entspringt natürlich der Einführungsvorlesung in die Quantenmechanik. Angeregt wurde sie durch Norbert Straumanns Beweis des fraglichen Satzes[1]. Das Ziel ist die Elimination des etwas komplizierten Beweises p. 322 ff in I.M. Gel'fand und N.J. Wilenkins "Verallgemeinerte Funktionen IV" (Berlin 1964), auf den Straumann sich stützt. Alles was der Leser zu wissen braucht (und viel mehr als das) findet er im sehr lesbaren Buch von M. Reed und B. Simon[2].

Die folgende Erörterung des von Neumannschen Satzes über die Eindeutigkeit der irreduziblen Darstellung der Heisenberg - Weylschen Vertauschungsrelationen (V R)[3] beruht im wesentlichen auf dem bekannten Satz von H. Lebesgue[4], dass jede monotone Funktion $\mathbb{R} \to \mathbb{R}$ L- fast überall eine Ableitung besitzt. Im übrigen verwendet sie nur die primitivsten Elemente der Operatorrechnung[2].

p und q bedeuten Impuls und Koordinate (eines eindimensionalen quantenmechanischen Systems) $\hbar$ ist 1 gesetzt. $U(a) = \exp ipa$ $a \in \mathbb{R}$ definiere eine stetige einparametrige unitäre Gruppe. Jedem stetigen und beschränkten $f : \mathbb{R} \to \mathbb{R}$ sei ein beschränkter selbstadjungierter Operator $f(q)$ zugeordnet, wobei die Zuordnung die üblichen Eigenschaften der Operatorrechnung besitze. Insbesondere sei $||f(q)|| \leq \sup_{x \in \mathbb{R}} |f(x)|$.

Als VR postulieren wir

Enz/Mehra (eds.), Physical Reality and Mathematical Description. 234–238.

$$U(a)\ f(q)\ U(-a) = f(q + a), \ \forall\, a \in \mathbb{R}.$$

Diese Gleichung ist (unter den Voraussetzungen) völlig äquivalent zu den Weylschen VR, in denen $V(b) = \exp i\, b\, q$, $b \in \mathbb{R}$ die Rolle von $f(q)$ spielen. $V(b)$ bestimmt wieder eine stetige unitäre Gruppe.

Wenn wir zeigen, dass jede zyklische Darstellung der VR sich in Schrödingersche Darstellungen zerlegen lässt, dann haben wir unser Ziel erreicht, denn jede Darstellung ist direkte Summe zyklischer Darstellungen und die Schrödingersche Darstellung ist offenbar irreduzibel.

Von nun an bedeuten U, V, f die entsprechenden Darsteller in einer zyklsichen Darstellung zum Vektor ϕ_0 ($\neq 0$). Weiter sei $E(x)$, $x \in \mathbb{R}$ das Spektralmass zu q, so dass

$$f(q) = \int f(x)\ d\,E(x).$$

Der Hilbertraum $\mathfrak{H}$ unserer Darstellung ist <u>separabel</u>, denn die Menge $\{U(r)\ V(s)\ \phi_0 | r \in \mathbb{Q},\ s \in \mathbb{Q}\}$ ($\mathbb{Q}$ die rationalen Zahlen) ist in $\mathfrak{H}$ total. $\mathfrak{H}$ besitzt also eine (abzählbare) Orthonormalbasis $\mathfrak{B}$: $= \{U_k | k \in \mathbb{N}\}$. Wir betrachten nun ($i = \sqrt{-1}$) die abzählbar vielen beschränkten monotonen Funktionen

$$(u_k + i^\nu\, u_\ell,\ E(x)\ (u_k + i^\nu\, u_\ell)\ ) = \mu_{k\ell\nu}(x),$$

$\nu \in \mathbb{Z}/\,4\mathbb{Z}, k \geq \ell$ aus $\mathbb{N}$. Aus diesen Funktionen lassen sich alle Matrixelemente von $E(x)$ in der Basis $\mathfrak{B}$ durch Polarisierung bestimmen. $N_{k\ell\nu}$ sei der L-Nullmenge, auf welcher $\mu_{k\ell\nu}$ nicht differenzierbar ist.

$$N = \bigcup_{k,\ell,\nu} N_{k\ell\nu}$$

ist dann ebenfalls eine L-Nullmenge und zwar mit der folgenden Eigenschaft :

<u>Falls</u> $N = N_0$ <u>ist und</u> N' <u>eine beliebige L-Nullmenge ist, dann gilt</u>

$$\int_{N'} d\,E(x) = \int_{N_0 \cap N'} d\,E(x).$$

Unter den Mengen N_0, welche der letzten Gleichung genügen, gibt es offenbar eine kleinste - und eben diese werde fortan mit

N_o bezeichnet. N_o ist eine unitäre Invariante von q. Aber aus den VR folgt

$$U(a)\ E(x)\ U(-a) = E(x - a),$$

also ist N_o translationsinvariant und daher leer. So haben wir das

Lemma : Für jede zyklische Darstellung gilt

$$\forall\ u \in \mathfrak{h}\ ,\ \forall\ v \in \mathfrak{h}$$

ist (u, E(x) v) absolut stetig und von beschränkter Variation.

Das Lemma erlaubt es die Ableitung d E / d x = e(x) wenigstens als Bilinearform einzuführen durch

$$(u,\ e(x)\ v) := \frac{d}{dx}\ (u,\ E(x)\ v).$$

Im Sinn der Bilinearformen gilt kraft der VR

$$U(a)\ e(x)\ U(-a) = e(x - a)$$

$$e(x)\ f(q) = f(q)\ e(x) = f(x)\ e(x).$$

Falls u und v C^∞-Vektoren von U sind, dann ist

$$(U(x)u,\ e(o)\ U(x)\ v) = (u,\ e(x)\ v)$$

für alle x definiert und C^∞. Aber die

$$(u,\ e(x)\ v) := \langle u(x),\ v(x)\rangle$$

haben für festes $x \in \mathbb{R}$ alle Eigenschaften eines Skalarproduktes [Bi- (Sesqui-) Linearität, Positivität] und definieren daher Halme $\mathfrak{f}_x$ von Hilberträumen in einem Feld von Hilberträumen über $\mathbb{R}$. Alle $\mathfrak{f}_x$ sind unitär äquivalent und werden durch U ineinander unitär transformiert. Wir können sie mit Hilfe von U miteinander und mit $\mathfrak{f}_{(o)}$ identifizieren. Dadurch stellt sich ein beliebiger Vektor $u \in \mathfrak{h}$ jetzt dar als Abbildung

$$\mathfrak{h} \ni u : \mathbb{R} \to \mathfrak{f}\ ,\ x \to u(x)$$

mit der Eigenschaft, dass $\langle u'(x), u(x)\rangle$ in x integrabel ist und gilt

$$\int \langle u'(x),\ u(x)\rangle\ dx = \int (u',\ e(x)\ u)\ dx = (u',u).$$

Mit der einleitenden Bemerkung haben wir den

Satz : Zu jeder Darstellung der VR gehört ein Hilbertraum $\mathfrak{h}$ mit Skalarprodukt < , >. Die Vektoren des Darstellungsraumes $\mathfrak{H}$ sind Abbildungen u : $\mathbb{R} \to \mathfrak{h}$ mit den Eigenschaften

$$(u',u) = \int \langle u'(x) \rangle \, dx$$

$$(f(q)\, u)\,(x) = f(x)\, u(x)$$

$$(U(a)\, u)(x) = u(x + a).$$

Das ist der v. Neumannsche Satz.

Die Verallgemeinerung der Überlegungen auf N Freiheitsgrade ($N < \aleph_0$!) ist evident. Ebenso dürfte die Verallgemeinerung auf die Mackey-Weylschen* Vertauschungsrelationen keine wesentlichen Schwierigkeiten aufweisen.

Was nun den anfänglich erwähnten Beweis von Gel'fand und Wilenkin angeht, so scheint sich dieser bequem der hier durchgeführten Diskussion des v. Neumannschen Satzes unterzuordnen. Es gilt der

Satz : Sei μ ein σ -finites (reguläres) Mass auf $\mathbb{R}$, das zu allen seinen Translatierten äquivalent ist. Dann ist μ zum L-Mass äquivalent.

In der Tat : betrachten wir auf $L_2(\mathbb{R}, d\mu)$ die unitäre Gruppe U und das Spektralmass E definiert durch

---

*) Darunter verstehen wir das folgende : Sei G eine lokal kompakte abelsche Gruppe und $\hat{G}$ die (mit der entsprechenden Topologie versehene) duale Gruppe. Sei V eine stetige unitäre Darstellung von G und $\hat{V}$ eine ebensolche von $\hat{G}$. Jedes Element $\hat{g} \in G$ bestimmt einen (unitären) Charakter: $\hat{g} : G \to \mathbb{C}$, den wir mit $(\hat{g},\cdot)$ bezeichnen. Die Mackey-Weylschen VR verlangen nun zusätzlich auf $\hat{G} \times G$

$$\hat{V}(\hat{g})\; V\,(g) = (\hat{g},g)\; V(g)\; \hat{V}(\hat{g}).$$

Für sie gilt ein dem Satz von v. Neumann analoger Eindeutigkeitssatz[5].

$$(U(a)\ \phi)(x') = \phi\ (x' + a) \sqrt{\frac{d\mu_a}{d\mu}\ (x')}$$

$$(E(x)\ \phi)(x') = \Theta\ (x - x')\ \phi\ (x'),$$

wobei μ_a das um $a \in \mathbb{R}$ translatierte Mass μ und Θ die charakteristische Funktion von $[o,\infty)$ ist, dann sind die Vertauschungsrelationen $U(a)\ E(x)\ U(-a) = E(x - a)$ erfüllt. Also gilt das Lemma (zu dessen Herleitung die Stetigkeit von U auf $\mathbb{R}$ nicht benutzt wurde, d.h. für jedes $\phi \in L_2(\mathbb{R}, d\mu)$ ist

$$(\phi,\ E(x)\ \phi) = \int_{-\infty}^{x} |\phi(x')|^2\ d\mu\ (x')$$

in x absolut stetig, was nur möglich ist, wenn μ eine Dichte besitzt. Diese Dichte kann wegen der Äquivalenz aller Translatierten als strikte positiv angenommen werden.

Verallgemeinerungen erscheinen automatisch.

REFERENZEN

1) N. Straumann, "A New Proof of von Neumanns Theorem ..." Helv. Phys. Acta 40 (1967) 518.

2) M. Reed and B. Simon, Functional Analysis Academic Press N.Y. and London 1972.

3) J. von Neumann, Collected Works Vol. II p. 221-229.

4) z.B. F. Riesz et B. Sz. Nagy, "leçons d'analyse fonctionnelle" Budapest 1952 p. 3-10.

5) Etwa G.W. Mackey : "Infinite Dimensional Group Representations and their Applications", Forschungsinstitut für Mathematik ETHZ 1971 p. 28.

REAL VERSUS COMPLEX REPRESENTATIONS AND LINEAR-ANTILINEAR COMMUTANT(+).

R.Ascoli, G.Teppati *
C.Garola, L.Solombrino †

* Istituto di Matematica dell'Università - Parma
† Istituto di Fisica dell'Università - Lecce

A proposition of Autonne, Frobenius and Schur concerning the equivalence of any unitary finite-dimensional complex group representation to a real representation and a related classification of such complex representations are both generalized to every family U of arbitrary mappings in any vector space over a (commutative or non-commutative) field endowed with a conjugation (non-identical involutory automorphism) j. The subject is connected with the group of the linear and "antilinear" (that is semilinear with respect to j) invertible mappings that commute with U.

1. INTRODUCTION

A necessary and sufficient condition for a unitary finite-dimensional complex group representation to be equivalent to a real representation was first given by Autonne (1902) [1] and afterwards simplified by Frobenius and Schur (1906) [2].

A more modern version of this statement may be found in several books on group theory [3], especially those written for physicists, like the books of Wigner [4], Hamermesh [5] and Lyubarskii [6].

In this latter form the condition may essentially be formulated as follows.

Theorem 0.- *Let* $D : g \to D(g)$ *be a unitary finite-dimensional*

(+) Research supported by C.N.R. (Italy).

Enz/Mehra (eds.), Physical Reality and Mathematical Description. 239–251.

complex (1) *matrix representation of a group G. Let us call* $\overline{D}$ *the complex conjugate of D.*

Then a necessary and usfficient condition for D to be equivalent to a real representation is that $\overline{D}$ *be equivalent to D through a matrix S such that* (2) (3) $S\overline{S} = E$.

The statement of Theorem 0 is clearly related to the following classification of complex group representations, which has been introduced by Frobenius and Schur in 1906 [2] :

1) representations which are equivalent to a real representation,

2) representations which are equivalent to the complex conjugate representation, but not to a real representation,

3) representations which are not equivalent to the complex conjugate representation.

From this point of view Theorem 0 connects this classification with other properties of the representations.

In order to permit an easier comparison with our work, (especially with Theorem 3 below,) let us first restate Theorem 0 in terms of linear mappings instead of matrices.

Theorem 0'.- *Let* $U : g \to U(g)$ *be a unitary representation of a group G in a finite-dimensional complex scalar-product space X.*

For any orthogonal basis $\mathcal{E}$ *in X let us call J the mapping of X that transforms the components of any vector into their complex conjugates and for any linear mapping A of X let us call* $\overline{A} = JAJ^{-1}$ *the linear mapping in which A is transformed under the mapping J of X.*

---

(1) The peculiarity of the complex field, that is used in the current proofs, is to be an algebraically closed field, endowed with an involutary automorphism (the complex conjugation $\alpha \to \overline{\alpha}$).

(2) Here and in the following we call E the identity mapping.

(3) In the formulations of this theorem that are quoted above, the representations are also assumed to be irreducible. We have dropped this assumption because unitarity of the representation implies complete reducibility of the reducibility of the representation : it is then easy to go over from the statement concerning irreducible representations to the one given here, in analogy to the original treatment of Frobenius and Schur that is quoted above.

Then the following conditions are equivalent (2) *:*

i) a basis exists in which $\overline{U} = U$,

ii) a basis exists in which $\overline{U}$ *is equivalent to* U *through a linear mapping* S *such that* $S\overline{S} = E$.

iii) in any basis $\overline{U}$ *is equivalent to* U *through a linear mapping* S *such that* $S\overline{S} = E$.

This paper generalizes the above treatment from several points of view (4):

1) the group G is generalized to be any set S,

2) the complex field is generalized to ba any field K(commutative or non-commutative), endowed with a non-identical involutory automorphism (conjugation) j : $\alpha \to \overline{\alpha}$, which takes the place of the complex conjugation,

3) the vector space X is not bound to be finite-dimensional,

4) no scalar product is assumed on X, so that the representation is not required to be unitary,

5) the representing linear mappings are generalized to be arbitrary mappings,

6) the representation is neither assumed to be irreducible, nor completely reducible.

Moreover the properties of the representations that are considered by the above classification of Frobenius and Schur, and especially by Theorem 0', are here connected with properties of the group $\mathcal{U}^c$ of the linear ($\mathcal{U}^{c\ell}$) and "antilinear"(5) ($\mathcal{U}^{ca}$) invertible mappings of X that commute with the given representations(6)

---

(4) To this purpose an essentially new proof of Theorem 0' is required. In fact all the current proofs use complete reducibility and derive first the statement for irreducible representations; moreover they require the existence of the square root of any unitary operator, hence, at least with reference to the usual proofs, the fundamental theorem of algebra and the finite-dimensionality of the space.

(5) That is semilinear with respect to the "conjugation" j of K.

(6) The consideration of antilinear mappings has a remarkable interest in Quantum Physics, where symmetries are described by antilinear as well as linear mappings of a complex vector space [7][8]. The mathematical reason is that in Quantum Physics the relevant structure is the projective geometry of the vector subspaces; the endomorphisms of such a structure are known to be described by the semilinear mappings : so in the complex case some further continuity requirement restricts the consideration to the linear and the antilinear mappings.

(from this point of view the main results of this paper are summarized in the Table attached to Theorem 4 below).

At the end the connection between operator representations and matrix representations is briefly discussed for representations consisting of semilinear mappings (Section 4). This connection turns out to be particularly simple whenever the representation consists of linear and antilinear mappings only.

So the main interest of this paper is perhaps in its application to groups of linear and antilinear mappings[6].

2. CONJUGATION OF REPRESENTATIONS WITH RESPECT TO A BASIS

We use the following definition of representation, in which the representing mappings are not required to be linear.

Definition 1.- *Let X be a vector space*[7] *over a (commutative or non-commutative) field K and let $\mathcal{S}$ be a set.*

We call representation space *of $\mathcal{S}$ over K any pair (X,U) of the vector space X and a mapping U from $\mathcal{S}$ into the set of all the mappings of X. The mapping U will be said to be a* representation *of the set $\mathcal{S}$ in the vector space X.*

If (X,U), (Y,V) are representation spaces of the same set $\mathcal{S}$, U and V are said to be equivalent *whenever (X,U) and (Y,V) are isomorphic, that is whenever an isomorphism $S : X \to Y$ exists such that :*

$$\forall\, s \in \mathcal{S} \qquad V(s) = S\, U(s)\, S^{-1} \quad .$$

The next definition generalizes the usual notion of the conjugate of a complex representation with respect to a basis.

Definition 2.- *We call* field K with a conjugation j *any field K (commutative or non-commutative), endowed with a non-identical involutory*[8] *automorphism $j : \alpha \to \overline{\alpha}$.*

Let X be a vector space over a field K with a conjugation j.

We call antilinear mapping *any mapping in X which is semilinear with respect to j.*

---

(7) We remark that Definitions 1,2,3, Theorem 2, the implications ii)$\Rightarrow$i)$\Rightarrow$iii) of Theorem 1 and Lemma 2 require only a module structure on X.

(8) This Section, as well as Section 4, does not depend on the assumption that j be non-identical and involutory: *Sections 2 and 4 hold for any automorphism j of K.*

Let $\mathfrak{E}$ *be a basis in X.*

We call $J_{\mathfrak{E}}$ *the (obviously invertible) antilinear mapping in X that leaves the elements of* $\mathfrak{E}$ *invariant. We say that* $J_{\mathfrak{E}}$ *is a* conjugation in X *and specifically the conjugation associated with the basis* $\mathfrak{E}$ *. The index* $\mathfrak{E}$ *is omitted whenever there is no possibility of confusion.*

For any mapping A of X we define the conjugate mapping $\overline{A} = J A J^{-1}$ *(more specifically the conjugate mapping* $\overline{A}^{\mathfrak{E}}$ *of A with respect to the basis* $\mathfrak{E}$ *).*

In analogy for any representation U in X (Def.1) we define the conjugate representation $\overline{U} = J U J^{-1}$ *with respect to the basis*(9) (10) $\mathfrak{E}$.

Theorem 1.- *Let (X,U) be a representation space of a set* $\mathcal{S}$ *over a field K with a conjugation*(8) *j (Def.2).*

Then the following conditions are equivalent :

i) an antilinear invertible mapping A, exists which commutes with U,

ii) a basis exists in which $\overline{U}$ *is equivalent to U,*

iii) in any basis $\overline{U}$ *is equivalent to U.*

More specifically, if i) occurs, then the equivalence in ii) and iii) is implemented by $S = A J^{-1}$*, where J is the conjugation in X associated to the basis that is being considered (Def.2). Conversely, if S implements the equivalence in ii) or iii) then* $A = S J$ *satisfies condition i).*

---

(9) The conjugate of any representation with respect to a basis $\mathfrak{E}$ depends on the basis $\mathfrak{E}$ but conjugate representations of U with respect to different bases are equivalent(8). In fact let $\mathfrak{E}$, $\mathfrak{F}$ be two bases in X. It is $\overline{U}^{\mathfrak{F}} = J_{\mathfrak{F}}\, U\, J_{\mathfrak{F}}^{-1} = J_{\mathfrak{F}}\, J_{\mathfrak{E}}^{-1} J_{\mathfrak{E}}\, U\, J_{\mathfrak{E}}^{-1} J_{\mathfrak{E}}\, J_{\mathfrak{F}}^{-1} = (J_{\mathfrak{F}}\, J_{\mathfrak{E}}^{-1})\, \overline{U}^{\mathfrak{E}} (J_{\mathfrak{F}}\, J_{\mathfrak{E}}^{-1})^{-1}$. Hence $\overline{U}^{\mathfrak{F}}$ is equivalent to $\overline{U}^{\mathfrak{E}}$ through the linear bijection $J_{\mathfrak{F}}\, J_{\mathfrak{E}}^{-1}$.

(10) Let us call $K_0 = \{\alpha \in K \mid \overline{\alpha} = \alpha\}$ the subfield of K consisting of the self-conjugate elements (8). Then the reference to a basis may be avoided in Def.2, by referring instead to the concept of a K_0-structure Ξ in X, for which we refer to Bourbaki, ref. [9], §8 no 1. Specifically Def.2 may be restated by defining J_{Ξ} to be the antilinear mapping in X that leaves invariant the vectors of a K_0-structure Ξ (Ξ is the K_0-linear span of $\mathfrak{E}$; conversely $\mathfrak{E}$ is any K_0-basis in Ξ).

Proof.- The implication iii) ⇒ ii) is an obvious consequence of the existence of bases in vector spaces. Let us first prove ii) ⇒ i), then i) ⇒ iii) (at the same time the last specifications of the statement are also proved).

For a given basis $\mathcal{E}$ of X, let S implement the equivalence of $\bar{U} = J\,U\,J^{-1}$ and U, where J is the conjugation associated to the basis $\mathcal{E}$ (Def.2) ; then

$$U = S\,\bar{U}\,S^{-1} = S\,J\,U\,J^{-1}\,S^{-1} = S\,J\,U\,(SJ)^{-1}.$$

Hence A = SJ, which is obviously an antilinear invertible mapping, commutes with U. So the implication ii) ⇒ i) is proved.

Let now A be an antilinear invertible mapping which commutes with U. Let $\mathcal{E}$ be any basis of X and J the associated conjugation. We define $S = AJ^{-1}$. It is A = SJ. Then, according to the formula above, commutativity of A with U implies equivalence of U to $\bar{U}$ through $S = AJ^{-1}$. This is the implication : i) ⇒ iii). Thus the theorem is proved.

3. SELF-CONJUGATE REPRESENTATIONS AND LINEAR-ANTILINEAR CENTRALIZER

Definition 3.- *Let X be a vector space over a field K with a conjugation j (Def.2). For any set* $\mathcal{U}$ *of mappings of X, we call* $\mathcal{U}^{\ell}$ *the subset of the linear mappings of* $\mathcal{U}$, $\mathcal{U}^{a}$ *the subset of the antilinear mappings of* $\mathcal{U}$, $\mathcal{U}^{c}$ (linear-antilinear centralizer *of* $\mathcal{U}$) *the group of the linear or antilinear invertible mappings of X which commute with all the mappings of* $\mathcal{U}$. *Hence,* $\mathcal{U}^{c\ell}$ (linear centralizer *of* $\mathcal{U}$) *is the group of the linear invertible mappings of X which commute with* $\mathcal{U}$, $\mathcal{U}^{ca}$ (antilinear centralizer) *is the set of the antilinear invertible mappings of X which commutes with* $\mathcal{U}$ *(obviously* $\mathcal{U}^{c} = \mathcal{U}^{c\ell} \cup \mathcal{U}^{ca}$*).*

Theorem 2.- *Let (X,U) be a representation space of a set* $\mathcal{S}$ *over a field K with a conjugation j (Def.2).*

Then the linear centralizer $\mathcal{U}^{c\ell}$ *of* $\mathcal{U} = \mathcal{U}(\mathcal{S})$ *is a normal subgroup of the linear-antilinear centralizer* $\mathcal{U}^{c}$ *and* $\mathcal{U}^{c}$ *either coincides with* $\mathcal{U}^{c\ell}$, *or is an extension of* $\mathcal{U}^{c\ell}$ *through the two-element group* G_2 *: that is, either* $\mathcal{U}^{c} = \mathcal{U}^{c\ell}$ *(equally* $\mathcal{U}^{ca} = \emptyset$*), or we have the short exact sequence :*

$$1 \longrightarrow \mathcal{U}^{c\ell} \underset{\varphi_1}{\longrightarrow} \mathcal{U}^{c} \underset{\varphi_2}{\longrightarrow} G_2 \longrightarrow 1$$

where φ_1 is the natural injection and φ_2 maps any linear mapping of $\mathcal{U}^c$ onto the identity of G_2 and any antilinear mapping of $\mathcal{U}^c$ onto the generator of G_2.

In the second case the short exact sequence splits (or equally $\mathcal{U}^c$ is isomorphic to a semidirect product of $\mathcal{U}^{c\ell}$ and G_2) if and only if an antilinear involutory element A exists within $\mathcal{U}^c$.

More specifically, whenever there exists an antilinear involutory element A within $\mathcal{U}^c$, let us call $\tilde{G}_2$ the group $\{E,A\}$: then $\mathcal{U}^c$ is canonically identical(11) to the semidirect product $\mathcal{U}^{c\ell} \circledS_\phi \tilde{G}_2$ of $\mathcal{U}^{c\ell}$ and $\tilde{G}_2$, defined by the mapping ϕ of $\tilde{G}_2$ into the group of the automorphisms of $\mathcal{U}^{c\ell}$ that makes the identity automorphism of $\mathcal{U}^{c\ell}$ correspond to E and the automorphism of $\mathcal{U}^{c\ell}$: $L \to ALA^{-1}$ to A.

Proof.- For any $L \in \mathcal{U}^{c\ell}$ and $A \in \mathcal{U}^c$, ALA^{-1} is a linear mapping which belongs to $\mathcal{U}^c$, hence it also belongs to $\mathcal{U}^{c\ell}$. So $\mathcal{U}^{c\ell}$ is a normal subgroup of $\mathcal{U}^c$.

As $\mathcal{U}^c = \mathcal{U}^{c\ell} \cup \mathcal{U}^{ca}$, and $\mathcal{U}^{c\ell} \cap \mathcal{U}^{ca} = \emptyset$, either $\mathcal{U}^c = \mathcal{U}^{c\ell}$ or $\exists\, A \in \mathcal{U}^{ca}$; in this latter case, clearly $\mathcal{U}^c = \mathcal{U}^{c\ell} \cup A\mathcal{U}^{c\ell} =$ $= \mathcal{U}^{c\ell} \cup \mathcal{U}^{c\ell}A$, the unions being disjoint, so that $\mathcal{U}^c/\mathcal{U}^{c\ell}$ is a two-element group.

The statements concerning the splitting of the short exact sequence occur as particular cases of the standard connection between splitting sequences and semidirect products. Thus the proof of the theorem is concluded.

The least trivial statement to be proved concerns the existence, under suitable conditions, of a basis in which $\bar{U} = U$. The next lemma includes the main part of the proof.

Lemma 1.- *Let X be a vector space over a field K with a conjugation j (Def.2).*

Then every antilinear involutory mapping A of X is a conjugation in X (Def.2) (that is a basis $\mathfrak{E}$ exists in X whose elements are invariant under A, so that $A = J_{\mathfrak{E}}$)(12).

---

(11) Through the identification $(A,B) \to AB$.

(12) An equivalent statement is the following. Let $\mathfrak{F}$ be any basis in X : a linear mapping S of X satisfies $S\,\bar{S}^{\mathfrak{F}} = E$ if and only if a linear invertible mapping T of X exists such that $S = T^{-1}\bar{T}^{\mathfrak{F}}$. Let in fact $J_{\mathfrak{F}}$ be the conjugation of X associated with the basis $\mathfrak{F}$(Def.2) and let us put $A = S\,J_{\mathfrak{F}}$. Then we

Proof.- As j is non-identical, an element $\alpha \in K$ exists such that $\bar{\alpha} \neq \alpha$. So $(\alpha - \bar{\alpha})^{-1}$ exists and we have identically

$$\forall\, x \in X \quad x = (\alpha - \bar{\alpha})^{-1} (\alpha x + \bar{\alpha} A x - \bar{\alpha}(x + Ax)).$$

How it is easily checked that for any $x \in X$ and $\beta \in K$ the vector $\beta x + \bar{\beta} A x$ is invariant under A : in fact,

$$A(\beta x + \bar{\beta} A x) = \bar{\beta} A x + \beta A^2 x = \beta x + \bar{\beta} A x.$$

Thus, in particular, both the vectors $\alpha x + \bar{\alpha} A x$ and $x + Ax$ are invariant under A.

It follows from the above identity that the set of the vectors invariant under A generates X.

Hence, according to a fundamental theorem on the bases of vector spaces *(13)*, a basis $\mathfrak{E}$ of vectors which are invariant under A exists in X.

So, according to Def.2, A and $J_{\mathfrak{E}}$ coincide on $\mathfrak{E}$. As both A and $J_{\mathfrak{E}}$ are antilinear, it follows $A = J_{\mathfrak{E}}$.

Thus the lemma is proved.
The main theorem may now be easily derived.

Theorem 3.- *Let (X,U) be a representation space of a set* $\mathcal{S}$ *over a field K with a conjugation j (Def.2).*
Then the following statements are equivalent *(14)* :

i) a basis exists in which $\bar{U} \equiv U$,
ii) a basis exists, in which $\bar{U}$ *is equivalent to U through S such that* *(2)* $S\bar{S} = E$,

---

(12)cont. observe : i) $A^2 = E$ if and only if $S\,\bar{S}^{\mathfrak{J}} = E$; ii) A is a conjugation in X if and only if a linear invertible mapping T exists such that $S = T^{-1}\bar{T}^{\mathfrak{J}}$ (because $S = T^{-1}\,\bar{T}^{\mathfrak{J}} \Longleftrightarrow A = T^{-1} J_{\mathfrak{J}}\, T \Longleftrightarrow A = J_{T^{-1}\mathfrak{J}}$). Hence the above statement is equivalent to Lemma 1. An alternative proof has been found by D.Lenzi, ref. [10].

(13) See for instance ref. [9], §7 no 1 Th.2.

(14) Corresponding to footnote *(10)* the statements of Theorem 3 may be formulated with reference to K_0-structures instead of bases. In particular the statement i) may be expressed as follows : a K_0-structure Ξ exists in X which is stable under $U(\mathcal{S})$ (that is $U(\mathcal{S})$ is rational on K_0, ref. [9], §8, no 3, Def.3).

iii) in any basis, $\bar{U}$ is equivalent to U through S such that $S\bar{S} = E$,

iv) an involutory antilinear mapping A exists, which commutes with U,

v) the linear-antilinear centralizer $\mathcal{U}^c$ of U (Def.3) is isomorphic to a semidirect product of the linear centralizer $\mathcal{U}^{cl}$ and the two-element group G_2.

Proof.- It is immediately seen that the equivalence of the statements ii),iii),iv) follows from Theorem 1 by further observing that : a) every involutory mapping is invertible; b) if A=SJ then A is involutory if and only if $S\bar{S} = E$. This latter statement follows from the identity (which uses the assumption of involutory j)

$$A^2 = SJSJ = SJSJ^{-1} = S\bar{S} .$$

Let us now prove the equivalence of i) and iv). The implication i)$\Longrightarrow$iv) is evident : in fact, if i) is assumed, then, according to Def.2, it is $J\ U\ J^{-1} = U$, hence $J\ U = U\ J$, so that iv) is satisfied by choosing A = J.

The implication iv)$\Longrightarrow$i) is the least trivial, but it follows easily from Lemma 1. In fact, if iv) is assumed, then, according to Lemma 1, there exists a basis $\mathcal{E}$ such that A = J, where J is the conjugation in X associated with the basis $\mathcal{E}$. As A commutes with U, it is $A\ U\ A^{-1} = U$. It follows that $J\ U\ J^{-1} = U$, that is, $\bar{U} = U$ in the basis $\mathcal{E}$. This is the statement i).

Finally iv) and v) are equivalent according to one of the statements of Theorem 2.

Thus the proof of Theorem 3 is completed.

We finally collect in the next proposition the main statements of Theorems 1, 2, and 3 so as to generalize Frobenius and Schur's classification of the complex group representations in three different types (see Introduction).

Theorem 4.- *Let (X,U) be a representation space of a set $\mathcal{S}$ over a field K with a conjugation j(Def.2).*

Then one of the three mutually exclusive cases described by the three lines of the next Table occurs, the statements on each line being equivalent. (The semidirect product occurring in the last statement of line 1 is specified in the last part of Theorem 2).

Table (Ref. Theorem 4)
(the symbol $\sim$ means equivalence of representations, E is the identity mapping)

<table>
<tr><th rowspan="2">Type</th><th colspan="3">Representation U
($\bar{A}$, the conjugate of A with reference to a basis $\mathcal{E}$, is defined in Def.2)</th><th colspan="2">Antilinear Centralizer $\mathcal{U}^{ca}$ (Def.3)</th><th colspan="2">Linear-antilinear centralizer $\mathcal{U}^{c}$ (Def.3)
($\mathcal{U}^{c\ell}$ is the linear centralizer, G_2 is the two-elements group, Ⓢ means semidirect product)</th></tr>
<tr><th></th><th colspan="2">(The statements do not depend on the basis $\mathcal{E}$)</th><th colspan="2"></th><th colspan="2"></th></tr>
<tr><td>1</td><td>$\exists$ basis such that $\bar{U} = U$</td><td rowspan="2">$\bar{U} \sim U$</td><td>$\bar{U} \sim U$ implemented by S such that $S\bar{S} = E$</td><td rowspan="2">$\mathcal{U}^{ca} \neq \emptyset$</td><td>$\exists A \in \mathcal{U}^{ca}$ such that $A^2 = E$</td><td rowspan="2">$\mathcal{U}^{c}$ is an extension of $\mathcal{U}^{c\ell}$ through G_2</td><td>$\mathcal{U}^{c} \sim \mathcal{U}^{c\ell}$ Ⓢ G_2</td></tr>
<tr><td>2</td><td rowspan="2">$\nexists$ basis such that $\bar{U} = U$</td><td>$\bar{U} \sim U$ implemented by no S such that $S\bar{S} = E$</td><td>$\mathcal{U}^{ca} \neq \emptyset$ but $\nexists A \in \mathcal{U}^{ca}$ such that $A^2 = E$</td><td>$\mathcal{U}^{c}$ is an extension of $\mathcal{U}^{c\ell}$ through G_2 but $\mathcal{U}^{c} \not\sim \mathcal{U}^{c\ell}$ Ⓢ G_2</td></tr>
<tr><td>3</td><td>$\bar{U} \not\sim U$</td><td>$\bar{U} \not\sim U$</td><td>$\mathcal{U}^{ca} = \emptyset$</td><td>$\mathcal{U}^{ca} = \emptyset$</td><td>$\mathcal{U}^{c} = \mathcal{U}^{c\ell}$</td><td>$\mathcal{U}^{c} = \mathcal{U}^{c\ell}$</td></tr>
</table>

4. LINEAR-ANTILINEAR REPRESENTATIONS AND MATRIX DESCRIPTION.

The original treatments of Autonne, Frobenius and Schur and its more modern versions which are quoted in the Introduction concern representations consisting of linear mappings and more specifically matrix representations rather than operator representations. This latter point is especially relevant with reference to the definition of $\bar{U}$ as the representation that is obtained from U through the conjugation of the matrix elements.

It is known that more generally semilinear mappings may be described by means of matrices[15]. In this Section we consider briefly the connection between operator representations and corresponding matrix representations for representations consisting of semilinear mappings and more particularly of linear and antilinear mappings only[6].

The next proposition[8] shows that, whenever the representation U consists of linear and antilinear mappings, the conjugate representation $\bar{U}$ with respect to a basis $\mathfrak{E}$, as it is introduced in Def.2, coincides with the representation that is obtained from U through the conjugation of its matrix elements with respect to the basis $\mathfrak{E}$. This statement holds more generally if U consists of semilinear mappings, but then the automorphism of K that is associated with any mapping of $\bar{U}$ is generally not the same as the one that is associated with the corresponding mapping of U.

Lemma 2.-*Let X be a vector space over a field K with a conjugation*[8] *j (Def.2) and let* $\mathfrak{E} = (e_\lambda)$ *be any basis in X.*

For any semilinear mapping A of X let us call j_A *its associated automorphism of K and let us define in the usual way*[14] *the corresponding matrix* $M(A) = (A_{\mu\lambda})$ *where for any* μ *and* λ $A_{\mu\lambda} = (Ae_\lambda)_\mu$, x_λ *meaning the* λ *component of any vector x.*

Then

$$j_{\bar{A}} = j\, j_A j^{-1} \quad , \quad M(\bar{A}) = \bar{M}(A).$$

In particular for any linear or antilinear A it is $j_{\bar{A}} = j_A$.

---

(15) Ref. [9],§10, no 6. It must be stressed that any semilinear mapping A is singled out by the pair of its associated automorphisms j_A of K and its matrix M(A) : when going over from operators to matrices the product operation is not described by the product of matrices, a more involved operation being required ; it is $j_{BA} = j_B\, j_A$ and, in the right vector spaces, $M(BA) = M(B)\, M(A)^{j_B}$, where $M(A)^{j_B}$ is the image of M(A) under j_B.

Proof.- The statements concerning $j_{\bar{A}}$ follow immediately from the definition $\bar{A} = J\ A\ J^{-1}$. It remains to prove that $M(J\ A\ J^{-1}) = M(\bar{A})$.

We have first $M(A\ J^{-1}) = M(A)$: in fact, using the definition of J, we get

$$\forall\ \lambda,\mu \quad (A\ J^{-1})_{\mu\lambda} = (A\ J^{-1}\ e_\lambda)_\mu = (Ae_\lambda)_\mu = A_{\mu\lambda}.$$

Moreover, we get $M(J\ A) = \bar{M}(A)$: in fact, using again the definition of J, for any vector x we get $\forall\ \lambda\ (J\ x)_\lambda = \bar{x}_\lambda$ hence

$$\forall\ \lambda,\mu \quad (J\ A)_{\mu\lambda} = (J\ Ae_\lambda)_\mu = \overline{(Ae_\lambda)}_\mu = \bar{A}_{\mu\lambda}.$$

Combining the two results, we obtain

$$M(J\ A\ J^{-1}) = \bar{M}(A),$$

so that the Lemma is proved.

Concerning linear-antilinear representations, and, more generally, semilinear representations, we observe finally that, inspite of Lemma 2, equivalence of operator representations does obviously not coincide with equivalence of the corresponding matrix representations.

In fact, the operator equivalence

$$\forall\ s \in \mathcal{S} \quad V(s) = S\ U(s)\ S^{-1}$$

is expressed in terms of matrices (in the case of a right vector space) by

$$\forall\ s \in \mathcal{S} \quad j_{V(s)} = j_{U(s)}, \quad M(V(s)) = M(S)\ M(U(s))(M(S)^{-1})^{j_{U(s)}}$$

where, for any s, $(M(S)^{-1})^{j_{U(s)}}$ is the image of $M(S)^{-1}$ under the automorphism $j_{U(s)}$ of K that is associated to the semilinear mapping U(s).

This latter formula expresses equivalence of matrices only in some special cases, which include the case of linear representations considered in the literature. When generalizing the treatment to semilinear representations, according to the preceding sections, the relevant concept is not equivalence of matrix representations, at least in its usual meaning, but rather equivalence of operator representations.

REFERENCES

[1] L.Autonne, Bull.de la Soc.Math.de France 30, 121-134 (1902).

[2] G.Frobenius, I.Schur, Berl.Berichte, 186-208 (1906).

[3] V.I.Smirnov, Linear Algebra and Group Theory, edited by R.A.Silberman (McGraw-Hill, NewYork, 1961), Ch.8, Problems,p.378. "These Problems are mostly By J.S.Lomont..." according to a footnote on page 375.

[4] E.P.Wigner, Group Theory(Academic Press, New York, 1959), Sec.24, p.285.

[5] M.Hamermesh, Group Theory, (Addison-Wesley,Reading,Mass.,1962) Ch.5, §5.

[6] G.Ya.Lyubarskii, The Application of Group Theory in Physics (Pergamon, Oxford, 1960) Sec.23.

[7] G.Emch,C.Piron, Journ.of Math.Phys. 4, 469-473 (1963).

[8] J.M.Jauch, Foundations of Quantum Mechanics (Addison-Wesley, Reading,Mass.,1968) Ch.9.

[9] N.Bourbaki, Eléments de Mathématique, Algèbre,Ch.2, Algèbre linéaire(Hermann,Paris,1962).

[10] D.Lenzi, Un teorema di decomposizione per alcune matrici quadrate a elementi in un corpo. To be published in : Atti Acc.Naz.dei Lincei, Rendiconti Cl.di Sc.Mat. Fis.Nat. (1973).

Part III

SCATTERING THEORY AND FIELD THEORY

APPROCHE ALGEBRIQUE DE LA THEORIE NON-RELATIVISTE DE LA DIFFUSION A CANAUX MULTIPLES

W.O.Amrein(*), V.Georgescu(*)(†) et Ph.A.Martin(**)(†)

(*)Département de Physique Théorique, Université de Genève. (**)Laboratoire de Physique Théorique de l'Ecole Polytechnique Fédérale de Lausanne.

§1. Introduction

Celui qui fait une expérience de diffusion a dans son champ d'observation des objets, particules ou agglomérats de particules, dont les mouvements lui apparaissent indépendants les uns des autres et qu'il qualifie "mouvements libres". Il détermine alors quelles corrélations existent entre les mouvements libres des particules avant et après le moment de leur interaction. Une théorie de la diffusion doit permettre de prédire ces corrélations à partir d'un modèle dynamique de l'interaction. A son point de départ, on postule une condition asymptotique qui exprime le fait que le mouvement des particules apparait libre en dehors de la région d'interaction mutuelle. La condition asymptotique en mécanique quantique met en jeu deux éléments : le groupe d'évolution dynamique du système et une notion de mouvement libre ou d'observations asymptotiques accessibles à l'expérimentateur. La formulation précise de cette notion et l'expression mathématique de la condition asymptotique déterminent la structure de la théorie, c'est-à-dire la forme mathématique des formules à utiliser pour comparer les observations au modèle dynamique.

La formulation communément admise de la condition asymptotique pour la diffusion quantique non-relativiste est bien connue. Soit $\mathcal{H}$ l'espace de Hilbert des états et f_t un état du système au temps t. Le mouvement f_t sera dit libre si $f_t=U_tf$ où $U_t=\exp(-iH_ot)$

(†)Travail financé partiellement par le Fonds National Suisse de la Recherche scientifique.

Enz/Mehra (eds.), Physical Reality and Mathematical Description. 255–276.

est le groupe d'évolution engendré par l'opérateur d'énergie cinétique H_o. Désignons par V_t le groupe d'évolution totale du système incluant l'interaction et par $g_t=V_t g$ un état de diffusion au temps t. La condition asymptotique requiert que g_t approche un mouvement libre lorsque $t\to\pm\infty$ dans la topologie forte de l'espace de Hilbert, propriété qu'on exprime habituellement de la façon suivante : les limites fortes des suites d'opérateurs $\underset{t\to\pm\infty}{\text{s-lim}}\ V_t{}^*U_t=\Omega_\pm$ existent sur $\mathcal{H}$.

Cette formulation en termes mathématiques précis de la condition asymptotique et de la définition des opérateurs d'onde $\Omega_\pm$ a été avancée par Jauch [1] en 1958 et Cook [2] et généralisée par la suite aux systèmes de diffusion à canaux multiples par Jauch [3] et Hack [4]. Dans [1] Jauch reconnait aussi que l'opérateur de diffusion $S=\Omega_+{}^*\Omega_-$ est unitaire si et seulement si les buts des opérateurs d'onde Ω_+ et Ω_- sont identiques, et que cette propriété doit être postulée indépendamment et en plus de la condition asymptotique. Complétée par une analyse de la relation entre l'opérateur de diffusion et la section efficace [1], la théorie accomplit donc le dessein qui lui est assigné.

Parmi les publications de cette époque sur la théorie de diffusion, le travail [1] de Jauch se remarque par la clarté conceptuelle des notions qu'il présente. L'auteur y reconnait la nécessité d'employer un langage rigoureux aussi bien pour la discussion des idées physiques que pour la cohérence de leur expression mathématique. De nombreux exemples ont montré par la suite combien la formulation de la théorie de la diffusion a gagné en précision et en clarté par l'emploi des élégants outils de l'analyse fonctionnelle. Il est aussi très remarquable que cette forme simple de la condition asymptotique ait également joué un rôle important dans le développement d'un chapitre entier de l'analyse fonctionnelle, à savoir l'étude de la similitude des opérateurs autoadjoints à spectre continu sur un espace de Hilbert (puisque les opérateurs d'onde sont aussi des opérateurs d'entrelacement entre les deux groupes $\{V_t\}$ et $\{U_t\}$). Nous en prenons à temoin les premiers travaux sur ce sujet par Friedrichs [5] et Kato [6] ainsi que l'abondante littérature qui s'y rattache (voir Kuroda [7] pour les références antérieures à 1959 et le dernier chapitre du livre de Kato [8]). Nous pensons que nous trouvons ici l'exemple d'une rencontre particulièrement fructueuse entre le physicien et le mathématicien qui atteste bien, pour le physicien théoricien, la justesse et la généralité de son point de vue.

C'est le choix d'une topologie sur l'ensemble des états du système, au moyen de laquelle s'exprime la condition asymptotique, qui définit précisément ce que l'on entend par "mouvement asymptotiquement libre". On pouvait se demander si la topologie forte de l'espace de Hilbert était bien motivée, ou s'il ne convenait pas de considérer sur l'ensemble des états la topologie définie par la convergence des valeurs moyennes de toutes les observables, qui est physiquement plus transparente. C'est la question qui est examinée dans le travail de Jauch, Misra et Gibson [9]. Ces auteurs concluent que la condition asymptotique est vérifiée dans cette dernière topologie seulement si pour tout t il existe un facteur scalaire ξ_t tel que les limites fortes s-lim $V_t{}^*U_t\xi_t$ lorsque $t\to\pm\infty$, existent sur $\mathcal{H}$. Il apparait donc clairement qu'en exprimant la notion de mouvement asymptotique libre à l'aide de la nouvelle topologie, on obtient une généralisation de la condition asymptotique usuelle, et de là un élargissement de la classe des interactions admises. Toutefois, les travaux de Dollard [10] ont révélé que ni la condition asymptotique usuelle ni cette extension subséquente ne pouvait inclure la diffusion coulombienne, si bien que ce cas important restait encore exclu d'un formalisme général.

A ce point, il était naturel de penser qu'il fallait encore affaiblir la topologie en n'exigeant la convergence asymptotique que pour une sous-algèbre engendrée par certaines observables particulières. En effet, dans une expérience de diffusion, l'expérimentateur n'attribue aux particules entrantes et sortantes les caractéristiques d'un mouvement libre que par l'observation de certaines grandeurs, comme leur impulsion par exemple; il n'en mesure certainement pas toutes les observables possibles. C'est dans cet esprit qu'on a cherché dans [11] à obtenir une nouvelle généralisation de la condition asymptotique qui puisse s'appliquer au cas des forces à longue portée. Une étude mathématique montre que ne requérir la convergence des valeurs moyennes que pour un sous-ensemble strict d'observables ne suffit pas en général à assurer l'existence d'opérateurs d'onde. Mais on établit dans [11] qu'en ajoutant quelques conditions naturelles il devient possible de construire des opérateurs d'onde. Ceux-ci résultent alors d'un isomorphisme entre les algèbres prises au temps $t=0$ et $t=\pm\infty$ des observables qui sont asymptotiquement constantes dans le processus de diffusion, d'où le terme de théorie algébrique de la diffusion (voir aussi les travaux de Combes [12], Corbett [13] et Lavine [14]).

La comparaison avec les formulations précédentes [1] et [9] se fait facilement quand on sait que la dernière forme [11] de la condition asymptotique équivaut à postuler l'existence d'une famille d'opérateurs $\hat{U}_t$ tels que les limites fortes s-lim $V_t{}^*U_t\hat{U}_t$ lorsque $t\to\pm\infty$, existent, où les $\hat{U}_t$ ne sont fonction que de l'énergie libre. La généralisation des résultats de [9] est claire : le choix du facteur de convergence ξ_t n'est plus limité aux seuls scalaires, mais il peut maintenant avoir une dépendance de l'énergie; c'est précisément ce qu'il faut pour pouvoir traiter le cas coulombien et les forces à plus longue portée encore. Nous pensons que, pour la diffusion quantique non-relativiste, les conditions que l'on impose dans [11] et dans l'étude qui suit sont minimales si on veut avoir une théorie avec opérateurs d'onde. On peut naturellement en affaiblir encore certaines et concevoir un formalisme sans opérateurs d'onde, mais dans lequel l'opérateur de diffusion serait seul défini. Pour aller plus loin, on peut imaginer une condition asymptotique qui n'impliquerait pas même l'existence d'un opérateur de diffusion, comme cela a été récemment proposé [15]. Faudrait-il encore trouver dans une telle théorie quels objets mathématiques peuvent être définis pour servir utilement au calcul des quantités mesurables.

Le présent exposé offre un prolongement des diverses études de la condition asymptotique mentionnées plus haut. Nous y donnons une version algébrique de la condition asymptotique pour une théorie de diffusion à plusieurs canaux qui est une extension naturelle du travail [11]. Bien que les postulats soient similaires avec ceux qu'on lit dans [11] avec les adaptations requises par cette situation plus complexe, plusieurs preuves que nous proposons ici sont nouvelles, et souvent plus directes. Certaines parties des démonstrations qui sont identiques à celles qu'on peut trouver dans [11] ne seront pas reproduites. Nous exposons tout d'abord le formalisme général de manière à en dégager la structure abstraite. Nous donnons ensuite des exemples et mentionnons quelques problèmes propres à une théorie de diffusion à canaux multiples et qui se posent d'une façon particulière dans le contexte de cette étude.

§2. La condition asymptotique pour un système de diffusion à canaux multiples

L'ensemble des états des N particules forme un espace de Hilbert $\mathcal{H}$ et le mouvement de ces particules est décrit par le groupe d'opérateurs unitaires $\{V_t\}, -\infty<t<+\infty$. Nous écrirons toute la théorie dans le système de coordonnées relatives au centre de masse total, celui-ci étant gardé fixé. (Pour N particules sans spin, $\mathcal{H} = L^2(R^{3N-3})$.) La notion de canal est définie de la manière usuelle : nous appelons un fragment un sous-ensemble strict des N particules susceptible de former un état lié. Un canal α est une partition des N particules en un certain nombre de fragments dont les états liés sont spécifiés. L'ensemble des états dynamiques des fragments d'un canal constitue un sous-espace fermé $\mathcal{H}_\alpha$ de $\mathcal{H}$ et nous dénotons par E_α l'opérateur de projection qui lui correspond. Sur chaque canal α, nous avons également le groupe d'évolution libre U_t^α engendré par l'opérateur H_α d'énergie cinétique des fragments (voir [16] et [3] pour de plus amples détails).

Le fait caractéristique d'une expérience de diffusion est que certaines observables des fragments (leurs impulsions par exemple) approchent des valeurs constantes lorsque $t \to \pm\infty$. A chaque canal nous attachons donc une algèbre d'opérateurs $\mathcal{A}_\alpha$. $\mathcal{A}_\alpha$ comprend les observables des fragments qui servent à spécifier les états initiaux et finaux du processus de diffusion et qui deviennent asymptotiquement stationnaires au cours de l'évolution. Nous ne préciserons pas ici $\mathcal{A}_\alpha$ davantage (voir les exemples donnés au §3). En tous les cas, $\mathcal{A}_\alpha$ a les propriétés mathématiques suivantes :

(i) $\mathcal{A}_\alpha$ est une algèbre de von Neumann agissant sur le sous-espace $\mathcal{H}_\alpha$ du canal α qui est comprise dans l'ensemble des constantes du mouvement libre : $\mathcal{A}_\alpha \subset \{U_t{}^\alpha\}'|_{E_\alpha \mathcal{H}}$.

(ii) $\mathcal{A}_\alpha$ contient au moins une sous-algèbre $\mathcal{B}_\alpha$ maximale abélienne sur $\mathcal{H}_\alpha$, de sorte que $\mathcal{A}_\alpha$ possède suffisamment d'observables pour distinguer tous les états du canal α.

Remarquons que cette dernière propriété implique que $\mathcal{A}_\alpha$ a un commutant abélien : $\mathcal{A}'_\alpha \subset \mathcal{A}_\alpha$. Il sera commode de considérer que l'algèbre $\mathcal{A}_\alpha$ agit sur l'espace de Hilbert $\mathcal{H}$ entier en posant $A_\alpha = A_\alpha E_\alpha$ pour tout A_α appartenant à $\mathcal{A}_\alpha$. C'est ce que nous entendrons dorénavant lorsque nous composerons un élément d'une algèbre $\mathcal{A}_\alpha$ avec un opérateur agissant sur $\mathcal{H}$ entier. Nous avons ainsi défini les objets de la théorie qui sont
(1) le groupe d'évolution du système $\{V_t\}$,
(2) la famille $\{E_\alpha\}$ de projecteurs sur les sous-espaces des canaux,
(3) la famille $\{\mathcal{A}_\alpha\}$ d'algèbres d'observables des canaux ayant les propriétés (i),(ii).

Il nous faut maintenant caractériser les états de diffusion du système : un état de diffusion correspondant au canal α est un état pour lequel les opérateurs $A_\alpha(t) = V_t{}^* A_\alpha V_t$ deviennent stationnaires lorsque $t \to \pm\infty$, dans la topologie forte de $\mathcal{H}$, ceci pour tous les éléments A_α de l'algèbre $\mathcal{A}_\alpha$ du canal. La condition asymptotique est essentiellement l'affirmation de l'existence d'états de diffusion en suffisance pour chaque canal. Plus précisement, nous dirons que nous avons une théorie de diffusion à canaux multiples si le système $\{\mathcal{A}_\alpha, V_t\}$ satisfait aux conditions suivantes :

I. Existence d'états de diffusion

Pour chaque canal α, il existe deux sous-espaces (fermés) $M^\pm_\alpha = P^\pm_\alpha \mathcal{H}$ de $\mathcal{H}$ $(P^\pm_\alpha = P^{\pm *}_\alpha = P^{\pm 2}_\alpha)$ pour lesquels les limites

(i) $\quad \underset{t\to\pm\infty}{\text{s-lim}} \; V_t{}^* A_\alpha V_t P^\pm_\alpha = \mu^\pm_\alpha(A_\alpha)$ existent pour tout $A_\alpha \in \mathcal{A}_\alpha$.

Les sous-espaces $M^\pm_\alpha$ sont supposés invariants sous le groupe d'évolution V_t ainsi que sous l'action des observables asymptotiques $\mu^\pm_\alpha(A_\alpha)$, c'est-à-dire

(ii) $[P_\alpha^\pm, V_t]=0$

(iii) $[P_\alpha^\pm, \mu_\alpha^\pm(A_\alpha)]=0$ pour tout $A_\alpha \in \mathcal{A}_\alpha$.

II. Abondance des états de diffusion

(i) A tout état f_α arbitrairement préparé dans le canal α à $t\to-\infty$ correspond un état de diffusion g_α^- dans M_α^-, et par symétrie sous le renversement du temps, nous imposons la même condition à $t\to+\infty$. Pour tout $f_\alpha \in \mathcal{H}_\alpha$, il existe $g_\alpha^\pm \in M_\alpha^\pm$ tels que

$$\lim_{t\to\pm\infty} (V_t g_\alpha^\pm, A_\alpha V_t g_\alpha^\pm) = (f_\alpha, A_\alpha f_\alpha), \text{ pour tout } A_\alpha \in \mathcal{A}_\alpha.$$

(ii) Les observables du canal α permettent de caractériser complètement tous les états de diffusion qui lui sont associés. Il existe une algèbre de von Neumann maximale abélienne $\mathcal{B}_\alpha$ dans $\mathcal{A}_\alpha$ dont les images $\mu_\alpha^\pm(\mathcal{B}_\alpha)$ engendrent des sous-algèbres maximales abéliennes sur $M_\alpha^\pm$.

III. Compatibilité et indépendance des canaux

Les observables $\{E_\alpha\}$ qui mesurent l'appartenance aux différents canaux doivent être asymptotiquement compatibles entre elles:

$$[\mu_\alpha^\pm(E_\alpha), \mu_\beta^\pm(E_\beta)] = 0 \text{ pour tout } \alpha,\beta.$$

De plus, l'attribution d'états de diffusion à des canaux doit être univoque, c.à.d. $M_\alpha^\pm \cap M_\beta^\pm=\{0\}$ si $\alpha\neq\beta$.

Nous voulons montrer que les conditions I et II sont suffisantes pour assurer l'existence des opérateurs d'onde de la théorie. La condition III impliquera alors l'orthogonalité des buts des opérateurs d'onde correspondant à des canaux différents (pour le même signe du temps). Remarquons en premier lieu que les applications $\mu_\alpha^\pm$ définies dans I sont des isomorphismes entre les algèbres des canaux $\mathcal{A}_\alpha$ et leurs images asymptotiques $\mu_\alpha^\pm(\mathcal{A}_\alpha)$: le fait que $\mu_\alpha^\pm$ soient linéaires, multiplicatifs et préservent l'adjoint se vérifie immédiatement à partir de I(i)-(iii). De plus $\mu_\alpha^\pm(A_\alpha)=0$ avec un élément $A_\alpha\neq0$ contredit certainement II(i), si bien que $\mu_\alpha^\pm$ doit être injectif en vertu de cette condition. Le second point, qui est considérablement plus délicat, est de s'assurer que les algèbres asymptotiques $\mu_\alpha^\pm(\mathcal{A}_\alpha)$ sont elles-mêmes des algèbres de von Neumann sur les sous-espaces $P_\alpha^\pm$. La preuve, qui met en jeu les propriétés de continuité des applications $\mu_\alpha^\pm$, en est donnée dans l'appendice (lemme 1). Une fois ces résultats acquis, il est facile de construire les opérateurs d'onde et c'est ce que nous faisons dans la proposition 1.

Proposition 1 : Si $\{\mathcal{A}_\alpha, V_t\}$ vérifie les conditions I et II, il existe pour chaque canal α deux isométries partielles $\Omega_\alpha^\pm$ telles que

(a) $\mu_\alpha^\pm(A_\alpha) = \Omega_\alpha^\pm A_\alpha \Omega_\alpha^{\pm *}$ pour tout $A_\alpha \in \mathcal{A}_\alpha$

(b) $\Omega_\alpha^{\pm *}\Omega_\alpha^\pm = E_\alpha$, $\Omega_\alpha^\pm \Omega_\alpha^{\pm *} = P_\alpha^\pm$

Les $\Omega_\alpha^\pm$ sont déterminés à un facteur unitaire près dans le centre $\mathcal{A}_\alpha'$ de $\mathcal{A}_\alpha$.

<u>Démonstration</u> : Soient e_α un vecteur cyclique pour $\mathcal{A}_\alpha$ sur $E_\alpha\mathcal{H}$ (un tel vecteur existe puisque $\mathcal{A}_\alpha$ contient une algèbre $\mathcal{B}_\alpha$ maximale abélienne sur $E_\alpha\mathcal{H}$) et g_α le vecteur dans $P_\alpha\mathcal{H}$ qui lui est associé par II(i). Alors g_α est cyclique pour $\mu_\alpha(\mathcal{A}_\alpha)$ sur $P_\alpha\mathcal{H}$. Remarquons tout d'abord qu'en vertu de II(ii) $\mu_\alpha(\mathcal{B}_\alpha)$ est maximale abélienne sur $P_\alpha\mathcal{H}$ et on a

$$\mu_\alpha(\mathcal{A}_\alpha)' \subset \mu_\alpha(\mathcal{B}_\alpha)' = \mu_\alpha(\mathcal{B}_\alpha) \subset \mu_\alpha(\mathcal{A}_\alpha) \qquad (1)$$

Selon une propriété bien connue ([17] ,p.6) il suffit de vérifier que g_α est séparateur pour le commutant $\mu_\alpha(\mathcal{A}_\alpha)'$ de $\mu_\alpha(\mathcal{A}_\alpha)$. C'est en effet le cas : si Q est un élément du centre $\mu_\alpha(\mathcal{A}_\alpha)'$ de $\mu_\alpha(\mathcal{A}_\alpha)$, Q est l'image $\mu_\alpha(R)$ d'un élément R du centre $\mathcal{A}_\alpha'$ de $\mathcal{A}_\alpha$, en vertu de (1) et du fait que μ_α est un isomorphisme des algèbres; et on a $Q^*Q=\mu_\alpha(R^*R)$. On obtient de II(i) $\|Qg_\alpha\|^2=(g_\alpha,\mu_\alpha(R^*R)g_\alpha)=\|Re_\alpha\|^2$.

Donc $Qg_\alpha=0$ entraine $Re_\alpha=0$, d'où R=0 puisque e_α est séparateur pour $\mathcal{A}_\alpha'$. Par conséquent $Q=\mu_\alpha(R)=0$, ce qui établit que g_α est cyclique pour $\mu_\alpha(\mathcal{A}_\alpha)$.

Nous pouvons définir une application linéaire Ω_α sur l'ensemble de vecteurs $\{A_\alpha e_\alpha, A_\alpha \in \mathcal{A}_\alpha\}$, dense dans $E_\alpha\mathcal{H}$, par

$$\Omega_\alpha A_\alpha e_\alpha = \mu_\alpha(A_\alpha)g_\alpha \qquad A_\alpha \in \mathcal{A}_\alpha$$

Cette application est isométrique :

$$\|\Omega_\alpha A_\alpha e_\alpha\|^2 = \|\mu_\alpha(A_\alpha)g_\alpha\|^2 = (g_\alpha,\mu_\alpha(A_\alpha^*A_\alpha)g_\alpha) = \|A_\alpha e_\alpha\|^2$$

et elle s'étend par continuité à une isométrie de $E_\alpha\mathcal{H}$ sur $P_\alpha\mathcal{H}$ puisque l'ensemble des vecteurs $\{\mu_\alpha(A_\alpha)g_\alpha\}$ est dense dans $P_\alpha\mathcal{H}$.

On vérifie (a) à partir de la définition de Ω_α : Si A_α et B_α appartiennent à $\mathcal{A}_\alpha$, on a

$$\mu_\alpha(A_\alpha)\mu_\alpha(B_\alpha)g_\alpha = \mu_\alpha(A_\alpha B_\alpha)g_\alpha = \Omega_\alpha A_\alpha B_\alpha e_\alpha$$

$$= \Omega_\alpha A_\alpha \Omega_\alpha^*\Omega_\alpha B_\alpha e_\alpha = \Omega_\alpha A_\alpha \Omega_\alpha^* \mu_\alpha(B_\alpha)g_\alpha$$

d'où $\mu_\alpha(A_\alpha) = \Omega_\alpha A_\alpha \Omega_\alpha^*$. Si Ω_α' est une autre isométrie qui jouit des mêmes propriétés, on voit de (b) que $U_\alpha=\Omega_\alpha^*\Omega_\alpha'$ est un opérateur unitaire sur $E_\alpha\mathcal{H}$ qui appartient à $\mathcal{A}_\alpha'$, et $\Omega_\alpha'=\Omega_\alpha U_\alpha$. ∎

Puisque $\mu_\alpha^\pm(E_\alpha)=P_\alpha^\pm$, nous avons

Corollaire : Si $\{\mathfrak{A}_\alpha, V_t\}$ vérifie les conditions I,II et III, alors $P_\alpha^\pm P_\beta^\pm = \delta_{\alpha\beta}P_\alpha^\pm$.

L'évolution au cours du temps d'un état de diffusion $\Omega_\alpha^- f_\alpha$ initialement préparé dans le canal α est donnée par $V_t\Omega_\alpha^- f_\alpha$. On sait que pour un système de diffusion avec forces à courte portée cet état se comporte asymptotiquement comme $U_t^\alpha f_\alpha$ lorsque $t\to-\infty$, et comme la superposition $\Sigma_\beta U_t^\beta S_{\beta\alpha} f_\alpha$ de mouvements libres dans les différents canaux lorsque $t\to+\infty$. $S_{\beta\alpha}$ est ici la matrice de diffusion définie par

$$S_{\beta\alpha} = \Omega_\beta^{+*}\Omega_\alpha^- \tag{2}$$

Nous allons montrer qu'avec nos conditions nous obtenons un comportement très similaire avec la différence que le mouvement asymptotique est décrit, dans chaque canal, à l'aide d'une famille d'opérateurs d'évolution libre modifiée T_t^α qui peut différer de U_t^α. Leurs propriétés sont données par l'énoncé de la proposition 2, dans la démonstration de laquelle les opérateurs T_t^α sont explicitement construits.

Proposition 2 : Supposons que $\{\mathfrak{A}_\alpha, V_t\}$ vérifie les conditions I et II, et soit $\Omega_\alpha^\pm$ un choix d'opérateurs d'onde correspondants. Il existe pour chaque canal α deux familles $\{T_t^{\alpha\pm}\}$ d'opérateurs linéaires et fermés sur $E_\alpha\mathcal{H}$ jouissant des propriétés suivantes :

(a) $T_t^{\alpha\pm}$ est affilié à $\mathfrak{A}_\alpha'$.

Il existe un domaine $\mathcal{D}_\alpha$ dense dans $E_\alpha\mathcal{H}$, invariant sous $\mathfrak{A}_\alpha$, sur lequel

(b) $\lim_{t\to\pm\infty} \|T_t^{\alpha\pm}h\| = \|h\|$, $h\in\mathcal{D}_\alpha$

(c) $\text{s-lim}_{t\to\pm\infty} V_t^* T_t^{\alpha\pm}h = \Omega_\alpha^\pm h$, $h\in\mathcal{D}_\alpha$.

Démonstration : Soit à nouveau un vecteur e_α cyclique pour $\mathfrak{A}_\alpha$ dans $E_\alpha\mathcal{H}$ et soit $\mathcal{D}_\alpha$ l'ensemble dense de vecteurs de la forme $\{A_\alpha e_\alpha, A_\alpha\in\mathfrak{A}_\alpha\}$. Nous considérons le projecteur C_α sur le sous-espace engendré par les vecteurs de la forme $\{A_\alpha e_\alpha, A_\alpha\in\mathfrak{A}_\alpha'\}$. Il est clair que $C_\alpha e_\alpha=e_\alpha$ et que C_α appartient à $\mathfrak{A}_\alpha''=\mathfrak{A}_\alpha$. De plus, $A_\alpha e_\alpha=0$ pour un $A_\alpha\in\mathfrak{A}_\alpha$ entraine $A_\alpha C_\alpha=0$. En effet, si $A_\alpha e_\alpha=0$, on a aussi $A_\alpha B_\alpha e_\alpha= B_\alpha A_\alpha e_\alpha=0$ pour tous les vecteurs de la forme $B_\alpha e_\alpha, B_\alpha\in\mathfrak{A}_\alpha'$, qui engendrent le sous-espace $C_\alpha\mathcal{H}$.

Nous définissons la famille $\{T_t^\alpha\}$ avec domaine commun $\mathcal{D}_\alpha$ comme suit :

$$T_t^\alpha h = A_\alpha C_\alpha V_t \Omega_\alpha e_\alpha \quad \text{avec } h = A_\alpha e_\alpha \tag{3}$$

T_t^α est un opérateur linéaire bien défini, puisque $h=A_\alpha e_\alpha=0$ entraine $A_\alpha C_\alpha=0$ et par suite $T_t^\alpha h=0$, et on a $\|T_t^\alpha h\| \leq \|A_\alpha\| \, \|e_\alpha\|$. Si $B_\alpha \in \mathcal{a}_\alpha$ et $h \in \mathcal{D}_\alpha$

$$T_t^\alpha B_\alpha h = T_t^\alpha B_\alpha A_\alpha e_\alpha = B_\alpha A_\alpha C_\alpha V_t \Omega_\alpha e_\alpha = B_\alpha T_t^\alpha h \tag{4}$$

ce qui montre que T_t^α commute avec $\mathcal{a}_\alpha$ sur $\mathcal{D}_\alpha$. Pour établir le premier point de la proposition, il reste à montrer que chaque T_t^α possède une extension fermée sur $E_\alpha\mathcal{H}$ qui commute avec $\mathcal{a}_\alpha$ sur son domaine. Le lecteur trouvera cette partie plus technique de la démonstration dans l'appendice (lemme 2). Quant aux points (b) et (c), ils suivent immédiatement de la définition de T_t^α. Si $h=A_\alpha e_\alpha \in \mathcal{D}_\alpha$, nous avons

$$\begin{aligned}\|(\Omega_\alpha - V_t{}^*T_t^\alpha)h\| &= \|(\Omega_\alpha A_\alpha C_\alpha - V_t{}^*A_\alpha C_\alpha V_t \Omega_\alpha)e_\alpha\| \\ &= \|(\Omega_\alpha A_\alpha C_\alpha \Omega_\alpha{}^* - V_t{}^*A_\alpha C_\alpha V_t)\Omega_\alpha e_\alpha\|\end{aligned}$$

qui converge vers zéro lorsque $t\to\infty$ puisque $A_\alpha C_\alpha \in \mathcal{a}_\alpha$. (b) est donc prouvé. Pour (c) nous considérons

$$\begin{aligned}|(T_t^\alpha h,T_t^\alpha h)-(h,h)| &= |((T_t^\alpha - V_t\Omega_\alpha)h, V_t\Omega_\alpha h) + (T_t^\alpha h,(T_t^\alpha - V_t\Omega_\alpha)h)| \\ &\leq \|(T_t^\alpha - V_t\Omega_\alpha)h\| \left[\|h\| + \|T_t^\alpha h\|\right]\end{aligned}$$

qui converge également vers zéro lorsque $t\to\infty$ puisque $\|T_t^\alpha h\|$ est une fonction bornée de t pour $h \in \mathcal{D}_\alpha$. ∎

Remarquons qu'une réciproque de la proposition 2 est aussi vraie dans le sens suivant. Si pour chaque canal α nous avons deux familles $T_t^{\alpha\pm}$ vérifiant (a),(b) et (c) et telles que les buts des isométries partielles $\Omega_\alpha^\pm$ définies par (c) soient invariants sous V_t, alors I et II sont vrais avec $P_\alpha^\pm = \Omega_\alpha^\pm \Omega_\alpha^{\pm *}$.

Examinons quelques-unes des propriétés des évolutions modifiées $T_t^{\alpha\pm}$. T_t^α ne constitue pas en général un groupe, mais sa dépendance du temps ne peut cependant pas être arbitraire. Pour le voir, nous définissons pour chaque canal deux groupes unitaires $W_t^{\alpha\pm}$ sur $E_\alpha\mathcal{H}$ par

$$W_t^{\alpha\pm} = \Omega_\alpha^{\pm *} V_t \Omega_\alpha^\pm \tag{5}$$

Comme l'algèbre $\mu_\alpha^\pm(\mathcal{a}_\alpha)$ est contenue dans la restriction du commutant $\{V_t\}'_{P_\alpha^\pm}$ au sous-espace $P_\alpha^\pm\mathcal{H}$, nous avons $\{V_t\}''_{P_\alpha^\pm} \subset \mu_\alpha^\pm(\mathcal{a}_\alpha)' = \mu_\alpha^\pm(\mathcal{a}_\alpha')$.

Ceci montre que $W_t^{\alpha\pm}$ appartient au centre $\mathcal{a}_\alpha'$ de $\mathcal{a}_\alpha$. Les opérateurs d'onde entrelacent donc l'hamiltonien total H avec les générateurs des groupes $W_t^{\alpha\pm}$ dont l'interprétation est de donner l'énergie asymptotique des fragments dans chaque canal. Nous supposerons dans la suite que l'énergie asymptotique des fragments est simplement leur énergie cinétique H_α, c'est-à-dire que

$$W_t^{\alpha+} = W_t^{\alpha-} = U_t^\alpha \tag{6}$$

Dans le cas contraire nous aurions une théorie plus générale avec des renormalisations d'énergie possibles dans les différents canaux que nous ne voulons pas considérer ici.

Il existe une relation simple entre l'évolution asymptotique modifiée T_t^α et le groupe U_t^α qui se déduit facilement de (c) ou de la définition (3). Elle s'écrit

$$\lim_{t\to\pm\infty} \|(T_{t+\tau}^{\alpha\pm} - U_\tau^\alpha T_t^{\alpha\pm})h\| = 0, \quad h \in \mathcal{D}_\alpha \tag{7}$$

Cette équation signifie que l'évolution asymptotique modifiée est asymptotiquement comparable à l'évolution libre U_τ^α sur des intervalles de temps finis. Soit $\hat{U}_t^{\alpha\pm}=U_t^{\alpha}{}^*T_t^{\alpha\pm}$ la modification qu'il faut apporter à U_t^α pour obtenir la convergence forte de $V_t{}^*U_t^\alpha\hat{U}_t^{\alpha\pm}E_\alpha$. La propriété (7), qui s'écrit encore

$$\lim_{t\to\pm\infty} \|(\hat{U}_{t+\tau}^{\alpha\pm} - \hat{U}_t^{\alpha\pm})h\| = 0$$

indique que ce facteur ne peut osciller trop rapidement lorsque $t\to\pm\infty$. Plus précisément, il existe un sous-ensemble dense $\mathcal{D}'_\alpha$ de $\mathcal{D}_\alpha$ tel que pour $h \in \mathcal{D}'_\alpha$ on ait

$$\operatorname*{s-lim}_{t\to\pm\infty} \frac{d}{dt} \hat{U}_t^{\alpha\pm}h = 0 \tag{8}$$

Pour le voir, supposons que le vecteur cyclique e_α utilisé dans la définition (3) appartienne à $D(H_\alpha)$, et soit F_k^α le projecteur spectral de H_α (considéré comme agissant dans le sous-espace $E_\alpha\mathcal{H}$) correspondant à l'intervalle $(-k,+k]$, $k \in \mathbb{R}^+$. Soit

$$\mathcal{D}'_\alpha = \{h \in \mathcal{D}_\alpha | h=F_k^\alpha A_\alpha e_\alpha \quad \text{avec } A_\alpha \in \mathcal{a}_\alpha \text{ et } k < \infty\}$$

En utilisant (3), (5) et (6) nous avons

$$-i \frac{d}{dt} \hat{U}_t^\alpha h = U_t^\alpha{}^*H_\alpha F_k^\alpha A_\alpha C_\alpha V_t\Omega_\alpha e_\alpha - U_t^\alpha{}^*F_k^\alpha A_\alpha C_\alpha V_t\Omega_\alpha H_\alpha e_\alpha$$

Il en découle que

$$\left\|\frac{d}{dt} \hat{U}_t^\alpha h\right\| \leq \|A_\alpha C_\alpha\| \, \|V_t{}^*H_\alpha F_k^\alpha V_t\Omega_\alpha e_\alpha - V_t{}^*F_k^\alpha V_t\Omega_\alpha H_\alpha e_\alpha\|$$

et lorsque $t\to\infty$ le membre de droite converge vers

$$\|A_\alpha C_\alpha\| \|\Omega_\alpha H_\alpha F_k^\alpha\Omega_\alpha{}^*\Omega_\alpha e_\alpha - \Omega_\alpha F_k^\alpha\Omega_\alpha{}^*\Omega_\alpha H_\alpha e_\alpha\| = 0$$

Dans toutes les applications connues il est possible de trouver des opérateurs T_t^α unitaires satisfaisant les conditions de la proposition 2 (ce choix peut différer de celui de la définition (3)).

Ceci ne découle pas de la condition asymptotique telle que nous la considérons ici. Les conditions supplémentaires nécessaires et suffisantes pour garantir l'existence des T_t^α unitaires sont données dans un travail de Mourre [18].

Pour déduire l'orthogonalité des buts des différents opérateurs d'onde dans la théorie de diffusion à canaux multiples usuelle, Jauch [3] a utilisé la condition suivante d'indépendance des évolutions libres U_t^α :

$$U_t^\alpha f = U_t^\beta f \quad \text{pour tout } t \text{ et } \alpha\neq\beta \Longrightarrow f = 0$$

Dans notre théorie également cette condition, combinée avec la propriété (8) des facteurs de distorsion, permet d'établir l'orthogonalité des canaux III. Pour le montrer nous avons néanmoins besoin de quelques propriétés supplémentaires de régularité des évolutions modifiées qui ne découlent pas de la condition asymptotique algébrique, à savoir la possibilité de choisir des modifications $\hat{U}_t^\alpha$ unitaires pour lesquelles (8) a lieu sur des domaines convenables. Nous nous basons sur le lemme suivant dont la démonstration se trouve dans l'appendice :

<u>Lemme 3</u> : Soit K un opérateur autoadjoint et $\{G(t)\}$, $t \in R^+$, une famille d'opérateurs uniformément bornée dans un espace de Hilbert $\mathcal{H}$. Supposons que Kf=0 n'ait que la solution f=0 et qu'il existe un sous-ensemble $\mathcal{M}$ dense dans $\mathcal{H}$ tel que pour tout $f \in \mathcal{M}$ la fonction $t\mapsto G(t)f$ est fortement différentiable et s-lim $\frac{d}{dt}G(t)f=0$ lorsque $t\to+\infty$. Alors pour tout $g,h \in \mathcal{H}$:

$$\lim_{T\to\infty} \frac{1}{T}\int_0^T dt(G(t)g, \exp(-iKt)h) = 0 \tag{9}$$

Supposons maintenant les $\hat{U}_t^\alpha$ unitaires, et soit $f_\alpha \in E_\alpha\mathcal{H}$, $f_\beta \in E_\beta\mathcal{H}$. Alors

$$(\Omega_\alpha^+ f_\alpha, \Omega_\beta^+ f_\beta) = \lim_{t\to\infty} (V_t{}^* T_t{}^{\alpha+} f_\alpha, V_t{}^* T_t^{\beta+} f_\beta)$$

$$= \lim_{T\to\infty} \frac{1}{T}\int_0^T dt\ (T_t^{\alpha+} f_\alpha, T_t^{\beta+} f_\beta) = \lim_{T\to\infty} \frac{1}{T}\int_0^T dt\ (U_t^\alpha \hat{U}_t^{\alpha+} f_\alpha, U_t^\beta \hat{U}_t^{\beta+} f_\beta) \tag{10}$$

Les opérateurs $U_t^\alpha, \hat{U}_t^\alpha$ ne sont définis que sur le sous-espace $E_\alpha\mathcal{H}$. Dans des applications particulières ce sont des fonctions des opérateurs d'impulsion relative des fragments du canal α. On peut alors étendre leur domaine de définition à $\mathcal{H}$ tout entier en les définissant partout par ces mêmes fonctions des opérateurs d'impulsion. Dans ces conditions nous pouvons considérer tous les opérateurs apparaissant dans le dernier membre de (10) comme étant définis sur $\mathcal{H}$, et tous ces opérateurs commuteront entre eux (ils sont tous des fonctions des opérateurs d'impulsion). Il suit alors que

$$(\Omega_\alpha^+ f_\alpha, \Omega_\beta^+ f_\beta) = \lim_{T\to\infty} \frac{1}{T} \int_0^T dt\ (\mho_t^{\beta+} * \mho_t^{\alpha+} f_\alpha, \exp\left[-it(H_\beta - H_\alpha)\right] f_\beta) \quad (11)$$

Il convient maintenant de distinguer deux cas (voir [16]):

(α) Si les fragments du canal α et du canal β sont identiques et $\alpha \neq \beta$, $E_\alpha \mathcal{H}$ est orthogonal à $E_\beta \mathcal{H}$ et l'intégrand dans l'avant-dernier membre de (10) est égal à zéro.

(β) Si les fragments du canal α et du canal β sont différents, $H_\beta - H_\alpha$ est une forme quadratique réelle des opérateurs d'impulsion relative des N particules qui n'est pas identiquement nulle (nous considérons ici le cas non-relativiste). L'opérateur $K = H_\beta - H_\alpha$ apparaissant dans (11) a donc un spectre absolument continu et la condition d'indépendance de Jauch est vérifiée, de même que celle du lemme 3. Celui-ci implique alors l'orthogonalité de M_α^+ et M_β^+ pourvu que la dérivée de $G(t) \equiv \mho_t^{\beta+} * \mho_t^{\alpha+}$ converge fortement vers zéro sur un ensemble dense $\mathcal{M}_{\alpha\beta}$ lorsque $t \to \infty$.

Pour des interactions coulombiennes les fonctions $\mho_t^\alpha(\vec{p}_1, \ldots, \vec{p}_{N-1})$, $\vec{p}_i \in R^3$, ont été données explicitement par Dollard [10] ($\vec{p}_1, \ldots, \vec{p}_{N-1}$ sont des impulsions relatives linéairement indépendantes entre les N particules). On voit que dans ce cas

$$\frac{d}{dt} \mho_t^\alpha(\vec{p}_1, \ldots, \vec{p}_{N-1}) = \frac{1}{|t|} \mho_t^\alpha(\vec{p}_1, \ldots, \vec{p}_{N-1}) S^\alpha(\vec{p}_1, \ldots, \vec{p}_{N-1})$$

où S^α est continue dans $R^{3(N-1)}$ à l'exception d'un ensemble fini de hypersurfaces σ_i^α. Il suffit de prendre pour $\mathcal{M}_{\alpha\beta}$ les fonctions à support compact appartenant à $L^2(R^{3N-3})$ qui s'annulent dans un voisinage de $(\bigcup_i \sigma_i^\alpha) \cup (\bigcup_j \sigma_j^\beta)$.

On peut maintenant définir l'opérateur de diffusion $S_{\beta\alpha}$ entre le canal α et le canal β. Si l'on se donne un état initial f_α dans le canal α ($f_\alpha \in E_\alpha \mathcal{H}$), on peut lui associer un état de diffusion $\Omega_\alpha^- f_\alpha$, le choix de Ω_α^- dans la classe d'équivalence (voir la proposition 1) n'influant guère sur les valeurs moyennes asymptotiques des observables de $\mathcal{A}_\alpha$ lorsque $t \to -\infty$. Si $t \to +\infty$, l'évolution de $\Omega_\alpha^- f_\alpha$ se rapproche de

$$V_t \Omega_\alpha^- f_\alpha \sim \sum_\beta T_t^{\beta+} \Omega_\beta^+ * \Omega_\alpha^- f_\alpha \quad (12)$$

Ceci suggère de définir $S_{\beta\alpha}$ par la formule (2). $S_{\beta\alpha}$ n'est pas uniquement déterminé (voir la proposition 1), mais tout choix possible de $S_{\beta\alpha}$ permet de prédire les valeurs moyennes des observables de $\mathcal{A}_\beta$ après la diffusion si on suppose connues celles des observables de $\mathcal{A}_\alpha$ avant la diffusion. Remarquons encore qu'il découle de (5) et (6) que

$$H_\beta S_{\beta\alpha} = S_{\beta\alpha} H_\alpha \tag{13}$$

La décomposition dans (12) du vecteur $\Omega_\alpha^- f_\alpha$ dans une somme de vecteurs appartenant chacun à un M_β^+ est unique, les M_β^+ étant orthogonaux deux à deux. Pourtant l'affirmation (12) n'est juste que si l'on suppose encore que

$$M_\alpha^- \subseteq \bigoplus_\beta M_\beta^+ \tag{14}$$

Normalement on s'attend à avoir une propriété plus forte que (14), à savoir la complétude asymptotique de la théorie, c.à.d.

$$\bigoplus_\alpha M_\alpha^- = \bigoplus_\alpha M_\alpha^+ = \mathcal{H}_p(H)^\perp \tag{15}$$

$\mathcal{H}_p(H)^\perp$ étant le complément orthogonal de l'ensemble des vecteurs propres de l'hamiltonien total H.

L'argument donné pour démontrer l'orthogonalité des canaux peut en principe aussi être appliqué pour démontrer l'unicité des deux familles $\{M_\alpha^\pm\}$ satisfaisant I,II,III,(6) et (15). En effet, si par exemple pour $t\to+\infty$ il existe deux familles $\{M_\alpha^{+(1)}\}$ et $\{M_\alpha^{+(2)}\}$ vérifiant toutes ces conditions, on aurait comme dans (10) (nous omettons l'indice +) :

$$(\Omega_\alpha^{(1)} f_\alpha, \Omega_\beta^{(2)} f_\beta) = \lim_{T\to\infty} \frac{1}{T} \int_0^T dt (U_t^\alpha \hat{U}_t^{\alpha(1)} f_\alpha, U_t^\beta \hat{U}_t^{\beta(2)} f_\beta)$$

En supposant que les domaines des dérivées des $\hat{U}_t^{\alpha(i)}$ soient tels que le lemme 3 s'applique, on en déduit que $M_\alpha^{(1)} \perp M_\beta^{(2)}$ si $\alpha\neq\beta$. Ceci implique avec (15) que $M_\alpha^{(1)} = M_\alpha^{(2)}$ pour tout α. Il serait satisfaisant de pouvoir prouver l'unicité des sous-espaces $\{M_\alpha^\pm\}$ des états de diffusion, dont l'existence est postulée par la condition asymptotique, indépendamment de toute hypothèse supplémentaire.

§3. Exemples et remarques

L'ensemble des observables des fragments qui deviennent asymptotiquement constantes au cours du processus de diffusion dépend en général de la nature des interactions. Plus cet ensemble sera petit, c'est-à-dire moins $\mathfrak{A}_\alpha$ contiendra d'éléments, plus son centre $\mathfrak{A}_\alpha'$ sera grand, puisque $\mathfrak{A}_\alpha$ inclut son commutant $\mathfrak{A}_\alpha'$. Comme l'opérateur d'onde Ω_α est déterminé à une fonction unitaire près dans $\mathfrak{A}_\alpha'$, la classe des opérateurs d'onde équivalents, et de là la classe des opérateurs de diffusion $S_{\beta\alpha}$, en sera d'autant plus grande. Pour que la théorie ait un pouvoir de prédiction, il faut naturellement que les quantités qui sont finalement comparées aux données expérimentales, soient libres de toute ambiguité, c'est-à-dire indépendantes du choix de l'opérateur de diffusion dans sa classe d'équivalence. Il est donc important de choisir judicieusement les algèbres $\mathfrak{A}_\alpha$, d'une part suffisamment restreintes pour que les limites postulées par la condition asymptotique existent, d'autre part suffisamment grandes pour que l'indétermination des opérateurs d'onde soit réduite au minimum.

Pour pouvoir donner une discussion complète de cette question, il serait d'abord nécessaire de faire l'étude des sections efficaces dans le cadre algébrique de la théorie de la diffusion. Nous pouvons admettre que la formule de diffusion dans les cônes proposée par Dollard [19] est une première étape raisonnable sur cette voie. La formule de Dollard affirme que la probabilité $P(f_\alpha;C_1,\ldots,C_N)$ que les N particules 1,2,...,N émergent dans les cônes $C_1,\ldots,C_N$ (dans R^3) lorsque $t\to+\infty$, et pour un processus de diffusion initié dans un état f_α du canal α, est donné par

$$\sum_\beta \int_{I_1^\beta \times\ldots\times I_{n(\beta)}^\beta} d^3p_1^\beta\ldots d^3p_{n(\beta)}^\beta \left|\left[(I\otimes S_{\beta\alpha})f_\alpha\right](\vec{p}_1^\beta,\ldots,\vec{p}_{n(\beta)}^\beta)\right|^2 \qquad (16)$$

$1,2,\ldots,n(\beta)$ indexe les éléments de la partition des N particules en $n(\beta)$ fragments qui forment le canal β et dont les impulsions totales sont $\vec{p}_1^\beta,\ldots,\vec{p}_{n(\beta)}^\beta$. $I_1^\beta,\ldots,I_{n(\beta)}^\beta$ désignent les intersections dans R^3 de ceux des cônes $C_1,\ldots,C_N$ qui correspondent aux particules composant les différents fragments du canal β. Le produit tensoriel dans (16) signifie que nous avons rajouté le mouvement du centre de masse afin de pouvoir appliquer directement les résultats du travail de Dollard [19]. On peut considérer que cette formule, qui vaut pour les forces à courte portée aussi bien que pour une classe d'interactions à longue portée, donne (avec (12))l'interprétation des opérateurs de diffusion $S_{\beta\alpha}$.

Nous décrivons toujours les N particules dans un système de coordonnées lié au centre de masse, et $\mathcal{H}$ sera l'espace de Hilbert correspondant. Le choix d'un canal α détermine le sous-espace $E_\alpha\mathcal{H}$ qui peut être décrit de la manière suivante : on note $\vec{x}_1^\alpha,\ldots,\vec{x}_{n(\alpha)-1}^\alpha$ les coordonnées des centres de masse des fragments $1,\ldots,n(\alpha)-1$ relativement au centre de masse du dernier fragment $n(\alpha)$, $\vec{Q}_1^\alpha,\ldots,\vec{Q}_{n(\alpha)-1}^\alpha$ les opérateurs de position correspondants et $\vec{P}_1^\alpha,\ldots,\vec{P}_{n(\alpha)-1}^\alpha$ leurs opérateurs conjugués (c.à.d. les opérateurs d'impulsion relative entre les fragments i et $n(\alpha)$,$i=1,\ldots,n(\alpha)-1$). Si tous les fragments sont à spin zéro, $E_\alpha\mathcal{H}$ est isomorphe à $L^2(R^{3n(\alpha)-3})$. Cet isomorphisme est uniquement déterminé par la condition que $\vec{P}_1^\alpha,\ldots,\vec{P}_{n(\alpha)-1}^\alpha$ deviennent les opérateurs de multiplication par les variables indépendantes dans $L^2(R^{3n(\alpha)-3})$. Nous désignons par m_i^α la masse du fragment i et par $\vec{V}_{ij}^\alpha$ l'opérateur de vitesse relative entre les fragments i et j du canal α. En particulier $\vec{V}_{in(\alpha)}^\alpha=(\mu_i^\alpha)^{-1}\vec{P}_i^\alpha$ avec $(\mu_i^\alpha)^{-1} = (m_i^\alpha)^{-1} + (m_{n(\alpha)}^\alpha)^{-1}$.

Nous envisageons dans la suite trois choix possibles des algèbres $\mathfrak{A}_\alpha$.

(a) Le premier, qui est minimal, consiste à prendre pour $\mathcal{A}_\alpha$ l'algèbre maximale abélienne sur $E_\alpha\mathcal{H}$ engendrée par $\vec{P}_1^\alpha,\ldots,\vec{P}_{n(\alpha)-1}^\alpha$.

(b) Pour le problème à deux corps on avait pris dans [11] l'algèbre donnée par le commutant de l'opérateur de vitesse relative $|\vec{V}_{12}|$ des deux particules, qui coincide dans ce cas avec l'ensemble de toutes les constantes du mouvement libre. La généralisation de ce choix au problème à N corps est de définir $\mathcal{A}_\alpha$ par le commutant sur $E_\alpha\mathcal{H}$ de toutes les vitesses relatives $|\vec{V}_{ij}^\alpha|$ des fragments du canal α les uns par rapport aux autres. Dans la représentation que nous avons choisie pour $E_\alpha\mathcal{H}$, on a donc

$$\mathcal{A}_\alpha = \{|\frac{\vec{P}_i^\alpha}{m_i^\alpha} - \frac{\vec{P}_j^\alpha}{m_j^\alpha}|,\ |P_i^\alpha|\}'$$

(c) Finalement nous considérons le cas où $\mathcal{A}_\alpha$ comprend toutes les constantes du mouvement libre dans le canal α, c'est-à-dire $\mathcal{A}_\alpha=\{U_t^\alpha\}'|_{E_\alpha\mathcal{H}}$, et son centre $\mathcal{A}'_\alpha$ est l'ensemble des fonctions de H_α. Remarquons que dans la représentation choisie, on a ([21],p.192) :

$$H_\alpha = \sum_{i=1}^{n(\alpha)-1} \frac{\vec{P}_i^{\alpha 2}}{2\mu_i^\alpha} + \sum_{i\neq j<n(\alpha)} \frac{\vec{P}_i^\alpha \cdot \vec{P}_j^\alpha}{2m_{n(\alpha)}^\alpha} + \varepsilon_\alpha$$

ce qui peut être écrit comme

$$H_\alpha = \sum_{i=1}^{n(\alpha)-1} \frac{\mu_i^\alpha}{2}|\vec{V}_i^\alpha|^2 + \sum_{i\neq j<n(\alpha)} \frac{\mu_i^\alpha \mu_j^\alpha}{4m_{n(\alpha)}^\alpha}\left[|\vec{V}_i^\alpha|^2+|\vec{V}_j^\alpha|^2-|\vec{V}_{ij}^\alpha|^2\right]+\varepsilon_\alpha$$

(ε_α est la somme des énergies des états liés des $n(\alpha)$ fragments). Il s'en suit que l'algèbre définie sous (b) est bien contenue dans le commutant de H_α, condition que nous avons requise au début du §2.

Le premier exemple est celui d'un système de particules dont le potentiel des forces décroit à l'infini plus rapidement que le potentiel de Coulomb. Dans ce cas les limites fortes s-lim $V_t^* U_t^\alpha = \Omega_\alpha^\pm$ $(t\to\pm\infty)$ existent sur tout $E_\alpha\mathcal{H}$. Les algèbres des canaux peuvent être définies comme au point (c), et les parties I et II de la condition asymptotique se vérifient immédiatement avec $P_\alpha^\pm=\Omega_\alpha^\pm\Omega_\alpha^{\pm *}$. En vertu de (13), $S_{\alpha\beta}$ est un opérateur d'entrelacement des centres $\mathcal{A}'_\alpha$ et $\mathcal{A}'_\beta$ puisque ceux-ci se réduisent aux seules fonctions de H_α et H_β respectivement. Si $\tilde{\Omega}_\beta^+=\Omega_\beta^+ U_\beta(H_\beta)$ et $\tilde{\Omega}_\alpha^-=\Omega_\alpha^- U_\alpha(H_\alpha)$ sont deux paires d'opérateurs d'onde équivalents, nous avons donc

$$\tilde{S}_{\beta\alpha} = \tilde{\Omega}_\beta^{+*}\tilde{\Omega}_\alpha^- = U_\beta^*(H_\beta)S_{\beta\alpha}U_\alpha(H_\alpha) = U_\beta^*(H_\beta)U_\alpha(H_\beta)S_{\beta\alpha}$$

si bien que deux opérateurs de diffusion équivalents ne diffèrent que d'une fonction multiplicative à gauche. Il est clair que cette indétermination est sans effet dans (16) puisqu'elle équivaut à modifier la fonction $[(I\otimes S_{\beta\alpha})f_\alpha](\vec{p}_1^\beta,\ldots,\vec{p}_{n(\beta)}^\beta)$ par une fonction de phase. Les probabilités données par (16) sont donc uniques.

La seconde illustration est la diffusion coulombienne. Dollard a donné explicitement dans ce cas une famille d'évolutions modifiées T_t^α pour lesquelles les limites fortes $V_t^*T_t^\alpha$ existent [10]. L'examen de ces formules montre que les T_t^α sont des fonctions des vitesses relatives $|\vec{V}_{ij}^\alpha|$ des fragments du canal α. C'est donc le choix (b) des algèbres $\mathfrak{a}_\alpha$ qui convient. Avec celui-ci, les points (a) (b) et (c) de la proposition 2 sont vérifiés et de là la condition asymptotique. La question de l'unicité des probabilités (16) est ici plus délicate. Comme les centres des algèbres sont constitués des fonctions des vitesses relatives, deux opérateurs de diffusion équivalents sont dans la relation

$$\tilde{S}_{\beta\alpha} = U_\beta^*(|\vec{V}_{ij}^\beta|)S_{\beta\alpha}U_\alpha(|\vec{V}_{ij}^\alpha|),U_\beta^*(|\vec{V}_{ij}^\beta|)\in\mathfrak{a}_\beta',U_\alpha(|\vec{V}_{ij}^\alpha|)\in\mathfrak{a}_\alpha'.$$

Nous savons que $S_{\beta\alpha}$ entrelace les fonctions de H_β et de H_α, mais nous ne savons pas en général si $S_{\beta\alpha}$ entrelace aussi $\mathfrak{a}_\beta'$ et $\mathfrak{a}_\alpha'$. Nous ne pouvons donc pas conclure à première vue que les fonctions $[(I\otimes S_{\beta\alpha})f_\alpha](\vec{p}_1^\beta,\ldots,\vec{p}_{n(\beta)}^\beta)$ et $[(I\otimes \tilde{S}_{\beta\alpha})f_\alpha](\vec{p}_1^\beta,\ldots,\vec{p}_{n(\beta)}^\beta)$ ne diffèrent que d'un facteur de phase et que les probabilités (16) sont déterminées uniquement. Il y a cependant un cas d'importance pratique où l'ambiguité est résolue, c'est celui où le canal initial α ne comprend que deux fragments. La raison en est que l'algèbre $\mathfrak{a}_\alpha=\{|\vec{V}_{12}^\alpha|\}'|_{E_\alpha\mathcal{H}}$ coincide avec l'ensemble des constantes du mouvement $\{H_\alpha\}'|_{E_\alpha\mathcal{H}}$ puisque $H_\alpha=\frac{1}{2}\mu_{12}^\alpha|\vec{V}_{12}^\alpha|^2+\varepsilon_\alpha$. Par conséquent l'argument donné plus haut pour les forces à courte portée s'applique encore à ce cas particulier. Si le canal initial a plus de deux fragments, la question reste irrésolue. Notons que l'ambiguité pourrait être levée pour d'autres raisons. Si par exemple nous pouvions prouver l'analogue du théorème 2 de [20] pour le problème à N corps avec forces à longue portée, comme Dollard [19] a essayé de le faire, nous aurions la certitude que les probabilités $P(f_\alpha;C_1,\ldots,C_N)$ sont uniques puisque celles-ci pourraient être calculées à partir d'une formule qui ne met en jeu aucune fonction indéterminée. La théorie pourrait aussi se faire avec le choix (a) des algèbres $\mathfrak{a}_\alpha$, c'est-à-dire les fonctions des impulsions.Mais l'indétermination de l'opérateur de diffusion serait encore accrue, et les problèmes que nous venons de mentionner se poseraient avec d'autant plus d'acuité.

<u>Appendice</u>

<u>Lemme 1</u> : Soit $\mathfrak{a}$ une algèbre de von Neumann, H un opérateur

autoadjoint, $V_t=\exp(-iHt)$ et P un projecteur tel que $[P,V_t]=0$ dans un espace de Hilbert séparable $\mathcal{H}$. Supposons que les limites faibles

$$\mu(A) = \underset{t\to\infty}{\text{w-lim}}\ PV_t{}^*AV_tP$$

existent pour tout $A\in\mathcal{A}$. Alors l'application $\mu : \mathcal{A}\to\mathcal{B}(P\mathcal{H})$ est ultrafaiblement continue.

Remarque : La démonstration de la continuité ultrafaible de $\mu_\pm$ donnée dans la proposition 1 dans [11] est incomplète. En effet l'applicabilité du théorème 1 du travail de Feldman et Fell ne se justifie pas par la continuité absolue du spectre de H_o mais plutôt par la propriété que le spectre de H_o est à multiplicité infinie. Cette démonstration ne s'appliquerait donc pas si $\mathcal{A}$ était le commutant d'un opérateur à multiplicité spectrale finie (par exemple si $\mathcal{A}$ était maximale abélienne). Une autre démonstration fût donnée par Zachary [22] qui ne met en jeu que la positivité de l'hamiltonien libre H_o.

Nous indiquons dans la suite une preuve très générale de ce lemme qui utilise la théorie de la dualité entre espaces vectoriels topologiques. Plus précisément nous utiliserons les faits suivants (voir [17],Ch.I.3.3 ou [23],Chs.1.1,1.8 et 1.13) :

(i) toute algèbre de von Neumann est l'espace dual d'un espace de Banach $\mathcal{A}_*$, (ii) la topologie faible $\sigma(\mathcal{A},\mathcal{A}_*)$ induite sur $\mathcal{A}$ par cette dualité coincide avec la topologie ultrafaible sur les opérateurs, (iii) le prédual $\mathcal{B}(\mathcal{H})_*$ de l'ensemble $\mathcal{B}(\mathcal{H})$ des opérateurs bornés peut être identifié avec l'ensemble des opérateurs nucléaires dans $\mathcal{H}$ muni de la norme trace $\|\cdot\|_1$ ([23],Thm.1.15.3).

Démonstration : Soit $\rho\in\mathcal{B}(P\mathcal{H})_*$, c.à.d.

$$\rho = \sum_{i=1}^{\infty} \lambda_i |f_i\rangle\langle g_i|$$

avec $\{f_i\}$, $\{g_i\}$ deux familles orthonormées d'éléments de $P\mathcal{H}$, $\lambda_i\geq 0$ et $\Sigma_i\lambda_i<\infty$ ([24],Thms.VI.17 et VI.21). Soit

$$\rho_t = V_t\rho V_t{}^* = \sum_{i=1}^{\infty} \lambda_i |V_tf_i\rangle\langle V_tg_i|$$

Pour $A\in\mathcal{A}$, calculons $\alpha_{\rho_t}(A) \equiv \mathrm{Tr}(\rho_tA)$ dans une base obtenue en complétant $\{V_tf_i\}$. Il vient

$$|\mathrm{Tr}(\rho_tA)| = |\sum_i \lambda_i(V_tg_i,AV_tf_i)| \leq \sum_i \lambda_i\|A\| < \infty . \tag{17}$$

Par hypothèse, chaque terme de la somme dans (17) converge, et le théorème de Lebesgue sur la convergence dominée implique que

$$\lim_{t\to\infty} \alpha_{\rho_t}(A) = \sum_i \lambda_i(g_i,\mu(A)f_i) \tag{18}$$

α_{ρ_t} est une forme linéaire ultrafaiblement continue sur $\mathfrak{A}$ et appartient donc à $\mathfrak{A}_*$ ([17],ch.I.3.3,Thm .1(ii),(iii)). (18) signifie que $\{\alpha_{\rho_t}\}$ est une suite de Cauchy dans la topologie faible $\sigma(\mathfrak{A}_*,\mathfrak{A})$. Comme $\mathfrak{A}_*$ est $\sigma(\mathfrak{A}_*,\mathfrak{A})$-séquentiellement complet ([23],Ch.1.8,Rem.1), il suit que $\alpha_\rho \equiv \lim_{t\to\infty} \alpha_{\rho_t}$ appartient à $\mathfrak{A}_*$.

Nous avons ainsi obtenu une application linéaire $\mu_* : \mathcal{B}(P\mathcal{H})_* \to \mathfrak{A}_*$ définie par $\mu_*(\rho)=\alpha_\rho$. μ_* est aussi continue en norme. En effet si $A\in\mathfrak{A}$:

$$|[\mu_*(\rho)](A)| = \lim_{t\to\infty} |\mathrm{Tr}(V_t\rho V_t^*A)| \leq \lim_{t\to\infty} \|V_t\rho V_t^*\|_1 \|A\| = \|\rho\|_1 \|A\|$$

et par conséquent

$$\|\mu_*(\rho)\|_{\mathfrak{A}_*} \equiv \sup_{\substack{A\in\mathfrak{A}\\ \|A\|=1}} |[\mu_*(\rho)](A)| \leq \|\rho\|_1 = \|\rho\|_{\mathcal{B}(P\mathcal{H})_*}$$

La continuité en norme de μ_* implique que, pour tout $A\in\mathfrak{A}$, l'application $\varphi_A : \rho\mapsto[\mu_*(\rho)](A)$ est une fonctionnelle linéaire sur $\mathcal{B}(P\mathcal{H})_*$ continue en norme, c.à.d. appartenant au dual de $\mathcal{B}(P\mathcal{H})_*$. Ainsi, par définition de la topologie faible, φ_A est aussi $\sigma(\mathcal{B}(P\mathcal{H})_*,\mathcal{B}(P\mathcal{H}))$-continue. Vu la définition de φ_A, ceci est équivalent à dire que μ_* est continue pour les topologies faibles $\sigma(\mathcal{B}(P\mathcal{H})_*,\mathcal{B}(P\mathcal{H}))$ et $\sigma(\mathfrak{A}_*,\mathfrak{A})$.

Considérons maintenant le transposé $\mu_*^T : \mathfrak{A}\to\mathcal{B}(P\mathcal{H})$ de μ_* qui est défini comme

$$[\mu_*^T(A)](\rho) \equiv [\mu_*(\rho)](A) \tag{19}$$

Comme nous venons de le voir, $\mu_*^T(A)$ appartient à $\mathcal{B}(P\mathcal{H})$, le dual de $\mathcal{B}(P\mathcal{H})_*$, donc μ_*^T est bien défini. Le fait que μ_* est continue pour les topologies faibles implique que μ_*^T est continu pour les topologies $\sigma(\mathfrak{A},\mathfrak{A}_*)$ et $\sigma(\mathcal{B}(P\mathcal{H}),\mathcal{B}(P\mathcal{H})_*)$ ([25],Ch.II,Cor. de la Prop. 12), c.à.d. μ_*^T est ultrafaiblement continu. Pour prouver le lemme, il ne nous reste donc qu'à montrer que $\mu=\mu_*^T$.

Pour ceci, soit $A\in\mathfrak{A}$ et $g\in\mathcal{B}(P\mathcal{H})_*$. On aura avec (19) et (18)

$$[\mu_*^T(A)](\rho) = \alpha_\rho(A) = \lim_{t\to\infty} \alpha_{\rho_t}(A) = \mathrm{Tr}[\rho\mu(A)].$$

D'autre part, avec la dualité entre $\mathcal{B}(P\mathcal{H})$ et $\mathcal{B}(P\mathcal{H})_*$:

$$\left[\mu_*^T(A)\right](\rho) = \mathrm{Tr}\left[\rho\mu_*^T(A)\right].$$

Ainsi

$$\mathrm{Tr}\left[\rho\{\mu_*^T(A) - \mu(A)\}\right] = 0 \quad \text{pour tout } \rho \in \mathcal{B}(P\mathcal{H})_*.$$

En prenant $\rho=|f><g|$, $f,g\in P\mathcal{H}$, il suit que $\mu_*{}^T(A)=\mu(A)$ pour tout $A \in \mathcal{a}$, donc $\mu_*^T=\mu$. ∎

Nous pouvons maintenant prouver que les images $\mu_\alpha^\pm(\mathcal{a}_\alpha)$ des algèbres $\mathcal{a}_\alpha$ satisfaisant la condition I sont des algèbres de von Neumann. En effet les hypothèses du lemme 1 se vérifient immédiatement à partir de I, donc $\mu_\alpha^\pm$ sont ultrafaiblement continues. D'autre part nous avons déjà remarqué que ce sont des *-homomorphismes de $\mathcal{a}_\alpha$ dans $\mathcal{B}(P_\alpha^\pm\mathcal{H})$. En répétant la partie(iv) de la démonstration de la proposition 1 dans [11] ou en appliquant le théorème 1.16.2 de [23], on déduit que $\mu_\alpha^\pm(\mathcal{a}_\alpha)$ sont ultrafaiblement fermées, donc des algèbres de von Neumann.

<u>Lemme 2</u> : L'opérateur T_t^α défini dans (3) possède une extension fermée $\overline{T}_t^\alpha$ sur $E_\alpha\mathcal{H}$ qui commute avec $\mathcal{a}_\alpha$ sur $D(\overline{T}_t^\alpha)$.

<u>Démonstration</u> : (i) Montrons d'abord que e_α appartient au domaine de l'adjoint $T_t^{\alpha *}$ de T_t^α. Pour ceci, considérons la fonctionnelle linéaire suivante sur $\mathcal{D}_\alpha$:

$$\phi(h) = (e_\alpha, T_t^\alpha h), \ h\in\mathcal{D}_\alpha$$

On a

$$\|\phi\|_{\mathcal{D}_\alpha} = \sup_{\substack{h\in\mathcal{D}_\alpha \\ h\neq 0}} \frac{|\phi(h)|}{\|h\|} = \sup_{\substack{A\in\mathcal{a}_\alpha \\ Ae_\alpha\neq 0}} \frac{|(e_\alpha, T_t A e_\alpha)|}{\|Ae_\alpha\|}$$

$$= \sup_{\substack{A\in\mathcal{a}_\alpha \\ Ae_\alpha\neq 0}} \frac{|(C_\alpha A^* e_\alpha, V_t \Omega_\alpha e_\alpha)|}{\|Ae_\alpha\|}$$

$$\leqslant \sup_{\substack{A\in\mathcal{a}_\alpha \\ Ae_\alpha\neq 0}} \frac{\|C_\alpha A^* e_\alpha\|}{\|Ae_\alpha\|}\|e_\alpha\| \tag{20}$$

Soit $\{f_i\}$ une base orthonormée dans $C_\alpha\mathcal{H}$ de la forme $f_i=B_ie_\alpha$ avec $B_i \in \mathcal{a}_\alpha'$. (Une telle base existe:on peut choisir un ensemble totalisateur de vecteurs linéairement indépendants G_ie_α, $G_i \in \mathcal{a}_\alpha'$, que l'on peut orthogonaliser selon la méthode de Schmidt. $\mathcal{a}_\alpha'$ étant une algèbre, on voit que tout vecteur de la base obtenue ainsi est de la forme B_ie_α pour un $B_i\in \mathcal{a}_\alpha'$.)

On a

$$(B_i{}^*e_\alpha, B_j{}^*e_\alpha) = (B_j e_\alpha, B_i e_\alpha) = \delta_{ij}$$

Ainsi les vecteurs $\{B_i{}^*e_\alpha\}$ forment une partie d'une base orthonormée de $C_\alpha\mathcal{H}$. En utilisant ceci, il vient

$$\|C_\alpha A^* e_\alpha\|^2 = \sum_i |(A^*e_\alpha, f_i)|^2 = \sum_i |(B_i{}^*e_\alpha, Ae_\alpha)|^2 \leq \|Ae_\alpha\|^2.$$

Si nous insérons cette inégalité dans (20), nous obtenons

$$\|\phi\|_{\mathcal{D}_\alpha} \leq \|e_\alpha\|.$$

Ainsi on peut étendre ϕ à une fonctionnelle linéaire bornée sur $E_\alpha\mathcal{H}$, et selon le théorème de Riesz il existe un vecteur $e_t \in E_\alpha\mathcal{H}$ tel que

$$(e_\alpha, T_t^\alpha h) = (e_t, h) \qquad \text{pour tout } h \in \mathcal{D}_\alpha.$$

Ceci montre que $e_\alpha \in D(T_t^{\alpha *})$.

(ii) Soit $h \in \mathcal{D}_\alpha$, $g \in D(T_t^{\alpha *})$ et $A \in \mathcal{a}_\alpha$. En utilisant (4) on trouve

$$(Ag, T_t^\alpha h) = (g, A^* T_t^\alpha h) = (g, T_t^\alpha A^* h) = (AT_t^{\alpha *} g, h).$$

Ainsi

$$g \in D(T_t^{\alpha *}) \Rightarrow Ag \in D(T_t^{\alpha *}) \qquad \text{pour tout } A \in \mathcal{a}_\alpha \tag{21}$$

$$\text{et} \quad AT_t^{\alpha *} \subseteq T_t^{\alpha *} A \qquad \text{pour tout } A \in \mathcal{a}_\alpha \tag{22}$$

Nous pouvons utiliser cela de deux façons :

(α) Puisque $e_\alpha \in D(T_t^{\alpha *})$, (21) implique que $Ae_\alpha \in D(T_t^{\alpha *})$ pour tout $A \in \mathcal{a}_\alpha$. Ainsi $D(T_t^{\alpha *})$ est dense dans $E_\alpha\mathcal{H}$, ce qui implique que T_t^α possède une extension fermée $\overline{T}_t^\alpha = T_t^{\alpha **}$ ([24],Thm .VIII.1).

(β) En répétant le raisonnement qui nous menait à (21) et (22) pour l'opérateur $T_t^{\alpha *}$ à la place de T_t^α et en utilisant (22) au lieu de (4), on conclut que $g \in D(T_t^{\alpha **}) \Rightarrow Ag \in D(T_t^{\alpha **})$ pour tout $A \in \mathcal{a}_\alpha$ et $AT_t^{\alpha **} \subseteq T_t^{\alpha **} A$. ∎

<u>Démonstration du lemme 3</u> : Comme $\|\exp(-iKt)\| = 1$ et $\|G(t)\| \leq M$ pour tout $t \geq 0$, il suffit de vérifier (9) pour $g \in \mathcal{M}_1$ et $h \in \mathcal{M}_2$, $\mathcal{M}_1$ et $\mathcal{M}_2$ étant deux sous-ensembles denses de $\mathcal{H}$. Nous prenons $\mathcal{M}_1 = \mathcal{M}$ et $\mathcal{M}_2 = D(K^{-1})$, et nous poserons $X_t = \exp(-iKt)$. Alors

$$\frac{d}{dt}(G(t)g, X_t K^{-1} h) = (G'(t)g, X_t K^{-1} h) - i(G(t)g, X_t h).$$

Par conséquent

$$\left|\frac{1}{T}\int_0^T dt(G(t)g, X_t h)\right| =$$

$$\left|\frac{i}{T}\left[(G(T)g,X_T K^{-1}h) - (G(0)g,K^{-1}h)\right] - \frac{i}{T}\int_0^T dt(G'(t)g,X_t K^{-1}h)\right|$$

$$\leq \frac{2}{T} M\|g\| \, \|K^{-1}h\| + \|K^{-1}h\| \frac{1}{T}\int_0^T dt \, \|G'(t)g\|$$

ce qui converge vers zéro lorsque $T\to\infty$. ∎

Références

[1] J.M.Jauch, Helv. Phys. Acta 31, 127 (1958)

[2] J.M.Cook, J. Math. and Phys. 36, 82 (1957)

[3] J.M.Jauch, Helv. Phys. Acta 31, 661 (1958)

[4] M.N.Hack, Nuovo Cim. 13, 231-236 (1959)

[5] K.O.Friedrichs, Comm. Pure Appl. Math. 1, 361 (1948)

[6] T.Kato, J. Math. Soc. Japan 9, 239 (1957) and Proc. Japan Acad. 33, 260 (1957)

[7] S.T.Kuroda, Nuovo Cim. 12, 431 (1959)

[8] T.Kato, "Perturbation Theory for Linear Operators", Springer, New York (1966)

[9] J.M.Jauch, B.Misra and A.G.Gibson, Helv. Phys. Acta 41, 513 (1968)

[10] J.D.Dollard, J. Math. Phys. 5, 729 (1964)

[11] W.O.Amrein, Ph.A.Martin and B.Misra, Helv. Phys. Acta 43, 313 (1970)

[12] J.M.Combes, "An Algebraic Approach to Scattering Theory", CNRS Marseille, preprint

[13] J.V.Corbett, Phys. Rev. D1, 3331 (1970)

[14] R.Lavine, J. Funct. Anal. 5, 368 (1970)

[15] H.Ekstein, Ann. Phys. 74, 303 (1972)

[16] W.Hunziker, in "Lectures in Theoretical Physics", Vol.XA, W.E.Brittin ed., 1, Gordon and Breach, New York (1968)

[17] J.Dixmier, "Les algèbres d'opérateurs dans l'espace Hilbertien", Gauthier-Villars, Paris (1969)

[18] E.Mourre, Ann. Inst. Henri Poincaré 18 (1973)

[19] J.D.Dollard, J. Math. Phys. 14, 708 (1973)

[20] J.D.Dollard, Comm. Math. Phys. 12, 193 (1969)

[21] B.Simon, "Quantum Mechanics for Hamiltonians Defined as Quadratic Forms", Princeton University Press (1971)

[22] W.W.Zachary, Helv. Phys. Acta 46, 327 (1973)

[23] S.Sakai, "C*-Algebras and W*-Algebras", Springer, Berlin (1971)

[24] M.Reed and B.Simon, "Functional Analysis", Academic Press, New York (1972)

[25] A.P.Robertson and W.J.Robertson, "Topological Vector Spaces", Cambridge Univ. Press, Cambridge (1964).

FOURIER SCATTERING SUBSPACES

Karl Gustafson*

Department of Mathematics
University of Colorado
Boulder, Colorado

In the literature of quantum mechanical scattering one finds that the scattering states are often being thought of as "the orthogonal complement of the bound states", or as "the outgoing states", or as "the asymptotic states", etc.; somewhat more precisely, as the physically realizable evolutions "possessing incoming and outgoing asymptotic free states", or a little more precisely, as evolutions whose initial values (at $t = 0$) are in the range of a wave operator or operators (e.g., in $\mathfrak{R}^+ \cap \mathfrak{R}^-$), etc. Recently there has been considerable interest in seeking a more fundamental understanding of scattering states; in this paper we wish to show and to emphasize, albeit in a brief and not yet complete manner, the use of the Fourier theory as an essential ingredient in such investigations. In particular, we will obtain from a mathematical point of view some orderings of important types of scattering subspaces now under consideration.

Let $H_1 = H_0 + V$ and H_0 be self-adjoint operators in a separable complex Hilbert space $\mathfrak{K}$, and let H denote a general self-adjoint operator in $\mathfrak{K}$. It is useful, before discussing the choice mathematically of a scattering subspace $\mathfrak{M}$ for the pair H_1, H_0, to first recall that there are at least three different viewpoints from which one can begin; by a scattering subspace $\mathfrak{M}$ we will presume that we are speaking of some subspace of $\mathfrak{K}$ consisting of initial values of certain types of evolutions.

* Partially supported by NSF GP15239 A1.

Enz/Mehra (eds.), Physical Reality and Mathematical Description. 277–285. *All Rights Reserved.*

From the time-dependent approach, in one of the earliest basic papers on the subject Jauch [1] used $\mathfrak{M} = \mathfrak{H}_c(H_0)$, the orthogonal complement of the subspace $\mathfrak{H}_p(H_0)$ of eigenfunctions of H_0. This subspace may be regarded as one of the "larger" choices of $\mathfrak{M}$ if one wishes to leave open the possibility of existence, for a reasonable class of potentials V, of the wave operators

$W_\pm = s - \lim_{t \to \pm\infty} e^{itH_1} e^{-itH_0} P$, where P is the orthogonal projection onto $\mathfrak{M}$. On the other hand, choosing $\mathfrak{M}$ smaller may provide a better likelihood for the existence of the wave operators, as pointed out by Kato [2, p. 532, Remark 3.6], where $\mathfrak{M} = \mathfrak{H}_{ac}(H_0)$ is taken to be the absolutely continuous subspace of H_0. Alternately, one can insist that the time-dependent wave operators exist by choosing $\mathfrak{M}$ to be the subspace of $\mathfrak{H}_c(H_0)$ (or of $\mathfrak{H}$) on which the above strong limits exist, as is done for example by Prugovecki [3]; however, as noted in [3, p. 415, 422], such an $\mathfrak{M}$ may not be of the proper size to describe the physics.

It seems that, although certainly known, it is not widely realized that in general, in the time-dependent framework, $\mathfrak{M}$ can be taken to be any reducing subspace of H_0. As a point of general information let us therefore state this fact somewhat informally as follows, the details (see [4]) following as in [2].

<u>Lemma 1</u>. Let $\mathfrak{M}$ be any reducing subspace of H_0; then if W_+ exists, the properties of [2, pp. 529-532] hold; also the sufficient condition [2, p. 533] for the existence of W_+ holds (on $\overline{\mathfrak{M}}$ rather than the D of [2]).

<u>Remark 1</u>. Thus for example candidates for $\mathfrak{M}$ could be smaller than $\mathfrak{H}_{ac}(H_0)$ so that the final set $\mathfrak{M}^+$ (see [2]) is within $\mathfrak{H}_{ac}(H_1)$ or smaller than $\mathfrak{H}_c(H_0)$ so that $\mathfrak{M}^+$ is in $\mathfrak{H}_c(H_1)$. In the other direction, one may consider $\mathfrak{M}$ larger than $\mathfrak{H}_{ac}(H_0)$ or $\mathfrak{H}_c(H_0)$; for example, in order to include certain eigenfunctions (as observed in [3]), or, and possibly more interesting for applications, as a localized form of $\mathfrak{H}_{ac}$ (or of $\mathfrak{H}_c$ if desired) as indicated in Gustafson and Johnson [5, Remark 2]. Let us elaborate somewhat what w intended in the aforementioned remark of [5]; let $\mathfrak{H}_{ac}(H,\Sigma)$, Σ an open set in the reals $\mathfrak{R}$, be the intersection of the subspace $\mathfrak{H}_{ac}(H,(\alpha,\beta))$ over all subintervals (α,β) in Σ, where $\mathfrak{H}_{ac}(H,(\alpha,\beta)) = \{\phi \in \mathfrak{H} \mid \langle E(\lambda)\phi,\phi\rangle$ is absolutely continuous on (α,β), i.e., equivalently, $\mathfrak{H}_{ac}(H,(\alpha,\beta)) = \mathfrak{H}_{ac}(H|_{E(\alpha,\beta)\mathfrak{H}})$. One may similarly introduce $\mathfrak{H}_s(H,(\alpha,\beta)) = \{\phi \in \mathfrak{H} \mid \langle E(\lambda)\phi,\phi\rangle$ is constant on $(-\infty,\alpha]$ and $[\beta,\infty)$, singular on (α,β). Then as in [2] $\mathfrak{H}_{ac}(\alpha,\beta)$ and $\mathfrak{H}_s(\alpha,\beta)$ are seen to be orthogonal closed subspaces of $\mathfrak{H}$; and it also fol-

lows that they reduce H. The $\mathcal{H}_{ac}(H,\Sigma)$ thus enlarge $\mathcal{H}_{ac}(H)$; and clearly $\bigcap_{(\alpha,\beta)} \mathcal{H}_{ac}(\alpha,\beta) = \mathcal{H}_{ac}(H)$.

Secondly, from the time-independent or stationary approach, the wave operators and corresponding scattering subspaces depend on considerations involving boundary values of analytic functions. Thus for example in [5, Remark 1] $\mathfrak{M} = \mathcal{H}_{ac}(H)$ is shown to consist of exactly those vectors ϕ for which the imaginary part $\mathrm{Im}\langle R_z\phi,\phi\rangle$ of the resolvent operator (denoted here by $R_z \equiv R_z(H)$, $z = x + iy$) matrix elements attains its boundary value in the sense of $\mathcal{L}^1 \equiv \mathcal{L}^1(-\infty,\infty)$ norm convergence as $y\to 0^+$. Usually some type of stronger convergence to boundary values is sought in establishing the existence of the wave operators; for example, see Kato [6], Howland [7], and Kato and Kuroda [8]. Let us summarize here some subspaces characterizable in this (or related) manner.

Lemma 2.

$$E_{\sigma_{ess}(H)}\mathcal{H} \supset \mathcal{H}_c(H) = \{\phi \mid \langle E(\lambda)\phi,\phi\rangle \text{ is cont.}\} = E_{\sigma_c(H)}\mathcal{H}$$

$$\supset E_{\sigma(H_c)}\mathcal{H} \supset \mathcal{H}_{ac}(H) = \{\phi \mid \langle E(\lambda)\phi,\phi\rangle \text{ is abs.cont.}\} = E_{\sigma(H_{ac})}\mathcal{H}$$

$$= \{\phi \mid \mathrm{Im}\langle R_z\phi,\phi\rangle \text{ converge in } \mathcal{L}^1 \text{ as } y\to 0^+\}$$

$$= \{\phi \mid \mathrm{Im}\langle R_z\phi,\phi\rangle \text{ conv. in } \mathcal{L}^1_{loc}\}$$

$$\supset \{\phi \mid \langle E(\lambda)\phi,\phi\rangle \text{ is loc. Hölder cont.}\}$$

$$\supset \{\phi \mid \langle E(\lambda)\phi,\phi\rangle \text{ is loc. Lipschitz}\}$$

$$\supset \{\phi \mid \|R_z\phi\| = O_{loc}(y^{-\frac{1}{2}}),\ y\to 0^+\}$$

$$= \{\phi\in\mathcal{H}_{ac} \mid g(\lambda) \equiv d\langle E(\lambda)\phi,\phi\rangle/d\lambda \in \mathcal{L}^\infty_{loc}\}$$

$$\supset \{\phi\in\mathcal{H}_{ac} \mid g\in\mathcal{L}^\infty\} = \{\phi \mid \|R_z\phi\| = O(y^{-\frac{1}{2}})\}$$

$$\supset \{\phi \mid g\in\mathcal{L}^\infty\cap C^0,\ g(\pm\infty) = 0\}$$

$$\supset \{\phi \mid g\in\mathcal{S}\} \supset \{\phi \mid g\in C_0^\infty(R)\}.$$

In Lemma 2 $\sigma_{ess}(H)$ denotes the (Weyl) essential spectrum, consisting of the spectrum $\sigma(H)$ minus all isolated eigenvalues of finite (algebraic and geometric) multiplicity, σ_c denotes continuous spectrum, H_c the continuous part of H, etc.; for further details see [2] and Gustafson [9]. Aspects of this approach, although in other contexts, have been used for determining absolute continuity for example in Weidmann [10], Rejto [11], and Pincus [12], among others.

Remark 2. (i) Let us mention that the resolvent growth rate condition sufficient for absolute continuity of [5] may be local and moreover can depend on individual vectors, and that the subspace $\mathfrak{M} = \{\phi \mid \| R_z \| = O(y^{-\frac{1}{2}}), y \to 0^+\}$ is shown in [5] to be dense in $\mathfrak{H}_{ac}(H)$. As such it could be a good test subspace for $\mathfrak{H}_{ac}$; on the other hand, the extent of the smaller dense test subspaces for $\mathfrak{H}_{ac}$ seems to be unclear.

(ii) It is natural in this approach that one might suggest using the Hardy space analytic function theory; and for closely related considerations some spaces of Hardy type have recently been introduced by Van Winter [13]. However let us observe here that the situation where the (complex valued) matrix element $\langle R_z\phi, \phi\rangle$ is in the Hardy space $\mathfrak{H}^p$ (upper half plane) for some $1 \leqslant p \leqslant \infty$, although certainly a sufficient condition for $\phi \in \mathfrak{H}_{ac}(H)$, never occurs.

Thirdly, and probably closer to the physical point of view, scattering states may be defined in terms of their ergodic behavior. Thus by Wiener's Theorem one has the characterization of $\mathfrak{H}_c(H)$ as those ϕ for which the expectation $T^{-1} \int_0^T |\langle e^{itH}\phi, \phi\rangle|^2\, dt \to 0$ as $T \to \infty$. For the use of this approach in recent work in scattering theory and decay phenomena, see for example Lax and Phillips [14], Ruelle [15], Horwitz, La Vita, and Marchand [16], Sinha [17], Wilcox [18], and Amrein and Georgescu [19]. Jauch [20] has recognized and emphasized the use of correlation functions $\langle U_t\phi, \phi\rangle$ in studying general fluctuating dissipative systems.

Let us recall in the following lemma some of the subspaces that have been recently introduced from this point of view, some of their properties as used in some of the above mentioned references, and some related properties.

Lemma 3.

$$\text{(i)}\quad \mathfrak{H}_c(H) = \{\phi \mid \int_0^T |\langle e^{itH}\phi, \phi\rangle|^2\, dt = o(T)\}$$

$$\supset \overline{\mathfrak{M}^{\pm}_{\infty}} = \{\phi \mid \int_0^T \| F_r\, e^{itH}\phi \|^2\, dt = o(T)\}$$

$$\supset \mathfrak{M}_\infty^\pm = \{\phi \mid \| F_r . e^{itH}\phi\|^2 \to 0\} = \mathcal{H}^s.$$

(ii) $\{\phi \mid \langle e^{it_n H}\phi,\phi\rangle \to 0\} \supset \mathcal{H}_c(H)$

$$\supset \mathfrak{M}_{RL} \equiv \{\phi \mid \langle e^{it\ H}\phi,\phi\rangle \to 0\}$$

$$\supset \mathcal{H}_{ac}(H) \supset \{\phi \mid \langle e^{it\ H}\phi,\phi\rangle\ \mathcal{L}^2\}$$

$$\supset \{\phi \in \mathcal{H}_{ac} \mid g \in \mathcal{L}^\infty\}.$$

(iii) $\mathcal{H}_{ac}(H) \supset \{\phi \mid \langle e^{it\ H}\phi,\phi\rangle = O(|t|^{-\frac{1}{2}-\varepsilon}),\ \varepsilon>0$; or

$$\int_{-\infty}^{\infty} |t|^p |\langle e^{it\ H}\phi,\phi\rangle| dt<\infty, p\geq 0; \text{ or}$$

$$\langle e^{it\ H}\phi,\phi\rangle = O(e^{-\varepsilon|t|}),\ \varepsilon>0\}\ .$$

(iv) $\{\phi \mid \forall\ t_0>0$ the angle between $e^{it\ H}\phi$ and ϕ remains bounded uniformly away from zero $\forall\ t\geq t_0\} \supset \mathcal{H}_{ac}(H)$.

For the subspaces $\overline{\mathfrak{M}_\infty^\pm}$ and $\mathfrak{M}_\infty^\pm$ and the inclusion $\mathcal{H}_c \supset \overline{\mathfrak{M}_\infty^\pm}$, see [18,19], the inclusion $\overline{\mathfrak{M}_\infty^\pm} \supset \mathfrak{M}_\infty^\pm$ having been observed for example in [9]; in [18] the notation $\mathcal{H}^s$ rather than $\mathfrak{M}_\infty^\pm$ was used. The F_r, $r>0$, denotes a family of orthogonal projections converging strongly to the identity; under certain relative compactness conditions of the F_r with respect to H, in [15,18,19] the inclusion $\mathfrak{M}_\infty^\pm \supset \mathcal{H}_{ac}(H)$ is shown.

The inclusions of (ii) are all proper and well-known, the last mentioned being the Kato-Rosenbloom Lemma. We comment that in some of these characterizations one may (and sometimes should,(e.g., as in $\mathfrak{M}_{RL}$, in order to guarantee weak convergence and hence a subspace rather than just a subset)) replace $\langle e^{it\ H}\phi,\phi\rangle$ by $\langle e^{it\ H}\phi,f\rangle$ $\forall\ f \in \mathcal{H}$, but we omit the detailing of this property.

The first condition in (iii) guarantees that $\langle e^{it\ H}\phi,\phi\rangle \in \mathcal{L}^2$, the second describes convergence stronger than needed for $\langle e^{itH}\phi,\phi\rangle \in \mathcal{L}^1$, the third occuring only when g is analytic. Sharpness statements for these (strong) conditions for absolute continuity are also known.

<u>Remark 3</u>. The requirement for absolute continuity in (iv) is mentioned here to bring out an interpretation (not usually discus-

sed) of the degree of orthogonality from the initial state that must be immediately attained and maintained by the evolution.

In the following lemma we give a further ordering (from the ergodic viewpoint) of some scattering subspaces.

<u>Lemma 4.</u>

$$\mathcal{H}_c(H) = \{\phi \mid T^{-1}\int_0^T |\langle e^{itH}\phi,\phi\rangle|^2\, dt \to 0\}$$

$$= \{\phi \mid T^{-1}\int_0^T |\langle e^{itH}\phi,\phi\rangle|\, dt \to 0\}$$

$$\supset \mathcal{H}_{ac}(H)$$

$$\supset \{\phi \mid T^{-1}\int_0^T |\langle e^{itH}\phi,\phi\rangle|^2\, dt = O(T^{-\alpha}),\ 0 < \alpha < 1\}$$

$$= \{\phi \mid \langle E(\lambda)\phi,\phi\rangle \text{ is loc. Hölder cont.}\}$$

$$\supset \{\phi \mid T^{-1}\int_0^T |\langle e^{itH}\phi,\phi\rangle|^2\, dt = O(T^{-1})\}$$

$$= \{\phi \mid \langle e^{itH}\phi,\phi\rangle \in \mathcal{L}^2\} = \{\phi \in \mathcal{H}_{ac} \mid g \in \mathcal{L}^2\}$$

$$\supset \{\phi \mid \langle e^{itH}\phi,\phi\rangle \in \mathcal{L}^p,\ 1 < p < 2\}$$

$$\supset \{\phi \mid T^{-1}\int_0^T |\langle e^{itH}\phi,\phi\rangle|\, dt = O(T^{-1})\}$$

$$= \{\phi \mid \langle e^{itH}\phi,\phi\rangle \in \mathcal{L}^1\}\ .$$

<u>Remark 4.</u> (i) It would be desirable to have in the hierarchy of Lemma 4 an ergodic characterization of $\mathcal{H}_{ac}(H)$ of the same type, which perhaps exists but of which we are not aware.

(ii) Let us also mention as a matter of related interest that the nature of the last mentioned set in Lemma 4 is a long-standing open problem in the Fourier Theory, namely, to determine the subspace of $g \in \mathcal{L}^1$ which have their transforms in $\mathcal{L}^1$.

Final Remark. From Lemmas 1-4 and Remarks 1-4 above it seems to us that it must be admitted that neither $\mathfrak{H}_{ac}(H)$ nor the meaning of scattering subspace of states is yet completely understood. (1) The disparity between the subspaces associated with the time dependent and the time-independent theories lingers. (2) The nature of the dense subspaces of $\mathfrak{H}_{ac}(H)$ needs clarification. (3) The "localizing" or "compactifying" interpretations implicit in the use of the projections F_r are not a priori consistent with correct probabilistic or geometric interpretations of the covariance function $\langle e^{itH}\phi,\phi\rangle$. (4) The lack of a satisfactory ergodic description of $\mathfrak{H}_{ac}(H)$ renders its use possibly more coincidental than fundamental.

It is hoped that these reflections will all be rendered trivial in the near future. (1) The time-independent theory, through proper dressing and screening arguments, may embrace in a completely satisfactory way from the physical viewpoint the Coulomb potential. (2) By sufficient mathematical effort (e.g., using molifiers) it can certainly be determined whether or not some of the good test function subspaces are dense in $\mathfrak{H}_{ac}(H)$. (3) The distinctions between all mentioned subspaces between $\mathfrak{H}_{ac}(H)$ and $\mathfrak{H}_p(H)^{\perp}$ may vanish via a determination that in the foundations of quantum mechanics at a crucial point, due to symmetry and/or measurement assumptions, one has implicitly required that all probabilities are associated with Borel sets in a nonsingular (i.e., allowing only counting and Lebesgue measures) way. (4) A study of physical phenomena inducing truly singular-continuous measures may clarify the ergodic behavior of scattering subspaces.

REFERENCES

1. J.M.Jauch, Theory of the scattering operator, Helv. Phys. Acta. 31 (1958), 127-158.

2. T.Kato, Perturbation theory for linear operators, Springer, Berlin, (1966).

3. E. Prugovecki, Quantum mechanics in Hilbert space, Academic Press, New York, (1971).

4. K.Gustafson and K.Jörgens (unpublished); this observation is due principally to Jörgens.

5. K.Gustafson and G.Johnson, On the absolutely continuous subspace of a self-adjoint operator, Helv. Phys. Acta. (to appear).

6. T.Kato, Wave operators and similarity for non-self-adjoint operators, Math. Ann. 162 (1966), 258-279.

7. J.S.Howland, Banach space techniques in the perturbation theory of self-adjoint operators with continuous spectra, J. Math. Anal. Appl. 20(1967), 22-47.

8. T.Kato and S.T.Kuroda, The abstract theory of scattering, Rocky Mt. J. Math. 1(1971), 127-171.

9. K.Gustafson, Candidates for σ_{ac} and H_{ac}, Proc. NATO. Adv. Study Inst. on Scattering Theory in Mathematics and Physics, J. La Vita and J.P.Marchand, Reidel, Doortrecht (to appear).

10. J.Weidmann, Oszillationsmethoden für Systeme gewöhnlicher Differentialgleichungen, Math. Z. 119(1971), 349-373.

11. P.Rejto, On partly gentle perturbations I, J. Math. Anal. and Applic. 17(1967), 435-462.

12. J.Pincus, Commutators, generalized eigenfunction expansions and singular integral operators, Trans. Amer. Math. Soc. 121(1966), 358-377.

13. C.Van Winter, Fredholm equations on a Hilbert space of analytic functions, Trans. Amer. Math. Soc. 162(1971), 103-139.

14. P.Lax and R.S.Phillips, Scattering theory, Academic Press, New York, (1967).

15. D.Ruelle, A remark on bound states in potential scattering theory, Nuovo Cim. 61A (1969), 655-662.

16. L.Horwitz, J.La Vita, J.P.Marchand, The inverse decay problem, J. Math. Phys. 12(1971), 2537-2543.

17. K.Sinha, On the decay of an unstable particle, Helv. Phys. Acta. 45(1972), 619-628.

18. C.Wilcox, Scattering states and wave operators in the abstract theory of scattering, J. Funct. Anal. 12(1973), 257-274.

19. W.O.Amrein and V.Georgescu, On the characterization of bound states and scattering states in quantum mechanics, Helv. Phys. Acta. (to appear).

20. J.M.Jauch, Informal remarks on some recent work of Prigogine and others, petit seminaire, Institut de physique théorique, Genève (1972).

N-PARTICLE SCATTERING RATES (*)

Roger G. Newton and Roman Shtokhamer

Physics Department, Indiana University,
Bloomington

Starting from time-dependent scattering theory, we derive expressions for the observable scattering rates of arbitrary numbers of colliding particles.

1. Introduction

A few years ago Josef Jauch discussed with one of the authors (R.G.N.) the unsatisfactory state of the usual derivation of the expression for the scattering cross section in terms of the T-matrix, starting from the fundamentals of time-dependent scattering theory. Part of the result of that discussion was a joint paper[1] on a generalization of Dollard's[2] "scattering-into-cones" theorem. The present paper is a further response to the questions raised in that discussion. We believe that it may answer most of them, although perhaps not in as rigorous a mathematical fashion as one would like. In fact, it opens up some further questions that need mathematical investigation.

In this paper we derive expressions for observable scattering rates for any finite number of particles from time-dependent nonrelativistic scattering theory. The main points that characterize the method of this derivation are the use of Dollard's scattering-into-cones theorem, and of an impact-parameter summation in the incident beams. The latter is, of course, not new. It has been used, for example, by Hunziker[3] in an N-particle context, and by Taylor[4]

(*) Work supported in part by the National Science Foundation and the U.S. Army Research Office, Durham, North Carolina.

Enz/Mehra (eds.), Physical Reality and Mathematical Description. 286–312.

in two-particle scattering. The present paper, however, carries it further and, to make the transition from coordinate space, where measurements are performed, to the momentum-space T-matrix elements, combines it with Dollard's theorem.

In its general form, the scattering-into-cones theorem states that if $\psi(x,t)$ is a wave function on the n-dimensional coordinate space and it developes according to a free Hamiltonian $H_o(p^2)$ which is such that its derivative with respect to p^2 is positive definite, then for any infinite cone C with apex at the origin it is true that

$$P^C(\infty) = \lim_{t\to\infty} P^C(t) = \lim_{t\to\infty} \int_C d^n x \; |\psi(x,t)|^2$$

$$= \int_C d^n k |\tilde{\psi}(k)|^2 \tag{1.1}$$

where $\tilde{\psi}(k)$ is the Fourier transform of $\psi(x,o)$, i.e., the wave function in momentum space. In other words, the probability that a particle will be found in the infinite future in C is equal to the probability that its momentum lies in the same cone. This is, of course, physically, an eminently plausible proposition and it is usually taken for granted in one's intuitive thinking about scattering.

There is another step that needs to be taken before the probability P^C becomes relevant to scattering experiments with counters. Let A be the intersection of C with a spherical surface of radius R, with center at the vertex of the cone; it divides C into C', between A and the vertex, and C". Let B be the surface of C, separated by the circumference of A into B' and B", the surfaces of C' and C", respectively. We have [5)]

$$P^{C'}(\infty) = 0 \tag{1.2}$$

since C' is a finite region, and we assume that

$$P^C(-\infty) = 0. \tag{1.3}$$

(This, of course, depends on the fact that $\psi(x,t)$ does not develop according to H_o all the time, so that there is scattering. If C does not contain the direction of indicence then (1.3) follows, for example, from a stationary-phase argument.) Furthermore, we have

$$P^C(t) = P^{C'}(t) + P^{C''}(t). \tag{1.4}$$

It then follows that

$$P^{C}(\infty) = P^{C''}(\infty) = \int_{-\infty}^{\infty} dt \, \frac{d}{dt} P^{C''}(t). \tag{1.5}$$

Now probability-flux conservation implies that for any region V

$$\frac{d}{dt} P^{V}(t) = -\int_{S} d\underset{\sim}{s} \cdot \underset{\sim}{f}$$

if S is the surface of V and $\underset{\sim}{f}$ is the flux density. Consequently, (1.5) says that

$$P^{C}(\infty) = \eta(A) - \eta(B'') \tag{1.6}$$

where

$$\eta(A) = \int_{-\infty}^{\infty} dt \int_{A} d\underset{\sim}{s} \cdot \underset{\sim}{f}$$

with the surface normal of A pointing away from the vertex, and similarly for B''.

Let us now let R tend to infinity. It is then to be expected that

$$\lim_{R\to\infty} \eta(B'') = 0 \tag{1.7}$$

although a rigorous proof of this is lacking. If (1.7) is correct then (1.6) implies that

$$P^{C}(\infty) = \lim_{R\to\infty} \eta(A). \tag{1.8}$$

The outgoing-wave boundary condition on the wave function implies that at large distances all particles that cross A do so in the outward direction. Hence, the right-hand side of (1.8) can be interpreted as the <u>probability that the particle will, at some time, cross the distant surface A</u>. This is exactly what a counter will measure. It is, therefore, the combination of (1.8) and (1.1) that gives the right-hand side of (1.1) its physical significance in terms of measurements by counters.

The use of impact parameters is essentially the following. One first derives an expression for the probability of finding particles in a given infinite cone on the assumption that the wave packet has an impact parameter b. Then one assumes that the beam is incoherently made up of a uniform distribution of impact parameters, as one would classically, and this generates the final expression which includes energy and momentum conserving delta-functions. The entire derivation is both physically reasonable and mathematically relatively clean (although our presentation does not pretend to be rigorous).

In order to give the ideas in their simplest context, we first derive the elastic scattering cross section and the optical theorem for particles scattered by a fixed scattering center, in Sec. 2. In Sec. 3 we give a similar derivation for the realistic situation of scattering by a foil of many identical scatterers. In that case a mixed beam is unnecessary. Section 4 contains the general derivation of the counting rate for m bound systems of particles colliding and giving rise to n bound systems, in an arbitrary coordinate system.

2. Scattering of a Particle by a Single Center

Let a wave packet[6)]

$$\psi_{in}^{(b)}(\underset{\sim}{r},t) = (2\pi)^{-3/2}\int(d\underset{\sim}{k})\, g(\underset{\sim}{k})\, e^{i\underset{\sim}{k}\cdot(\underset{\sim}{r}-\underset{\sim}{b})-itk^2/2m} \tag{2.1}$$

be sent toward the scattering center at the origin. We assume that $g(\underset{\sim}{k}) = g_o(\underset{\sim}{k}-\underset{\sim}{k}_o)$ is peaked about $\underset{\sim}{k}_o$ with a width $\Delta \ll k_o$, and that the impact parameter $\underset{\sim}{b}$ is a vector orthogonal to $\underset{\sim}{k}_o$. The function $g(\underset{\sim}{k})$ is normalized so that

$$\int(d\underset{\sim}{r})\,|\psi_{in}^{(b)}(\underset{\sim}{r},t)|^2 = \int(d\underset{\sim}{k})\,|g(\underset{\sim}{k})|^2 = 1. \tag{2.2}$$

The full wave function $\psi_b(\underset{\sim}{r},t)$, which strongly approaches $\psi_{in}^{(b)}(\underset{\sim}{r},t)$ as $t \to -\infty$, is then of the form

$$\psi_b(\underset{\sim}{r},t) = (2\pi)^{-3/2}\int(d\underset{\sim}{k})\, g(\underset{\sim}{k})\, \psi^{(+)}(\underset{\sim}{k},\underset{\sim}{r})\, e^{-i\underset{\sim}{k}\cdot\underset{\sim}{b}-itk^2/2m} \tag{2.3}$$

where $\psi^{(+)}(\underset{\sim}{k},\underset{\sim}{r})$ is the usual time-independent full outgoing-wave scattering wave function.

The probability of finding the particle in the infinite cone C in the distant future is given by

$$P_b = \lim_{t\to\infty}\int_C(d\underset{\sim}{r})\,|\psi_b(\underset{\sim}{r},t)|^2 = \lim_{t\to\infty}\int_C(d\underset{\sim}{r})\,|\psi_{out}(\underset{\sim}{r},t)|^2. \tag{2.4}$$

By Dollard's theorem (1.1) this equals

$$P_b = \int_C (d\underset{\sim}{k}) |\psi_{out}(\underset{\sim}{k})|^2 \tag{2.5}$$

where $\tilde{\psi}_{out}(\underset{\sim}{k})$ is the Fourier transform of $\psi_{out}(\underset{\sim}{r},0)$ and hence

$$\tilde{\psi}_{out}(\underset{\sim}{k}) = \int (d\underset{\sim}{k}') \, S(\underset{\sim}{k},\underset{\sim}{k}') \, g(\underset{\sim}{k}') \, e^{-i\underset{\sim}{k}'\cdot\underset{\sim}{b}} . \tag{2.6}$$

Here $S(\underset{\sim}{k},\underset{\sim}{k}')$ is the S-matrix,

$$S(\underset{\sim}{k},\underset{\sim}{k}') = \delta(\underset{\sim}{k}-\underset{\sim}{k}') - 2\pi i \, \delta(E-E') \, T_E(\hat{\underset{\sim}{k}},\hat{\underset{\sim}{k}}'). \tag{2.7}$$

If the cone C does not intersect a cone about the forward direction $\underset{\sim}{k}_o$, of opening angle $\Delta/\underset{\sim}{k}_o$, then the first term on the right-hand side of (2.7) does not contribute to the integral in (2.5).

Therefore,

$$P_b = (2\pi)^2 \int_C (d\underset{\sim}{k}) \int (d\underset{\sim}{k}')(d\underset{\sim}{k}'') \, \delta(E-E'') \, g(\underset{\sim}{k}') \, g^*(\underset{\sim}{k}'') \, T_E(\hat{\underset{\sim}{k}},\hat{\underset{\sim}{k}}') \, T_E^*(\hat{\underset{\sim}{k}},\hat{\underset{\sim}{k}}'') \, e^{i(\underset{\sim}{k}'-\underset{\sim}{k}'')\cdot\underset{\sim}{b}} . \tag{2.8}$$

We assume that in the region of the peak of g, T_E is essentially constant. One may therefore write :

$$P_b = \int_C (d\underset{\sim}{k}) |T_{E_o}(\hat{\underset{\sim}{k}},\hat{\underset{\sim}{k}}_o)|^2 \, h_b(\underset{\sim}{k}), \tag{2.9}$$

$$h_b = (4\pi m)^2 \int (d\underset{\sim}{k}')(d\underset{\sim}{k}'') \, \delta(k^2-k'^2) \, \delta(k'^2-k''^2) \, g(\underset{\sim}{k}') \, g^*(\underset{\sim}{k}'') \, e^{i(\underset{\sim}{k}'-\underset{\sim}{k}'')\cdot\underset{\sim}{b}} . \tag{2.10}$$

The next step is to shift both $\underset{\sim}{k}'$ and $\underset{\sim}{k}''$ by $\underset{\sim}{k}_o$, so that

$$k'^2-k''^2 \to (\underset{\sim}{k}'+\underset{\sim}{k}_o)^2 - (\underset{\sim}{k}''+\underset{\sim}{k}_o)^2 = k'^2-k''^2+2\underset{\sim}{k}_o\cdot(\underset{\sim}{k}'-\underset{\sim}{k}'').$$

Since $g_o(\underset{\sim}{k})$ is sharply peaked about zero with a width $\Delta \ll k_o$, we may replace

$$\delta(k'^2-k''^2) \to \delta\left[2\underset{\sim}{k}_o\cdot(\underset{\sim}{k}'-\underset{\sim}{k}'')\right] = (2k_o)^{-1} \, \delta(k_z'-k_z'')$$

if we choose the z-axis along $\underset{\sim}{k}_o$. Thus

$$h_b = \tfrac{1}{2}(4\pi m)^2 \int (d\underset{\sim}{k}')(d\underset{\sim}{k}'')\ k_o^{-1}\ \delta(k^2-k_o^2)\ \delta(k_z'-k_z'')\ g_o(\underset{\sim}{k}')\ g_o^*(\underset{\sim}{k}'')\ e^{i(\underset{\sim}{k}'-\underset{\sim}{k}'')\cdot\underset{\sim}{b}}\ . \qquad (2.11)$$

Suppose now that the initial state is mixed and described by a density matrix

$$\rho = \sum_b p_b\ \Psi_b\ \Psi_b^\dagger$$

where p_b is the probability for the impact parameter $\underset{\sim}{b}$. Then the probability of finding the particle in the cone C as $t \to \infty$ is given by

$$P = \lim_{t\to\infty} \int_C (d\underset{\sim}{r})\ \rho(\underset{\sim}{r},\underset{\sim}{r}) = \sum_b p_b\ P_b .$$

We take the "beam" to be uniform, with a cross sectional area a, centered at the origin. Then we may write

$$P = (1/a) \int_{|b|\leq B} (d\underset{\sim}{b})\ P_b \qquad (2.12)$$

with $a = \pi B^2$.

We must now make the important assumption that the forces are microscopic and the "beam" macroscopic. This means, specifically, that P_b, as a function of b, rapidly drops to zero at values of b much smaller than B. We may then allow the b-integral effectively to extend to infinity, and set

$$\int (d^2\underset{\sim}{b})\ e^{i(\underset{\sim}{k}'-\underset{\sim}{k}'')\cdot\underset{\sim}{b}} = (2\pi)^2\ \delta^2(\underset{\sim}{k}_\perp'-\underset{\sim}{k}_\perp'') \qquad (2.13)$$

$\underset{\sim}{k}_\perp$ being perpendicular to $\underset{\sim}{k}_o$ because that is the plane over which $\underset{\sim}{b}$ varies.

Using (2.9), (2.11), (2.13), and (2.2) in (2.12) we obtain

$$P = (2\pi)^4\ m^2 a^{-1} \int_C (d\hat{\underset{\sim}{k}})\,|T_{E_o}(\hat{\underset{\sim}{k}},\hat{\underset{\sim}{k}}_o)|^2 \qquad (2.14)$$

Now if the original "beam" contains N particles then it follows from (2.14) that the number of particles found in C is given by

$$N_C = (2\pi)^4 m^2 dL \int_C (d\hat{k}) |T_{E_o}(\hat{k},\hat{k}_o)|^2$$

where $d = (N/aL)$ is the density of the "beam" of length $L = v_o T$, if T is the duration of the "beam". We have $Ld = TF$, F being the flux. Hence the number of particles counted in C per unit time, or by (1.8), the number of particles crossing a distant surface intersecting C, per unit time, is given by

$$N_C/T = F \int_C (d\hat{k})\ d\sigma/d\Omega \tag{2.15}$$

with

$$d\sigma/d\Omega = (2\pi)^4 m^2 |T_{E_o}|^2. \tag{2.16}$$

On the other hand, let C be a narrow cone that includes the forward direction. Then the first term on the right-hand side of (2.7), used in (2.6), gives "1" in (2.5) and in (2.12). The second term, used alone in (2.6) gives (2.10) and (2.14). But the cross-term between the first and second term on the right of (2.7) in (2.5) gives rise to the following:

$$P_b^{int} = 4\pi \int_C (dk) \int (dk')\ \mathrm{Im}\left[g^*(k)\ g(k')\ \delta(E-E')\ e^{i(k'-k)\cdot b}\ T_E(k,k')\right]$$

$$= 4\pi m k_o^{-1}\ \mathrm{Im} \int (dk)\ T_{E_o}(k,k_o) \int (dk')\ g_o^*(k)\ g_o(k')\ \delta(k_z-k_z')\ e^{i(k'-k)\cdot b}$$

by the same arguments that lead to (2.11). The integration in (2.12) then leads, via (2.13), to

$$P^{int} = 2(2\pi)^3 m k_o^{-1} a^{-1}\ \mathrm{Im} \int_C (dk)\ T_{E_o}(k,k_o) |g_o(k)|^2.$$

If we assume that the opening angle of C is narrow but large compared to Δ/k_o, this may be evaluated as

$$P^{int} = (4\pi/k_o)\ \mathrm{Im}\left[(2\pi)^2\ m\ T_{E_o}(k_o,k_o)\right]\ a^{-1}$$

because of (2.2). At the same time, making the cone narrow, makes (2.14) very small. Hence, we get

$$P = 1 + (4\pi/k_o)\ \mathrm{Im}\left[(2\pi)^2 m\ T_{E_o}(k_o,k_o)\right]\ a^{-1}.$$

The probability of finding the particle not in the narrow forward cone is therefore,

$$1 - P = -(4\pi/k_o)\ \mathrm{Im}\left[(2\pi)^2 m\ T_{E_o}(\underset{\sim}{k}_o, \underset{\sim}{k}_o)\right] a^{-1}$$

and that must equal the probability of being scattered anywhere which, according to (2.14) is equal to

$$a^{-1} \int (d\hat{\underset{\sim}{k}}) |(2\pi)^2 m\ T_{E_o}(\hat{\underset{\sim}{k}}, \hat{\underset{\sim}{k}}_o)|^2 .$$

This is the optical theorem.

3. Scattering by Many Centers

The wave function of a particle scattered by many identical centers located at $\{\underset{\sim}{r}_\alpha\}$, satisfies the integral equation

$$\psi^{(+)}(\underset{\sim}{k}, \underset{\sim}{r}) = e^{i\underset{\sim}{k}\cdot\underset{\sim}{r}} + \sum_\alpha \int (d\underset{\sim}{r}')\ G_o^+(k, \underset{\sim}{r}-\underset{\sim}{r}')\ V(\underset{\sim}{r}'-\underset{\sim}{r}_\alpha)\psi^{(+)}(\underset{\sim}{k}, \underset{\sim}{r}') \tag{3.1}$$

which at large distance from all the scatterers has the asymptotic value

$$\psi^{(+)}(\underset{\sim}{k}, \underset{\sim}{r}) = e^{i\underset{\sim}{k}\cdot\underset{\sim}{r}} + r^{-1} e^{ikr}\ A^{coll}(k\hat{\underset{\sim}{r}}, \underset{\sim}{k}) + O(r^{-1})$$

with

$$A^{coll}(\underset{\sim}{k}', \underset{\sim}{k}) = -(1/4\pi) \sum_\alpha \int (d\underset{\sim}{r})\ e^{-i\underset{\sim}{k}'\cdot\underset{\sim}{r}}\ V(\underset{\sim}{r}-\underset{\sim}{r}_\alpha)\ \psi^{(+)}(\underset{\sim}{k}, \underset{\sim}{r}) .$$

Setting

$$\psi_\alpha(\underset{\sim}{r}) = \psi^{(+)}(\underset{\sim}{k}, \underset{\sim}{r}+\underset{\sim}{r}_\alpha),$$

we can write

$$A^{coll}(\underset{\sim}{k}', \underset{\sim}{k}) = \sum_\alpha A_\alpha(\underset{\sim}{k}', \underset{\sim}{k})\ e^{-i\underset{\sim}{k}'\cdot\underset{\sim}{r}_\alpha} \tag{3.2}$$

where

$$A_\alpha(\underset{\sim}{k}', \underset{\sim}{k}) = -(1/4\pi) \int (d\underset{\sim}{r})\ e^{-i\underset{\sim}{k}'\cdot\underset{\sim}{r}}\ V(\underset{\sim}{r})\ \psi_\alpha(\underset{\sim}{r}).$$

If the potential has the range R, then we need here $\psi_\alpha(r)$ only for $r \leq R$. Let us assume that for $\alpha \neq \beta$, $r_{\alpha\beta} = |\underset{\sim}{r}_\alpha - \underset{\sim}{r}_\beta| >> R$, that is, the scattering centers are well outside each other's range. Then we may approximate the Green's function in the integral equation

$$\psi_\alpha(\underset{\sim}{r}) = e^{i\underset{\sim}{k}\cdot(\underset{\sim}{r}+\underset{\sim}{r}_\alpha)} + \sum_{\beta\neq\alpha} \int (d\underset{\sim}{r}')\, G_o^+(k;\underset{\sim}{r}-\underset{\sim}{r}'+\underset{\sim}{r}_{\alpha\beta})\, V(\underset{\sim}{r}')$$

$$\psi_\beta(\underset{\sim}{r}') + \int (d\underset{\sim}{r}')\, G_o^+(k,\underset{\sim}{r}-\underset{\sim}{r}')\, V(\underset{\sim}{r}')\, \psi_\alpha(\underset{\sim}{r}')$$

by

$$G_o^+(k,\underset{\sim}{r}+\underset{\sim}{r}_{\alpha\beta}-\underset{\sim}{r}') \simeq -(1/4\pi)\, e^{ikr_{\alpha\beta}}\, r_{\alpha\beta}^{-1}\, e^{i\underset{\sim}{k}_{\alpha\beta}\cdot(\underset{\sim}{r}-\underset{\sim}{r}')}$$

where $\underset{\sim}{k}_{\alpha\beta} = k\hat{\underset{\sim}{r}}_{\alpha\beta}$. Hence for $r < R$

$$\psi_\alpha(\underset{\sim}{r}) \simeq e^{i\underset{\sim}{k}\cdot(\underset{\sim}{r}+\underset{\sim}{r}_\alpha)} + \int (d\underset{\sim}{r}')\, G_o^+(k,\underset{\sim}{r}-\underset{\sim}{r}')\, V(\underset{\sim}{r}')\, \psi_\alpha(\underset{\sim}{r}')$$

$$+ \sum_{\beta\neq\alpha} r_{\alpha\beta}^{-1}\, e^{ikr_{\alpha\beta}}\, e^{i\underset{\sim}{k}_{\alpha\beta}\cdot\underset{\sim}{r}}\, A_\beta(\underset{\sim}{k}_{\alpha\beta},\underset{\sim}{k}). \qquad (3.3)$$

This is an integral equation for $\psi_\alpha(\underset{\sim}{r})$ in the region $r < R$. Let $\phi^{(+)}(\underset{\sim}{k},\underset{\sim}{r})$ be the solution of the one-center equation

$$\phi^{(+)}(\underset{\sim}{k},\underset{\sim}{r}) = e^{i\underset{\sim}{k}\cdot\underset{\sim}{r}} + \int (d\underset{\sim}{r}')\, G_o^+(k,\underset{\sim}{r}-\underset{\sim}{r}')\, V(\underset{\sim}{r}')\, \phi^{(+)}(\underset{\sim}{k},\underset{\sim}{r}').$$

We then obtain from (3.3) for $r < R$

$$\psi_\alpha(\underset{\sim}{r}) \simeq e^{i\underset{\sim}{k}\cdot\underset{\sim}{r}_\alpha}\, \phi^+(\underset{\sim}{k},\underset{\sim}{r}) + \sum_{\alpha\neq\beta} r_{\alpha\beta}^{-1}\, e^{ikr_{\alpha\beta}}\, A_\beta(\underset{\sim}{k}_{\alpha\beta},\underset{\sim}{k})\, \phi^+(\underset{\sim}{k}_{\alpha\beta},\underset{\sim}{r}).$$

Writing $A(\underset{\sim}{k}',\underset{\sim}{k})$ for the one-center scattering amplitude,

$$A(\underset{\sim}{k}',\underset{\sim}{k}) = -(1/4\pi) \int (d\underset{\sim}{r})\, e^{-i\underset{\sim}{k}'\cdot\underset{\sim}{r}}\, V(\underset{\sim}{r})\, \phi^{(+)}(\underset{\sim}{k},\underset{\sim}{r})$$

we get in the approximation $r_{\alpha\beta} >> R$

$$A_\beta(\underset{\sim}{k}',\underset{\sim}{k}) \simeq e^{i\underset{\sim}{k}\cdot\underset{\sim}{r}_\alpha}\, A(\underset{\sim}{k}',\underset{\sim}{k}) + \sum_{\alpha\neq\beta} r_{\alpha\beta}^{-1} e^{ikr_{\alpha\beta}}\, A_\alpha(\underset{\sim}{k}_{\beta\alpha},\underset{\sim}{k})\, A(\underset{\sim}{k}',\underset{\sim}{k}_{\beta\alpha}) .$$

On the assumption that not only $r_{\alpha\beta} >> R$ but that $r_{\alpha\beta} >> |A|$, that is, the inter-center distance is large compared to the "scattering diameter", then the above equations can be solved by iteration and the first term can be expected to be an excellent approximation to this rapidly converging multiple-scattering series. In this approximation then the scattering amplitude for the collection of scatterers is given by (3.2) as

$$A^{coll}(\underset{\sim}{k}',\underset{\sim}{k}) \simeq A(\underset{\sim}{k}',\underset{\sim}{k}) \sum_{\alpha} e^{i(\underset{\sim}{k}-\underset{\sim}{k}')\cdot \underset{\sim}{r}_{\alpha}} \tag{3.4}$$

in terms of the single-center scattering amplitude A. The same relation therefore holds for the T-matrix:

$$T_E^{coll}(\hat{\underset{\sim}{k}}',\hat{\underset{\sim}{k}}) \simeq T_E(\hat{\underset{\sim}{k}}',\hat{\underset{\sim}{k}}) \sum_{\alpha} e^{i(\underset{\sim}{k}-\underset{\sim}{k}')\cdot \underset{\sim}{r}_{\alpha}}. \tag{3.5}$$

We now evaluate the probability P for finding the particle in the cone C in the distant future, assuming that it was sent in as a wave packet (2.3), which in the infinite part was of the form (2.1), normalized as in (2.2). Since it will not matter in the future, we will set b = 0.

If the cone C does not include the forward direction we then get, by the same arguments that lead to (2.9), and by (3.5),

$$P = \int_C (d\underset{\sim}{k}) \; |T_{E_o}(\hat{\underset{\sim}{k}},\hat{\underset{\sim}{k}}_o)|^2 \; h(\underset{\sim}{k}) \tag{3.6}$$

where

$$h = (4\pi m)^2 \; \delta(k^2-k_o^2) \int (d\underset{\sim}{k}')(d\underset{\sim}{k}'') \; \delta(k'^2-k''^2) \; g(\underset{\sim}{k}') \; g^*(\underset{\sim}{k}'')$$
$$\sum_{\beta} e^{i(\underset{\sim}{k}'-\underset{\sim}{k}'')\cdot \underset{\sim}{r}_{\beta}} \sum_{\alpha} e^{i(\underset{\sim}{k}'-\underset{\sim}{k})\cdot \underset{\sim}{r}_{\alpha\beta}} . \tag{3.7}$$

If the scattering centers form a regular array, as in a crystal, the function h leads to regular interference patterns with Bragg angles. In general the term $e^{-i\underset{\sim}{k}\cdot \underset{\sim}{r}_{\alpha\beta}}$ for $\alpha\neq\beta$ will lead to interference fringes that are rapidly favying functions of the angles of $\hat{\underset{\sim}{k}}$ if $k_o r_{\alpha\beta} >> 1$. Suppose then that the angular opening Θ of the cone C is such that it contains many of these interference fringes, that is, that although $\Theta << 1$, we have for $\alpha\neq\beta$

$$\Theta k_o r_{\alpha\beta} >> 1.$$

Then in the main region of the $\underset{\sim}{k}'$-integral the $\underset{\sim}{\hat{k}}$-integral oscillates to zero. More specifically,

$$I = \int_C (d\hat{k})\ e^{i\lambda \underset{\sim}{\hat{k}} \cdot \underset{\sim}{n}} = \Theta^2\ O\left[(\lambda\Theta)^{-\frac{1}{2}}\right].$$

when $\lambda\ \Theta >> 1$. (We have written here $\lambda = -kr_{\alpha\beta}$, $\underset{\sim}{\hat{n}} = \underset{\sim}{\hat{r}}_{\alpha\beta}$). This is easily seen by a stationary-phase argument.

In our case we have effectively $\lambda = -k_o r_{\alpha\beta}$ and as a result the terms in (3.7) with $\alpha \neq \beta$ are of order $(k_o r_{\alpha\beta})^{-\frac{1}{2}}$. Thus, the terms with $\alpha=\beta$ dominate and we get to an excellent approximation

$$h = (4\pi m)^2\ \delta(k^2-k_o^2) \int (d\underset{\sim}{k}')(d\underset{\sim}{k}'')\ \delta(k'^2-k''^2)\ g(\underset{\sim}{k}')\ g^*(\underset{\sim}{k}'')$$

$$\sum_\beta e^{i(\underset{\sim}{k}'-\underset{\sim}{k}'')\cdot \underset{\sim}{r}_\beta}.$$

Let us now shift variables of integration as in the transition from (2.10) to (2.11). That leads to

$$h = 2\ (2\pi m)^2\ k_o^{-1}\ \delta(k^2-k_o^2) \int (d\underset{\sim}{k}')(d\underset{\sim}{k}'')\ \delta(k_z'-k_z'') g_o(\underset{\sim}{k}')\ g_o^*(\underset{\sim}{k}'')$$

$$\sum_\beta e^{i(\underset{\sim}{k}_\perp'-\underset{\sim}{k}_\perp'')\cdot \underset{\sim}{r}_\beta}.$$

The vectors with subscripts $_\perp$ are projections onto the x,y-plane. (These arise from $\delta(k_z'-k_z'')$). We now replace the sum over β by an integral,

$$\sum_\beta e^{i(\underset{\sim}{k}_\perp'-\underset{\sim}{k}_\perp'')\cdot \underset{\sim}{r}_{\perp\beta}} \rightarrow d \int (d\underset{\sim}{r}_\perp)\ e^{i(\underset{\sim}{k}_\perp'-\underset{\sim}{k}_\perp'')\cdot \underset{\sim}{r}_\perp}$$

$$= (2\pi)^2\ d\ \delta(\underset{\sim}{k}_\perp'-\underset{\sim}{k}_\perp'')$$

where d is the number of scatterers per unit area perpendicular to the direction $\underset{\sim}{k}_o$ of the incident beam. Hence, because of (2.2)

$$h = 2(4\pi^2 m)^2\ k_o^{-1}\ \delta(k^2-k_o^2),$$

and (3.6) gives

$$P = d(2\pi)^4 m^2 \int_C (d\underset{\sim}{\hat{k}}) |T_{E_o}(\underset{\sim}{\hat{k}}, \underset{\sim}{\hat{k}}_o)|^2 \tag{3.8}$$

with the integral now running over the solid angle of C only.

4. N-Body Reactions

We shall now consider the general situation of N particles, initially bound in m clusters. We start from an abstract formulation of time-dependent scattering theory. It should be kept in mind that we are always using normalizable state vectors; in other words, wave packets.

A scattering state $\Psi^{(\pm)}_{\alpha j}(t)$ satisfies the Schrödinger equation

$$i \frac{\partial}{\partial t} \Psi^{(\pm)}_{\alpha j}(t) = H \Psi^{(\pm)}_{\alpha j}(t) \tag{4.1}$$

and is uniquely specified by its "preparation" as an H_α-developing state $\Psi_{\alpha j}(t)$ in the distant past or future. That is,

$$i \frac{\partial}{\partial t} \Psi_{\alpha j}(t) = H_\alpha \Psi_{\alpha j}(t) \tag{4.2}$$

and

$$\lim_{t \to \mp\infty} \| \Psi^{(\pm)}_{\alpha j}(t) - \Psi_{\alpha j}(t) \| = 0 \ . \tag{4.3}$$

We assume both states normalized to unity:

$$\| \Psi_{\alpha j}(t) \| = \| \Psi^{(\pm)}_{\alpha j}(t) \| = 1 \ . \tag{4.4}$$

The label α specifies the "arrangement channel", i.e., which particles are bound to which others, and j labels the specific state of the fragments. H_α is the conventional channel Hamiltonian, and the space spanned by all $\Psi_{\alpha j}(t)$ with fixed α is called R_α. It is physically interesting to note that with (4.4), Eq.(4.3) is equivalent to the demand

$$\lim_{t \to \mp\infty} (\Psi_{\alpha j}(t), \Psi^{(\pm)}_{\alpha j}(t)) = 1 \tag{4.5}$$

i.e., that the probability of finding the particles in the state $\Psi_{\alpha j}(t)$ in the infinite past or future, respectively, be unity. The wave operators[7] map $\Psi_{\alpha j}(t)$ to $\Psi^{(\pm)}_{\alpha j}(t)$:

$$\Psi_{\alpha j}^{(\pm)}(t) = \Omega_{\alpha}^{(\pm)} \Psi_{\alpha j}(t) \tag{4.6}$$

and vanish on the orthogonal complement of R_α.

The probability of finding the particles in the state $\Psi_{\beta f}(t)$ in the distant future, knowing that in the distant past they were in the state $\Psi_{\alpha i}(t)$, is given by the squared magnitude of

$$(\Psi_{\beta f}^{(-)}(t), \Psi_{\alpha i}^{(+)}(t)) = (\Psi_{\beta f}(t), S_{\beta\alpha} \Psi_{\alpha i}(t)) = (\Psi_{\beta f}(0), S_{\beta\alpha} \Psi_{\alpha i}(0)) \tag{4.7}$$

where

$$S_{\beta\alpha} = \Omega_{\beta}^{(-)\dagger} \Omega_{\alpha}^{(+)} \tag{4.8}$$

is a scattering operator.

The wave operators $\Omega_{\alpha}^{(-)}$ are partially isometric, and

$$Q_{\alpha}^{(-)} = \Omega_{\alpha}^{(-)} \Omega_{\alpha}^{(-)\dagger}$$

is the orthogonal projection onto the range $R_{\alpha}^{(-)}$ of $\Omega_{\alpha}^{(-)}$. As is well-known, the spaces $R_{\alpha}^{(-)}$ and $R_{\beta}^{(-)}$, for $\alpha \neq \beta$, are orthogonal to one another and to the space R_B spanned by the bound states of the N-particle system. Moreover,

$$\sum_{\alpha} Q_{\alpha}^{(-)} = 1 - \Lambda, \tag{4.9}$$

where Λ is the orthogonal projection onto R_B.

Now $Q_{\beta}^{(-)}\Psi_{\alpha i}^{(+)}(t)$ is a vector in $R_{\beta}^{(-)}$; hence, as $t \to +\infty$, it must strongly approach a Ψ_β-state, say $\Psi'_\beta(t)$:

$$\lim_{t\to+\infty} \| Q_{\beta}^{(-)} \Psi_{\alpha i}^{(+)}(t) - \Psi'_{\beta}(t) \| = 0. \tag{4.10}$$

It follows that for any arbitrary state $\Psi_{\beta}^{(-)}(t)$ in $R_{\beta}^{(-)}$

$$(\Psi_\beta(0), S_{\beta\alpha} \Psi_{\alpha i}(0)) = \lim_{t\to+\infty} (\Psi_{\beta}^{(-)}(t), Q_{\beta}^{(-)} \Psi_{\alpha i}^{(+)}(t)) = (\Psi_\beta(0), \Psi'_\beta(0)).$$

Hence for all $\Psi_\beta \in R_\beta$ and Ψ''_β defined by

$$\Psi''_\beta = S_{\beta\alpha}\,\Psi_{\alpha i}(0) - \Psi'_\beta(0)$$

we have

$$(\Psi_\beta, \Psi''_\beta) = 0.$$

Specifically, we may choose $\Psi_\beta = \Psi''_\beta$ and then obtain $||\Psi''_\beta|| = 0$. Therefore,

$$\Psi'_\beta(0) = S_{\beta\alpha}\,\Psi_{\alpha i}(0) \tag{4.11}$$

and consequently from (4.10)

$$\lim_{t\to+\infty} ||Q^{(-)}_\beta\,\Psi^{(+)}_{\alpha i}(t) - S_{\beta\alpha}\,\Psi_{\alpha i}(t)|| = 0 \tag{4.12}$$

because it follows from (4.2), (4.6), and (4.8) that (4.11) implies:

$$\Psi'_\beta(t) = S_{\beta\alpha}\,\Psi_{\alpha i}(t).$$

Eq. (4.9) together with (4.12) implies

$$\lim_{t\to+\infty} ||\Psi^{(+)}_{\alpha i}(t) - \sum_\beta S_{\beta\alpha}\,\Psi_{\alpha i}(t)|| = 0 . \tag{4.13}$$

In a specific representation (4.13) is not very useful because the elements of the S-matrix are not matrix elements of $S_{\beta\alpha}$ on one and the same basis. However, (4.12) can be represented in a straight-forward manner in a representation that uses the bound-state fragment wave functions in the β-channel together with a momentum representation for the motion of the centers of mass of the fragments. Let us first subject (4.12) to further projection onto the f-channel in the β-arrangement channel (i.e., the f-level of the fragments)

$$Q^{(-)}_{\beta f}\Psi^{(+)}_{\alpha i}(t) \underset{t\to+\infty}{\overset{s}{\to}} \Psi'_{\beta f}(t) = P_{\beta f}\,S_{\beta\alpha}\,\Psi_{\alpha i}(t) \tag{4.14}$$

where $Q^{(-)}_{\beta f}$ is the orthogonal projection onto the space spanned by the vectors $\Psi^{(-)}_{\beta f}(t)$ for fixed β and f, and $P_{\beta f}$ is the orthogonal projection onto the βf-channel.

The initial wave function in a mixed α-channel momentum representation is given by

$$\tilde{\psi}_{\alpha i}(k,t) = g(k)\, e^{-i[E(k) + \varepsilon_i]t}$$

where g is normalized to unity, ε_i is the bound-state energy of the fragments in their ith bound state, k is an m-dimensional momentum vector comprising the momenta of the centers of mass of the m fragments in the α-channel, and E(k) is the energy of this fragment motion. This means that the initial wave packet in the (partial) coordinate representation is given by

$$\psi_{\alpha i}(x,t) = (2\pi)^{-3m/2} \int d^{3m}k\, g(k)\, e^{-i[E(k)+\varepsilon_i]t + ik\cdot x} .$$

Similarly, the projection of the final state on the βf-channel in a partial momentum representation will be called $\psi'_{\beta f}(k',t)$. In the (partial) coordinate representation it is then given by

$$\psi'_{\beta f}(x,t) = (2\pi)^{-3m/2} \int d^{3m} k'\, \psi'_{\beta f}(k',t)\, e^{ix'\cdot k'} .$$

The transcription of (4.14) in the partial momentum representation now reads

$$\psi'_{\beta f}(k',t) = \int d^{3m} k\, S_{\beta\alpha}(f,k';i,k)\, g(k)\, e^{-i[E(k) + \varepsilon_i]t}$$

$$= e^{-i[E(k') + \varepsilon_f]t} \int d^{3m} k\, S_{\beta\alpha}(f,k';i,k)\, g(\underset{\sim}{k}),$$

$S_{\beta\alpha}(f,k';i,k)$ being of the form

$$S_{\beta\alpha}(f,k';i,k) = \delta_{\beta\alpha}\, \delta_{fi}\, \delta^{3m}(k-k')$$

$$-2\pi i\, \delta\left[E(k) + \varepsilon_i - E(k') - \varepsilon_f\right] \delta^3 (\underset{\sim}{K}-\underset{\sim}{K}')T_E(\beta f,k';\alpha i,k) \qquad (4.15)$$

where $\underset{\sim}{K}$ is the total 3-dimensional momentum of the fragments

$$\underset{\sim}{K} = \sum_{j=1}^{m} \underset{\sim}{k}_j ,\quad \underset{\sim}{K}' = \sum_{j=1}^{n} \underset{\sim}{k}'_j .$$

The probability of finding the system in the distant future

in the βf-channel, with the centers of mass of the fragments in the cone C, is given by

$$P = \lim_{t\to\infty} \| Q_C \; Q_{\beta f} \; \Psi^{(+)}_{\alpha i}(t) \|^2$$

$$= \lim_{t\to\infty} \int_C d^{3n}x' \, |\psi_{\beta f}(x',t)|^2$$

if Q_C is the projection onto the cone C in coordinate space. The "scattering-into-cones" theorem (1.1) tells us that

$$P = \int_C d^{3n} \, k |\tilde{\psi}_{\beta f}(k,t)|^2$$

$$= \int_C d^{3n}k | \int d^{3m} \, k' \; S_{\beta\alpha}(f,k;i,k') \; g(k')|^2 \; . \qquad (4.16)$$

If C does not include the forward direction then the first term in (4.15) does not contribute. If, furthermore, g(k) is sharply peaked at $k = k_o$ and T_E is essentially constant in that peak, then T_E may be taken outside the k' - integral in (4.16) and we get[8)]

$$P = \int_C d^{3n}k |T_{E_o}(f,k;i,k_o)|^2 \; h(k) \qquad (4.17)$$

where

$$h = (2\pi)^2 \int d^{3m}k' d^{3m}k'' \; g(k') g^*(k'')$$

$$\delta(E'-E'') \; \delta(E-E'+\varepsilon_f-\varepsilon_i) \; \delta^3(K'-K'') \; \delta^3(K'-K) \quad . \qquad (4.18)$$

We must now be more specific about the incoming wave packets. Suppose they have impact parameter b. Then we will lable P and h by that and call them P_b and h_b. We must examine the notion of an impact parameter for an N-particle system with some care[9)].

For a single particle of momentum $\underset{\sim}{k}$ the impact parameter relative to a center (say, the origin) is a displacement vector orthogonal to $\underset{\sim}{k}$; for a wave packet of one particle, with the central momentum $\underset{\sim}{k}^o$, it is orthogonal to $\underset{\sim}{k}^o$. For m fragments with central momenta $\underset{\sim}{k}_i^o$ there are two things to note: The first is that we must not displace the center of mass; the second is that even a displacement of a particle along its track may have an impact parameter-like effect in the collision. If all particles are displaced along their tracks in proportion to their (central) velocities, on the other hand, the result is merely a displacement of the time of collision.

In view of these considerations it is clear that the impact parameter space can be taken to be the orthogonal complement (in a convenient metric) of the four-dimensional space that is the direct sum of the three-dimensional space in which each particle is located at the center of mass, and the one-dimensional space generated by the (central) particle velocities relative to the center of mass.

We implement these ideas as follows.

Let $\underset{\sim}{k}_i^o$, $i = 1, \ldots, m$ be the particle momenta (at the center of the wave packet);

$$\underset{\sim}{K}^o = \sum_{i=1}^{m} \underset{\sim}{k}_i^o$$

the total momentum;

$$\underset{\sim}{V}_i^o = \frac{\underset{\sim}{k}_i^o}{m_i} - \frac{\underset{\sim}{K}^o}{M}, \quad i = 1, \ldots,$$

the velocities relative to the center of mass, with

$$M = \sum_{i=1}^{m} m_i,$$

and

$$E^o = \sum_{i=1}^{m} \frac{\underset{\sim}{k}_i^{o2}}{2m_i} - \frac{\underset{\sim}{K}^{o2}}{2M} = \sum_{i=1}^{m} \tfrac{1}{2} m_i \underset{\sim}{V}_i^{o2}$$

the energy in the center of mass system. Let $\xi = \{\underset{\sim}{\xi}_i\}_{i=1}^{m}$ be any 3M-dimensional vector made up out of m 3-vectors.

We first transform to a new set of components defined by

$$\tilde{\underset{\sim}{\xi}}_i = m_i^{\frac{1}{2}} \underset{\sim}{\xi}_i, \quad i = 1, \ldots, m \quad .$$

Next we perform a rotation to obtain the components

$$\bar{\xi}_i = M^{-\frac{1}{2}} \sum_{j=1}^{m} m_j^{\frac{1}{2}} \tilde{\xi}_j^{(i)} \quad , \; i = 1, 2, 3$$

$$\bar{\xi}_4 = (2E^o)^{-\frac{1}{2}} \sum_{j=1}^{m} m_j^{\frac{1}{2}} \underset{\sim}{V}_j^o \cdot \underset{\sim}{\tilde{\xi}}_j$$

and the remaining components $\bar{\xi}_i$, $i = 5, \ldots, 3m$ arbitrarily related to the $\tilde{\xi}_j$, except that the entire transformation be orthogonal. We then have

$$\sum_{i=1}^{3m} \bar{\xi}_i^2 = \bar{\xi} \cdot \bar{\xi} = \sum_i m_i \; |\xi_i|^2$$

and

$$\frac{\partial(\bar{\xi}_1, \ldots, \bar{\xi}_{3m})}{\partial(\xi_1, \ldots, \xi_{3m})} = (\prod_i m_i)^{3/2} \; . \tag{4.19}$$

We shall set

$$\xi^F = \{\xi_1, \xi_2, \xi_3, \xi_4, \; 0, \; 0, \; 0, \; \ldots\}$$

and

$$\xi^{\#} = \bar{\xi} - \xi^F .$$

The volume element is given by

$$d^{3m} \xi = (\prod_i m_i)^{-3/2} \, d^4 \xi^F \, d^{3m-4} \xi^{\#} \tag{4.20}$$

because of (4.19). It follows that

$$\delta^{3m}(\xi) = (\prod_i m_i)^{3/2} \, \delta^4(\xi^F) \, \delta^{3m-4}(\xi^{\#}) \; .$$

Now

$$\delta(\bar{\xi}_1)\,\delta(\bar{\xi}_2)\delta(\bar{\xi}_3) = M^{3/2}\,\delta^3(\sum_{i=1}^{m} m_i\,\underset{\sim}{\xi}_i)$$

and

$$\delta(\bar{\xi}_4) = (2E^o)^{\frac{1}{2}}\,\delta(\sum_{i=1}^{m} m_i\,\underset{\sim}{V}_i^o\cdot\underset{\sim}{\xi}_i).$$

This means that in the case of velocities

$$\delta(\bar{v}_1 - \bar{v}_1')\,\delta(\bar{v}_2 - \bar{v}_2')\,\delta(\bar{v}_3 - \bar{v}_3') = M^{3/2}\,\delta(\underset{\sim}{K} - \underset{\sim}{K}')$$

and

$$\delta(\bar{v}_4 - \bar{v}_4') = (2E^o)^{\frac{1}{2}}\,\delta[\sum_{i=1}^{m} \underset{\sim}{V}_i^o\cdot(\underset{\sim}{k}_i - \underset{\sim}{k}_i')]$$

where $\underset{\sim}{k}_i = m_i\,\underset{\sim}{v}_i$ and $\underset{\sim}{K} = \sum_i \underset{\sim}{k}_i$. If $\underset{\sim}{k}_i$ and $\underset{\sim}{k}_i'$ are both restricted to be close to $\underset{\sim}{k}_i^o$ then one finds that with $E = \sum_i k_i^2/2m_i$

$$\delta^4(v_i^F - v_i^{F'}) = (2E^o)^{\frac{1}{2}}\,M^{3/2}\,\delta^3(\underset{\sim}{K}-\underset{\sim}{K}')\,\delta(E-E'). \qquad (4.21)$$

Now the probability of finding the fragments in the cone C in the distant future, if the incoming impact parameter was b, is, by (4.17),

$$P_b = \int_C d^{3n}k\,|T_{E_o}(f,k;i,k_o)|^2\,h_b(k) \qquad (4.22)$$

where by (4.21)

$$h_b = (2\pi)^2 \int dk^{3m'}\,d^{3m}\,k''\,g(k')\,g^*(k'')\,e^{-i\sum_j b_j\cdot(k_j'-k_j'')}$$

$$\delta(E'-E'')\,\delta(E-E'+\varepsilon_f-\varepsilon_i)\,\delta^3(\underset{\sim}{K}'-\underset{\sim}{K}'')\,\delta^3(\underset{\sim}{K}'-\underset{\sim}{K})$$

$$= (2\pi)^2\,M^{-3/2}\,(2E^o)^{-\frac{1}{2}} \int d^{3m}k'\,d^{3m}k''\,g(k')\,g^*(k'')$$

$$\delta(E-E'+\varepsilon_f-\varepsilon_i)\,\delta^4(v^{F'}-v^{F''})\,\delta^3(\underset{\sim}{K}-\underset{\sim}{K}')\,\exp[-ib^{\#}\cdot(v^{\#'}-v^{\#''})]\,. \qquad (4.23)$$

If the initial state is mixed with impact parameter b_α having the probability p_α then

$$P = \sum_\alpha p_\alpha P_{b_\alpha} = \int d^{3m-4} b^{\#} \mu(b^{\#}) P_b$$

or, by (4.22)

$$P = \int_C d^{3n}k |T_{E_o}(f,k;i,k_o)|^2 \; h(k) \tag{4.24}$$

$$h(k) = \int d^{3m-4} b^{\#} \mu(b^{\#}) \; h_b(k), \tag{4.25}$$

where $\mu(b^{\#})$ is the density in impact-parameter space, i.e., the probability density for the particles to have their impact parameter in $d^{3m-4} b^{\#}$. We shall assume that the ith "beam" has the density ρ_i. If we think of ρ_i as depending on the variables $\underset{\sim}{v}_i^o(t)$ and $\underset{\sim}{b}_i$ with $\underset{\sim}{b}_i \cdot \underset{\sim}{v}_i^o = 0$, then we further assume that $\rho_i \neq 0$ only when $t_i \leq t \leq t_i + T_i$ and $\pi |b_i|^2 \leq a_i$. When these conditions are satisfied then ρ_i is assumed to be constant and hence it must have the value

$$\rho_i^o = 1/V_i$$

where

$$V_i = a_i \; v_i^o \; T_i,$$

so that V_i is the volume of the ith pulse, a_i its cross sectional area, T_i its duration, and $v_i^o \; T_i$ its length. We then have

$$\mu(b^{\#}) = (\prod_i \rho_i)/\beta$$

where

$$\beta = \int d^{3m-4} b^{\#} \prod_i \rho_i . \tag{4.26}$$

We now have to make two assumptions that in a realistic situation usually are extremely well satisfied. Assumption (Aa) is that P_b be a sufficiently rapidly decreasing function of b so that in the calculation of P the integral over $b^{\#}$ may be extended to infinity without appreciable error. In fact, we must assume (Ab) that the radii a_i^2/π of the beams are so large that the impact parameters that contribute to the scattering are negligible compared to the maximal impact parameters contributed by the beams. This should be no more than the recognition that the beams are macroscopic and the ranges of the forces involved, microscopic. There is, however, the following caveat.

One of the notorious troubles besetting the N-particle T-matrix, for N > 2, comes from the possibility of on-energy-shell multiple scattering. In the case of three incident fragments, this would be real double scattering.[10] Such processes, in which, for example, particle 1 is scattered by particle 2, conserving energy and total momentum, and then it is scattered by particle 3, are effectively of infinite range. In other words, they involve macroscopic impact parameters even though the forces have microscopic ranges. Therefore, in order for assumption (A) to be satisfied we must assume that the cone C does not include those momenta which satisfy the multiple-scattering conditions. This, however, is necessary to assume anyway, because the T-matrix becomes infinite at those momenta, and in such a way that the k-integral in (4.24) would diverge. Hence, C must stay away from them for that reason, too. That will be part of what we shall call assumption (A).

On assumption (Ab), then, of the 3m-dimensional macroscopic volume occupied by the beams, only that part for which the impact parameters are negligible really matters and we loose nothing by restricting the density μ to that.

Assumption (B) is that each T_i is so large that certain "edge effects" that occur at the beginning and at the end of the beam intersection may be neglected.

Using assumption (Aa), the sharp peaking of g(k) at k_o, and the normalization of g, we get

$$h = (2\pi)^{3m-2}\ \beta^{-1}\ M^{-3/2}\ (\prod_i \rho_i^o\ m_i^{3/2})\ (2E^o)^{-\frac{1}{2}}$$

$$\int d^{3m}k'\ d^{3m}k''\ \ g(k')\ g^*(k'')\ \delta(E-E'+\varepsilon_f-\varepsilon_i)\ \delta^3(\underset{\sim}{K}-\underset{\sim}{K}')\ \delta^{3m}(k'-k'')$$

$$=(2\pi)^{3m-2}\ \beta^{-1}\ M^{-3/2}\ (\prod_i \rho_i^o\ m_i^{3/2})(2E^o)^{-\frac{1}{2}}\ \delta^3(\underset{\sim}{K}-\underset{\sim}{K}_o)\ \delta(E-E_o+\varepsilon_f-\varepsilon_i). \tag{4.27}$$

We must now calculate β. This we do by calculating the 3m-dimensional integral $\int d^{3m}x\ \rho_1 \ldots \rho_m$. It is important to recognize that in this integration we may take $x^{\#}$ to be negligible because of assumption (Ab). This very much simplifies the region of integration. When $x^{\#} = 0$, then

$$\underset{\sim}{x}_i = \frac{1}{M} \Sigma\, m_j\, \underset{\sim}{x}_j + \frac{1}{2E^o} \Sigma\, m_j \left(\underset{\sim}{V}_j^o \cdot \underset{\sim}{x}_j \right) \underset{\sim}{V}_i^o .$$

If we set

$$\underset{\sim}{x}_i = \underset{\sim}{b}_i + \underset{\sim}{v}_i^o\,(t - t_o)$$

with $\underset{\sim}{b}_i \cdot \underset{\sim}{v}_i^o = 0$ and

$$t_o = -\Sigma\, m_i\, \underset{\sim}{b}_i \cdot \underset{\sim}{K}^o / 2ME^o, \quad \underset{\sim}{x} = \Sigma\, m_i\, \underset{\sim}{b}_i / M + t_o\, \underset{\sim}{K}^o / M,$$

then we obtain

$$\underset{\sim}{x}_i = \underset{\sim}{x} + \underset{\sim}{v}_i^o\, t. \tag{4.28}$$

With these values of $\underset{\sim}{x}_i$ we have

$$\bar{x}_1 = xM^{\frac{1}{2}} + M^{-\frac{1}{2}}\, K_x^o\, t$$

$$\bar{x}_2 = yM^{\frac{1}{2}} + M^{-\frac{1}{2}}\, K_y^o\, t$$

$$\bar{x}_3 = zM^{\frac{1}{2}} + M^{-\frac{1}{2}}\, K_z^o\, t$$

$$\bar{x}_4 = (2E^o)^{\frac{1}{2}}\, t$$

and therefore

$$d\bar{x}_1 \ldots d\bar{x}_4 = (2E^o)^{\frac{1}{2}}\, M^{3/2}\, dx\, dy\, dz\, dt.$$

Using assumption (B) that the time T during which the beams all intersect fully is long compared to the transition time during which the beams intersect partially,[11] it follows from (4.28) that we can well approximate the integral

$$\int dx\ dy\ dz\ dt\ \prod_i \rho_i \simeq T\, V_{int} \prod_i \rho_i$$

where

$$T = \min\{t_i + T_i\} - \max\{t_i\}$$

is the duration of the beam intersection and V_{int} is the volume of the spacial region of intersection of all the beams.

Hence

$$\int d\bar{x}_1 \ldots d\bar{x}_4 \prod_i \rho_i = (2E^o)^{\frac{1}{2}} M^{3/2} V_{int}\, T \prod_i \rho_i^o$$

and therefore effectively by (4.26) and (4.20)

$$\int d^{2m}x \prod_i \rho_i = (\prod_i m_i)^{-3/2} \int d\bar{x}_1 \ldots d\bar{x}_4 \int d^{3m-4}x^{\#} \prod_i \rho_i$$

$$= \beta(\prod_i m_i)^{-3/2} M^{3/2} V_{int}\, T(2E^o)^{\frac{1}{2}}.$$

But the left-hand side equals $\prod_i \rho_i^o V_i = 1$. Therefore we conclude that

$$\beta^{-1} = (\prod_i m_i)^{-3/2} M^{3/2} V_{int}\, T(2E^o)^{\frac{1}{2}}. \tag{4.29}$$

Consequently the probability of finding the particles in C is given by (4.24), (4.27), and (4.29) as

$$P = (2\pi)^{3m-2} V_{int}\, T(\prod_i \rho_i^o) \int_C d^{3n}k |T_{E_o}(f,k;i,k_o)|^2$$

$$\delta(E-E_o+\varepsilon_f-\varepsilon_i)\ \delta^3(\underset{\sim}{K}-\underset{\sim}{K}_o). \tag{4.30}$$

Note that $\prod_i \rho_i^o = 1/\prod_i V_i$. Now if we send in N_i fragments of kind i then the number of m-tuples of particles that can be expected to be found in C is $(\prod_i N_i)$ P and since $\prod_i \rho_i^o N_i = \prod_i (N_i/V_i) = \prod_i d_i$, where d_i is the fragment-density of the ith beam, the rate at which fragments will be found scattered into C per unit time, or, by (1.8), the rate at which they will be found crossing a distant surface intersecting C, is given by

$$R = V_{int} \left(\prod_i d_i\right) \sigma \tag{4.31}$$

where

$$\sigma = (2\pi)^{3m-2} \int_C d^{3n}k \, |T_{E_o}(f,k;i,k_o)|^2 \, \delta(E-E_o+\varepsilon_f-\varepsilon_i) \, \delta^3(\underset{\sim}{K}-\underset{\sim}{K}_o) \tag{4.32}$$

may be called the <u>specific scattering rate</u>.

The quantity σ has the dimensions $(\text{distance})^{3m-3}$ $(\text{time})^{-1}$. In the most common case of two incident particles, this becomes $(\text{distance})^3$ $(\text{time})^{-1}$. Thus, dividing it by the incident relative velocity gives an <u>area</u>, the scattering cross section. For more than two incident particles there is no natural way to obtain an area. One may again divide by a velocity and get a quantity of dimensions $(\text{distance})^{3m-4}$ but only at the expense of losing the symmetry of (4.32). Another possibility would be to divide (4.32) by the total energy. That preserves its symmetry and (σ/E_o) has the dimensions $(\text{distance})^{3m-3}$.

In order to relate (4.32) to physical measurements the 3n-dimensional cone C has to be made up of n three-dimensional cones, that is, the integral in (4.32) extends over the cones C_j, $j = 1, \ldots, n$, each in three dimensions:

$$\sigma = (2\pi)^{3m-2} \int_{C_1} (d\underset{\sim}{k}_1) \ldots \int_{C_n} (d\underset{\sim}{k}_n) |T_{E_o}(f,k;i,k_o)|^2 \, \delta(E-E_o+\varepsilon_f-\varepsilon_i) \, \delta^3(\underset{\sim}{K}-\underset{\sim}{K}_o). \tag{4.33}$$

One may be tempted to allow the cones $C_1, \ldots C_n$ to become all of three-space and thereby to define a total scattering rate. The difficulty is that (4.33) has been derived only for cones that do not include the multiple scattering momenta (see p. 22). What is more, at these multiple-scattering momenta $|T_{E_o}|^2$ can in general be expected to be infinite in a non-integrable way, as it is in the three-particle case[10]. Thus, the total scattering rate calculated by integrating (4.33) is generally infinite.

How could measurements of a scattering rate such as (4.33) have to be actually performed? If the experiment consisted of many repetitions of sending in an m-tuple of initial fragments and in each case waiting to see if the n specified final fragments arrive in their n given cones, and thus to measure the probability (4.30), then it would only be necessary to put n specific fragment sensitive counters in arbitrary position, and wait a sufficiently long time between repetitons to catch even the slow fragments. In reality, however, the initial fragments arrive in steady streams. It is therefore not enough simply to count fragments at specified angular positions. In order to be sure that fragments registering in counters 1, 2, ... n come from the same collision it is necessary tu put the counters at distances from the beam intersection that are proportional to the (central) particle velocities, and then to make coincidence measurements. Let us be more specific and for simplicity take a three-particle system. We want to measure the differential scattering rate

$$d\sigma = (2\pi)^7 \, (d\underset{\sim}{k}_1)(d\hat{k}_2)\,|T_{E_o}(f,k;i,k_o)|^2$$

where $\underset{\sim}{k}_3$ and k_2 are determined by conservation of energy and momentum.

Our arrangement is to mount counters #1 and #2 at distances D_1 and D_2, respectively, from the beam intersections in arbitrarily given directions $\hat{\underset{\sim}{k}}_1$ and $\hat{\underset{\sim}{k}}_2$, with counter #1 sensitive to particles 1 in the momentum interval between k_1 and $k_1 + dk_1$, and counter #2, to particles 2.

The single collisions of particle 1 in which one of the other two particles does not collide at all, and the double collisions in which particle 1 collides only once, and first, are both characterized by the same momentum magnitude k_1 determined by energy and momentum conservation in the binary collision. Consequently, the momentum sensitivity of counter #1 can be set so as to rule out these single and double collisions.

Now energy conservation in the three-particle collision determines k_2 if $\underset{\sim}{k}_1$ and the direction $\hat{\underset{\sim}{k}}_2$ are given. Hence, the momentum sensitivity of counter #1 also serves to rule out the double collisions in which particle 2 collides first, and only once.

Finally, since measurement of $\underset{\sim}{k}_1$ and $\hat{\underset{\sim}{k}}_2$ determines $\underset{\sim}{k}_2$, $\underset{\sim}{k}_3$ is determined by overall momentum conservation. Therefore, the double collisions in which particle 3 collides first and only once can also be ruled out by proper choice of k_1.

Determination of $\underset{\sim}{k}_1$ and $\hat{\underset{\sim}{k}}_2$ therefore is enough to rule out all double collisions, provided that the particles counted come from the same collision. The remaining phenomena to be garded against are events in counters #1 and 2 that come from different collisions. Most of these would be from single collisions of particle 2 with 1 or 3, accompanied by a three-particle collision of particle 1 (either a "genuine" three-particle collision or a double collision) in which particle 2 is undetected. Some of them would come from separate three-particle collisions of particles 1 and 2. These could be eliminated as follows.

If events in counters #1 and 2 come from the same collision and their distances from the beam intersection are such that $D_1/D_2 = m_2k_1/m_1k_2$, then they should arrive simultaneously. So we arrange for an event in counter #1 to sensitize counter #2 from the arrival time t_1 of particle 1 to $t_1 + \Delta t$. The time interval Δt should be large compared to the "natural" arrival time spread $m_1D_1k_1^{-2}dk_1$ of particle 2 associated with particle 1 whose momenta are spread by dk_1. On the other hand, Δt must be so small that the number of accidental coincidences[12] is small compared to the number of real events.

In principle, this coincidence arrangement would serve to insure that particles detected in counters #1 and 2 come from the same collision. However, the binary collisions of particles 2 and 3 are so much more frequent than three-particle events that it would be important to eliminate them separately. This can be done by making counter #2 momentum sensitive and to allow it to accept only particles in the momentum interval from k_2 to $k_2 + \Delta k_2$. Here Δk_2 should be large but small enough to eliminate all binary collisions of particle 2. With the use of such a counter #2 the number of accidental coincidences during Δt would be drastically reduced to the results of separate three-particle collisions of particles 1 and 2.

For more than three intersecting beams, or for two beams and more than two final fragments, similar considerations apply. In principle, then, the reaction rate (4.31) is a measurable quantity.

REFERENCES

1) J.M. Jauch, R. Lavine, and R.G. Newton, Helv. Phys. Acta 45, 325 (1972).

2) J.D. Dollard, Comm. Math. Phys. 12, 193 (1969) and J. Math. Phys. 14, 708 (1973).

3) W. Hunziker, in Lectures in Theoretical Physics, vol. X-A, edi-

ted by A.O. Barut and W.E. Brittin (Gordon and Breach, New York, 1968).

4) J.R. Taylor, *Scattering Theory* (J. Wiley & Sons, New York, 1972), pp. 49-51.

5) This follows from the Riemann-Lebesgue lemma.

6) Our notation is to use $\mathbf{k}$ for a vector in R^3 and k for its magnitude; $\hat{\mathbf{k}}$ is the unit vector $\mathbf{k}/k$. The three dimensional volume element is $(d\mathbf{k})$ and a solid-angle integration is indicated by $(d\hat{\mathbf{k}})$.

7) We are here, of course, merely outlining well-known matters that are based largely on J.M. Jauch, Helv. Phys. Acta 31, 661 (1958).

8) We now use a somewhat simplified notation in which i and f stand for the combinations α, i and β, f.

9) The following is merely an elaboration of ideas contained in Ref. 3.

10) E. Gerjuoy, Proc. Phys. Soc. London, J. Phys. B3, L92 (1970) and Phil. Trans. Royal Soc. London A 270, 197 (1971); J.Nuttall, J. Math. Phys. 12, 1896 (1971); S.P. Mercuriev, Teor. i Mat. Fiz. 8, 235 (1971); R.G. Newton, Ann. Phys. (NY) 74, 324 (1972) and ibid 78, 561 (1973) and preprint, "The Three-Particle S-Matrix," 1973.

11) Note that this implies that the angles of intersection of the beams are not too small. Their lower limit depends on T.

12) At this writing we cannot give a prescription for the computation of this number other than to measure it by alternately closing off counters #1 and 2. We do not know how to calculate it theoretically, because it is not known how to calculate the "inclusive" reaction rate. [See the remarks following Eq. (4.33)].

CROSS SECTIONS IN THE QUANTUM THEORY OF SCATTERING (*)

F. Coester

Argonne National Laboratory,
Argonne, Illinois 60439

In any physical theory emphasis and direction are shaped by the question: What are the observable quantities? Most empirical information about atomic nuclei and subnuclear particles is derived from scattering and reaction experiments. The quantum theory of scattering thus occupies a place of central importance. The observables in scattering experiments are cross sections, but prevailing fashion has been to treat the S matrix, or by abstraction the S operator, as the quantity that is in principle observable. The computation of cross sections from the S matrix is then a technical detail without theoretical significance. However, the requirement that a sensible theory should yield scattering cross sections imposes conditions on the theory that are not manifestly satisfied as soon as the existence of wave operators has been demonstrated.

A careful review of the definition of scattering cross sections should serve to emphasize the theoretical implications of this requirement. For my purposes it will be sufficient to consider the scattering of spinless particles by fixed short-range potentials.

The measurement of a cross section requires an ensemble of scattered particles. Since the notion of a scattering cross section is in essence classical, let us at first consider an ensemble of classical particles described by the phase-space density $f(\vec{x},\vec{k},t)$. The total number of particles is

$$N = \int d^3x \int d^3k \; f(\vec{x},\vec{k},t). \tag{1}$$

(*) Work supported under the auspices of the U.S. Atomic Energy Commission.

Enz/Mehra (eds.), Physical Reality and Mathematical Description. 313–320. *All Rights Reserved.*

Momentum and space distributions are given by

$$\tilde{\rho}(\vec{k},t) := \int d^3x\ f(\vec{x},\vec{k},t), \tag{2}$$

and

$$\rho(\vec{x},t) := \int d^3k\ f(\vec{x},\vec{k},t). \tag{3}$$

The particle-current density is

$$\vec{j}(\vec{x},t) := \int d^3k(\vec{k}/m)f(\vec{x},\vec{k},t). \tag{4}$$

In the absence of any forces the time dependence of f is given by

$$f(\vec{x},\vec{k},t) = f(\vec{x}-\vec{k}t/m,\vec{k}) \tag{5}$$

where

$$f(\vec{x},\vec{k}) := f(\vec{x},\vec{k},0). \tag{6}$$

For scattering processes Eq.(5) holds asymptotically for large times. There exist initial and final phase space distributions $f_a(\vec{x},\vec{k})$ and $f_b(\vec{x},\vec{k})$ such that

$$\lim_{t\to-\infty} \int d^3x \int d^3k \left| f(\vec{x},\vec{k},t) - f_a(\vec{x}-\vec{k}t/m,\vec{k})\right| = 0, \tag{7}$$

$$\lim_{t\to+\infty} \int d^3x \int d^3k \left| f(\vec{x},\vec{k},t) - f_b(\vec{x}-\vec{k}t/m,\vec{k})\right| = 0. \tag{8}$$

This description appears to be appropriate for scattering experiments in which a burst of particles sweeps over the target and the observer counts the total number of particles traveling in a specified direction after the collision. An example would be the case in which the data consist of a set of bubble chamber tracks.

The definition of the differential cross section, appropriate for this experiment is

$$d\sigma := d\Omega_k \int dk k^2\ \rho_b(\vec{k}) \Big/ \left[\int_{-\infty}^{+\infty} dt\ \vec{n}\cdot j_a(\vec{x}^0,t)\right] \tag{9}$$

where $\vec{n}$ is a unit vector in the direction of $\vec{k}^0$,

$$\vec{k}^0 := \int d^3k\ \vec{k}\rho_a(\vec{k}), \tag{10}$$

and $\vec{x}^0$ is the point where the incident flux is monitored.

In a stationary scattering experiment a beam of particles impinges on a target continuously and a counter records the rate at which particles are scattered into a solid angle $d\Omega_x$. In that case $f_a(\vec{x}-\vec{k}t/m,\vec{k})$ and $f(\vec{x},\vec{k},t)$ are time-independent in the experimental area during the experiment. The appropriate expression for the differential cross section is then

$$d\sigma := d\Omega_x |\vec{x}|\vec{x}\cdot j(\vec{x}) \Big/ \vec{n}\cdot j_a(\vec{x}^0). \tag{11}$$

A particle ensemble is described by the density matrix ρ in either the space or momentum representation, $(\vec{x}'|\rho|\vec{x}')$, or $(\vec{k}'|\tilde{\rho}|\vec{k}')$. They are related by Fourier transform

$$(\vec{k}'|\tilde{\rho}|\vec{k}'') = (2\pi)^{-3}\int d^3x' \int d^3x''\ (\vec{x}'|\rho|\vec{x}'')\ e^{i\vec{k}''\cdot\vec{x}''}\ e^{-i\vec{k}'\cdot\vec{x}'}. \tag{12}$$

The function $f(\vec{x},\vec{k})$ defined by *)

$$f(\vec{x},\vec{k}) = (2\pi)^{-\frac{3}{2}}\int d^3x' \int d^3x''\ \delta\left[\vec{x} - \tfrac{1}{2}(\vec{x}'+\vec{x}'')\right] (\vec{x}'|\rho|\vec{x}'')\ e^{i\vec{k}\cdot(\vec{x}'-\vec{x}'')} \tag{13}$$

goes over into the classical phase space density in the classical limit. From the definition it follows that

$$\int d^3x\ f(\vec{x},\vec{k}) = (\vec{k}|\tilde{\rho}|\vec{k}) = \tilde{\rho}(\vec{k}), \tag{14}$$

$$\int d^3k\ f(\vec{x},\vec{k}) = (\vec{x}|\rho|\vec{x}) = \rho(\vec{x}), \tag{15}$$

and

$$\int d^3k(\vec{k}/m)f(\vec{x},\vec{k}) = \vec{j}(\vec{x}) = (2mi)^{-1}\left[\nabla'(\vec{x}'|\rho|\vec{x}) - \nabla(\vec{x}'|\rho|\vec{x})\right]_{\vec{x}'=\vec{x}} \tag{16}$$

*) This phase-space distribution was first introduced by Dirac[1] in the context of the Thomas-Fermi model for atoms and independently by Wigner[2] in the context of statistical mechanics. Its properties were developed in detail by Moyal[3].

The expressions (9) and (11) for the differential cross section remain applicable without change. The time-dependent theory calls for the definition (9)+,++). The definition (11) appropriate for the time-independent scattering theory is found in every textbook. Next we need to express both cross sections in terms of the scattering operator.

The time dependence of the density matrix ρ is given by

$$\rho(t) = e^{-iHt}\rho\, e^{iHt} . \tag{17}$$

If the wave operators $\Omega_\pm$ defined by strong operator limit

$$\Omega_\pm : = \operatorname*{s-lim}_{t\to\pm\infty} e^{iHt} e^{-iH_0 t} \tag{18}$$

exist, and we define

$$\rho_a(t) = \Omega_-^\dagger \rho(t)\Omega_- = e^{-iH_0 t}\rho_a e^{iH_0 t} , \tag{19}$$

and

$$\rho_b(t) = \Omega_+^\dagger \rho(t)\Omega_+ = e^{-iH_0 t}\rho_b e^{iH_0 t} , \tag{20}$$

then it follows that

$$\lim_{t\to-\infty} [\rho(t) - \rho_a(t)] = 0, \tag{21}$$

and

$$\lim_{t\to+\infty} [\rho(t) - \rho_b(t)] = 0 \tag{22}$$

in analogy to Eqs. (7) and (8). Since

$$S: = \Omega_+^\dagger \Omega_- \tag{23}$$

we have

$$\rho_b = S\rho_a S^+ . \tag{24}$$

+) See Sec. XII of ref. 4.
++) See Sec. 6 of ref. 5.

In order to proceed further we need the properties of the S matrix as a momentum space kernel.*) Let

$$V: = H - H_0 \tag{25}$$

and

$$T: = V\Omega_- \ . \tag{26}$$

We have then+)

$$(S-1)\psi = s - \lim_{\varepsilon \downarrow 0} \int \frac{-2i\varepsilon}{(H_o-\lambda)^2 + \varepsilon^2} \ T \ dE_\lambda^0 \psi \tag{27}$$

for any ψ in the dense domain D on which both H and H_0 are self-adjoint. Assume that the Schwartz test functions are included in D, $\mathcal{S} \subset D$. It follows that for any pair f, $g \in \mathcal{S}$

$$\langle f,(S-1)g\rangle = \lim_{\varepsilon\downarrow 0} \int \langle f, \frac{-2i\varepsilon}{(H_0-\lambda)^2 + \varepsilon^2} \ T \ dE_\lambda^0 \ g\rangle$$

$$= \lim_{\varepsilon\downarrow 0} \int d^3k' \int d^3k \ \overset{*}{f}(\vec{k}') \ \frac{-2i\varepsilon}{(\omega(k')-\omega(k))^2 + \varepsilon^2} \ (\vec{k}'|T|\vec{k}) \ g(\vec{k}) \tag{28}$$

where $\omega(k)$ is the kinetic energy and the kernel $(\vec{k}'|T|\vec{k})$ is defined as a generalized function on $\mathcal{S}' \times \mathcal{S}'$. Equation (28) is usually written in the form

$$(\vec{k}'|S|\vec{k}) = \delta(\vec{k}'-\vec{k}) - 2\pi i \ \delta[\omega(k') - \omega(k)] \ (\vec{k}'|T|\vec{k}). \tag{29}$$

The meaning of both equations is precisely the same. In order to obtain cross sections we must require that the kernel $(\vec{k}'|T|\vec{k})$ be a continuous function of $\vec{k}'$ and $\vec{k}$. This property does not follow from the existence of the wave operators. In a deductive theory it must be proved from the known properties of the Hamiltonian++).

In the time-independent scattering theory the kernel $(\vec{k}'|T|\vec{k})$ is often defined as the solution of the Lippman-Schwinger equation

*) See Sec. V.5, p. 143 of ref. 6.
+) See Theorem 7 of ref. 7.
++) See Sec. V.5, p. 143 of ref. 6.

$$(\vec{k}'|T|\vec{k}) = (\vec{k}'|V|\vec{k}) - \lim_{\varepsilon\downarrow 0} \int d^3k'' \frac{(\vec{k}'|V|\vec{k}'')(\vec{k}''|T|\vec{k})}{\omega(k'') - \omega(k') - i\varepsilon} \tag{30}$$

in which it is essential that the limit exists for fixed $\vec{k}$. The necessary smoothness of the kernel depends on the properties of V. The principal virtue of Eq.(30) is that it is suitable for numerical solution, which cannot be said of the defining equations (26) and (18). From the Hilbert space version of the Lippman-Schwinger equation Ω_- 7,8,9) *)

$$(\Omega_- -1)\psi = - \operatorname{s-lim}_{\varepsilon\downarrow 0} \int (H_0-\lambda-i\varepsilon)^{-1} V\Omega_- \, dE^0_\lambda\psi, \tag{31}$$

for $\psi \in D$, it follows that

$$T\psi = V\psi - \operatorname{s-lim}_{\varepsilon\downarrow 0} \int V(H_0 - \lambda - i\varepsilon)^{-1}\, T\, dE^0_\lambda\psi. \tag{32}$$

However, Eq.(32) is insufficient to derive Eq.(30). For purposes of numerical computation Eq.(32) is just as useless as the definition

$$T\psi = \operatorname{s-lim}_{t\to-\infty} V e^{iHt} e^{-iH_0t}\psi. \tag{33}$$

If the incoming beam is well collimated and sufficiently broad in the lateral direction then $(\vec{k}'|\rho_a|\vec{k}'')$ vanishes if the direction of $\vec{k}'$ or $\vec{k}''$ differs substantially from that of $\vec{k}^0$. The differential cross section (9) may then be written in the form

$$d\sigma = (2\pi)^4 d\Omega_k \frac{\int k^2 dk \langle\delta(\omega-\omega'')(\vec{k}|T|\vec{k}')(\vec{k}|T|\vec{k}'')\rangle}{\langle\vec{n}\cdot(\vec{k}'+\vec{k}'')/2m\rangle} \tag{34}$$

where $\langle f(\vec{k}',\vec{k}'')\rangle := \int d^3k' \int d^3k'' f(\vec{k}',\vec{k}'')(\vec{k}'|\rho_a|\vec{k}'')\cdot\delta(\omega'-\omega'')$. If $(\vec{k}'|\rho_a|\vec{k}'')$ is different from zero only if $\vec{k}'$ and $\vec{k}''$ are both in a sufficiently small neighborhood of $\vec{k}^0$, then we may factor out the integrations abbreviated by the brackets < ... > in numerator and denominator and we are left with the usual expression

$$d\sigma = (2\pi)^4 \, d\Omega_k \, m^2 \{|(\vec{k}|T|\vec{k}_0)|^2\}_{|\vec{k}| = |\vec{k}_0|}. \tag{35}$$

It is only in the limit just described that the cross section be-

*) See Theorem 6 of ref. 7.

comes independent of details of the density matrix ρ_a that are not controllable in practice. The physical significance of the restriction on ρ_a is that the spatial extent of the incident wave packets must be sufficiently large. Exactly how large is sufficient depends on the properties of the kernel T. But crudely the requirement is that the wave packets be larger than the scattering region.

The evaluation of the cross section (11) follows familiar lines. For wave packets that are sharp in momentum the result is again Eq.(35). A brief sketch will suffice. In order to evaluate the right-hand side of Eq.(11) we need the density matrix

$$(\vec{x}'|\rho|\vec{x}'') = (\vec{x}'|\Omega_-\rho_a\Omega_-^+|\vec{x}'') . \tag{36}$$

From Eq.(31) it follows that

$$(\vec{k}'|\Omega_-|\vec{k}) = \delta(\vec{k}'-\vec{k}) - \lim_{\varepsilon\downarrow 0} (\omega(k')-\omega(k)-i\varepsilon)^{-1}(\vec{k}'|T|\vec{k}) \tag{37}$$

in analogy to Eq.(29). The scattering wave function $(\vec{x}|\Omega_-|\vec{k})$ is obtained from Eq.(37) by Fourier transform. The formal derivation of the asymptotic form for large $|\vec{x}|$ is straightforward. The important point is as before the requirement that the function $(\vec{k}'|T|\vec{k})$ be sufficiently smooth.

The main result of this survey is the emphasis placed on the smoothness properties of the kernel $(\vec{k}'|T|\vec{k})$. They are essential for the link between the abstract theory and observable numbers.

REFERENCES

1) P.A.M. Dirac, Proc. Cambr. Phil. Soc. 26, 376 (1930).

2) E.P. Wigner, Phys. Rev. 40, 749 (1932).

3) J.E. Moyal, Proc. Cambr. Phil. Soc. 45, 99 (1949).

4) H. Ekstein, Phys. Rev. 101, 880 (1956).

5) J.M. Jauch, Helv. Phys. Acta 31, 127 (1958).

6) B. Simon, Quantum Mechanics for Hamiltonians Defined As Quadratic Forms, Princeton University Press, 1961.

7) W.O. Amrein, V. Georgescu and J.M. Jauch, Helv. Phys. Acta 44, 407 (1971).

8) E. Prugovecki, Nuovo Cim. 63B, 569 (1969).

9) D.B. Pearson, Nuovo Cim. 2A, 853 (1971).

ON LONG-RANGE POTENTIALS

I. Zinnes

Physics Department
Fordham University, New York

I.- In the rigorous scattering theory so neatly formulated by Professor Jauch in two classic papers[1] one of the problems by-passed for some time was the problem of scattering in a Coulomb field as well as a fortiori in fields decreasing more slowly than the Coulomb field at infinity. For these cases it was known that the famous Ω did not exist. There has always been an implication that, since quantum mechanics reflects nature completely, one should not expect it to provide answers to non-physical problems, something which classical physics can do quite readily. So, the work by Dollard[2] in 1964 which extended Jauch's original formulation to include scattering in a Coulomb potential was an important breakthrough.

If $H_o = -\Delta/2\mu$ and $H = H_o + e_1e_2/r$ are the conventional free and full Hamiltonians, respectively, for single channel (or two body) scattering, then Dollard has shown that operators $\Omega^{(\pm)}$ exist in the present case if we define

$$\Omega^{(\pm)} = \underset{t\to\pm\infty}{\text{s-lim}}\ V_t^* U_c(t) \tag{1}$$

where

$$\begin{aligned} &(a)\quad V_t = e^{-iHt} \quad ; \quad (b)\ U_c(t) = e^{-iH_{oc}(t)} \\ &(c)\quad H_{oc}(t) = H_o t + \frac{\epsilon(t)\mu e_1 e_2}{(-\Delta)^{\frac{1}{2}}}\,\ell n\left(-\frac{2|t|\Delta}{\mu}\right) \quad ; \\ &\qquad \epsilon(t) = \pm I \text{ for } t \gtrless 0. \end{aligned} \tag{2}$$

In fact, he succeeded in showing that the limit in (1) also exists when the full Hamiltonian includes a short-range potential which we shall define as one which is either (1) square-integrable or (2) locally square-integrable and vanishes at infinity more rapidly than r^{-1}.

The identification of $\Omega^{(\pm)}$ in (1) with the desired wave-operator is supported by at least three properties which it possesses.

(a) For $h \in L^2(\mathbb{R}^3)$,

$$\lim_{t\to\pm\infty} \left\| U_c(t)h(\underset{\sim}{r}) - \left(\frac{\mu}{it}\right)^{3/2} \phi_c(\underset{\sim}{r})\hat{h}\left(\frac{\mu \underset{\sim}{r}}{t}\right)\right\| = 0 , \tag{3}$$

where $\hat{h}$ is the Fourier transform[3] of h, and ϕ_c is a function such that $|\phi_c| = I$. This relation implies that for large $|t|$ the probability density $|U_c(t)h(\underset{\sim}{r})|^2$ of the modified free packet $U_c(t)h$ can be approximated by the density $|\frac{\mu}{t}|^3 \cdot |\hat{h}(\frac{\mu \underset{\sim}{r}}{t})|^2$, a result one might expect classically.

(b) With the interpretation of $U_c(t)$ as a possible "free" propagator justified, the definition of $\Omega^{(\pm)}$ in (1) becomes just an extension of Professor Jauch's original definition of the wave-operator :

$$\Omega_o^{(\pm)} = \underset{t\to\pm\infty}{\text{s-lim}}\ V_t^* U_t \ , \text{ where } U_t = \exp(-iH_o t), \tag{4}$$

for short-range potentials, since $U_c(t) = U_t$ when $e_1 = e_2 = 0$.

(c) Finally, definition (1) is consistent with what we shall call the Ikebe relation : If $f \in L^2$, $\hat{f}$ is its Fourier transform, and $\psi^{(\pm)}(\underset{\sim}{k},\underset{\sim}{r})$ the conventional time-independent scattering states, then f, $\hat{f}$, $\psi^{(\pm)}$ and $\Omega^{(\pm)}$ are related through the following relations

$$\begin{aligned} &(a) \quad f(\underset{\sim}{r}) = (2\pi)^{-3/2}\int d\underset{\sim}{k}\ e^{i\underset{\sim}{k}\cdot\underset{\sim}{r}}\,\hat{f}(\underset{\sim}{k}) \\ &(b) \quad f^{(\pm)}(\underset{\sim}{r}) = (\Omega^{(\pm)}f)(\underset{\sim}{r}) = (2\pi)^{-3/2}\int d\underset{\sim}{k}\ \psi^{(\pm)}(\underset{\sim}{k},\underset{\sim}{r})\hat{f}(\underset{\sim}{k}) \ . \end{aligned} \tag{5}$$

It is evident that in (5) $\hat{f}$ is both the Fourier transform of f and the $\psi^{(\pm)}$ - transform of the scattering states $\Omega^{(\pm)}f$. This relation was proved by Ikebe[4] for (super) short-range[5] potentials and by Dollard[1] for pure Coulomb potentials. The $\Omega^{(\pm)}$ obtained through the relations (5) satisfy (4) for Ikebe's potentials, but of course not for the Coulomb potential. For short-range potentials the kernels of (5) are related through the conventional space-asymptotic condition :

$$\psi^{(-)}(\underset{\sim}{k}_o\underset{\sim}{r}) \simeq e^{i\underset{\sim}{k}_o\cdot\underset{\sim}{r}} + f(\vartheta)e^{ikr}/r \ ; \ \psi^{(+)}(\underset{\sim}{k},\underset{\sim}{r}) = \psi^{(-)^*}(-\underset{\sim}{k},\underset{\sim}{r}) \quad (6)$$

It is evident that the kernel in (5a) is the "plane-wave" part of the kernel in (5b). Thus the Ikebe relation connects the time-dependent and time-independent theories.

Since the pure Coulomb time-independent scattering state does not satisfy (6), it is not surprising that one can replace (5) by

$$\begin{aligned} &(a)\ f(\underset{\sim}{r}) = (2\pi)^{-3/2}\int d\underset{\sim}{k}\ e^{i\underset{\sim}{k}\cdot\underset{\sim}{r}}\,\hat{f}(\underset{\sim}{k}) \\ &(b)\ f_{co}^{(\pm)}(\underset{\sim}{r}) \equiv K(\lambda)f(\underset{\sim}{r}) \equiv \int dk\ \phi_{co}^{(\pm)}(\underset{\sim}{k},\underset{\sim}{r})\hat{f}(\underset{\sim}{k}) \\ &(c)\ f_c^{(\pm)}(\underset{\sim}{r}) \equiv (\Omega_c^{(\pm)}f)(\underset{\sim}{r}) \equiv (2\pi)^{-3/2}\int d\underset{\sim}{k}\ \psi_c^{(\pm)}(\underset{\sim}{k},\underset{\sim}{r})\hat{f}(\underset{\sim}{k}) \ , \end{aligned} \quad (7)$$

where $\phi_{co}^{(\pm)}(\underset{\sim}{k},\underset{\sim}{r})$ is the plane-wave part of the space-asymptotic form of $\psi_c^{(\pm)}(\underset{\sim}{k},\underset{\sim}{r})$:

$$\psi_{co}^{(\pm)}(\underset{\sim}{k},\underset{\sim}{r}) = \exp i\left[\underset{\sim}{k}\cdot\underset{\sim}{r} \mp (\lambda/k)\ln(kr \pm \underset{\sim}{k}\cdot\underset{\sim}{r})\right] \ , \quad (8)$$

and $\lambda = \mu e_1 e_2$. It has been shown that as expected[6)]

$$||V_t f_c^{(\pm)} - K(\lambda)U_t f|| \to 0 \ , \ \text{as } t \to \pm\infty \ , \quad (9)$$

i.e., that there exists an operator $\Omega_c^{(\pm)}$ such that

$$\Omega_c^{(\pm)} = \underset{t\to\pm\infty}{\text{s-lim.}}\ V_t^* K(\lambda)U_t \ . \quad (10)$$

From (5) and (9) it is clear that, since $\Omega^{(\pm)}f$ of (5) is just the f_c of (9), the wave-operators in (1) and (10) are identical. With respect to the operator $K(\lambda)$ defined implicitly in (7) one may regard it as a dressing operator with the property that $K(0) = 1$. In an interesting recent paper Chandler and Gibson have extended the result (10) to the many-channel case and found agreement with Dollard's many-channel result[1)].

In section III we propose an extension of the result (9) to potentials $V(r) = O(r^{-\beta})$ as $r \to \infty$, where $\beta > 0$. First however we shall prove a generalized form of the well-known theorem that (4) does not exist for long-range potentials.

II.-Theorem 1. If H_1 and H_2 are self-adjoint operators, where

$$H_1(t) = H_0 t + \nu(H_0,t) \;;\; H_0 = -\Delta/2\mu \;; \qquad (11)$$

$$\nu(\lambda,t) \text{ is real and } \lim_{t\to\pm\infty} \int_{\Delta\lambda} \exp\left[i\nu(\lambda,t)\right] d\lambda = 0$$

over any interval $\Delta\lambda$ of the real line; furthermore, if $\exp(iH_2 t)\exp\left[-H_1(t)\right] \equiv \Omega(t)$ converges strongly as $t \to \pm\infty$ to an operator $\Omega^{(\pm)}$ which is complete[8], then $\exp(iH_2 t)\exp(-iH_0 t) \equiv \Omega_0(t) \xrightarrow{\omega} 0$ and therefore, since $||\Omega_0(t)|| = 1$, $\Omega_0(t)$ does not converge strongly as $t \to \pm\infty$.

Thus, since the Coulomb potential satisfies the conditions of the theorem, the existence of the modified wave-operator according to (1) precludes the existence of $\Omega_0^{(\pm)}$ defined in (4). We can prove this theorem swiftly with aid of two lemmas following the procedure outlined by Dollard[10] for the Coulomb case.

Lemma 1. If A, $\{A_n\}$, B, $\{B_n\}$ is a set of bounded operators such that $A_n \xrightarrow{\omega} A$ and $B_n \xrightarrow{s} B$, as $n \to \infty$, then $A_n B_n \xrightarrow{\omega} AB$ as $n \to \infty$.

Proof. According to a well-known theorem, $A_n \xrightarrow{\omega} A \Longrightarrow ||A_n|| \leqslant c$, independent of n. Therefore, for $f, g \in \mathcal{H}$

$$\left|\left((A_n B_n - AB)f,g\right)\right| \leqslant \left|\left((A_n - A)Bf,g\right)\right| + \left|\left(A_n(B_n-B)f,g\right)\right|$$

$$\leqslant \left((A_n - A)\, Bf,g\right) + c\, ||B_n - B|| \cdot ||f|| \cdot ||g|| \to 0, n \to \infty.$$

Lemma 2. With $H_1(t)$ and H_0 as defined in theorem 1, we have the result that

$$\exp(iH_1(t))\exp(-iH_0 t) \xrightarrow{\omega} 0. \qquad (12)$$

Proof. Using the spectral theorem with $E^{(o)}(\lambda)$ the spectral family for H_0, we find for $g,h \in \mathcal{H}$ that

$$(g, e^{-iH_1(t)} e^{-iH_0 t} h) = \int_{-\infty}^{\infty} e^{i\nu(\lambda,t)} d\sigma(\lambda) = \int_{-\infty}^{\infty} e^{i\nu(\lambda,t)} \frac{d\sigma}{d\lambda}\, d\lambda \qquad (13)$$

where $\sigma(\lambda) = (g, E^{(0)}(\lambda)h)$ and the last step follows from an application of the Radon-Nikodym theorem to the absolutely continuous measure $\sigma(\lambda)$. Now, from the properties of $\nu(\lambda,t)$ postulated and an application of a generalized version[11] of the Riemann-Lebesgue lemma, the rhs. of (13) vanishes when $t \to \pm\infty$.

We can now prove theorem 1. An arbitrary element $f \in \mathcal{H}$ can be written as $f = f_s + f_{ac}$, where $f_s \in \mathcal{H}_{2s}$ and $f_{ac} \in \mathcal{H}_{2,ac}$. However, it can be shown that, since the spectrum of H_0 is absolutely continuous, if the strong limits of $\Omega_o(t)$ exist, then $\mathcal{R}(\Omega_o^{(\pm)}) \subset \mathcal{H}_{2,ac} \perp \mathcal{H}_{2s}$. So, in considering the weak convergence of $\Omega_o(t)$, it is sufficient to show that $(f_{ac}, \Omega_o(t)g) \xrightarrow[t\to\pm\infty]{} 0$ for $g \in \mathcal{H}$. But this is indeed true, since

$$(f_{ac}, \Omega_o(t)g) = (f_{ac}, e^{iH_2t}e^{-iH_1(t)}e^{iH_1(t)}e^{-iH_0t}g) \tag{14}$$

$$= (e^{iH_0t}e^{-iH_1(t)} \cdot e^{iH_1(t)}e^{-iH_2t}f_{ac}, g) .$$

By lemma 2, $\exp(iH_0t)\exp\left[-iH_1(t)\right] \xrightarrow[t\to\pm\infty]{\omega} 0$; by the hypotheses, $\exp\left[iH_1(t)\right]\exp(-iH_2t)f_{ac} \xrightarrow[t\to\pm\infty]{s} \Omega^{(\pm)*}f_{ac}$; and by lemma 1, the limit on the r.h.s. of (14) is zero.

III.-We now return to the original problem of extending the asymptotic condition presented in (9) to more general potentials. Of course, extensions of Jauch's original asymptotic condition to long-range potentials have been recently formulated by several authors[13,14], but all these formulations lead to non-uniquely defined wave-operators (14). Insofar as it is desirable to associate a precise scattering state[15] $\psi^{(\pm)}$ with a precisely *prepared* state ψ_o, which association is accomplished in the case of short-range potentials through the formula $\psi^{(\pm)} = \Omega_o^{(\pm)}\psi_o$, and the physically 'reasonable' definition of $\Omega_o^{(\pm)}$ in (4), the non-uniqueness of the wave-operator is troublesome[16]. It is precisely this problem which the formulation given in (9) may overcome, i.e., by tying the time-dependent asymptotic condition to the space-asymptotic condition of conventional time-independent scattering theory. Specifically, we shall motivate a theorem which extends (9) to spherically symmetric potentials (in three dimensions) which vanish (almost) arbitrarily slowly at infinity.

Consider the partial-wave expansion of the state ψ :

$$\psi(\underset{\sim}{r}) = \sum_{\ell=0}^{\infty} \frac{\phi_\ell(r)}{r} P_\ell(\cos\nu) , \tag{15}$$

where $\phi_\ell(r)$ satisfies the radial Schrödinger equation :

$$\phi'' + \left[k^2 - \frac{\ell(\ell+1)}{r^2} - U(r)\right]\phi = 0 , \tag{16}$$

and $U(r) = 2mV(r)$, $\hbar = 1$, $\phi(0) = 0$. We now quote a basic theorem concerning the asymptotic behaviour of linear equations of the type (16).

Theorem 2.[17)] In the (matrix) equation

$$\hat{x}' = \left[\hat{A} + \hat{B}(t) + \hat{C}(t)\right]\hat{x} \tag{17}$$

let $\hat{x}$ be an n-rowed column vector; $\hat{A}$, an nxn constant matrix with eigenvalues μ_j, $j = 1,2,\ldots n$, all distinct. Let the matrix $\hat{B}$ be differentiable and satisfy

$$\int_a^\infty |\hat{B}'(t)|dt < \infty \ , \ a > 0 \ ; \tag{18}$$

and let $\hat{B}(t) \to 0$ as $t \to \infty$. Let the nxn matrix $\hat{C}(t)$ be integrable and let

$$\int_a^\infty |\hat{C}(t)| \ dt < \infty \ . \tag{19}$$

Let the roots of $\det(\hat{A}+\hat{B}(t) - \lambda E) = 0$ be $\lambda_j(t)$, $j = 1,2,\ldots n$, and ordered so that $\lim_{t\to\infty}\lambda_j(t) = \mu_j$. Define $D_{kj}(t) = \mathrm{Re}(\lambda_k(t) - \lambda_j(t))$ and suppose that for fixed k and all j, $1 \leq j \leq n$,

$$\int_{t_1}^{t_2} D_{kj}(\tau)d\tau < K \ , \ a \leq t_1 \leq t_2, \ K = \text{const.} \tag{20}$$

Furthermore, let $\hat{p}_k$ be an eigenvector of $\hat{A}$ so that $\hat{A}\hat{p}_k = \mu_k\hat{p}_k$. Then, there is a solution of (17) and a t_0, $a \leq t_0 \leq \infty$ such that

$$\lim_{t\to\infty} \hat{a}_k(t)\exp\left[-\int_{t_0}^{t} \lambda_k(\tau)d\tau\right] = \hat{p}_k \ . \tag{21}$$

We apply this theorem to (16) by converting this equation to a system of first order equations in the usual way. Let $\hat{u}$ be a 2-rowed column vector with $u_1 = \phi$, $u_2 = \phi'$, and let

$$U(r) = U_1(r) + U_2(r), \text{ where} \tag{22}$$

$$\begin{aligned} &\text{(a)} \quad U_1(r) = \Lambda/r^\beta \ ; \qquad \Lambda, \text{ a constant; } \beta > 0^{18} \\ &\text{(b)} \quad U_2(r): \int_a^\infty |U_2(r)|dr < \infty \ . \end{aligned} \tag{23}$$

Then (16) takes the form

$$\frac{d\hat{u}}{dr} = \Bigg[\underbrace{\begin{pmatrix} 0 & 1 \\ -k^2 & 0 \end{pmatrix}}_{\hat{A}} + \underbrace{\begin{pmatrix} 0 & 0 \\ U_1(r) & 0 \end{pmatrix}}_{\hat{B}(r)} + \underbrace{\begin{pmatrix} 0 & 0 \\ U_2(r) + \frac{\ell(\ell+1)}{r^2} & 0 \end{pmatrix}}_{\hat{C}(r)}\Bigg] \hat{u} \, . \quad (24)$$

The eigenvalues of $\hat{A} + \hat{B}(r)$ are

$$\lambda_{\pm}(r) = \pm\, i\sqrt{k^2 - U_1(r)} \quad (25)$$

and the eigen-elements when $r \to \infty$ are

$$\hat{p}_+ = \begin{pmatrix} 1 \\ ik \end{pmatrix}, \ \lambda_+ = ik \ ; \ \hat{p}_- = \begin{pmatrix} 1 \\ -ik \end{pmatrix}, \ \lambda_- = -ik. \quad (26)$$

It is straightforward to show that $\hat{B}(r)$ and $\hat{C}(r)$ satisfy the requirements[19)] of theorem 2, so that we are guaranteed the existence of two linearly independent solutions of (16), ϕ_+, ϕ_- which have the asymptotic forms

$$\phi_{\pm}(k,r) \simeq \exp\left[\pm i \int_{r_o}^{r} \sqrt{k^2 - U_1(s)}\, ds\right] , \quad (27)$$

and evidently may be called generalized Jost solutions. The dependence on ℓ in (27) comes about through the arbitrary multiplicative constants which may depend on ℓ.

If one chooses $\beta = 1/n$ in (23), n a positive integer, so that

$$U_1(r) = \Lambda r^{-1/n} \ ; \quad (n = 1,2,\ldots) \ , \quad (28)$$

one can evaluate the r.h.s. of (27) in closed form and obtain the result

$$\phi_{\pm}(k,r,1/n) \simeq \exp\left[\pm i(kr + \omega(k,r,1/n))\right] \quad (29)$$

where $\omega(k,r,1/n) = c_1(k,1/n)\ln r + c_2(k,r,1/n)$ and $c_2(k,r,1/n)$ is a linear combination of a finite number of positive fractional powers of r. For example

$$\begin{aligned} &(a) \quad \omega(k,r,1) = -\frac{\Lambda}{2k}\,\ell n\, 2\,k\,r \\ &(b) \quad \omega(1,r,\tfrac{1}{2}) = -\frac{\Lambda}{k}\, r^{\frac{1}{2}} - \frac{\Lambda^2}{8k^3}\,\ell n\, k\, r \end{aligned} \quad (30)$$

(c) $\omega(k,r,1/3) = -\frac{3}{4}\frac{\Lambda}{k} r^{2/3} - \frac{\Lambda^3}{16k^5} \ln k r$

(d) $\omega(k,r,\infty) = 0$

Relation (30d) shows that when U(r) is given only by $U_2(r)$ in (23) we have the short-range case for which according to (29)

$$\phi_{\pm}(k,r,\infty) \simeq \exp(\pm ikr) . \tag{31}$$

The solution of the scattering problem now consists in inserting the proper linear combinations of $\phi_{\pm}$ -- this brings in the phase shift -- into (15). However, since we are interested only in the plane-wave part $\psi_{PW}(\underset{\sim}{r})$ of the asymptotic form of $\psi(r)$ in (15) -- the part which corresponds to exp $(i\underset{\sim}{k}\cdot\underset{\sim}{r})$ in (6) -- we can obtain $\psi_{PW}^{(\pm)}$ directly from (29) by generalizing the conventional procedure used in the short-range case whereby one passes from (15) to (6). The result is

$$\psi_{PW}^{(\pm)}(k,r,1/n) = \exp i(\underset{\sim}{k}\cdot\underset{\sim}{r} \pm \omega(k,r,1/n), \quad (n = 0,1,2,..). \tag{32}$$

The discussion leading up to the relations (7) - (9) suggests that if we replace $f_{co}^{(\pm)}$ and $\phi_{co}^{(\pm)}$ in (7) by

$$\text{(a)} \quad f_{co}^{(\pm)} \longrightarrow f_{c}^{(\pm)}(\underset{\sim}{r}) \equiv K_{\beta}^{(\pm)}(\lambda)f(\underset{\sim}{r}) \equiv \int d\underset{\sim}{k}\,\phi_{\beta o}^{(\pm)}(\underset{\sim}{k},\underset{\sim}{r})\hat{f}(\underset{\sim}{k})$$

$$\text{(b)} \quad \phi_{co}^{(\pm)} \longrightarrow \phi_{\beta o}^{(\pm)}(\underset{\sim}{k},\underset{\sim}{r}) \equiv \psi_{PW}^{(\pm)}(\underset{\sim}{k},\underset{\sim}{r},\beta) \;, \tag{33}$$

[20]

where $\lambda = \Lambda/2$ and $\psi_{PW}^{(\pm)}(\underset{\sim}{k},\underset{\sim}{r},\beta)$ is defined in (32) with $1/n \to \beta$, we max expect the following conjecture to be a valid extension of these relations.

If $H = H_0 + V(r)$, $H_0 = -\Delta/2\mu$, V(r) is locally square-integrable, and for r large enough, i.e. for $r > R$

$$V(r) = V_1(r) + V_2(r)$$

$$V_1(r) = \alpha/r^{\beta} \;, \; \beta > 0, \; \alpha = \text{const.} \tag{34}$$

$$\int_R^{\infty} |V_2(r)| \, dr < \infty \;,$$

then, with $\lambda = \mu\alpha$,

$$||e^{-iHt}f_\beta^{(\pm)} - K_\beta^{(\pm)}(\lambda)e^{-iH_ot}f|| \to 0, \text{ as } t \to \pm\infty , \tag{35}$$

where

$$f_\beta^{(\pm)}(\underset{\sim}{r}) = (2\pi)^{-3/2}\int d\underset{\sim}{k}\ \psi_\beta^{(\pm)}(\underset{\sim}{k},\underset{\sim}{r})\hat{f}(\underset{\sim}{k}) \tag{36}$$

and $\psi_\beta^{(\pm)}$ are given alternatively by (a) : (15) with $\phi_\ell(r)$ given by appropriate linear combinations of $\phi_\pm(k,r,\beta)$ defined in (29); or (b) by solutions of $H\psi_\beta^{(\pm)} = \frac{k^2}{2\mu}\psi_\beta^{(\pm)}$ satisfying $\binom{\text{incoming}}{\text{outgoing}}$ boundary conditions as $r \to \infty$.

References

(1) J.M. Jauch, Helv. Phys. Acta 31, 127, 661 (1958), Theory of the Scattering Operator. I, II.

(2) J.D. Dollard, J. Math. Phys. 6, 729 (1964), Asymptotic Convergence and the Coulomb Interaction. Also, thesis, Princeton, 1963.

(3) We shall use the circumflex from now on to indicate the Fourier transform in the symmetric form shown in (5a) below.

(4) T. Ikebe, Archiv. Ratl. Mech. Anal. 5, 1, (1960).

(5) Specifically, it is required that (1) V be real and locally Holder continuous except for a finite number of singularities; (2) $V \in L^2$; and (3) $V(\underset{\sim}{r}) = O(r^{-2-\varepsilon})$ for $\varepsilon > 0$ and $r \to \infty$.

(6) D. Mulherin and I.I. Zinnes, J. Math. Phys. 11, 1402 (1970), Coulomb Scattering. I. Single Channel.

(7) Colston Chandler and A.G. Gibson, J. Math. Phys. (to appear).

(8) $\Omega^{(\pm)}$ will be said to be complete[9] if $\mathcal{R}(\Omega^{(\pm)}) = \mathcal{H}_{2,ac}$ where $\mathcal{R}$ is the range and $\mathcal{H}_{2,ac}$ is the absolutely continuous part of the Hilbert space with respect to H_2.

(9) T. Kato, Perturbation Theory for Linear Operators, Springer-Verlag, Berlin, 1966; ch. 10.

(10) J.D. Dollard, J. Math. Phys. 7, 802 (1966), Adiabatic Switching in the Schrödinger Theory of Scattering. See also E. Prugovečki and J. Zorbas, J. Math. Phys. 14, 1398 (1973).

(11) E.J. McShane, Integration. Princeton University Press, 1944, pp. 231 - 3.

(12) For the significance of the notation here and below, see ref. 9.

(13) R.B. Lavine, J. Funct. Anal. 5, 368 (1970); V.S. Buslaev and V.B. Matveev, Teor. Mat. Fiz. 2, 367 (1970).

(14) W.O. Amrein, Ph. A. Martin and B. Misra, Helv. Phys. Acta 43, 313 (1970), On the Asymptotic Condition of Scattering Theory; P. Alsholm and T. Kato, Proc. of Symp. in Pure Math. 23, 393 (1973), A.M.S., Scattering with Long Range Potentials.

(15) In the short-range case the scattering and the prepared states are just the states ψ, ψ_o, respectively, which are related through $\psi = \Omega_o\psi_o$. When long-range potentials are involved it was found convenient in ref. 6 to introduce the notion of an asymptotic state which plays the role of a modified free state such as $K(\lambda)f$ in (9).

(16) There is no great problem when one is interested only in the scattering matrix. See ref. 14.

(17) E.A. Coddington and N. Levinson, Theory of Ordinary Differential Equations. McGraw-Hill, 1955, p. 92.

(18) Note that for the Coulomb case $\Lambda = 2\mu\alpha$, where $V(r) = \alpha/r$.

(19) It is evident that the condition on U_1 given in (23a) can be relaxed.

(20) This relation together with (32) and (30) imply that when $\beta = 1$ (corresponding to the pure Coulomb potential) $\phi^{(\pm)}_{1,0}(k,r) = \exp i[\underset{\sim}{k}\cdot\underset{\sim}{r} \pm (\lambda/k) \ln 2kr]$, which differs in the ln - term from $\phi^{(\pm)}_{co}$ defined in (8). Detailed calculations show that both functions are equally effective in defining the modified free wave.

PHENOMENOLOGICAL ASPECTS OF LOCALIZABILITY*

L.P. Horwitz

Department of Physics and Astronomy
University of Tel Aviv
Ramat Aviv, Israel

Recent experimental and theoretical advances have provided some additional insight into the value of local field theories for the description of actual phenomena. It is suggested, on the basis of some of these results, that there is evidence that the concept of the position of an elementary system may have phenomenological importance.

1. Introduction

The meaning of the position of an elementary system of non-vanishing mass, from the viewpoint of relativistic quantum theory, was clarified in a simple and beautiful way in the fundamental work of Newton and Wigner[1] in 1949. Jauch and Piron[2] later suggested a generalized notion of localizability for application to cases where an operator of the type constructed by Newton and Wigner does not exist (e.g., the photon and neutrino).

The relativistic quantum field theory of electrons and photons has successfully postulated a local interaction of the form

$$j^\mu(x)\, A_\mu(x) = \bar{\psi}(x)\gamma^\mu\psi(x)\, A_\mu(x) \tag{1.1}$$

The positive energy Dirac wave function is a function on the manifold of space-time, but it does not have the same interpretation, in terms of physical measurement, as the Schrödinger wave func-

*Research supported in part by the Israel Academy of Sciences.

tion in the non-relativistic theory. Newton and Wigner[1)] showed that a relativistically covariant wave function describing a localized elementary system of definite mass and spin is not a δ-function in coordinate space. The interaction (1.1) is therefore not "local" in the sense that electrons and photons are really expected to interact only when the corresponding particles overlap. Amrein showed in his thesis[3)] that the relation between the position of a particle and the corresponding energy density is not local. In a relativistic quantum field theory, every particle is associated with a particle field, which contains the energy of the particle and causes its interactions with other particles (it is this field which corresponds to the "coordinate space wave function" of Newton and Wigner[1)]). The notion of locality in axiomatic field theory refers to a property of the particle field, and not the particles themselves (as defined through position measurements). Local observables refer to the position of the particle field, and it is in this sense that we refer to the interaction (1) as local.

The general form of electromagnetic theory has been carried over for the description of other systems, i.e., those involving much weaker, or much stronger, interactions. These attempts have not been as successful, but the postulate of local interactions, analogous to (1.1), has been maintained for many years[4)].

As Jauch and Piron[2)] have pointed out, many new facts of elementary particle physics have been easily assimilated within this customary framework. The additional new quantum numbers of isotopic spin, hypercharge, and even the representation structure of SU(3), SU(6), etc. have been incorporated in the direct product sense in the Hilbert space of states, therefore suggesting little connection between the observed phenomena. As a partial motivation for their work, they suggested that the study of localizability might be fruitful in preparation for a general theory of elementary particles.

Recent experimental and theoretical advances have provided some additional insight into the value of local field theories for the description of actual phenomena, and in this essay, I shall review some of these results and suggest that there is evidence that the concept of the position of an elementary system may have phenomenological importance.

2. Historical Background.

Since the advent of SU(3)[5)6)] in 1961, the "existence" or "non-existence" of the lowest representation as a triplet of physically observable particles has been an issue. The strong results of SU(3), i.e., the classification of mesons into octets and baryons into an octet and a decuplet, and even the tensor properties of currents, do not depend on the existence of these lowest representations as particles, or even as unobservable fields. They were discreetly considered to be only abstract mathematical entities (tensor indices) used in the construction of physically observable particle representations, but they were called "quarks,"[7] in case some stronger evidence became available that would imply the utility of granting them some degree of physical reality.

The extraordinary success of Gell-Mann's current algebra[8] in explaining the ratio of axial to vector couplings in weak leptonic decay (G_A/G_V)[9], and in obtaining magnetic moment sum rules[10][11] could not be taken as strong evidence in favor of the existence of quark fields, even though the algebra could be derived by assuming that the SU(3) charges were the integrals over local currents carried by free quark fields:

$$F(\lambda_a) = \int d^3x \; j_o^a \; (x) \tag{2.1}$$

where

$$j_\mu{}^a(x) = \bar{q}(x)\gamma_\mu \frac{\lambda a}{2} q(x) \tag{2.2}$$

and q(x) is a local Dirac field in the triplet representation of SU(3). Although the algebra of currents was originally derived by means of (2.2) (with the assumption of free field anticommutation relations), the algebra could then be postulated without reference to its source, since no consequent results depended upon the existence of free quark fields. This process was called "abstraction", and is still in use, as we shall see, for example, in light cone physics.

The "discovery"[12] of SU(6) in 1964 added significantly to the utility of SU(3) in the classification of particle multiplets, and successfully predicted the ratio of neutron and proton magnetic moments, among other applications such as decay widths. The 35 dimensional negative parity representation of SU(6) provides a place for 8 pseudoscalar mesons, and 9 vector mesons, and the 56 dimensional representation, for 8 spin 1/2 baryons (including the nucleons) and 10 spin 3/2 baryons (including the famous $\frac{3}{2}\,\frac{3}{2}\,\Delta$ resonance).

Since SU(3) could successfully be made to say something about particle masses[13], and SU(6) included statements about spin, it was tempting to try to develop a relativistic theory which would non-trivially contain these "internal symmetries". These attempts[14] ended in failure, but a peripheral effort, which was less ambitious, will be of interest to us in our discussion.

Suppose we wish to characterize a free Dirac electron by its momentum, its mass, total angular momentum, and helicity, but we want the helicity to have the interpretation of a magnetic quantum number associated with an SU(2) group which is independent of the momentum. The difficulty with this program is that $\alpha_3(\gamma^5\sigma_3)$, the generator of Lorentz boosts in the 3 direction that appears formally in the non-unitary four dimensional Dirac representation, does not commute with σ_1 and σ_2, and hence the usual Pauli algebra of SU(2) generators cannot serve as a basis for a momentum independent SU(2). The three generators $(\beta\sigma_1, \beta\sigma_2, \sigma_3)$ do, however, form an SU(2) which commutes with α_3.
Multiplying these by the $\{\lambda_a\}$ of SU(3), we obtain an SU(6) called $SU(6)_W^{curr.}$; its charges are then of the form[15]

$$F(\beta\sigma_\perp \lambda_a) = \int q(x)^\dagger \beta\sigma_\perp \frac{\lambda_a}{2} q(x)\, d^3x \tag{2.3}$$

and

$$F(\sigma_3\lambda_a) = \int q(x)^\dagger \sigma_3 \frac{\lambda_a}{2} q(x)\, d^3x, \tag{2.4}$$

and integrals over local quark currents of the form (2.2).

At about the same time, Lipkin and Meshkov[16)17)] proposed a classification scheme for particle states according to a group $SU(6)_W^{str.}$; conservation laws for vertices and collinear processes were worked out, and were fairly successful. Moreover, Lipkin and Scheck[18], and Kokkedee and Van Hove[19] worked out relations among total hadron-hadron cross-sections as generalizations of the Levin-Frankfurt work[20], which rested upon a view (the "naive" quark model) of hadrons as being composed of two or three "quarks" subject to impulse approximation at high energies. Taking the Dashen-Gell-Mann suggestion[21] of classifying hadronic states according to $[U(6) \times U(6)]_\beta$, Cabibbo, Horwitz and Ne'eman proposed an algebraic scheme[22] for Regge couplings which produced all of the results of the "naive" quark model. In their theory, the Regge couplings were determined as matrix elements, in physical hadron states, of quark counting operators in this group.

It is interesting to note that the charges (2.3) and (2.4) (also contained in the scheme of CHN) do not commute with the free field Dirac Hamiltonian, and it is difficult to see how a classification scheme, based on these, could be successful. The SU(2) group which we proposed as a boost invariant classification scheme suffers from the same difficulty - the states would not be invariant in time to lowest order in perturbation theory. It is clear that the "quarks" of current algebra, and the "quarks" of supermultiplet schemes should not be considered to be the same, even if they do not have the physical properties of particles, if they are to provide a useful mathematical model with properties that can be "abstracted". It is in this distinction, as we shall see, that the concept of the position of an elementary system will arise.

In some of the most remarkable experiments done in recent years in elementary particle physics, involving deep inelastic production of hadrons by collision with very high energy incident electrons, the phenomenon of scaling was strikingly observed[23]. Bjorken[24] had predicted this phenomenon on the basis of what is apparently still another view of the structure of hadrons, the parton model of Feynman[25]. In this model, the hadron is thought of as constructed of a number (perhaps infinite) of sub-particles, each carrying an essentially localized charge. It is the local nature of the charges that gives rise to the observed scaling behavior. Frishman[26] and Fritzsch and Gell-Mann[27] have, however, been able to relate at least some of the results of the parton model to a new algebra of currents, assumed to be valid for light-like separated densities (instead of the equal time algebra of the older schemes discussed above). The model for this algebra was again taken to be that of local currents carried by free quark fields; the resulting scaling behavior was exactly that predicted by Bjorken on the basis of the parton model and strikingly observed experimentally. If we cast the "naive" quark model results in the form of the CHN algebra of Regge couplings, the problem we face is, therefore, to find some relation between the following disjoint ideas, each with their own special domain of phenomenological success:

a) Algebra of Regge residues
b) Supermultiplet symmetries
c) Current algebras
d) Parton models

3. Constituent quarks and current quarks.

As we have seen, it appears that the notion of quarks, and quark fields, has been useful in describing various features of hadron systems. There are, however, many contradictions that appear to be inherent in such a picture. We have already remarked that the expression of certain group operators in terms of local quark fields, as one-body observables in a field theory, immediately leads to the conclusion that the corresponding charges are not conserved under the evolution generated by the free field Hamiltonian. Hence the representations of the group cannot persist in time in lowest order perturbation theory. Furthermore, Coleman[28] has pointed out that if the operators of a classification group are represented by integrals over local currents, then they must correspond to an exact symmetry. A perhaps more serious criticism is that a symmetry model that provides a good D/F ratio for meson decays results in a much worse value for G_A/G_V in the coupling of the hadronic weak currents to leptons. The simple notion of two quarks in a meson and three in a baryon does not seem compatible with the parton model, which requires a large number of sub-particles in the hadron. Finally, we remark that in spite of extensive experimental searches, quarks have not yet been seen in asymptotic states.

What seems to emerge from this situation is a suggestion that was, in fact, implicit in the use of quark models since they were first introduced: There may be more than one kind of quark needed for the description of hadronic systems. Fritzsch and Gell-Mann[27] recently emphasized this point, and labelled as "constituent quarks" the subsystems characterizing supermultiplet classification schemes (as in, for example, the calculation of Regge coupling according to the rules of CHN, or the "naive quark model", or $SU(6)_W^{Str.}$ selection rules), and as "current quarks" the subsystems characterizing current algebra (the electromagnetic and weak currents). They suggested, furthermore, that there may be a transformation between them. A unitary transformation connecting the charges of current algebra (integrals of local current densities) to the charges of $SU(6)_W^{Str.}$ was then constructed, in a free field model, by Melosh[29]. The criterion used was that the transformed charges commute with the free field Hamiltonian. The resulting scheme removed many of the contradictions described above (we shall return to discuss the question of why quarks are not seen in the next section).

In what follows, I shall describe the connection, based on an extension of the Melosh transformation[29], recently proposed by Gomberoff, Horwitz and Ne'eman[30], between the first three schemes listed at the end of the previous section, and suggest, on this basis, a relation to the fourth. It is precisely in the concept of <u>localizability</u> that the distinctions can be drawn. The discussion will be almost entirely based on the theory of free fields. In a concluding section, recent progress in attempts to develop a theory of interacting fields, of a non-perturbative nature, which might eventually provide some justification for the use of free field models, will be described. In this context, one of the obvious, but most fundamental questions, seems also to be related to the notion of localizability - why don't we observe individual quarks in asymptotic states?

We start by remarking that the system of charges (2.1), (2.3) and (2.4), supplemented by

$$F(\beta\lambda_a) = \int d^3x\ q(x)^{+}\beta\,\frac{\lambda_a}{2}\,q(x) \qquad (3.1)$$

$$F(\sigma_{\perp}\lambda_a) = \int d^3x\ q(x)^{+}\,\sigma_{\perp}\frac{\lambda_a}{2}\,q(x) \qquad (3.2)$$

$$F(\beta\sigma_3\lambda_a) = \int d^3x\ q(x)^{+}\beta\sigma_3\,\frac{\lambda_a}{2}\,q(x) \qquad (3.3)$$

close on an algebra* which we shall call $\left[U(6) \times U(6)\right]^{currents}_{\beta W}$; except for $F(\lambda_a)$, these do not commute with the free field Hamiltonian

$$H_o = \int d^3x\ q(x)^{+}(-i\,\underset{\sim}{\alpha}\cdot\underset{\sim}{\nabla} + \beta m)\,q(x).$$

As we have remarked above, it is our objective to construct a unitary transformation that will carry the operators F to another set of operators, which we shall call W, which commute with the free field Hamiltonian and generate the group $\left[U(6) \times U(6)\right]^{strong}_{\beta W}$.

As a first step, we use the transformation $V = e^{iS}$ introduced by Melosh[29]), where

---

* Using free field anti-commutation relations for the fields

$$S = \tfrac{1}{2}\int d^3x\ q(x)^{+}\ \tan^{-1}\left(\frac{\underset{\sim}{\gamma}_{\perp}\cdot\underset{\sim}{\partial}_{\perp}}{m}\right) q(x) \qquad (3.4)$$

Applying this transformation to the $U(6)_W^{current}$ charges of (2.1), (2.3) and (2.4), in the sense $F \to VFV^{-1}$, we obtain the charges of $U(6)_W^{strong}$ which commute with H_o. To see this result in a simple way, let us apply the transformation to H_o in the inverse sense instead:

$$H_M = V^{-1}H_o\ V = \int d^3x\ q(x)^{+}(-i\alpha_3\partial_3+\beta\kappa)\ q(x) \qquad (3.5)$$

where $\kappa = \sqrt{(\underset{\sim}{\gamma}_{\perp}\cdot\underset{\sim}{\partial}_{\perp})^2 + m^2}$. Since α_3 and β commute with σ_3, $\beta\sigma_{\perp}$, it is clear that one may construct a supermultiplet symmetry with the $U(6)_W^{strong}$ charges, and still maintain contact with the system of currents.

Although only the very simple notions of free field commutation relations, and commutivity with the free field Hamiltonian, were used in Melosh's construction, some remarkable successes were achieved in comparison with experiment. The pseudo-scalar density, acting as a transition operator in meson decays, looks like $\int d^3x\ q(x)^{+}\beta\gamma^5\frac{\lambda_a}{2}q(x)$ at small momentum transfer, and $\beta\gamma^5$ commutes with $\underset{\sim}{\gamma}_{\perp}$; Melosh transformation therefore does not affect its matrix elements, and the reasonably good D/F prediction[31)] of $U(6)_W^{currents}$ remains valid in $U(6)_W^{strong}$. On the other hand, a symmetry prediction of G_A/G_V involves the expectation value of the axial charge $\int d^3x\ q(x)^{+}\gamma^5\frac{\lambda_a}{2}q(x)$; since γ^5 does not commute with $\underset{\sim}{\gamma}_{\perp}$, the Melosh transformation changes this quantity, reducing it, as shown by Melosh[29)], to a value less than $\frac{2}{3}$. More recently, it has been shown by Gilman and Kugler,[32)] and Meshkov, Gilman and Kugler[33)] , that all of the meson decay amplitudes are well reproduced by the Melosh transformation.

The form of the transformation (3.4) is reminiscent of the Foldy-Wouthuysen transformation, and is, in fact, a two-dimensional version of it. In a first-quantized theory, the operator representing a position measurement which has the form $\underset{\sim}{x}$ in the

"Melosh representation" (where the Hamiltonian is H_M), is an operator in the original representation which has the structure of the Newton-Wigner position operator in its transverse part.

The partially transformed first-quantized "Dirac equation",

$$(\alpha_3 p^3 + \beta\kappa)\, \psi_M(p) = E\, \psi_M(p) \tag{3.6}$$

has an interesting interpretation in terms of null plane dynamics (the "front form" of dynamics described by Dirac[35], and given modern expression by Kogut and Soper[36], Susskind[37] and Bjorken[38], and others, in their treatment of the infinite momentum frame and the parton model).

Define*

$$\alpha_\pm = \frac{1}{\sqrt{2}}(1 \pm \alpha_3) \qquad p^\pm = \frac{1}{\sqrt{2}}(p^0 \pm p^3)\ , \tag{3.7}$$

and let us call

$$\psi_\pm = \alpha_\pm\, \psi_M \tag{3.8}$$

Multiplying (3.6) by α_+ and α_- in turn, one obtains the coupled (two-component) equations

$$p^-\, \psi_+ = \frac{\beta\kappa}{\sqrt{2}}\, \psi_-$$

$$p^+\, \psi_- = \frac{\beta\kappa}{\sqrt{2}}\, \psi_+ \tag{3.9}$$

---

* This interpretation was discovered during discussions with Prof. L.C. Biedenharn during my visit to Duke University in June, 1973. I wish to thank Prof. Biedenharn for his hospitality and collaboration.

In the "front form" of dynamics[35], p^+ is a constant which plays the role of a Galilean mass, and p^- is the "Hamiltonian" which translates the null plane, corresponding to dynamical evolution. The resulting structure was called "Galilean subdynamics" by Biedenharn, Han and van Dam[39], where the significance of the Galilean algebra embedded in the Poincaré algebra is worked out and is shown to lead to Dirac's new equation[40]. (This work was further developed and applied to the construction of dual models by Biedenharn and van Dam[41].) Assuming that p^+ is constant, we may eliminate ψ_- in (3.9) to obtain

$$p^- \psi_+ = \frac{\kappa^2}{2p^+} \psi_+ \qquad (3.10)$$

the Galilean equation of motion for the null-plane initial value problem. It is this equation which forms the basis for the dynamics of parton models[42].

The connection between the usual field theoretical approach and null-plane dynamics is not, <u>a priori</u>, very clear. An infinite boost of p^o, generally believed to be the evolution operator in the usual formulation, brings it to p^+ which has no dynamical role on the null plane. The Galilean Hamiltonian p^- is introduced as the "residual" motion of the transverse degrees of freedom. In view of the result(3.10), it is natural to suggest that the connection between the usual space-time formulation and the parton model is made by a unitary transformation of the type (3.4); <u>the solutions of</u> (3.10) <u>correspond to wave functions in which the transverse coordinates have the Schrödinger interpretation of localization</u>.

We furthermore remark that these states do not correspond to the physical states we started out to construct at the beginning of this section. The transformation V , instead of being applied to the system of $U(6)_W^{current}$ charges, was applied to the Hilbert space of physical states and to the Hamiltonian; the resulting description is that of a world with an altered Hamiltonian (3.6), and in this world the operators F of $[U(6)]_W^{currents}$ are the generators of the classification group for forming multiplets. Hence, in the parton model, the "constituent quarks" play no role, and we deal only with "current quarks".

Let us now turn to the remainder of our task of constructing a "good" $[U(6) \times U(6)]_{\beta W}^{strong}$ algebra. The operators (3.1), (3.2) and (3.3) still do not commute with the transformed Hamiltonian H_M of (3.5). We must therefore find an additional transformation which commutes with (2.1), (2.3) and (2.4) (so that the final form of these charges will be the same as given by Melosh, and the results concerning G_A/G_V , meson decays, etc. remain), but takes H_M to a form which commutes with the operators (3.1), (3.2) and (3.3). The solution found by Gomberoff, Horwitz and Ne'eman is given by $V' = e^{iS'}$, where

$$S' = \frac{1}{2}\int d^3x \; q(x)^{+} \tan^{-1} \frac{\gamma^3 \partial_3}{\kappa} \; q(x). \tag{3.11}$$

the combined transformation

$$G = V \; V' \tag{3.12}$$

brings H_o to the Foldy-Wouthuysen form

$$G^{-1} \; H_o \; G = \int d^3x \; q(x)^{+} \; \beta \; E \; q(x) \; , \tag{3.13}$$

but it is not identical to the transformation used by Foldy and Wouthuysen. Conversely, we may carry out this transformation on all of the charges F to obtain

$$W = G \; F \; G^{-1} \tag{3.14}$$

The "good" (conserved) charges of $[U(6) \times U(6)]_{\beta W}^{strong} \equiv W_B$, with which we may construct a classification scheme for the physical states.

Defining the quark fields in the usual way,

$$q(x) = \int \frac{d^3p}{(2\pi)^{3/2}} \sqrt{\frac{m}{E}} \sum_s \left(b(p,s)u(p,s)e^{-ip\cdot x} + d^\dagger(p,s)v(p,s)e^{ip\cdot x}\right), \tag{3.15}$$

where s refers to unitary spin as well as spin, b(p,s) annihilates a quark and $d^\dagger(p,s)$ creates an antiquark, one finds that

$$W(\beta\lambda_o) = G\, F(\beta\lambda_o) G^{-1} = \frac{1}{\sqrt{6}} \sum_s \int d^3p \left\{ b^\dagger(p,s)b(p,s) + d^\dagger(p,s)d(p,s) \right\} \tag{3.16}$$

explicitly commutes with the boost operator M^{03}. From this it follows that the entire transformed algebra is boost invariant (3 direction), so we have obtained a classification scheme that is invariant under a non-trivial part of the Poincaré group*. We have therefore made a connection between the supermultiplet classification scheme of W_B (of which the proper diagonal subgroup is $U(6)_W^{strong}$) and the algebra of currents.

It is of some interest to see the effect of our transformation on the quark fields of which the group operators are constructed. Carrying out the transformation (3.14) on the quark field, we find

$$\hat{q}(x) = G\, q(x)\, G^{-1} = (E + i\beta\partial^o) \frac{E+\kappa + i\,\gamma^3\partial_3}{\sqrt{2E(E+\kappa)}\sqrt{2\kappa(\kappa+m)}}\, q(x) \tag{3.17}$$

---

* Some of the corresponding charges in the current system vanish as $p^3 \to \infty$; they do not in the strong system. It is shown in Ref. 30 that the mechanism for this effect can be found in the fact that S' does not have a well-defined operator limit as $p^3 \to \infty$, although its commutation relations with the quark fields do. This implies

a highly non-local transformation. Since q(x) satisfies the Klein-Gordon equation, it follows that

$$(E - i\,\beta\,\partial^{o})\,\hat{q}(x) = o$$

or

$$i\,\partial^{o}\,\hat{q}(x) = \beta E\,\hat{q}(x)\ , \qquad (3.18)$$

i.e., the transformed fields satisfy the Foldy-Wouthuysen equation. In first-quantized form, these "constituent" quark states therefore have the Schrödinger interpretation as the wave functions of localized particles (in three dimensions). Since electromagnetic coupling is local in terms of the original, covariant fields, we do not expect that constituent quarks will have minimal electromagnetic couplings. The non-existence of minimal electromagnetic coupling for Dirac's new equation[39),40)] can be understood in these terms. It is possible, however, that the strong interaction, which seems to select $[U(6)xU(6)]^{strong}_{\beta W}$ as the appropriate classification group, would be simpler to express in the constituent system, in which the localizability of particles is made explicit.

The connection to Regge couplings is made as follows. In analogy to the Compton-type amplitudes studied in deep-inelastic electron or neutrino scattering (cf., for example, ref. 27), we introduce spin-averaged absorptive parts of the form

$$M_{ab} = \int e^{ig\cdot x} \langle N(p)\mid \hat{J}_a(x),\ J_b(o)\mid N(p)\rangle\, d^4x\ , \qquad (3.19)$$

---

* (cont.)

that our unitary transformation does not exist as $p^3\to\infty$, and the limiting form of the transformed algebra is not unitarily related to the algebra of local currents (if it existed, G at $p^3\to\infty$ would be "time" independent).

where $\hat{J}_a(x)$, $\hat{J}_b(o)$ are "strong currents", and the hadron states $|N(p)\rangle$ are in small representations of $[U(6) \times U(6)]^{strong}_{\beta W}$. Using, for example, the currents

$$\hat{v}_c{}^o(x) = \hat{q}(x)^{+} \frac{\lambda_c}{2} \hat{q}(x) \qquad (3.20)$$

and the strong free quark anti-commutation relations

$$\left\{\hat{q}(x) , \hat{q}(x')^{+}\right\} = i \beta \partial_o \Delta_1(x-x') - \partial_o \Delta(x-x') \qquad (3.21)$$

derived from (3.17)($i\Delta=\Delta_+-\Delta_-$, $\Delta_1=\Delta_++\Delta_-$, $\Delta_\pm=\frac{1}{(2\pi)^3}\int\frac{d^3p}{2p^o}\, e^{\pm ip\cdot x}$) ,

we obtain

$$\left[\hat{v}_a{}^o(x), \hat{v}_b{}^o(o)\right] = -(\hat{v}_c{}^{(-)}(x,o)d_{abc} + i\, \hat{v}_c{}^{(+)}(x,o)f_{abc})\partial_o\Delta(x)$$

$$+ i\, (\hat{s}_c{}^{(+)}(x,o)\, d_{abc} + i\, \hat{s}_c{}^{(-)}(x,o)f_{abc})\partial_o\Delta_1(x) ,$$

(3.22)

where

$$\hat{v}_a{}^{(\pm)}(x,o) = \tfrac{1}{2}\, (\, \hat{v}_a(x,o) \pm \hat{v}_a(o,x))$$

$$\hat{s}_a{}^{(\pm)}(x,o) = \tfrac{1}{2}\, (\, \hat{s}_a(x,o) \pm \hat{s}_a(o,x)) \qquad (3.23)$$

and

$$\hat{v}_a(x,o) = \hat{q}(x)^{+} \frac{\lambda_a}{2} \hat{q}(o)$$

$$\hat{s}_a(x,o) = \hat{q}(x)^{+} \beta \frac{\lambda_a}{2} \hat{q}(o) \quad . \tag{3.24}$$

The presence of the $\partial_o \Delta_1$ terms reflects the non-local nature of the transformation G.

In the limit that $q^\mu q_\mu = q^2 \to -\infty$, and $\nu = p \cdot q \to \infty$ such that $\omega = \frac{2\nu}{-q^2}$ = const. (the Bjorken limit[24]), the integration (3.19) is restricted by the Riemann-Lebesque lemma to the neighborhood of $x^2=o$, the light cone; under these conditions, the contributions of the various terms of (3.22) to the integral (3.19) become functions of ω alone, and we obtain the phenomenon of scaling. Finally, for large ω, we expect the amplitude M_{ab} to be asymptotically approximated by Regge behavior. Following a procedure suggested by Cabibbo and Testa[43], we may then pick out the hadron-Regge coupling, and use this result to predict features of hadron-hadron scattering.

Since the operators of the classification group that we have constructed are independent of time (in the free field theory), the quasi-bilocal operators (3.23) are elements of two distinct tensor operators transforming like the adjoint representation of W_B. For the first of these,

$$\hat{v}_q^{(+)}(x,o) \sim W(\lambda_a)$$

$$\hat{s}_a^{(+)}(x,o) \sim W(\beta\lambda_a)$$

and for the second,

$$\hat{v}_a^{(-)}(x,o) \sim W(\beta\lambda_a)$$

$$\hat{s}_a^{(-)}(x,o) \sim W(\lambda_a) \quad . \tag{3.25}$$

We may therefore apply the Wigner-Eckart theorem to the matrix elements occuring in (3.19). We assume in the symmetry limit in which we are working that the singlet is associated with the same reduced matrix elements. It is shown, furthermore, in ref. 30, that between states of equal momentum these operators transform like the fourth component of a four vector under Lorentz boosts, and we do not include the less singular contribution proportional to x^o :

$$\langle N(p)|\hat{v}_a^{(+)}(x,o)|N(p)\rangle = C^{NN}_{v,a}\, f^{(+)}(x\cdot p)\, p^o$$

$$\langle N(p)|\hat{s}_a^{(+)}(x,o)|N(p)\rangle = C^{NN}_{s,a}\, f^{(+)}(x\cdot p)\, p^o$$

$$\langle N(p)|\hat{v}_a^{(-)}(x,o)|N(p)\rangle = i\, C^{NN}_{s,a}\, f^{(-)}(x\cdot p)\, p^o$$

$$\langle N(p)|\hat{s}_a^{(-)}(x,o)|N(p)\rangle = i\, C^{NN}_{v,a}\, f^{(-)}(x\cdot p)\, p^o \quad ,$$

(3.26)

where the coefficients $C^{NN}_{s,a}$, $C^{NN}_{v,a}$, are Clebsch-Gordon coefficients normalized according to

$$\frac{p^o}{m}\; C^{NN}_{v,a}\; \delta^3(p-p') = \langle N(p)|W(\lambda_a)|N(p)\rangle$$

$$\frac{p^o}{m}\; C^{NN}_{s,a}\; \delta^3(p-p') = \langle N(p)|W(\beta\lambda_a)|N(p')\rangle$$

(3.27)

Carrying out the integral (3.19), the spin-zero parts are given by

$$\int d^4x\; e^{iq\cdot x}\; p^o f\; (x\cdot p)\partial_o\Delta(x) \xrightarrow[Bj]{} -\pi\; F(\tfrac{1}{\omega})$$

$$\int d^4x\; e^{iq\cdot x}\; p^o f(x\cdot p)\; \partial_o\Delta_1(x) \xrightarrow[Bj]{} -i\pi F(\tfrac{1}{\omega})$$

(3.28)

where

$$f(x\cdot p) = \int e^{i\,\xi(x\cdot p)}$$

The amplitude (3.19) can then be written (for the vector-vector example (3.22))

$$(M_{ab})_{\text{spin zero}} \xrightarrow[B_j]{} \pi\{F^{(+)}(\tfrac{1}{\omega}) + i\,F^{(-)}(\tfrac{1}{\omega})\}\{C^{NN}_{s,c}\,d_{abc} + if_{abc}\,C^{NN}_{v,c}\} \qquad (3.29)$$

In the large ω limit, we assume Regge behavior for this amplitude,

$$F^{(+)}(\tfrac{1}{\omega}) + F^{(-)}(\tfrac{1}{\omega}) = \sum_n a_n\omega^n \sim \beta\,\omega^{\alpha} \qquad (3.30)$$

where the coefficients of the power series correspond to matrix elements of the Taylor expansion of the corresponding quasi-bi-local generators. Since these correspond to n^{th} rank tensors, they carry spin $\leq n$, and hence the Regge asymptotic behavior could be associated with a Van Hove type model[44]. Factorization of the residues implies that

$$\beta = \beta^{\text{curr}}.\ \beta^{\text{hadr.}} \quad ; \qquad (3.31)$$

and

$$\beta^{\text{curr}}\ f_{abc}$$

$$\beta^{\text{curr}}\ d_{abc} \qquad (3.32)$$

are the couplings of the currents to the Regge exchange, and

$$\beta^{\text{hadr}} C^{NN}_{s,c}$$

$$\beta^{\text{hadr}} C^{NN}_{v,c} \qquad (3.33)$$

are the couplings to charge conjugation positive and negative Regge exchanges, respectively, as originally proposed by CHN.

For the validity of the Wigner-Eckart theorem as $\omega \to \infty$ it was essential that the group W_B does not deform under infinite Lorentz boosts, and hence the rather delicate properties of S' as $p^3 \to \infty$ are necessary for the structure of the theory. For the group $[U(6)xU(6)]^{\text{currents}}_{\beta W}$, half of the generators go to zero at $p^3 \to \infty$.

By closing the algebra of bilocal operators (in the sense of distributions) at $p^3 \to \infty$, Reddy et al have shown that the Regge couplings are universal if high mass intermediate states (many particle) do not contribute strongly, i.e., the coefficient $\beta^{\text{hadr.}}$ is the same for meson and baryon couplings.

Having made a connection between the first three schemes mentioned at the end of the previous section, we now turn to the fourth. The parton model of Feynman refers to a picture of the hadron as made of many sub-hadronic constituents, which have the property of almost point interaction with the electromagnetic field. It is this local, practically unstructured interaction which leads to scaling in deep inelastic electron scattering from this point of view[24) 25)]. As we have seen, the "constituent" quarks from which the hadrons are built as small representations are non-local objects, fields subjected to a transformation closely related to that of Foldy and Wouthuysen, and hence they would not be expected to couple locally to electro-magnetism (except approximately, in a non-relativistic limit). Let us consider again the eigenvalue equations defining the hadronic states:

$$W \psi^{\text{constit.}} = w' \psi^{\text{constit.}} \quad ; \qquad (3.34)$$

according to (3.14), this is equivalent to

$$F\ (G^{-1}\ \psi^{\text{constit.}}) = w'\ (G^{-1}\ \psi^{\text{constit.}})\ , \qquad (3.35)$$

and we recognize that $G^{-1}\psi^{\text{constit.}}$ is an eigenstate of the charges in the current system. This state, $\psi^{\text{curr.}}$ may be constructed of local quark creation operators acting on the vacuum. Hence the physical hadron state

$$\psi^{\text{constit.}} = G\ \psi^{\text{curr.}} \qquad (3.36)$$

At $p^3 \to \infty$, the part of G corresponding to the Melosh transformation V contains no pair terms, and therefore only induces a rotation in the sector of the Fock space in which $\psi^{\text{curr.}}$ lies[46] (but adding an indefinite amount of spin). The operator S' contains only pair terms as $p^3 \to \infty$, however, and the expansion of the factor V' in a power series adds an indefinite number of quark pairs to the original state $\psi^{\text{curr.}}$; the resulting "cloud" of quarks is composed of local fields, which may interact locally with electromagnetism. To calculate the momentum distribution of quarks in this cloud, for example, some additional information is needed. We have so far been discussing a free quark model, and there is therefore no constraint on the momentum spectrum of the fields. Whatever else the strong interaction does when it is turned on, however, it must have the result of confining the quarks to a bounded volume in space. It is appealing to guess, to the lowest order of approximation, that this is the only effect of the strong interaction, and that the quark cloud (parton) momentum distribution could be calculated in a model of free quarks confined to a bounded region of space. Some authors have proposed models of this type; in this question we have come close to one of the most fundamental problems of providing a field theoretical substructure for the hadrons. It is closely tied to the notion of localizability.

4. Renormalization and localizability

In this concluding section, I shall review some recent attempts to construct models in which hadrons are considered to be constructed of sub-hadronic fields, and make some highly speculative remarks concerning the role of localizability.

In 1964, Wu and Yang[47)] revived an idea of Fermi and others, which conceived of the hadron as a liquid drop. This work was originally motivated by the observation that differential cross-sections seemed to drop off exponentially, reflecting a "finite size" for hadrons with mean square radii roughly the same (about $10/(\frac{GeV}{c})^2$) . This idea was further developed by Yang and Byers[48)] using eikonal methods, and was extended to a theory of inelastic scattering in the fragmentation model of Chou and Yang[49)]. In this model, the hadronic matter of two particles in collision interacts pointwise (through integration of the overlap densities), across the plane perpendicular to the direction of motion, to build the phase of an eikonal type amplitude. The local point interaction of hadronic matter (not fields) in the direction transverse to the motion reminds one of the localization induced by the Melosh transformation; the appropriate density in a quark field model would be of the form $\rho(x_\perp) = \int \psi_m^*(r)\psi_m(r)dz$. Asymptotically, ψ_+ (cf. (3.8)) carries all of the information, and we obtain a potential link between limiting fragmentation and the parton model.

An approach to the description of scaling from a very different point of view than that of the parton model is that of the light cone algebra, as discussed in the previous section. As we have remarked, the scaling behavior is a reflection of the leading singularity on the light cone, and this singularity is characteristic of a free field theory. It has been suggested that interacting fields may behave, under certain conditions, as if they are free. Using the method of the renormalization group[50)], Gross and Wilczek[51)] showed that theories with non-Abelian gauge fields (Yang Mills fields) can be "asymptotically free", i.e., zero effective coupling is an ultraviolet stable fixed point ($q^2 \to \infty$). In this case, the coefficient functions of the Wilson product expansion[52)] have no anomalous dimension, and we obtain Bjorken scaling (up to logarithmic corrections) and the tensor properties of free field theory. Coleman and Gross[53)] have pointed out that no renormalizable field theory without non-Abelian gauge fields can be asymptotically free. It therefore appears that a strongly interacting quark field theory with non-Abelian gauge fields can account for the observed scaling behavior (up to logarithmic corrections), although the validity of scaling behavior observed at relatively low energies still remains as somewhat of a mystery. It is a still deeper mystery why the fundamental fields are not seen as particles. If a theory exhibiting asymptotic freedom is to be consistent with experiment, a mechanism for inhibiting the production of field quanta in (actual) asymptotic states must be found. The required condition must refer to the localization of quanta within bounded hadronic states.

An interesting proposal was recently made by Amati and Testa[54], who suggest that the requirement of "quark imprisonment" is sufficient to determine the existence of strong interactions. To the generating functional Z of the Feynman path integral method (for obtaining n point functions) is added the requirement that the quark current carrying a second SU(3) (called "color"[55]), which couples to the non-Abelian gauge fields, vanishes pointwise. To the action corresponding to a theory of weak and electromagnetic interaction alone, they add the term

$\sum_i C^i_\mu A^{\mu i}$, where $C^i_\mu = \bar{q}\,\gamma^\mu \frac{\chi_i}{2}\, q$ is the color current (χ_i are formally the same as the λ_i of SU(3) , but mix "color" indices alone), and integrate over the variables $A^{\mu i}(x)$. This term has the effect of a δ-function, forcing the color current to vanish pointwise, and therefore assuming that (color-carrying) quarks cannot be found separated in asymptotic states. This expression coincides with the limit of infinite coupling of a theory in which $A^i_\mu(x)$ corresponds to a new dynamical field; in this limit, the kinematic terms in the action of $A^i_\mu(x)$ become negligible compared to the coupling term, and we recover the $C^i_\mu(x) = 0$ condition. The bilocal algebra deals with quark fields at separate space-time points, and the condition of zero color current imposes a polarization current localized on the light cone segment between them suggestive of the string structure of dual models. The viewpoint taken by Amati and Testa, that of using the condition of containment as a mechanism for the induction of strong interactions, suggests a more general question:[56] What is the structure of a field theory in which single quanta have no asymptotic states?

Another interesting model, the so-called "bag" model, has been proposed by a group at M.I.T.[57]. In this model, the action integral is defined over what is essentially a free field Lagrangian, but the space domain of integration is truncated beyond a finite region, the "bag." The resulting field equations are those for a free field, but homogeneous boundary conditions are imposed which assure that the net color charge within any hadronic system is zero. The interpretation we proposed at the end of the previous section is clearly consistent with this model; to calculate the quark momentum distribution, only the fact of boundedness is required. The model also appears to be fruitful as a basis for dual structures.

It appears at present, however, that the physical basis of a theory in which hadrons are composed of fundamental substructures, but these substructures are not separately observed, remains to be discovered. We have argued that there is some evidence that the concept of localizability may be an important ingredient in such a theory.

It is from this point of view, that concepts close to the foundations of the quantum theory may form the basis for the construction of theories describing actual phenomena, that Josef Jauch has exercised his scientific leadership in Geneva; this essay is dedicated to him.

REFERENCES

1. T.D. Newton and E.P. Wigner, Rev. Mod. Phys. 21, 400 (1949).
2. J.M. Jauch and C. Piron, Helv. Phys. Acta 40 561 (1967).
3. W.O. Amrein, Helv. Phys. Acta 42 149 (1968).
4. See, for example, S. Schweber, "An Introduction to Relativistic Quantum Field Theory", Harper & Row, New York, 1964.
5. M. Gell-Mann, Cal Tech Report CTSL-20 (1961), unpublished.
6. Y. Ne'eman, Nuc. Phys. 26 222 (1961).
7. G. Zweig, CERN Preprint Th/401 (1964), unpublished.
M. Gell-Mann, Phys. Lett. 8 214 (1964).
8. M. Gell-Mann, Phys. Rev. 125 1067 (1962).
9. W.I. Weisberger, Phys. Rev. Lett. 14 1047 (1965).
S.L. Adler, Phys. Rev. Lett. 14 1051 (1965).
10. N. Cabibbo and L.A. Radicati, Phys. Lett. 19 697 (1966).
11. R.F. Dashen and M. Gell-Mann, Phys. Lett. 17 145 (1965).
12. F. Gürsey and L.A. Radicati, Phys. Lett. 13 173 (1964).
A. Pais, Phys. Rev. Lett. 13 175 (1964).
F. Gürsey, A. Pais and L.A. Radicati, Phys. Rev. Lett. 13 299 (1964).
13. S. Okubo, Prog. Theor. Phys. 27 949 (1962); cf. also ref. 5.
14. See, for example, F.J. Dyson, "Symmetry Groups in Nuclear and Particle Physics", W.A. Benjamin, Inc. New York (1966), for a review and reprints of relevant papers.
15. K.J. Barnes, P. Carruthers and F. von Hippel, Phys. Rev. Lett. 14 82 (1965).
16. H.J. Lipkin and S. Meshkov, Phys. Rev. Lett. 14 670 (1965).
17. H. Harari, D. Horn, M. Kugler, H.J. Lipkin and S. Meshkov, Phys. Rev. 146 1052 (1966).

18. H.J. Lipkin and F. Scheck, Phys. Rev. Lett. 16 71 (1966).

19. J.J.J. Kokkedee and L. Van Hove, Nuovo Cimento 42 711 (1966).

20. E.M. Levin and L.L. Frankfurt, Zhur. Eksp. i Theort. Fiz. Pisma v Redak. 2 105 (1965) (English translation JETP Lett. 2 65 (1965).

21. R.F. Dashen and M. Gell-Mann, Phys. Lett. 17 142 (1965).

22. N. Cabibbo, L. Horwitz and Y. Ne'eman, Phys. Lett. 22 336 (1966) - To be called CHN in the following.

23. M. Breidenbach, et al, Phys. Rev. Lett. 23 935 (1969); G. Miller, Report No. SLAC 129 (1970).

24. J.D. Bjorken, Phys. Rev. 179, 1547 (1969).

25. R.P. Feynman, Phys. Rev. Lett. 23 1415 (1969) and Proc. of the Third High Energy Collision Conf. Stony Brook, p. 237, Gordon and Breach, New York (1970).

26. Y. Frishman, Phys. Rev. Lett. 25, 966 (1970).

27. H. Fritzsch and M. Gell-Mann, in "International Conference on Duality and Symmetry in Hadron Physics," ed. E. Gotsman, Weizmann Science Press, Jerusalem (1971).

28. S. Coleman, Phys. Lett. 19 144 (1965).

29. H.J. Melosh IV, "Quarks: Currents and Constituents", Thesis, California Institute of Technology, 1973.

30. L. Gomberoff, L.P. Horwitz and Y. Ne'eman, Phys. Lett. 45 B, 131 (1973); a more complete treatment is to be published in the Phys.Rev. (in press).

31.. See, for example, Y. Ne'eman, "Algebraic Theory of Particle Physics", p. 233, W.A. Benjamin, Inc. New York (1967).

32. F.J. Gilman and M. Kugler, Phys. Rev. Lett. 30 518 (1973).

33. F.J. Gilman, M. Kugler, and S. Meshkov, "The transformation between current and constituent quarks and transition between hadrons", Phys. Rev. (in press).

34. L.L. Foldy and J.A. Wouthuysen, Phys. Rev. 78, 29 (1950).

35. P.A.M. Dirac, Rev. Mod. Phys. 21 392 (1949).

36. J.B. Kogut and D.E. Soper, Phys. Rev. D1 2901 (1970).

37. L. Susskind, Phys. Rev. 165 1535 (1968).

38. J.D. Bjorken, in "International Conference on Duality and Symmetry in Hadron Physics", ed. E. Gotsman, Weizmann Science Press, Jerusalem (1971).

39. L.C. Biedenharn, M.Y. Han and H. Van Dam, "Generalization and Interpretation of Dirac's Positive Energy Relativistic Wave Equation," Duke University Preprint 1973.

40. P.A.M. Dirac, Proc. Roy. Soc. (London) A322 435 (1971).

41. L.C. Biedenharn and H. van Dam, "Galilean Subdynamics and the Dual Resonance Model," Duke University Preprint, July 1973.

42. J. Kogut and L. Susskind, Phys. Lett. 8C (Physics Reports) 75 (1973).

43. N. Cabibbo and M. Testa, Phys. Lett. 42B 369 (1972).

44. M. Testa, CERN preprint TH/1540, (1972).

45. V.J. Reddy, L. Gomberoff, L.P. Horwitz and Y. Ne'eman, "Algebra of Strenghts Generated by Bilocal Currents Acting on Constituent States II: Universality" Tel-Aviv University preprint TAUP-396-73 (1973).

46. S.P. deAlwis, Nuc. Phys. B55 427 (1973).

47. T.T. Wu and C.N. Yang, Phys. Rev. 137 B 708 (1965).

48. N. Byers and C.N. Yang, Phys. Rev. 142, 976 (1966).

49. J. Benecke, T.T. Chou, C.N. Yang and E. Yen, Phys. Rev. 188 2159 (1969).

50. M. Gell-Mann and F. Low, Phys. Rev. 95 1300 (1954).
N.N. Bogoliubov and D.V. Shirkov, "Introduction to the Theory of Quantized Fields," Interscience, New York (1959).

51. D.J. Gross and F. Wilczek, Phys. Rev. Lett. 30, 1343 (1973).

52. K. Wilson, Phys. Rev. 179, 1499 (1969).

53. S. Coleman and D.J. Gross, Phys. Rev. Lett. 31 851 (1973).

54. D. Amati and M. Testa, CERN preprint TH/1770, 31 Oct. 1973.

55. H. Fritzsch and M. Gell-Mann, Proc. of the XVI International Conference on High Energy Physics, Chicago, NAL-Sept.-6-13 (1972).

56. This question was discussed for a first-quantized non-Abelian gauge theory by L.P. Horwitz, Helv. Phys. Acta 39, 144 (1966).

The one-particle Hamiltonian was shown not to be an observable since its expectation value is not gauge invariant under gauge transformations of the first kind. A similar phenomenon occurs in electromagnetism under gauge transformations of the second kind; in this case it is clear that the energy of the field must be included (since it carries momentum). This mechanism is also responsible for the non-invariance of the Hamiltonian of a particle in a non-Abelian gauge field; the canonical momentum is altered by a gauge transformation of the first kind. Hence it is difficult to define the "particle" independently of the field. I wish to thank Dr.A. Casher for a recent discussion of this point.

57. A Chodos, R.L. Jaffe, K. Johnson, C.B. Thorn and V.F.Weisskopf, M.I.T. preprint CTP, 387, Nov. 1973.

CHARGE DISTRIBUTIONS FROM RELATIVISTIC FORM FACTORS

Gordon N. Fleming

Department of Physics
The Pennsylvania State University
University Park, Pennsylvania

Position eigenvector matrix elements and highly local expectation values of the charge density field are examined in the relativistic domain to establish a more satisfactory notion of charge distribution than is customarily employed.

1. INTRODUCTION

In this note I will examine the relation between the electromagnetic form factor and localized matrix elements of the charge density field for a relativistic spinless particle. The motivation for this study lies in the desire to clarify the status and utility of the concept of spatial charge distribution in the domain of relativistic quantum theory. At a time when rapid phenomenological input to highly conjectural and semipictorial models of the internal structure of hadrons is at the forefront of high-energy research[1] it is distressing that the relation between such familiar structural notions as form factors and charge distributions that is almost universally employed is the conceptually vague one of a three-dimensional Fourier transform in the Breit frame[2]. This relation does not emerge from any analysis based upon the general principles of relativistic quantum theory and thus can not be interpreted in their light. All one can say for the Breit frame calculation is that it formally yields the same result that one obtains in the nonrelativistic domain[3]. The nonrelativistic connection between form factors and charge distributions, however, <u>does</u> emerge simply from basic principles and is thereby rendered conceptually clear[4]. This nonrelativistic connection can be achieved

Enz/Mehra (eds.), Physical Reality and Mathematical Description. 357–374.

by analysing the position eigenvector matrix elements and local limits of expectation values of the charge density field and in the following pages I will carry out the corresponding analysis in the presence of relativistic kinematics and Lorentz invariance.

The results of the study provide no support for the Breit frame analysis and in fact two-dimensional Fourier transforms, ubiquitous in infinite momentum frame studies[1], emerge naturally as the principal machinery for constructing charge distributions from form factors.

If this note should provoke further studies along these lines eventually leading to an improvement in our ability to handle and interpret the current flux of ultra-high energy data[5] and through this to a better understanding of the relativistic quantum domain then it will be worthy as a tribute to the work and person of Professor Jauch.

In §2, I will briefly review the nonrelativistic case. The proportionality of the position eigenvector matrix elements of the charge density field to a delta function of the difference of position eigenvalues ensures the clean separation of the internal distribution of charge from the external distribution of the particle in a normalized state. This keeps the problem of interpretation simple. In §3 the Newton-Wigner position eigenvector[6] matrix elements of the relativistic charge density field and its moments are studied. The above mentioned delta function fails to dominate in both the higher moments and the charge density itself and this is interpreted in terms of Z-graphs[7] which modify the very charge distribution one is looking for. To minimize the influence of Z-graphs I consider taking local limits of expectation values of the charge density field in normalizable states. It is in this step that the two dimensional Fourier transform emerges as a consequence of the high momentum dominance of the local limit. However it is the combination of the local limit with relativistic kinematics that is important in producing the two dimensional Fourier transform connection between charge distribution and form factor. In the nonrelativistic domain the local limit modifies nothing (since there are no Z-graphs to eliminate).

The expression for the r.m.s. charge radius[2] is also changed in the local limit by the factor $\sqrt{2/3}$. This may seem strange since the charge radius is generally regard as measuring a static property of the particle and the nonrelativistic calculation should be valid. This argument is correct if the particle is not too highly localized. The Z-graphs then raise the charge radius to its nonrelativistic value.

Finally in §4 I present some calculations of the modifications which the local limit charge distribution displays over the Breit frame charge distribution when some popular form factors[8] are used.

Although I believe all the results herein can be derived with absolute rigor I have not hesitated to make unimpeded use of improper eigenvectors and generalized functions.

2. THE NONRELATIVISTIC CASE

Let $\rho(\vec{y})$ denote the charge density field at the time t = 0. Employing momentum eigenvectors with the normalization

$$\langle\vec{p}'|\vec{p}\rangle = \delta^3(\vec{p}'-\vec{p}) \tag{2.1}$$

we find from Euclidean and Galilean invariance that

$$\begin{aligned}\langle\vec{p}'|\delta(\vec{y})|\vec{p}\rangle &= e^{\frac{i}{\hbar}(\vec{p}-\vec{p}')\cdot\vec{y}}\langle\vec{p}'|\delta(0)|\vec{p}\rangle \\ &= e^{\frac{i}{\hbar}(\vec{p}-\vec{p}')\cdot\vec{y}} f(\vec{q}^2)\end{aligned} \tag{2.2}$$

where

$$\vec{q} \equiv \vec{p} - \vec{p}' \tag{2.3}$$

The function, $f(\vec{q}^2)$, is called the electromagnetic form factors. The momentum eigenvector matrix elements of the current density field, $\vec{j}(\vec{y})$, can also be expressed in terms of $f(\vec{q}^2)$ but that is not a matter of concern to us here.

The effective charge distribution for the particle in the state $|\psi\rangle$ is given by the expectation value

$$\frac{\langle\psi|\rho(\vec{y})|\psi\rangle}{\langle\psi|\psi\rangle}$$

and if,

$$|\psi\rangle = \int d^3x\, |\vec{x}\rangle\, \psi(\vec{x}) \quad , \tag{2.4}$$

where the position eigenvectors are given by

$$|\vec{x}\rangle = (2\pi\hbar)^{-3/2} \int d^3p \; e^{-\frac{i}{\hbar}\vec{p}.\vec{x}} \; |\vec{p}\rangle \quad , \qquad (2.5)$$

then

$$\langle\psi|\rho(\vec{y})|\psi\rangle = \int d^3x' \; d^3x \; \psi(\vec{x}')^* \langle\vec{x}'|\rho(\vec{y})|\vec{x}\rangle \; \psi(\vec{x}), \qquad (2.6)$$

and so it is in order to examine the position eigenvector matrix elements of $\rho(\vec{y})$.

Combining (2.5) and (2.2) we have

$$\langle\vec{x}'|\rho(\vec{y})|\vec{x}\rangle = \delta^3(\vec{x}'-\vec{x}) \int d^3q \; e^{\frac{i}{\hbar}\vec{q}.(\vec{y}-\vec{x})} \; f(\vec{q}^2) \qquad (2.7)$$

and so

$$\langle\psi|\rho(\vec{y})|\psi\rangle = \int d^3x |\psi(\vec{x})|^2 \; \bar{\rho}(\vec{y}-\vec{x}) \qquad (2.8)$$

where

$$\bar{\rho}(\vec{y}) \equiv \int d^3q \; e^{\frac{i}{\hbar}\vec{q}.\vec{y}} \; f(\vec{q}^2) \qquad (2.9)$$

is identified as the <u>internal</u> charge distribution of the particle. We see that the effective charge distribution differs from the internal charge distribution only by virtue of the smearing of the location of the center of the charge inherent in the use of normalizable states. We can, in fact, isolate the internal charge distribution by performing a norm preserving scale transformation on $|\psi\rangle$ and then taking the limit as the scaling parameter vanishes. We call this the <u>local limit</u> of the expectation value. Thus define,

$$|\psi_\lambda\rangle = \int d^3x \; \psi_\lambda(\vec{x})|\vec{x}\rangle \equiv \int d^3x \; \lambda^{-3/2} \; \psi\left(\frac{\vec{x}}{\lambda}\right) |\vec{x}\rangle \quad , \qquad (2.10)$$

and assume

$$\int d^3x |\psi(\vec{x})|^2 = 1 \quad , \qquad (2.11)$$

to get,

$$\lim_{\lambda\to 0} \langle\psi_\lambda|\rho(\vec{y})|\psi_\lambda\rangle = \lim_{\lambda\to 0} \int \frac{d^3x}{\lambda^3} \; \left|\psi\left(\frac{\vec{x}}{\lambda}\right)\right|^2 \; \bar{\rho}(\vec{y}-\vec{x})$$

$$= \lim_{\lambda\to 0} \int d^3z |\psi(\vec{z})|^2 \bar{\rho}(\vec{y}-\lambda\vec{z}) = \int d^3z |\psi(\vec{z})|^2 \bar{\rho}(\vec{y})$$

$$= \bar{\rho}(\vec{y}) \quad . \tag{2.12}$$

This local limit will be the crucial ingredient of our definition of the internal charge distribution in the relativistic domain.

For comparison with the relativistic results we list the position eigenvector matrix elements of the zeroth, first and second moments of the charge density field. We have,

$$<\vec{x}'| \int d^3y \; \rho(\vec{y}) |\vec{x}> = \delta^3(\vec{x}'-\vec{x})(2\pi\hbar)^3 \; f(0) \tag{2.13}$$

so that $(2\pi\hbar)^3 f(0)$ is the total charge of the particle. Also

$$<\vec{x}'| \int d^3y \; y_a \rho(\vec{y}) |\vec{x}> = \delta^3(\vec{x}'-\vec{x})(2\pi\hbar)^3 f(0) x_a \tag{2.14}$$

so that in a normalized state the expectation value of $\vec{x}$ is the effective center of the charge distribution. Finally

$$<\vec{x}'| \int d^3y \; y_z y_b \rho(\vec{y}) |\vec{x}> = \delta^3(\vec{x}'-\vec{x})\{(2\pi\hbar)^3 f(0) x_a x_b$$

$$- 2\hbar^2 \delta_{ab} (2\pi\hbar)^3 f'(0)\} \tag{2.15}$$

where

$$f'(0) \equiv \frac{d}{dq^2} f(q^2)\Big|_{q^2=0} \quad . \tag{2.16}$$

From this last result we characterize $- 6\hbar^2(2\pi\hbar)^3 f'(0)$ as the square of r.m.s. charge radius of the internal distribution. Indeed

$$\int d^3y \; \vec{y}^2 \bar{\rho}(\vec{y}) = - 6\hbar^2(2\pi\hbar)^3 f'(0) \quad . \tag{2.17}$$

To the extent that one can make sense of the spatial integral of $\rho(\vec{y})$ as an unbounded operator[9] (2.13) is a weak version of the much stronger assertion that $|\vec{x}>$ is an eigenvector of the total charge operator. For the other moment integrals the stronger eigenvector analogues will, in general, not hold since $\rho(\vec{y})$ can couple single particles to various other systems providing the mass superselection rule of nonrelativistic quantum theory[10] is satisfied.

3. THE CHARGE DISTRIBUTION IN THE RELATIVISTIC DOMAIN

In this case the charge density field will be denoted by $c^{-1}j_o(\vec{y})$ at time t = 0 where c is the speed of light in vacuum and $j_o(\vec{y})$ is the time-like component of the four-vector current density $j_\mu(\vec{y},y_o=0)$. Lorentz invariance and local charge conservation, $\partial^\mu j_\mu=0$, yield the result

$$<\vec{p}'|j_o(\vec{y})|\vec{p}> = e^{\frac{i}{\hbar}(\vec{p}-\vec{p}').\vec{y}} <\vec{p}'|j_o(0)|\vec{p}>$$

$$= e^{\frac{i}{\hbar}(\vec{p}-\vec{p}').\vec{y}} \frac{p_o + p_o'}{2\sqrt{p_o p_o'}} F(q^2) \qquad (3.1)$$

where

$$p_o \equiv \sqrt{\vec{p}^2 + m^2c^2} , \quad p_o' \equiv \sqrt{\vec{p}'^2 + m^2c^2} , \quad q_\mu \equiv p_\mu - p_\mu'$$

$$q^2 = q_o^2 - \vec{q}^2 \qquad (3.2)$$

and the eigenvectors, $|\vec{p}>$, have the same normalization as in (2.1).

In the nonrelativistic limit, $|\vec{p}|, |\vec{p}'| << mc$, we have

$$\frac{p_o + p_o'}{2\sqrt{p_o p_o'}} F(q^2) \to F(-\vec{q}^2) \qquad (3.3)$$

and so the connection with the form factor of the previous section is

$$c^{-1} F(-\vec{q}^2) \xrightarrow[|\vec{p}|,\ |\vec{p}'| \to 0]{} f(\vec{q}^2) \ . \qquad (3.4)$$

On the basis of this connection it has become traditional[2] to note that in the Breit frame, defined by $\vec{p}' = -\vec{p}$, we have,

$$\frac{p_o + p_o'}{2\sqrt{p_o p_o'}} F(\vec{q}^2) = F(-\vec{q}^2), \qquad (3.5)$$

and to <u>define</u> the charge distribution $\bar{\rho}_B(\vec{y})$ by

$$\bar{\rho}_B(\vec{y}) \equiv c^{-1} \int d^3q \; e^{\frac{i}{\hbar}\vec{q}\cdot\vec{y}} \; F(-\vec{q}^2) \quad . \qquad (3.6)$$

Aside from the ad hoc character of this definition we note that from the viewpoint of the laboratory frame $\vec{p} = 0$ the transformation to the Breit frame varies with $\vec{p}'_{lab}$ and so there is no unique frame of reference to which (3.6) refers. But in the relativistic domain a charge distribution is not an invariant quantity and must, to be well defined, refer to some particular inertial frame.

We therefore turn to the examination of Newton-Wigner position eigenvector matrix elements of $j_o(\vec{y})$ and its moments to establish a well-defined notion of charge distribution.

For elementary systems Newton and Wigner established the uniqueness of the position eigenvectors that carry their name[6] and I have previously explored the transformation properties under the Poincaré group and the physical interpretation of these eigenvectors.[11] On the basis of these studies it seems clear that these eigenvectors are the appropriate analogues of the nonrelativistic position eigenvectors for the purposes of this paper. In particular we note that since a Lorentz transformation of a Newton-Wigner eigenvector is not a Newton-Wigner eigenvector[6,11] it follows that such an eigenvector singles out a particular inertial frame as the frame in which the system is localized at a definite time. This is precisely the kind of reference that is needed to render the charge distribution well defined.

Now it turns out that in the relativistic case the lower moments of the charge density have much simpler local matrix elements than the charge density itself and so we consider the moments first.

Straightforward calculation employing (3.1) and (2.5) (where now $|\vec{x}\rangle$ is the Newton-Wigner eigenvector) yields

$$\langle\vec{x}'| \int d^3y \; j_o(\vec{y})|x\rangle = \delta^3(\vec{x}-\vec{x}')(2\pi\hbar)^3F(0) \qquad (3.7)$$

$$\langle\vec{x}'| \int d^3y \; y_a j_o(\vec{y})|x\rangle = \delta^3(\vec{x}-\vec{x}')(2\pi\hbar)^3F(0)x_a \qquad (3.8)$$

$$\langle\vec{x}'| \int d^3y \; y_a y_b j_o(\vec{y})|\vec{x}\rangle = \delta^3(\vec{x}-\vec{x}')(2\pi\hbar)^3\{x_a x_b F(0)$$

$$+ 2\hbar^2\delta_{ab}F'(0)\} + \frac{\pi^2\hbar^4}{4mc}\,F(0)\,\frac{\partial^2}{\partial x_b \partial x_a}\,e^{-\frac{mc}{\hbar}|\vec{x}-\vec{x}'|}$$

$$+ 4\pi^2\hbar^5 F'(0)\,\frac{\partial^2}{\partial x_b \partial x_a}\left[\frac{1}{|\vec{x}-\vec{x}'|}\,e^{-\frac{mc}{\hbar}|\vec{x}-\vec{x}'|}\right] . \qquad (3.9)$$

The first two results exactly duplicate the nonrelativistic case and again one expects that in <u>some sense</u>[9] the eigenvector $\langle x'|$ could be removed from (3.7) turning the equation into an eigenvalue equation. In (3.9), however, we encounter new contributions from the relativistic kinematics (notice that the new terms "disappear" in the limit $c \to \infty$). It seems reasonable to expect that, in a perturbation theoretic analysis of (3.9) in which $j_o(\vec{y})$ is expressed as an infinite series in the asymptotic free fields for the particle in question and any other particles the present one may couple to, the new nonlocal terms arise from the possibility of Z-graphs[7] as indicated in Fig.1 for the pion. The Z-graphs are rendered possible by the absence of

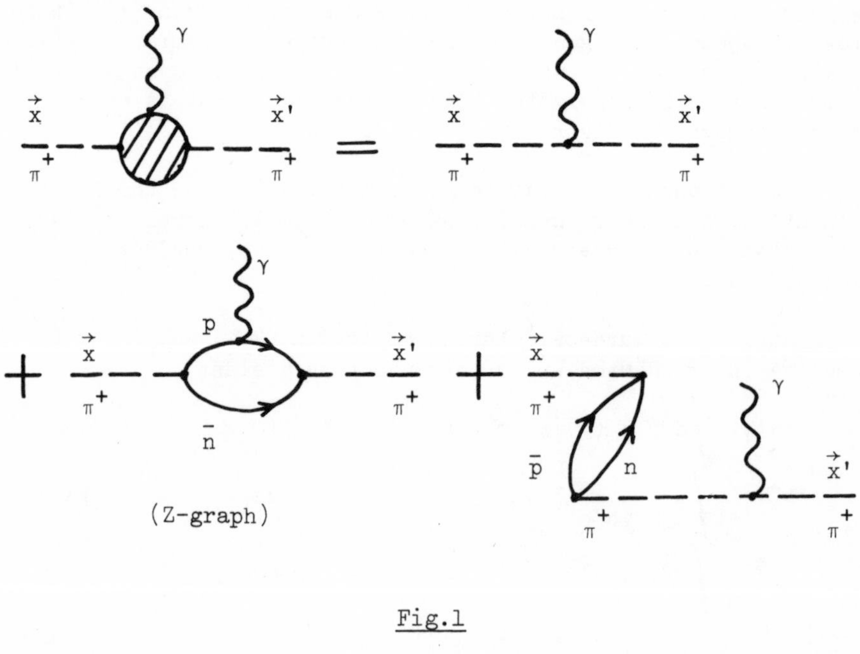

<u>Fig.1</u>

a super-selection rule on the mass[10] and by incompatibility between sharpness in $\vec{x}$, the Newton-Wigner eigenvalue, and sharpness in $\vec{y}$, the spatial position variable of the field $j_o(\vec{y})$. I am inclined to interpret the Z-graph contributions as representing a disturbance of an already present charge distribution by the very application of the charge density field to the position eigenvector. To the extent that such an interpretation is allowed the search for an intrinsic or internal charge distribution should employ some procedure which will tend to isolate only the strongest singularities at $\vec{x} = \vec{x}'$. We will see later that the local limit defined in the previous section has this feature.

I realize that in these last sentences I have indulged in a certain license in the use of language which brings me dangerously close to heresy vis-a-vis the so-called Copenhagen interpretation of quantum theory.[12] It is not my desire to foment heresies! It is my desire to understand results like (3.9) in some physical detail and the above suggestion may not be wholly erroneous.

If we actually calculate the indicated derivatives on the right hand side of (3.9) we obtain, putting $\vec{r} \equiv \vec{x} - \vec{x}'$,

$$\frac{\partial^2}{\partial x_b \partial x_a} e^{-\frac{mc}{\hbar} r} = \frac{mc}{\hbar r} \left\{ r_a r_b \left[\frac{1}{r^2} + \frac{mc}{\hbar r} \right] - \delta_{ab} \right\} e^{-\frac{mc}{\hbar} r} \tag{3.10}$$

and[13]

$$\frac{\partial^2}{\partial x_b \partial x_a} \left[\frac{e^{-\frac{mc}{\hbar} r}}{r} \right] = \left\{ \left(\frac{mc}{\hbar}\right)^2 \frac{\hat{r}_a \hat{r}_b}{r} - (\delta_{ab} - 3\hat{r}_a \hat{r}_b) \left[\frac{1}{r^3} + \frac{mc}{\hbar r^2} \right] \right\} e^{-\frac{mc}{\hbar} r}$$
$$- 4\pi \hat{r}_a \hat{r}_b \delta^3(\vec{r}) \ , \tag{3.11}$$

where $\hat{r} = \vec{r}/r$. The most striking feature is the delta function term on the right hand side of (3.11). Using the relation

$$\hat{r}_a \hat{r}_b \delta^3(\vec{r}) = \frac{1}{3} \delta_{ab} \ \delta^3(\vec{r}) \tag{3.12}$$

this term combines with the Kronecker delta term already explicit on the right hand side of (3.9). In this way the expression for the r.m.s. charge radius is altered from its nonrelativistic value. In fact upon taking a local limit we have

$$\lim_{\lambda\to 0} \langle\psi_\lambda| \int d^3y\, \vec{y}^2 j_o(\vec{y})|\psi_\lambda\rangle = 4\hbar^2(2\pi\hbar)^3F'(0) \ . \qquad (3.13)$$

The relation of this result to the nonrelativistic result (2.17) will become clearer as we consider the charge distribution itself.

The Newton-Wigner position-eigenvector matrix element of $j_o(\vec{y})$ can be written as

$$\langle\vec{x}'|j_o(\vec{y})|\vec{x}\rangle = (2\pi\hbar)^{-3}\frac{1}{8}\int d^3P\, d^3q\, e^{-\frac{i}{\hbar}P.\frac{\vec{r}}{2}}$$

$$e^{\frac{i}{\hbar}\vec{q}.(\vec{y}-\vec{R})} \left\langle \frac{\vec{P}-\vec{q}}{2}\right|j_o(0)\left|\frac{\vec{P}+\vec{q}}{2}\right\rangle \ , \qquad (3.14)$$

where we are using

$$\vec{r} \equiv \vec{x} - \vec{x}',\ \vec{R} = \frac{1}{2}(\vec{x}+\vec{x}')\ ,\ \vec{P} \equiv \vec{p} + \vec{p}'\ ,\ \vec{q} \equiv \vec{p} - \vec{p}' \ . \qquad (3.15)$$

If we combine (3.14) with (3.1) we see that all of the difference with the nonrelativistic results comes from the failure of the momentum eigenvector matrix element of $j_o(0)$ to decouple the $\vec{P}$ dependence and depend only on the 3-momentum transfer $\vec{q}$. Because of the $\vec{P}$ dependence in the matrix element we cannot expect (3.14) to be proportional to a delta function on $\vec{r}$ and this complicates the interpretation of (3.14) and provokes hapless authors to mumble about Z-graphs. At the same time it is clear that the most singular behavior of (3.14) for small $\vec{r}$ must come from the high $\vec{P}$ behavior of the integrand and perhaps some light can be shed by concentrating on this limit. In fact we will see that the local limit defined above has just the effect of extracting the dominant singularities from the high $\vec{P}$ behavior of the integrand in (3.14). Thus it behooves us to note that

$$\lim_{P\to\infty} \frac{p_o + p_o'}{2\sqrt{p_o p_o'}} F(q^2) = F(-\vec{q}_\perp^{\,2}) \qquad (3.16)$$

where

$$\vec{q}_\perp \equiv \vec{q} - \hat{P}(\hat{P}.\vec{q}) \ . \qquad (3.17)$$

This, of course, is the kind of result that infinite momentum frame calculators have known for some time.[1] Consequently we might expect two-dimensional Fourier transforms, which dominate space-time considerations in the infinite momentum frame, to play a role in the charge distribution we obtain from the local limit.

In this relativistic case it is simpler to take the local limit in terms of momentum space state functions and so we note that if

$$\tilde{\psi}(\vec{p}) \equiv (2\pi\hbar)^{-3/2} \int d^3x \; e^{-\frac{i}{\hbar}\vec{p}.\vec{x}} \; \psi(\vec{x}) \tag{3.18}$$

then

$$\tilde{\psi}_\lambda(\vec{p}) = \lambda^{3/2} \; \tilde{\psi}(\lambda\vec{p}) \tag{3.19}$$

and

$$|\psi_\lambda> = \int d^3p \; \tilde{\psi}_\lambda(\vec{p})|\vec{p}> \quad . \tag{3.20}$$

Therefore

$$\lim_{\lambda\to 0} <\psi_\lambda|j_o(\vec{y})|\psi_\lambda> = \lim_{\lambda\to 0} \int d^3p' \; d^3p \; \tilde{\psi}_\lambda(\vec{p}') \; e^{\frac{i}{\hbar}(\vec{p}-\vec{p}').\vec{y}} <\vec{p}'|j_o(0)|\vec{p}>\tilde{\psi}_\lambda(\vec{p})$$

$$= \lim_{\lambda\to 0} \frac{1}{8} \int d^3P \; d^3q \; \lambda^3 \; \tilde{\psi}^*\left(\lambda \; \frac{\vec{P}-\vec{q}}{2}\right) e^{\frac{i}{\hbar}\vec{q}.\vec{y}} \left< \frac{\vec{P}-\vec{q}}{2} \; |j_o(0)| \frac{\vec{P}+\vec{q}}{2} \right> \tilde{\psi}\left(\lambda \; \frac{\vec{P}+\vec{q}}{2}\right)$$

$$= \lim_{\lambda\to 0} \int d^3L \; d^3q \; \tilde{\psi}^*\left(\vec{L} - \frac{\lambda}{2}\vec{q}\right) \tilde{\psi}\left(\vec{L} + \frac{\lambda}{2}\vec{q}\right) \left< \frac{\vec{L}}{\lambda} - \frac{\vec{q}}{2} \; |j_o(0)| \frac{\vec{L}}{\lambda} + \frac{\vec{q}}{2} \right> e^{\frac{i}{\hbar}\vec{q}.\vec{y}}$$

$$= \int d^3L \; |\tilde{\psi}(\vec{L})|^2 \; d^3q \; e^{\frac{i}{\hbar}\vec{q}.\vec{y}} \; F(-\vec{q}_\perp^{\,2}) \quad . \tag{3.21}$$

But

$$\int d^3q \; e^{\frac{i}{\hbar}\vec{q}.\vec{y}} \; F(-\vec{q}_\perp^{\,2}) = 2\pi\hbar \; \delta(\hat{L}.\vec{y}) \int d^2q_\perp \; e^{\frac{i}{\hbar}\vec{q}_\perp.\vec{y}_\perp} \; F(-\vec{q}_\perp^{\,2})$$

$$= \delta(\hat{L}.\hat{y}) \frac{2\pi\hbar}{|\vec{y}|} \int d^2q_\perp \; e^{\frac{i}{\hbar}\vec{q}_\perp.\hat{y}_\perp|\vec{y}|} \; F(-\vec{q}_\perp^{\,2}) \tag{3.22}$$

and consequently

$$\lim_{\lambda\to 0} \langle\psi_\lambda|j_o(\vec{y})|\psi_\lambda\rangle = \frac{2\pi\hbar}{|\vec{y}|}\bar{\rho}_2(\vec{y}^2)\int d^3L|\tilde{\psi}(\vec{L})|^2\delta(\hat{L}.\hat{y}) \tag{3.23}$$

where

$$\bar{\rho}_2(\vec{y}^2) \equiv \int d^2q_\perp\, e^{\frac{i}{\hbar}\vec{q}_\perp.\hat{y}_\perp|\vec{y}|} F(-\vec{q}_\perp^2)$$

$$= \int_0^\infty q\,dq \int_0^{2\pi} d\phi\, e^{\frac{i}{\hbar}q|\vec{y}|\cos\phi} F(-q^2) \quad . \tag{3.24}$$

We note immediately that the local limit does not completely eliminate all dependence on the form of the initial normalized state $|\psi\rangle$. A residual <u>angular</u> dependence is determined by the <u>shape</u> of $\tilde{\psi}(\vec{L})$. The total charge, however, is not modified by this angular state dependence. Thus

$$\int d^3y \lim_{\lambda\to 0}\langle\psi_\lambda|j_o(\vec{y})|\psi_\lambda\rangle = 2\pi\hbar\int|\vec{y}|d|\vec{y}|\bar{\rho}_2(\vec{y}^2)\int d^3L\, |\tilde{\psi}(\vec{L})|^2\int d^2\hat{y}\, \delta(\hat{L}.\hat{y})$$

$$= 4\pi^2\hbar\int_0^\infty y\,dy\int_0^\infty q\,dq\int_0^{2\pi} d\phi\, e^{\frac{i}{\hbar}qy\cos\phi} F(-q^2)$$

$$= 4\pi^2\hbar\,(2\pi\hbar)^2\int_0^\infty q\,dq\,\delta^2(\vec{q})\,F(-q^2)$$

$$= (2\pi\hbar)^3 F(0) \tag{3.25}$$

and we similarly find that the first and second moment spatial integrals can also be interchanged with the local limit.

The angular dependence can also be expressed in terms of the shape of the <u>coordinate space</u> state function as follows. We have

$$\int d^3L\, |\tilde{\psi}(\vec{L})|^2\delta(\hat{L}.\hat{y}) = (2\pi\hbar)^{-3}\int d^3z'd^3z\, \psi^*(\vec{z}')\psi(\vec{z})$$

$$\times \int d^3L\, e^{\frac{i}{\hbar}\vec{L}.(\vec{z}'-\vec{z})}\, \delta(\hat{L}.\hat{y}) \qquad (3.26)$$

while

$$\int d^3L\, e^{\frac{i}{\hbar}\vec{L}.(\vec{z}'-\vec{z})}\, \delta(\hat{L}.\hat{y}) = \int d^2L_\perp\, e^{\frac{i}{\hbar}\vec{L}_\perp.(\vec{z}'-\vec{z})_\perp}\, L_\perp$$

$$= \frac{(2\pi\hbar)^3}{\pi^2} \frac{1}{|(\vec{z}'-\vec{z})_\perp|^3} \qquad (3.27)$$

so that

$$\lim_{\lambda\to 0} \langle\psi_\lambda|j_o(\vec{y})|\psi_\lambda\rangle = \frac{2\hbar}{\pi} \frac{\bar{\rho}_2(\vec{y}^2)}{|\vec{y}|} \int d^3z' d^3z \frac{\psi^*(\vec{z}')\psi(\vec{z})}{|(\vec{z}'-\vec{z})_\perp|^3} \quad . \qquad (3.28)$$

In this form, the angular dependence clearly invariant under the scale transformation employed in the local limit, displays the dominant inverse cube singularity of the position eigenvector matrix element of $j_o(\vec{y})$.

It appears then that in the relativistic domain one further step must be taken to extract a purely internal structural feature of the particle. The obvious step is to average (3.23) over all orientations of $|\psi_\lambda\rangle$. But should the isotropic average be performed first on $|\psi_\lambda\rangle$ and followed by the local limit or should the isotropic average be performed on the result of the local limit ? Fortunately, it doesn't make any difference and in either case we get

$$\overline{\lim_{\lambda\to 0} \langle\psi_\lambda|j_o(\vec{y})|\psi_\lambda\rangle} = \frac{\pi\hbar}{|\vec{y}|} \bar{\rho}_2(\vec{y}^2) \quad . \qquad (3.29)$$

This result I advance as the appropriate expression for the internal charge distribution of a spinless particle in the relativistic domain.

I will close this section with some comments on the relation between the nonrelativistic form factor and charge distribution described in §2 and an exact expectation value of $j_o(\vec{y})$ in a "mildly localized" state.

Let the momentum space state function for the state in question be $\tilde{\psi}(\vec{p})$. Then

$$<\psi|j_o(\vec{y})|\psi> = \int d^3p'd^3p\ \tilde{\psi}^*(\vec{p}')\tilde{\psi}(\vec{p})\ e^{\frac{i}{\hbar}(\vec{p}-\vec{p}')\cdot\vec{y}}$$

$$<\vec{p}'|j_o(0)|\vec{p}> \quad . \tag{3.30}$$

Now suppose that $\tilde{\psi}(\vec{p})$ is concentrated almost entirely in the region $|\vec{p}| << mc$ i.e.,

$$1 - \int_{|\vec{p}| \leq \frac{1}{10} mc} d^3p\ |\tilde{\psi}(\vec{p})|^2 << 1 \ . \tag{3.31}$$

Then the only part of the momentum eigenvector matrix element of $j_o(\vec{y})$ which gets to contribute to (3.30) is the low momentum portion for which

$$<\vec{p}'|j_o(0)|\vec{p}> \approx F(-\vec{q}^2) \quad . \tag{3.32}$$

Therefore

$$<\psi|j_o(\vec{y})|\psi> \approx \int d^3p'd^3p\ \tilde{\psi}^*(\vec{p}')\tilde{\psi}(\vec{p})\ e^{\frac{i}{\hbar}(\vec{p}-\vec{p}')\cdot\vec{y}}\ F\left(-(\vec{p}-\vec{p}')^2\right)$$

$$= \int d^3x|\psi(\vec{x})|^2 \int d^3q\ e^{\frac{i}{\hbar}\vec{q}\cdot(\vec{y}-\vec{x})}\ F(-\vec{q}^2) \tag{3.33}$$

which is the nonrelativistic result. For the second moment we get

$$<\psi|\int d^3y\ y_a y_b j_o(\vec{y})|\psi> \approx (2\pi\hbar)^3 F(0)\int d^3x\ |\psi(\vec{x})|^2 x_a x_b + \delta_{ab} 2\hbar^2 (2\pi\hbar)^3 F'(0) . \tag{3.34}$$

It is clear then that the nonrelativistic results can be reproduced only for states that are not dominated by high momentum and that includes all highly localized states. It is incorrect to regard the second term on the right of (3.34) as having fundamental significance because it is independent of $\psi(\vec{x})$. That independence is only approximate for a restricted set of sufficiently smeared state functions and upon increasing the localization of the particle the effective r.m.s. charge radius will gradually drop to the local limit value given by (3.13).

4. AN EXAMPLE

We will calculate the charge distributions corresponding to first order and second order pole terms in the time-like region of $F(q^2)$. In the case of the pion the former corresponds to vector meson dominance[14] while the latter follows the phenomenological dipole fit[15] suggested by nucleon form factors. Thus we consider

$$F_V(q^2) = \frac{F_V(0)}{1 - \frac{q^2}{m_V^2 c^2}} \tag{4.1}$$

and

$$F_D(q^2) = \frac{F_D(0)}{\left[1 - \frac{q^2}{m_D^2 c^2}\right]^2} \quad . \tag{4.2}$$

For the nonrelativistic or Breit frame charge distribution we obtain

$$\bar{\rho}_{BV}(\vec{y}) \equiv c^{-1} \int d^3q \, e^{\frac{i}{\hbar}\vec{q}\cdot\vec{y}} \, F_V(-\vec{q}^{\,2}) = c^{-1} F_V(0) \int d^3q \, \frac{e^{\frac{i}{\hbar}\vec{q}\cdot\vec{y}}}{1 + \frac{\vec{q}^{\,2}}{m_V^2 c^2}}$$

$$= 2\pi^2 \hbar \, m_V^2 c \, F_V(0) \, \frac{e^{-\frac{m_V c}{\hbar} y}}{y} \tag{4.3}$$

and

$$\bar{\rho}_{BD}(\vec{y}) = c^{-1} \int d^3q \, e^{\frac{i}{\hbar}\vec{q}\cdot\vec{y}} \, F_D(-\vec{q}^{\,2})$$

$$= c^{-1} F_D(0) \int d^3q \, \frac{e^{\frac{i}{\hbar}\vec{q}\cdot\vec{y}}}{\left[1 + \frac{\vec{q}^{\,2}}{m_D^2 c^2}\right]^2}$$

$$= \pi^2 m_D^3 c^2 \ F_D(0) \ e^{-\frac{m_D c}{\hbar} y} \ . \qquad (4.4)$$

The corresponding results for the local limit charge distributions can be expressed in several forms. I have chosen a representation that facilitates comparison with the Breit frame distributions;

$$\frac{\pi\hbar}{cy} \bar{\rho}_{2V}(y^2) = \frac{\pi\hbar}{cy} \int d^2q \ e^{\frac{i}{\hbar} \vec{q}\cdot\hat{y} \, y} \ \frac{F_V(0)}{\left[1 + \frac{q^2}{m_V^2 c^2}\right]}$$

$$= \frac{\pi\hbar}{cy} m_V^2 c^2 \ F_V(0) \int_{-\infty}^{\infty} dq_1 dq_2 \ \frac{e^{\frac{i}{\hbar} q_1 y}}{m_V^2 c^2 + q_1^2 + q_2^2}$$

$$= 2\pi^2\hbar \ m_V^2 c \ F_V(0) \int_{m_V}^{\infty} \frac{dm}{\sqrt{m^2 - m_V^2}} \left[\frac{e^{-\frac{mc}{\hbar} y}}{y}\right]$$

$$(4.5)$$

and

$$\frac{\pi\hbar}{cy} \bar{\rho}_{2D}(y^2) = \frac{\pi\hbar}{cy} \int d^2q \ e^{\frac{i}{\hbar} \vec{q}\cdot\hat{y} \, y} \ \frac{F_D(0)}{\left[1 + \frac{q^2}{m_D^2 c^2}\right]^2}$$

$$= \frac{\pi\hbar}{cy} m_D^4 c^4 \ F_D(0) \int_{-\infty}^{\infty} dq_1 dq_2 \ \frac{e^{\frac{i}{\hbar} q_1 y}}{[m_D^2 c^2 + q_1^2 + q_2^2]^2}$$

$$= \pi^2 \ m_D^3 c^2 \ F_D(0) \int_{m_D}^{\infty} \frac{dm}{\sqrt{m^2 - m_D^2}} \frac{m_D}{m} \left[1 + \frac{\hbar}{mcy}\right] e^{-\frac{mc}{\hbar} y} \ .$$

$$(4.6)$$

For both the simple pole and dipole form factors the local limit charge distribution involves a positive definite superposition of the forms comprising the Breit frame or nonrelativistic charge distributions. Of particular interest is the asymptotic behavior of the local limit charge distributions which are noticeably different from the Breit frame distributions. We find

$$\frac{\pi\hbar}{cy}\,\bar{\rho}_{2V}(y^2) \underset{y\to\infty}{\simeq} 2\pi^2\hbar\, m_V^2 c\, F_V(0)\,\frac{\pi}{2}\,\frac{e^{-\frac{m_V c}{\hbar}y}}{\left[\frac{m_V c}{\hbar}\right]^{1/2} y^{3/2}} \qquad (4.7)$$

$$\frac{\pi\hbar}{cy}\,\bar{\rho}_{2D}(y^2) \underset{y\to\infty}{\simeq} \pi^2\, m_D^3 c^2 F_D(0)\,\frac{\pi}{2}\,\frac{e^{-\frac{m_D c}{\hbar}y}}{\left[\frac{m_D c}{\hbar}y\right]^{1/2}}\,. \qquad (4.8)$$

While the asymptotic form is a bit unfamiliar it is compatible with our interpretation of the nonrelativistic charge distribution as an approximate result valid for particles that are not highly localized. One would expect the tail end of the charge distribution from a localized particle to fall off faster than that from a particle with significant uncertainty in the location of its center.

The extension of this study to the important case of spin $\frac{1}{2}$ particles and to mass distributions from form factors for the stress-energy-momentum tensor is presently under way.

REFERENCES

1. H.J.Lipkin, Phys.Rep. 8C, No.3 (August 1973); J.H.Schwartz, Phys.Rep. 8C, No.4 (Sept. 1973); J.Kogut and L.Sussking, Phys. Rep. 8C, No. 2 (June 1973).

2. Two exemplary references spanning a recent decade are S.Drell and F.Zachariasen "Electromagnetic Structure of Nucleons," Oxford University Press (1961), pp. 10-11; P.Roman, "Introduction to Quantum Field Theory," John Wiley and Sons, Inc. (1969), pp. 503-507.

3. The circumstance under which the Breit frame calculation yields justifiable results will be described later in §3.

4. E.Henley and W.Thirring, "Elementary Quantum Field Theory", McGraw-Hill Book Co., Inc. (1962), pp. 225-235.

5. As represented, for example, in the proceedings of the "XVI International Conference on High Energy Physics, Chicago-Batavia", v. 1-4, National Accel. Lab. (1973).

6. T.D.Newton and E.Wigner, Rev.Mod.Phys. 21, 400 (1949); see also A.S.Wightman, Rev.Mod.Phys. 34, 845 (1962).

7. See J.Kogut and L.Susskind, loc. cit.

8. R.P.Feynman, "Photon-Hadron Interactions," W.A.Benjamin Inc. (1972); R.Hofstadter, "Nuclear and Nucleon Structure," W.A.Benjamin, Inc. (1963).

9. This is a serious problem in the relativistic domain. See the review by C.Orzalesi, Rev.Mod.Phys. 42, 381 (1970) and H.Snellman, Journ.Math.Phys. 14, 1218 (1973).

10. V.Bargmann, Ann.Math. 59, 1 (1954); J.M.Jauch, Helv.Phys.Acta 33, 711 (1960); J.M.Jauch and B.Misra, Helv.Phys.Acta 34, 699 (1961); J.M.Levy-Leblond, Commun.Math.Phys. 4, 157 (1967).

11. G.N.Fleming, Phys.Rev. 137, B188 (1965), 139, B963 (1965), Journ.Math.Phys. 7, 1959 (1966); more recently see B.Durand, Journ.Math.Phys. 14, 921 (1973).

12. W.Heisenberg, "Physics and Philosophy," New York, Harper (1958); J.M.Jauch, "Foundations of Quantum Mechanics," Addison-Wesley Publ.Co. (1968), "Are Quanta Real?/A Galilean Dialogue", Indiana Univ. Press, Bloomington (1973); T.Bastin, Ed., "Quantum Theory and Beyond," Cambridge Univ. Press (1971).

13. I.M.Gelfand and G.E.Shilov, "Generalized Functions," Academic Press (1964), v. 1, pp. 71-74.

14. D.Schildknecht, "Vector Meson Dominance, Photo and Electroproduction from Nucleons," Springer Tracts on Modern Physics 63 (Springer-Verlag 1972), pp. 57-93; F.J.Gilman, Phys.Rep., 4C, No.3 (July/August 1972).

15. K.Huang, "Duality and the Pion Electromagnetic Form Factor," Springer Tracts in Modern Physics 62 (Springer-Verlag 1972), pp. 98-106; C.L.Hammer and T.Weber, Phys.Rev.Lett. 28, 1675 (1972), Phys.Rev. D, 5, 3087 (1972).

CHARGES AND CURRENTS IN THE THIRRING MODEL

R.F. Streater

Bedford College, London

We give an informal account of the recent theory of superselection rules due mainly to R. Haag, and we treat a simple model field theory from this point of view, following a paper of I.F. Wilde and the author. The model, similar to one studied by Skyrme in 1960, turns out to yield the Thirring model if suitable limits are taken, as is seen by comparing the equations with the work of Dell'Antonio, Frischman and Zwanziger.

Our treatment of the model shows that the charges and superselection rules, and to some extent the Fermi field operator and its equations of motion, are intrinsically determined by the observables, the currents, of the theory.

* * *

Until recent years, it had been taken for granted that a fundamental field theory of matter must contain, as given variables, fields carrying enough quantum numbers to generate all the desired particles - baryons, leptons, strange particles, charged particles etc. In particular, to describe Fermions it was thought essential to have spinor fields entering the fundamental equations, since, whereas Bosons can occur as bound states of Fermions, the converse is impossible. Given the usual connection between spin and statistics, this fact has a simple origin in the Clebsch-Gordon series for the group O(3); tensor products of integer-spin representations do not contain half-integer-spin representations as subrepresentations. It was natural for Heisenberg[1] to postulate the existence of a fundamental Fermi field in his theory of elementary

Enz/Mehra (eds.), Physical Reality and Mathematical Description. 375–386.

particles.

Skyrme, in 1960 [2], discussed a non-linear Boson field theory model in two-dimensional space-time, and was able to construct certain states of the field which behave as if they contained Fermions. This model was not mathematically rigorous, but a special case, that of the free field, has been treated in detail by the author and I.F. Wilde [3], who confirmed Skyrme's result for this case.

The idea that a Boson field might be able to describe Fermions was contained in a general way in the paper of Haag and Kastler [4], and in more detail by Borchers [5] and Haag, Doplicher and Roberts [6]. The Skyrme model discussed in [3] shows that these ideas can be realized, at least in two-dimensional space-time. In this paper we show that the resulting Fermion field theory can lead to the solution of the Thirring model in the form exhibited by Dell' Antonio, Frischman and Zwanziger [7]. Thus, the model gives an example where the existence of Fermions and the dynamics of the theory can be obtained from a knowledge of the observables only. Let us turn to the details.

The states of various charge, baryon number etc. can be labelled by operators Q,B,L.... States of different charge q or baryon number b, lepton number ℓ... are then orthogonal, and states with the same values form coherent subspaces $H_{q,b,\ell\ldots}$ between which superselection rules operate. If H is the Hilbert space of all states, we may write

$$H = \bigoplus_{q,b,\ell} H_{q,b,\ell\ldots}$$

and if $\mathfrak{A}$ is the C^*-algebra generated by the observables, this direct sum reduces $\mathfrak{A}$. Thus an operator $A \in \mathfrak{A}$ has a "block diagonal" form

$$A\psi = \begin{pmatrix} A_o & & & & \\ & A_1 & & & \\ & & \ddots & & \\ & & & A_q & \\ & & & & \ddots \end{pmatrix} \begin{pmatrix} \psi_o \\ \psi_1 \\ \vdots \\ \psi_q \\ \vdots \end{pmatrix}$$

where we have labelled the set of simultaneous eigenvalues by q. We use the operators A_q, as A runs over $\mathfrak{A}$ but for a fixed q, to generate an irreducible C^*-algebra $\mathfrak{A}_q$ acting on H_q.

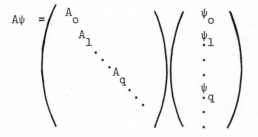

At this point, Haag and Kastler made the non-trivial assumption that all $\mathfrak{A}_q$ are isomorphic to each other, and that they are inequivalent representations of the same abstract algebra, $\mathfrak{A}$. Thus, q,b ... act as labels for inequivalent representations of $\mathfrak{A}$ - they are its Casimir operators! This gave Haag and Kastler the idea that one ought to be able to predict the charges and other conserved quantum numbers of a theory entirely from its algebra $\mathfrak{A}$; we just solve the mathematical problem of classifying its representations.

Let now $\mathfrak{A}$ denote the abstract algebra, and write $\pi_q(\mathfrak{A})$ for the algebra $\mathfrak{A}_q$ and $\pi_q(A)$ for A_q, where π_q denotes the representation on H_q. Let $\pi = \oplus_q \pi_q$. The representations π_q are mutually inequivalent, so that π is multiplicity free. In this case, $\pi(\mathfrak{A})'$, the commutant, is abelian, and conversely, if $\pi(\mathfrak{A})'$ is abelian, π is multiplicity free. Jauch and Misra [8] have shown that this means that $\pi(\mathfrak{A})$ has a complete set of commuting observables - which property is therefore equivalent to Wightman's hypothesis of commuting superselection rules (Q,B... all commute). Haag, Doplicher and Roberts [6] show that if $\pi(\mathfrak{A})'$ is not abelian, then parastatistics will occur for some of the particles, and that this is connected with non-abelian gauge groups.

We should not expect every state or representation to occur in a theory of elementary particles; all states made in the laboratory are obtained by making local perturbations of the vacuum, and inserting a finite amount of energy. These ideas are formulated as follows.

Let $\mathfrak{A}(\mathcal{O})$ be the C^*-algebra generated by the observables in $\mathcal{O} \subset \mathbb{R}^d$ = space-time, and $\mathfrak{A} = \overline{\cup_{\mathcal{O}} \mathfrak{A}(\mathcal{O})}$, (norm closed). We assume that $\mathfrak{A}$ is subject to the usual axioms of isotony and causality: if $\mathcal{O}_1 \subset \mathcal{O}_2$ then $\mathfrak{A}(\mathcal{O}_1) \subset \mathfrak{A}(\mathcal{O}_2)$, and if $\mathcal{O}_1$ is space-like to $\mathcal{O}_2$, then $\mathfrak{A}(\mathcal{O}_1)$ and $\mathfrak{A}(\mathcal{O}_2)$ commute. If we desire a relativistic theory, then the Poincaré group $\mathcal{P}$ acts on $\mathfrak{A}$ by automorphisms, $\tau(L)$, $L \in \mathcal{P}$. We express the finiteness of the energy of a state of $\mathfrak{A}$ by requiring it to lie in a covariant representation; we say π is covariant if there is a continuous projective representation, $L \to U(L)$, of $\mathcal{P}$, on the representation space of π, such that $\pi(\tau(L)A) = U(L)\pi(A)U^*(L)$ for all $L \in \mathcal{P}$ and $A \in \mathfrak{A}$, and such that the energy-momentum spectrum of U lies in the closed forward light-cone. Actually, this gives slightly less than finiteness of the energy- each state of a covariant representation is arbitrarily close, in norm, to a state of finite energy.

Let us assume that the vacuum representation, π_0 is a given covariant representation containing a state invariant under $\mathcal{P}$. Let

σ be any automorphism of $\mathfrak{A}$; an automorphism is the analogue, in a quantum theory, of a canonical transformation in a classical theory. We say that an automorphism, σ, of a C^*-algebra $\mathfrak{A}$, is implemented in a representation π if there exists a unitary operator, U, such that $\pi(\sigma(A)) = U\pi(A)U^{-1}$ for all $A \in \mathfrak{A}$. It is typical of systems with infinitely many degrees of freedom that there exist automorphisms and representations in which they are not implemented.

Suppose, then, that σ is an automorphism not implemented in the vacuum representation π_0. Define the representation π_σ of $\mathfrak{A}$ by $\pi_\sigma(A) = \pi_0(\sigma(A))$. Then π_σ is not equivalent to π_0; it acts on the same Hilbert space as π_0 but gives rise to a different set of states. It is not known whether every irreducible representation of a C^*-algebra can be obtained from one special one by a suitable automorphism in this way, but this does provide a physically interesting class of representations; π_σ, as well as π_0, provides an equally valid description of the system, since they are related by a canonical transformation. The states of π_σ should occur along with those of π_0. Doplicher, Haag and Roberts [6] consider those automorphisms σ that are localized in some region of space; that is, there exists a bounded set $\mathcal{O}$ such that $\sigma(A) = A$ for all A space-like to $\mathcal{O}$. This axiom expresses the fact that the states of π_σ are local perturbations of the vacuum. While it might be desirable to relax this condition somewhat if we wish electrodynamics to be included in the formalism, it will in fact hold in the model we discuss. Our model will only use automorphisms σ with the property that, for every open set $\mathcal{O}$, $\sigma(\mathfrak{A}(\mathcal{O})) \subset \mathfrak{A}(\mathcal{O})$, so σ does not move the space-time regions about. The general theory of Doplicher, Haag and Roberts does not need this restriction, but considers a more general class of automorphisms satisfying other properties instead. We shall not need to work from any particular set of axioms, since we have an explicit model and its properties can be analysed. However, it is worthwhile to see in general how the existence of the charge-raising operator is related to a general property of the representations π_σ considered by Borchers [5] - they are all strongly locally equivalent. This means that if $\mathcal{O} \subset \mathbb{R}^d$ is such that the set of points space-like to $\mathcal{O}$ has non-empty interior, then σ restricted to $\mathfrak{A}(\mathcal{O})$ is implemented in π_0 (though not implemented as an automorphism of the whole of $\mathfrak{A}$). If a field operator ψ^* creating charge did exist, and if it satisfies the usual property of commuting with the observables at space-like separation, then we have

$$\psi^* A = A\psi^*$$

if A is localized space-like to ψ. Rather, we should write $\pi(A)$

for A, where $\pi = \oplus_q \pi_q$. If ψ^* creates one unit of charge, this equation should be written

$$\psi^* \pi_q(A) = \pi_{q+1}(A)\psi^* \tag{1}$$

if ψ and A are space-like separated. Thus ψ^* is a <u>local intertwiner</u> of the two inequivalent representations π_q and π_{q+1}; that means that (1) holds, not for all A, but only for A space-like to the region of localization of ψ. It is not a far step from (1) to the statement that π_q and π_{q+1} are strongly locally equivalent. Indeed, for any $\mathcal{O}$ such that the set of points, $\mathcal{O}'$, space-like to $\mathcal{O}$, has non-empty interior, choose ψ localized in $\mathcal{O}'$, and then (1) will hold, showing that π_q and π_{q+1}, restricted to $\mathfrak{A}(\mathcal{O})$, are not disjoint. The idea behind the converse problem, that of getting the fields from the automorphisms, is that if π_q and π_{q+1} are strongly locally equivalent, then the field operator ψ^*, localized in $\mathcal{O}'$, is to be found among the intertwiners of $\pi_q \mid \mathfrak{A}(\mathcal{O})$ and $\pi_{q+1} \mid \mathfrak{A}(\mathcal{O})$. This is what happens in our model-certain limits ψ of such operators turn out to solve the Thirring model.

Let us recall a few details of these models.

In the free Skyrme model we start with a scalar massless field $\phi(x)$ in two-dimensional space-time; this satisfies

$$\Box\phi \equiv (\partial^0\partial_0 + \partial^1\partial_1)\phi(x,t) = 0$$

and is represented in the usual relativistic Fock space, so that

$$\phi(x,t) = \frac{1}{2\sqrt{2\pi}}\int (a^*(p)e^{ipx-i\omega t} + a(p)e^{-ipx+i\omega t})\,\frac{dp}{\sqrt{\omega}}$$

where (a^*,a) form the Fock representation, and $\omega = |p|$. For each t, $\int\phi(x,t)f(x)dx$ is a self-adjoint operator if $f \in L^2(\mathbb{R}, dp/|p|)$. It is well known that ϕ is not a Wightman field, because $L^2(\mathbb{R}, dp/|p|)$ does not contain $\mathcal{S}(\mathbb{R})$. This difficulty is avoided by choosing test-functions of the form df/dx, $f \in \mathcal{S}$. The extra factor of p in the Fourier transform cancels the singularity $1/|p|$. This is a special case of the idea of Schroer[9)] that the test-function class should be restricted, in preference to the introduction of an indefinite metric as in [10)].

We now define the local algebras of observables of the model. Let $\mathcal{O} \subset \mathbb{R}$ denote a bounded open interval; we regard this as a space-interval located at time t_o in space-time; that is, we consider the set of $(x,t_o) \in \mathbb{R}^2$ such that $x \in \mathcal{O}$, and denote by $\mathcal{O}'$ the set of points space-like to this set. The set of points in $\mathbb{R}^2$ that are space-like with respect to $\mathcal{O}'$, is denoted by $\mathcal{O}''$ and is called the double cone based on $\mathcal{O}$ at time t_o.

Let ξ be a solution of the wave-equation $\Box\xi(x,t) = 0$, such that $\xi(x,t_o) \in \mathcal{S}(\mathbb{R})$ and $\dot{\xi}(x,t_o) = df/dx$ for some $f \in \mathcal{S}(\mathbb{R})$. Then

$$W(\xi) = \exp i \int \{\phi(x,t_o)\, \dot{\xi}(x,t_o) - \dot{\phi}(x,t_o)\xi(x,t_o)\}dx$$

is a unitary operator on Fock space. If supp $\xi \subset \mathcal{O}$ and supp $f \subset \mathcal{O}$, we assign $W(\xi)$ to the algebra $\mathfrak{A}(\mathcal{O}'')$, which is defined to be the von Neumann algebra generated by these. The algebra of observables is taken to be the C^*-algebra generated by the $\mathfrak{A}(\mathcal{O}'')$, as $\mathcal{O}$ runs over all bounded open sets located at time t for any time.

This theory has two conserved currents, $j^\mu = \partial_\mu\phi$ and $j_5^\mu = \varepsilon^{\mu\nu}\partial_\nu\phi$. The corresponding charges, $Q = \int j^o(x,o)dx$ and $Q_5 = \int j_5^o(x,o)dx$ need care in their definition, since 1 is not a test-function for $\dot{\phi}$ or $\nabla\phi$. We choose a function $\theta(x)$, where $\theta(x) = 1$ on a large set, and zero outside a slightly larger set. For every such θ, define

$$Q(\theta) = \int j^o(x,o)\, \theta(x)dx, \quad Q_5(\theta) = \int j_5^o(x,o) \quad \theta(x)dx .$$

Let ξ be a solution of $\Box\xi = 0$ of the class considered above, and let Ψ_o denote the Fock vacuum. Then

$$\langle W(\xi)\, \Psi_o, Q(\theta)W(\xi)\, \Psi_o\rangle = \int \theta(x)\, \dot{\xi}(x,o)dx \qquad (1)$$

$$\langle W(\xi)\, \Psi_o, Q_5(\theta)W(\xi)\, \Psi_o\rangle = \int \nabla\theta(x)\xi(x)dx \qquad (2) .$$

Both of these go to zero as $\theta \to 1$, so we may say that these charges vanish in every "coherent" state $W(\xi)\, \Psi_o$. These states span Fock space, so Q and Q_5 vanish. But, far from being uninteresting, the charges Q and Q_5 are unitary invariants, acting as labels for the inequivalent representations. The values happen to be zero for the Fock representation, which contains the Poincaré invariant vector Ψ_o. We take this to be the representation denoted π_o in our discussion of the general theory.

The model under discussion has many similarities to a model in four dimensions, exhibiting a spontaneous breakdown of symmetry [11]. It should be mentioned, however, that the present model does <u>not</u> have a spontaneously broken symmetry. In four dimensions, the equation $\Box\phi = 0$ is invariant under the group $\phi(x,t) \to \phi(x,t) + \eta$ where $\eta \in \mathbb{R}$ is a constant function of (x,t). This symmetry group is spontaneously broken in the Fock representation. This means that the automorphisms generated by the group are not implemented in the Fock representation. There is a continuum of inequivalent representations π_η of the algebra of observables, each with a vacuum state Ψ_η and positive energy. These vacuum states can be distinguished by the expectation values of $\exp i \int\phi(\underline{x},t)f(\underline{x})d^3x$ where $\int f(\underline{x})d\underline{x} \neq 0$. In two-dimensional space-time, we still have the invariance of the wave-equation but operators $\int\phi(x,t)f(x)dx$ with $\int f(x)dx \neq 0$ do not exist. We must restrict ourselves to test-functions f such that $\int f(x)dx = 0$ in two-dimensions, because of the singularity at $p = 0$. Consequently, all the vacua ω_η coincide, all the π_η are equivalent, and the automorphism generated by $\phi \to \phi + \eta$, is the identity automorphism of the algebra of observables. So, while we have many features in common with a spontaneously broken symmetry, such a phenomenon does not occur here.

Returning to our model in two dimensions, we now describe how we get new representations from π_o by using automorphisms. Consider the automorphisms generated by

$$\phi(x,t) \to \phi(x,t) + \eta(x,t)$$

where $\Box\eta = 0$. Writing $\{\eta,\xi\}$ for $\int(\eta\dot{\xi} - \dot{\eta}\xi)dx$, this automorphism σ_η may be expressed as

$$\sigma_\eta(W(\xi)) = e^{i\{\eta,\xi\}}W(\xi)$$

which uniquely determines σ_η. Applying a theorem of Manuceau[12)], this automorphism is implemented in π_o if and only if $\eta(x,o)$ and $\dot{\eta}(x,o)$ satisfy $\eta(\infty,o) = \eta(-\infty,o)$ and $\int\dot{\eta}(x,o)dx = 0$. Indeed, in the representation $\pi_\eta = \pi_o \circ \sigma_\eta$, for any coherent state $\pi_\eta(W(\xi))\Psi_\eta$, we have

$$\langle\langle \int j^o(x,o)dx \rangle\rangle = \int \dot{\eta}(x,o)dx = Q(\eta)$$

$$\langle\langle \int j_5^o(x,o)dx \rangle\rangle = \eta(\infty,o) - \eta(-\infty,o) = Q_5(\eta) .$$

Here, Ψ_η is the state defined (up to a phase) by

$$\langle\Psi_\eta, \pi_\eta(A)\Psi_\eta\rangle = \langle\Psi_o, \sigma_\eta(A)\Psi_o\rangle$$

and $\langle\langle\cdots\rangle\rangle$ means expectation value in the state $\pi_\eta(W(\xi))\Psi_\eta$. The infinite integrals are to be understood in the sense described after Eq. (2).

One shows [3] that the representations π_η are covariant. The states Ψ_η are the coherent states corresponding to the classical field η, since $\langle\langle \phi(x,t) \rangle\rangle = \eta(x,t)$.

We can write down a formula for a unitary operator that transforms π_o to π_η, namely, $\exp i\{\phi,\eta\}$; this does not exist if $Q \neq 0$ or $Q_5 \neq 0$, consistent with the fact that π_o and π_η are inequivalent in this case. All the representations are strongly locally equivalent: if $\mathcal{O}'$ has open interior, we can, for any $\eta(x,t)$, find $\eta_1(x,t)$ such that η and η_1 differ by a constant on the interval $\mathcal{O}$, but with $\eta_1(\infty) = \eta_1(-\infty)$ and $\int\dot{\eta}_1 dx = 0$. Then σ_η and σ_{η_1} coincide on $\mathfrak{A}(\mathcal{O}'')$, since $\mathfrak{A}(\mathcal{O}'')$ contains only the gradient of the field. Then $\pi_\eta(\mathfrak{A}) \,|\mathfrak{A}(\mathcal{O}'')$ is equivalent, using the unitary operator $W(\eta_1)$, to $\pi_o(\mathfrak{A}) \,|\mathfrak{A}(\mathcal{O}'')$.

A meaning can be given to the operator $\exp i\{\phi,\eta\} = W(\eta)$ even when $Q(\eta)$ or $Q_5(\eta)$ is non-zero. The way to do this is to regard $W(\eta)$ as an operator on $\bigoplus_{q,q_5} H_{q,q_5}$, rather than on Fock space, $H_{o,o}$. The operator $W(\eta)$ increases q by $Q(\eta)$ and q_5 by $Q_5(\eta)$. Let us see how this happens.

It is convenient to introduce the related charges $q_\pm = q \pm q_5$ and the conserved currents $j^\mu_\pm = j^\mu \pm j^\mu_5$.

The general classical solution to $\eta = 0$ is the sum of a wave moving to the left and one moving to the right: $\eta = f(x+t)+g(x-t)$. Solutions with $q_+ = 0$ contain only the first, and those with $q_- = 0$ contain only the second. We can obtain all our states Ψ_η from the vacuum sector states by applying automorphisms, first of the form σ_η with $Q_+(\eta) = 0$, then of the form σ_η with $Q_-(\eta) = 0$. The class of f's and g's turns out to be $\{f : f \in C^\infty, f' \in \mathcal{S}\}$. Two f's and g's that differ by an overall constant lead to the same automorphism σ_η, and $Q_+(\eta) = 2\left[f(\infty) - f(-\infty)\right]$, $Q_-(\eta) = -2\left[g(\infty) - g(-\infty)\right]$. It suffices, then, to give meaning to $W(\eta)$ separately in the two cases $q_+ = 0$ and $q_- = 0$. We take our cue from the Weyl relations

$$W(\xi_1)W(\xi_2) = e^{i\{\xi_1,\xi_2\}/2}W(\xi_1 + \xi_2)$$

to define

$$W(\eta)(W(\xi)\,\Psi_o) = e^{\frac{i}{2}\{\eta,\xi\}}\,W(\xi + \eta)\Psi_o$$

where $W(\eta)\Psi_o$ is defined to be $\Psi_\eta = \sigma_\eta \Psi_o$. This defines a map from H_o to H_{q,q_5}, where $q = Q(\eta)$, $q_5 = Q_5(\eta)$. Similarly one defines $W(\eta)$ on $H_{q_1 q_5}$ for any $q_1 q_5$:

$$W(\eta)W(\eta_1)\Psi_o = e^{\frac{i}{2}\{\eta,\eta_1\}} W(\eta + \eta_1)\Psi_o .$$

These operators have relations

$$W(\eta_1)W(\eta_2) = e^{i\{\eta_1\eta_2\}}W(\eta_2)W(\eta_1) .$$

We seek states that are Boson or Fermi. We see that it is possible to have states in this model that are neither, if $e^{i\{\eta_1\eta_2\}}$ is not ± 1 for space-like separations of f_1 and g_1, f_2 and g_2.

If supp f_1' and supp g_1' are to the left of supp f_2' and g_2', then

$$\{\eta_1,\eta_2\} = \int(\dot{\eta}_1\eta_2 - \eta_2\dot{\eta}_1)dx = \eta(\infty)\int_{-\infty}^{\infty}\dot{\eta}_2\, dx = Q(\eta_1)Q_5(\eta_2) .$$

Thus, $W(\eta_1)$ and $W(\eta_2)$ anti-commute if $Q(\eta_1)Q_5(\eta_2) = (2n + 1)\pi$ for some integer n. That it is possible to have anti-commuting operators in a Boson theory was remarked by T.H.R. Skyrme[2]. Indeed, his model, in the linear case, is the same as that described here, except that he tries to work with step-functions $\theta(x-x_o)$ for f and g, calling the resulting $W(\eta)$ the Fermi-field $\psi(x_o)$. He argues that it satisfies the free Dirac equation; it has two components corresponding to $g = 0$ or $f = 0$ respectively.

A more rigorous approach to this question has been worked out by Dell'Antonio, Frischman and Zwanziger[7], apparently unaware of Skyrme's paper. These authors start with a theory, the Thirring model, in which there are Fermi fields ψ ab initio. Their point of view, then, is the conventional one, in which the superselection rules are built in, rather than derived, as here. Nevertheless, the calculations in [7] can be applied to our model, which shows that local Fermi operators $\psi(x)$ can be defined as limits in our model. Let us compare the notation of [7] with what we are using here.

The currents of the Thirring model are given by $j^\mu = \sqrt{c}\,\partial^\mu\phi$, $j_5^\mu = \sqrt{c}\,\varepsilon^{\mu\nu}j_\nu\phi$, where ϕ is the free field used here. The parameter c could be chosen to be 1 without altering the algebra of observables or the dynamics, and we shall do this. In that case, the parameters a and $\bar{a}$ of [7] are the same as our parameters $Q(\eta)$ and $Q_5(\eta)$, for one particular representation obtained from a pair $(\eta_o(x,o), \dot{\eta}_o(x,o))$ of a particular shape (determined by the function $\mathfrak{J}$ of [7]). Having chosen this shape (corresponding to the "standard" automorphism of [3]), the Thirring model exists on the Hilbert space $\bigoplus_{m,n=-\infty}^{\infty} H_{m,n}$, where $H_{m,n}$ carries the representation $\pi_{m,n}$ of $\mathfrak{A}$ belonging to charges $Q = ma$, $Q_5 = n\bar{a}$; that is, $\pi_{m,n}$ is the representation π_η where σ_η is the automorphism determined by the solution $\eta(x,t)$ whose Cauchy data are $(m\eta_o, n\dot{\eta}_o)$.

To define the Fermi operators, in the limit as η becomes a step-function, one proceeds as follows. Let η_1 denote the solution to the wave equation whose Cauchy data are $(\eta_o, 0)$ and η_2 the solution with Cauchy data $(0, \dot{\eta}_o)$. Then the representations of the canonical commutation relations given by

$$\xi \to \pi_{m+1,n}\,(W(\xi)\, e^{-i\{\eta_1,\xi\}})$$

is unitarily equivalent to $\pi_{m,n}$, and the representation

$$\xi \to \pi_{m,n+1}\,(W(\xi)\, e^{-i\{\eta_2,\xi\}})$$

is unitarily equivalent to $\pi_{m,n}$. It follows that there are unitary operators U,V, unique up to a phase (since $\pi_{m,n}$ is irreducible) such that

$$U\,\pi_{m,n}(W(\xi))U^{-1} = e^{-i\{\eta_1,\xi\}}\,\pi_{m+1,n}\,(W(\xi))$$

and

$$V\,\pi_{m,n}(W(\xi))V^{-1} = e^{-i\{\eta_2,\xi\}}\,\pi_{m,n+1}(W(\xi))\ .$$

The operators U,V increase the charges Q, Q_5 respectively, i.e. $U: H_{m,n} \to H_{m+1,n}$ and $V: H_{m,n} \to H_{m,n+1}$. Thus U,V are creation operators, creating the "standard" coherent states, with shape η_1 and η_2; the phase $e^{-i\{\eta_j,\xi\}}$, $j = 1,2$, necessary for unitary equivalence, does not alter the state defined by $U\Psi, \Psi \in H_{m,n}$ or by $V\Psi, \Psi \in H_{m,n}$.

Our candidate for the creation operator for a general shape η is $W(\eta)$, but this has no limit as η converges to a step function. Dell'Antonio, Frischman and Zwanziger choose a related operator, which is defined on $\bigoplus_{m,n} H_{m,n}$, and which does possess a limit. If $\dot{\eta}(x,o) = 0$, they define $\Psi^*(\eta) = : W(\eta-\eta_1)U:$ where: : means Wick ordering, but with U placed between the creation and annihilation operators. Wick ordering alters $W(\eta)$ only by a phase, so that the transformation induced by it is the same. We note that $W(\eta-\eta_1)$ actually exists in Fock space, and on each $H_{m,n}$, creating a coherent state $\eta-\eta_1$. The operator U creates the coherent state η_1, so $:W(\eta-\eta_1)U:$ creates the state Ψ_η. In 7) it is proved that Ψ^* converges as $\eta \to$ step-function, in the sense of strong convergence of distributions on a certain domain, leading to a Wightman theory. The sense $\eta \to$ step-function is a particular one, namely the step function, $\theta(x - x_o)$ is replaced by the momentum space cut-off function

$$\frac{1}{\sqrt{2\pi}} \int_{|k|\leq\Lambda} \tilde{\theta}(k)\, e^{ikx} dk, \text{ where } \tilde{\theta}(k) = \frac{1}{\sqrt{2\pi}} \int^{\infty} e^{-ikx} dx.$$

Then $\Lambda \to \infty$ as at the end. Similarly, $:W(\eta-\eta_2)V:$ leads to a Fermi Wightman field as $\dot{\eta}(x,o) \to \delta(x - x_o)$, if $\eta(x,o) = 0$. These two Fermi fields are the two components of the Thirring model. For the details, see 7).

There are several points one should make. The new representations $\pi_{m,n}$ are defined intrinsically from the vacuum representation, and exhibit Fermions if Q and Q_5 are suitably chosen. The Fermion appears as a condensed state of finite energy, whose equations of motion are non-linear and determined by the theory itself.

It would be interesting to try for a similar model in 3 or 4 space-time dimensions. The following features of the situation in two dimensions might suggest how the model might be generalized. The space $\mathbb{R}$, omitting a finite interval I, is not connected, whereas $\mathbb{R}^2$ or $\mathbb{R}^3$, omitting a compact set, is. This remark disguises the similarity between $\mathbb{R}$-I and $\mathbb{R}^2$ or $\mathbb{R}^3$ omitting a compact set, namely, that certain cohomology groups of the spaces are not trivial. Thus, $\mathbb{R}$-I, is a manifold on which there exists a 0-form f (a function), such that df = 0 but f is not constant (it is locally constant). Such an f is taken as the Cauchy data of our solution η_1 or η_2; it is because $f(+\infty) \neq f(-\infty)$ holds that we get new representations, whereas it is because df = 0 (outside I) that gives the localization, in I, or ψ^*. This suggests that, in 3 or 4 dimensions, automorphisms using non-exact closed forms might give other models predicting their own superselection rules.

REFERENCES

1) W. Heisenberg, Der derzeitige Stand der nichtlinearen Spinortheorie der Elementarteilchen, Act. Phys. Aus. 14, 328 (1961).

2) T.H.R. Skyrme, Proc. Roy. Soc. A262, 237-245 (1961).

3) R.F. Streater and I.F. Wilde, Fermion States of a Boson Field, Nuclear Phys. B24, 561-575 (1970).

4) R. Haag and D. Kastler, An Algebraic Approach to Quantum Field Theory, J. Mathematical Phys. 5, 848 (1964).

5) H. Borchers, Cargèse Lectures, 1965 (Ed. F. Lurçat), Gordon and Breach, N.Y. 1967.

6) S. Doplicher, R. Haag and J. Roberts, Comm. Math. Phys. 13, 1 (1969); 15, 173 (1969).

7) G.F.Dell'Antonio, Y. Frischman and D. Zwanziger, Phys. Rev. D, 6, 988 (1972). G.F. Dell'Antonio, Schladming Lectures, 1973 (Ed. P. Urban), Springer, Berlin (to appear).

8) J.M. Jauch and B. Misra, Supersymmetries & Essential Observables, Helv. Phys. Acta 34, 502, 699 (1961).

9) B. Schroer, Infrateilchen, Fortschr. der Phys. 11, 1 (1963).

10) J. Challifour, Time Ordered Products in Two-Dimensional Field Theories, J. Mathematical Phys. 9, 1137 (1968).

11) R.F. Streater, Spontaneous Breakdown of Symmetry in Axiomatic Theory, Proc. Roy. Soc. A287, 510-518 (1965).

THE NONLOCAL NATURE OF ELECTROMAGNETIC INTERACTIONS

F. Rohrlich

Syracuse University
Syracuse, New York

Ever since the beginning of our successful collaboration twenty years ago when I first began to know my friend and colleague Josef Jauch, he has continued to develop and deepen his mathematical understanding of our physical world, and he has succeeded admirably in enriching the scientific literature.

* * *

My thinking about electromagnetic interactions received considerable clarification from the study of classical point electrodynamics. Over the years I became convinced that electromagnetic interactions are intrinsically nonlocal, despite the fact that they are usually represented in terms of local, interacting electromagnetic potentials. This local description emerged more out of mathematical opportunism than out of a deep understanding of experimental facts.

In theoretical physics one sometimes encounters different formulations of a theory which all lead to identical predictions. But these formulations, although they lead to the same physical predictions, are in general not physically equivalent in the sense that the interpretation of their mathematical structures leads to different physical models. For example, the so-called "covariant" and "non-covariant" gauges of electrodynamics lead to the same cross sections and life times; but in one case the theory describes photons associated with the longitudinal and timelike modes of the field, while in the other case no such photons occur.

Enz/Mehra (eds.), Physical Reality and Mathematical Description. 387–402.

One can make a choice between such formulations either on mathematical grounds or on physical grounds. Computational convenience chooses the former; deeper understanding with a view toward generalization chooses the latter. It is here where my own preference suggests the use of Ockham's razor in favor of physics. Theoretical constructs which do not correspond to observable quantities should be minimized. It is my belief that if this philosophy is pursued, it will eventually pay off in providing a theory which is also mathematically more satisfactory.

The purpose of the following considerations is to look at electrodynamics (both classical and quantum) from this point of view. We shall find that this does not lead to the conventional formulation, although it is equivalent to it. It will lead to a separation of electromagnetic interactions into two nonlocal parts, characterized by the presence or absence of (observable) photons. This criterion takes precedent over the criterion of locality in the present formulation.

1. Two types of nonlocalities for classical fields

Everyone knows that electromagnetic interactions are nonlocal. In classical physics the retarded solutions of Maxwell's equations are selected; they relate in a unique manner the field at an arbitrary point x off the worldline of the charge to its point of origin on the world line, z_{ret}. Thus, $F^{\mu\nu}_{ret}$ depends not only on x but depends at least implicitly also on a point which may be far away. The well known convolution $A^{\mu}_{ret} = D_R * J^{\mu}$ expresses this nonlocal dependence on the source J^{μ}.

We shall call this nonlocality of the field: type I. Little more needs to be said about it. But there is another, less well known nonlocality. It appears most dramatically in radiation reaction.

According to Maxwell's equations the field produced by a point charge is infinite at the position of that charge. This is so quite independent of the asymptotic conditions which are chosen to specify the solution (e.g. retarded or advanced fields). Consequently, the self-interaction is also infinite.

However, upon closer inspection one finds that this self-interaction consists of two parts which separate covariantly, viz. a "Coulomb" self-interaction and a so-called "radiation reaction"; and only the former is divergent, as was first found by Wentzel[1]. This covariant separation of the observed field strength F_{ret} is just

$$F_{ret} = F_+ + F_- \qquad F_\pm \equiv \tfrac{1}{2}(F_{ret} \pm F_{adv}). \tag{1}$$

The "radiation reaction" Γ is the Lorentz self-force at the point of radiation emission, due to F_-,

$$\Gamma^\mu \equiv eF_-^{\mu\alpha}\, v_\alpha. \tag{2}$$

It is sometimes known as the Abraham fourvector.

Before proceeding, a word about advanced fields is in order: These fields do <u>not</u> describe a causality violating situation as is often believed. Rather, $F_{adv}(x)$ specifies that field at a point x which must be there so that it can be fully <u>absorbed</u> at a <u>later</u> point z_{adv} where the point charge's kinematics is given. This meaning of F_{adv} is not new but was first stated by Einstein in 1909[2].

Now the rate of energy and momentum emission of radiation is given by the timelike fourvector[3]

$$\frac{dP_R^\mu}{d\tau} = \frac{e^2}{6\pi}\,\dot{v}^\alpha \dot{v}_\alpha v^\mu. \tag{3}$$

But the <u>local</u> conservation laws require that the recoil momentum and energy due to this emission is just the negative of $dP_R^\mu/d\tau$. This is <u>not</u> the case because

$$dP_R^\mu/d\tau \neq -\Gamma^\mu. \tag{4}$$

Therefore Γ cannot be the true (local) radiation reaction. This fact is mathematically obvious since (2) is spacelike while (3) is timelike.

In fact, one finds

$$\frac{dP_R^\mu}{d\tau} = -\Gamma^\mu + \frac{e^2}{6\pi}\,\ddot{v}^\mu. \tag{5}$$

The extra term is the "mysterious" Schott term*).

---

*) One notes parenthetically that the Schott term cannot be reasonably regarded as a radiation reaction force since it does not have the structure of a Lorentz force. On the other hand, Γ, which does have this structure, (2), cannot really be the radiation reaction force because it does not vanish if and only if the radiation rate ($R \equiv (e^2/6\pi)\dot{v}^\alpha \dot{v}_\alpha > 0$) vanishes. For the global relations (6) and (7) below, these difficulties do not exist.

This balance between $dP^\mu_R/d\tau$ and Γ^μ which according to (4) does not hold locally, however does hold globally if the charge is asymptotically free, as is the case in scattering systems,

$$\Delta P^\mu_R \equiv \int_{-\infty}^{\infty} \frac{dP^\mu_R}{d\tau}\, d\tau = -\int_{-\infty}^{\infty} \Gamma^\mu d\tau. \tag{6}$$

This nonlocal conservation law is a manifestation of the nonlocality which we call: type II.

If the point charge is constrained by an external force to undergo periodic motion of period T one obtains instead of (6)

$$\int_\tau^{\tau+T} \frac{dP^\mu_R(\tau')}{d\tau'}\, d\tau' = -\int_\tau^{\tau+T} \Gamma^\mu d\tau \qquad \forall\ \tau, \tag{7}$$

another type II nonlocality relation.

This rather disquieting result of a lack of local conservation laws reappears in the equations of motion as will now be shown.

2. Equations of Motion

It is well-known that the Maxwell-Lorentz equation together with the integral conservation laws lead to the Lorentz-Dirac equation[4] for a point charge,

$$m\dot{v}^\mu = F^\mu + \Gamma^\mu \equiv F^\mu - \frac{e^2}{6\pi}\, \dot{v}^\alpha \dot{v}_\alpha v^\mu + \frac{e^2}{6\pi}\, \ddot{v}^\mu. \tag{8}$$

where F^μ is any external force. Because of the second time-derivative of v^μ this is not an equation of motion. An equation of motion must permit a unique solution on the basis of only two initial conditions, position and velocity. The problem can be resolved, as was first pointed out by Dirac, by adjoining to (8) an asymptotic condition. For example, one can require that the point charge be free in the distant future. A much more general condition however, suffices; one needs to require only that

$$\lim_{\tau\to\infty} e^{-\tau/\tau_0}\, \dot{v}^\mu(\tau) = 0 \tag{9}$$

where $\tau_o \equiv e^2/(6\pi m)$. Equations (8) and (9) are mathematically equivalent to the integro-differential equation

$$m\dot{v}^{\mu}(\tau) = \int_0^{\infty} K^{\mu}(\tau+\alpha\tau_o)e^{-\alpha}d\alpha \tag{10}$$

where

$$K^{\mu} \equiv F^{\mu} - \frac{e^2}{6\pi}\dot{v}^{\alpha}\dot{v}_{\alpha}v^{\mu}. \tag{11}$$

I have proposed this equation (10) as the equation of motion of a relativistic point charge[5]; it determines the world line of the particle from its initial position and velocity. Note that the Schott term has disappeared and the nonlocality of type II became explicit.

This equation answers the question raised before: the conservation laws are indeed nonlocal when expressed entirely in terms of particle variables. The inertial term on the left is not equal to the net force K^{μ}, the external force diminished by the (true) radiation reaction $dP_R^{\mu}/d\tau$, taken at the same time τ. Rather, the inertial term equals this net force integrated over a very steep tail from τ over the entire future. The corresponding nonlocal energy conservation law becomes explicit when the case $\mu = 0$ is considered.

While this result may seem rather unsatisfactory, it is an inescapable consequence of the very few generally accepted assumptions made above. The conservation laws in this form do show a nonlocal behavior.

That the matter is not as serious as it may seem can be seen as follows: the exponential tail in (10) extends effectively over only a few times the classical electron radius (divided by the velocity of light) while classical considerations cease to have physical importance below distances 137 times larger, viz. at the Compton wave length. Thus, one can argue that the nonlocality is outside the physical domain of applicability of this classical theory.

One might ask how a nonlocal conservation law can have emerged in a theory in which a local one, viz.

$$\partial_{\alpha}T^{\alpha\mu}(x) = F^{\mu\alpha}(x)\,J_{\alpha}(x) \tag{12}$$

is well known to exist. The answer lies in the observation that J_α is a nonlocal function of the world line of the point charge, $z(\tau)$,

$$J_\alpha(x) = e\int_{-\infty}^{\infty} \delta_4\Big(x-z(\tau)\Big)\, v_\alpha(\tau)d\tau. \tag{13}$$

If we now construct the mixed energy tensor

$$T_{+-}^{\mu\nu} \equiv F_+^{\mu\alpha}F_{-\alpha}{}^{\nu} + F_-^{\mu\alpha}F_{+\alpha}{}^{\nu} + \tfrac{1}{2}g^{\mu\nu}F_+^{\alpha\beta}F_{-\alpha\beta} \tag{14}$$

then (12) becomes

$$\partial_\alpha T_{+-}^{\alpha\mu} = F_-^{\mu\alpha}J_\alpha \tag*{(12)$_{+-}$}$$

and with (13) we have

$$\partial_\alpha T_{+-}^{\alpha\mu}(x) = \int_{-\infty}^{\infty} \delta_4\Big(x-z(\tau)\Big)\; eF_-^{\mu\alpha}(z)v_\alpha(\tau)d\tau = \int_{-\infty}^{\infty} \delta_4(x-z)\;\Gamma^\mu d\tau \tag{15}$$

where Γ^μ is given by (8). But without the τ-integration Γ^μ cannot be related to $T_{+-}^{\mu\nu}$.

Thus, the local conservation law (12) with respect to a field point x hides the nonlocal law which emerges when the particle variables are exhibited explicitly. The nonlocality is hidden in the boundary conditions for the field, i.e. in the subscripts ret and adv. This establishes a connection between the nonlocalities of type I and type II.

3. Action at a distance

The description of classical electromagnetic interactions in terms of the electromagnetic field is today practically the only one which is in common usage. However, it has been known that a very different description is also possible in which no such fields occur at all. This description[6)] is equally well in agreement with experiment. But it appears to many to be intuitively awkward.

This "action at a distance" formulation is based on the interaction between a system of n charges, characterized by the interaction Lagrangian

$$L_P = \frac{1}{2} \sum_{\substack{i=1 \\ (i \neq j)}}^{n} \sum_{j=1}^{n} e_i e_j \int v_i^\mu(\tau_i)\, D_P(z_i - z_j) v_{j\mu}(\tau_j) d\tau_i d\tau_j \tag{16}$$

The world lines of particle i is $z_i(\tau_i)$, its velocity $v_i(\tau_i)$, and the invariant function $D_P \equiv \frac{1}{2}(D_R + D_A)$ is given by a well-known principal value Fourier representation. This interaction is obviously <u>nonlocal</u>. It can be written as the arithmetic average of a retarded and an advanced interaction between each pair of charges. It does not violate causality. This is discussed in detail by Wheeler and Feynman[6].

The great virtue of (16) is its explicit mandate $i \neq j$ which exorcises the self-interactions and the associated divergence problems. The main objections are (i) the absence of electromagnetic fields which - at least as radiation and, in quantum mechanics, as photons - seem to demand just as much "reality" as the charges, and (ii) the need for an "absorber condition" to ensure the equivalence with the customary formulation.

One can meet both objections without giving up the great virtue. This can be done by treating (16) not as the total interaction, but as only the "$1/r^2$" interaction. The interaction via radiation ($1/r$ fields) is then described by additional terms which contain the fields explicitly. As will be shown below, this separation of the fields must be made in such a way that the Maxwell equations also separate. In this way one can generalize the one-particle Lorentz-Dirac equation (8) to a system of particles[7,8]. The interaction Lagrangian then contains L_P as well as the interaction with radiation,

$$L_R = \sum_{i=1}^{n} \int J_i^\mu(x)\, \bar{A}_\mu(x) d^4x \tag{17}$$

where

$$J_i^\mu(x) = e_i \int_{-\infty}^{\infty} \delta_4(x - z_i(\tau_i))\, v_i^\mu(\tau_i) d\tau_i, \qquad (13)_i$$

and $\bar{A}_\mu(x)$ is a <u>free</u> field. In this formulation there is <u>both</u> an action-at-a-distance and a field interaction. <u>But only free fields occur</u> in this interaction. No absorber condition is involved. Such a condition would be exactly equivalent to the requirement $L_R = 0$.

4. Free and bound fields

The electromagnetic field F of a point charge can be covariantly separated into a "$1/r^2$" and a "$1/r$" dependence, $F(1/\rho^2)$ and $F(1/\rho)$, for the retarded as well as for the advanced field[8]. The invariant ρ is the magnitude of the spacelike fourvector; it reduces to "the distance from the source point to the field point" in the rest frame of the source point.

The field $F(1/\rho^2)$ is not only the Coulomb field. It also includes the Biot-Savart field which is here defined as the magnetic field due to the velocity of the point charge; it vanishes only for a charge at rest. But in any case it is a field which seems attached to the source and which does not escape to infinity like the $F(1/\rho)$ field. It is therefore physically tempting to treat these "bound" fields separately from the radiation fields.

But this cannot be done ab initio. The reason is the following. One cannot replace Maxwell's equations by two other sets of equations with local sources which upon imposition of the boundary condition for retardation would yield $F_{ret}(1/\rho^2)$ and $F_{ret}(1/\rho)$, respectively.

Before resolving this difficulty we note that there exists a corresponding separation of the energy tensor[10]. If the part of $T^{\mu\nu}$ which contains only $F(1/\rho)$ is denoted by $T^{\mu\nu}_{II}$ while the rest is $T^{\mu\nu}_{I}$ then

$$\partial_\alpha T^{\alpha\mu}_{I}(x) = -\int_{-\infty}^{\infty} d\tau \; \delta_4(x-z) \left[m_{self}\dot{v}^\mu - \frac{e^2}{6\pi}\ddot{v}^\mu \right] \tag{18}$$

$$\partial_\alpha T^{\alpha\mu}_{II}(x) = -\int_{-\infty}^{\infty} d\tau \; \delta_4(x-z) \, \frac{e^2}{6\pi} \; \dot{v}^\alpha \dot{v}_\alpha v^\mu . \tag{19}$$

The last equation is of course similar to (15) and (6) but here applied to $F(1/\rho)$. We see from (18) that the bound fields give the mass renormalization term, and the interference term between the bound and the free (i.e. radiation) field give the Schott term. These relations are nonlocal.

If we wish a separation of the <u>field</u> into bound and free fields so that each field satisfies a covariant field equation, the characterization by powers of $1/\rho$ is clearly not possible.

An alternative separation is however provided by (1). It corresponds to a different characterization of "bound" and "free", as we shall see. But first we note that this separation is clearly nonlocal: neither F_+ nor F_- is a local field. This nonlocality is not only of type I, but involves corresponding nonlocal conservation laws (type II). In the quantized form we shall see that neither A_+^μ nor A_-^μ are local fields in the sense that neither satisfies a local commutation relation (type III).

The new characterization of bound and free fields can be seen from the corresponding Green functions. If

$$A = A_{in} + A_{ret} = A_{out} + A_{adv} \tag{20}$$

then

$$\left.\begin{aligned} A_+^\mu(x) &= \int D_P(x-x')\, J^\mu(x')d^4x' \\ A_-^\mu(x) &= \tfrac{1}{2}\int D(x-x')\, J^\mu(x')d^4x'. \end{aligned}\right\} \tag{21}$$

They satisfy inhomogeneous and homogeneous equations, respectively,

$$\left.\begin{aligned} \Box A_+^\mu &= -J^\mu \\ \Box A_-^\mu &= 0\ , \end{aligned}\right\} \tag{22}$$

with suitable boundary conditions which yield (21) uniquely, (we assume here the Lorentz condition). The Green function D_P has support only <u>off</u> the mass shell while D has support only <u>on</u> the mass shell. This corresponds to "virtual" and real photons, respectively.

Thus, the bound photons are off the mass shell, the free photons are on the mass shell. This characterization is obviously not only relativistic, but also quantum mechanically very useful.

The separation (1) is thus based on the mass shell as a criterion. And this same separation underlies the classical theory of a system of point charges[7,8]: the interaction consists of a part, L_P, eq.(16), which consists only of off-shell fields (note the appearance of D_P) and another part, L_R, eq.(17), which involves only physical photons, on-shell, free fields.

That this separation is also a separation into bound and free fields is seen as follows. The Lorentz force F_-, eq.(2), gives the radiation reaction in the nonlocal (global) sense (6). The interaction energy gives the action-at-a-distance interaction L_P, (16), when (13) and (21) are substituted. The interaction energy $J\cdot A_-$ combines with the interaction of the current with the incident field A_{in} to yield

$$J\cdot A_{in} + J\cdot A_- = J\cdot\tfrac{1}{2}(A_{in}+A_{out}) = J\cdot\bar{A}. \tag{23}$$

as follows from (20). Both A_{in} and A_{out} are free fields (on mass-shell). For a single particle (L_P=0) all electromagnetic effects are correctly described by this interaction[8].

It is interesting to observe that the integral over the divergence of the energy tensor constructed as it were from the interference of the bound and the free field gives the total momentum of radiation emission, as is seen from (15) and (6), i.e. again the global conservation law.

Finally, we repeat that the separation (1) based on the mass shell results in two nonlocal fields.

5. The free quantized electromagnetic field

The free electromagnetic field is well known to transform according to a representation of the Poincaré group involving zero mass. Consequently, as has been clear physically all along, it can have only two degrees of freedom. The characterization by A^μ, a four-component object, thus requires two conditions. But only one such condition exists which is manifestly covariant, viz. $\partial_\alpha A^\alpha = 0$. The manifestly covariant gauges therefore are forced to introduce two additional degrees of freedom ("longitudinal" and Coulomb photons) which are not observed and which are interrelated by the Lorentz condition. This leads necessarily to an indefinite Hilbert space[11].

Only if one is willing to give up manifest covariance and with it the physically unwarranted assumption that A^μ is a four-vector, is the road open to a deeper understanding. Since A^μ is not an observable, there is of course no reason why it should be a geometrical object (four-vector in this case).

The deeper understanding emerges from the physically natural second condition A^0=0 which relates perfectly with the transformation property of A^μ according to certain m=0 representations of the Lorentz group[12]. The condition A^0=0 is natural because the Coulomb field is an abstraction from the experimental facts involving two charges to a property of a single charge, for which it can never be observed. This abstraction is mathematically extremely desirable but it is inadmissible from a purist physical point of view (Ockham's razor again!). From this point of view the Coulomb interaction is an action-at-a-distance interaction, as described correctly by (16). There is no Coulomb field.

One is thus forced into the Coulomb (or radiation) gauge, a non-manifestly covariant gauge. Under a Lorentz transformation A^μ now transforms like a four-vector together with a gauge transformation, preserving the condition $A^0=0$ in every frame. The mathematical bonus of this description is a quantization in a positive definite Hilbert space[11].

This quantization must be performed consistent with

$$\Box A^\mu = 0 \tag{24}$$

$$\nabla \cdot \vec{A} = 0, \qquad A^0 = 0. \tag{25}$$

The commutation relations must therefore contain the projection to transverse modes*):

$$\left[A_k(x), A_\ell(x')\right] = -i\left(\delta_{k\ell} - \frac{\partial_k \partial_\ell}{\nabla^2}\right) D(x-x') \tag{26}$$

More precisely, the initial surface constraints (25) are augmented by the constraints on t = const:

$$\left[A_k(\vec{x},t), A_\ell(\vec{x}',t)\right] = 0 \tag{27}$$

$$\left[\dot{A}_k(\vec{x},t), A_\ell(\vec{x}',t)\right] = -i\left(\delta_{k\ell} - \frac{\partial_k \partial_\ell}{\nabla^2}\right) \delta_3(\vec{x}-\vec{x}'). \tag{28}$$

Now it would be incorrect to conclude that, because of (27), $A_k(x)$ is a local field; rather,

$$\left[A_k(f), A_\ell(g)\right] \neq 0 \quad \text{for} \quad \text{supp } f \sim \text{supp } g \tag{29}$$

---

*) The operator ∇^{-2} is defined by $\nabla^{-2} f(\vec{x}) \equiv -(4\pi)^{-1} \int f(\vec{x}') \times |\vec{x}-\vec{x}'|^{-1} d^3x'$ and can be applied to distributions as well as functions with suitable restrictions on their class.

where "$\sim$" indicates spacelike relationship. Thus, A_k is not local in the sense of "not microcausal" (type III nonlocality).

This assertion is easily verified by noting that

$$-2\pi\nabla^{-2}D(x) = \tfrac{1}{2}\varepsilon(t)\theta(|t|-r) + \frac{t}{2r}\,\theta(r-|t|). \tag{30}$$

Thus, $\nabla^{-2}D(x)$ does <u>not</u> vanish everywhere outside the light cone, but vanishes only on $t = 0$.

The physically natural description of a free field, correspon-ding to the absence of Coulomb fields, thus leads to a <u>nonlocal free field</u>.

6. The interacting electromagnetic field

The separation of the Maxwell equations into on and off mass-shell fields was already carried out in (22) and (23). In addi-tion, the off mass-shell fields are to be eliminated in favor of an action-at-a-distance interaction. The Dirac equation now must be written in this spirit. One finds in the Coulomb gauge

$$(\gamma\cdot\partial+m)\Psi = -ie\left[\vec{\gamma}\cdot(\bar{\vec{A}}_\perp + D_P * \vec{J}_\perp) + \gamma^0\nabla^{-2}J^0\right]\Psi \tag{31}$$

Here $\bar{\vec{A}}_\perp$ satisfies the free field equation and contains all such fields in the system. We see that in addition to the Coulomb in-teraction (last term) there is also the Biot-Savart interaction treated as action-at-a-distance, both containing only off mass-shell Green functions.

Eq.(31) can clearly be derived from a suitable Lagrangian if one wishes to cast the theory into a Lagrangian field theory.

If one omits $\bar{A}$ in (31) one obtains the quantum field theore-tical generalization of the Wheeler-Feynman theory, except that the self-interaction can no longer be eliminated; the system of n dis-crete particles has now become a field, characterized by Ψ and no discrete labels are available. As in the classical theory, the absence of $\bar{A}$ must be replaced by suitable asymptotic conditions ("absorber conditions") to ensure that no electromagnetic field can escape from the system. A quantum electrodynamics of this ty-pe was attempted by several authors recently[13]. Obviously, the theory based on (31) is capable of treating such a "pure" action-at-a-distance theory as a special case.

The free field $\bar{A}$ is actually the arithmetic mean of A_{in} and A_{out}, as indicated in (23). But A_{in} and A_{out} are not independent as follows from (20),

$$A_{out} = A_{in} + D{*}J_{\perp} = A_{in} + 2A_{-\perp} \quad . \tag{32}$$

Thus, the independent fields can be taken as $\vec{A}_{in}$ and Ψ, with the former having only two degrees of freedom as explained in the previous section.

Since (31) is nonlinear one can make it formally linear by defining a new field $A^{\mu} \equiv (A^{o},\vec{A})$,

$$\left.\begin{aligned} \vec{A} &\equiv \vec{\bar{A}}_{\perp} + D_{P}{*}\vec{J}_{\perp} = \vec{A}_{in} + \vec{A}_{-\perp} + D_{P}{*}\vec{J}_{\perp} = \vec{A}_{in} + D_{R}{*}\vec{J}_{\perp} \\ A^{o} &\equiv -\nabla^{-2}J^{o} \end{aligned}\right\} \tag{33}$$

The right side of (31) then becomes $-ie\gamma\cdot A\Psi$. Thus, one recovers the equations of the conventional theory.

However, if we wanted to recover the commutation relations of the conventional theory a rather non-trivial assumption must be made as we shall now exhibit. If A^{μ}_{in} is given by (24) to (26), it is not at all obvious whether A^{μ} is a local field, because its commutation relations are completely determined by those of A^{μ}_{in} and of Ψ. Even if A^{μ}_{in} is given in a covariant gauge, so that <u>it</u> is local, A^{μ} is not obviously local, since it would require that

$$\left[A^{\mu}_{in}(f),\, D_{R}{*}J^{\nu}(g)\right] + \left[D_{R}{*}J^{\mu}(f), A^{\nu}_{in}(g)\right] + \left[D_{R}{*}J^{\mu}(f), D_{R}{*}J^{\nu}(g)\right] = 0.$$

$$(\text{supp } f \sim \text{supp } g) \tag{34}$$

This condition can be satisfied if Bogoliubov causality is assumed and one works with operator derivatives, identifying the current in terms of the S-operator by

$$J_{\mu} = i\, S^{+} \frac{\delta S}{\delta A^{\mu}_{in}} \quad , \tag{35}$$

as was proven some time ago in asymptotic quantum field theory[14]. But the assumptions made here may appear too strong, and in any case are not generally accepted as necessary for a local A^{μ}.

Similarly, if the above assumptions are accepted but one starts with A^μ_{in} in the Coulomb gauge, one does not obtain a local field A^μ, but one which is nonlocal exactly as A^μ_{in}, satisfying (29). Otherwise, the commutator for $\vec{A}$ would not be the same as for $\vec{A}_{in}$ outside the light cone.

Thus, the usually assumed locality for the interpolating field $A^\mu(x)$ is a highly nontrivial assumption and indeed questionable from the present point of view.

We recall again that the use of the Coulomb gauge to which we have been led by physical considerations ensures an underlying Hilbert space which is positive definite[11].

7. The mass-shell criterion

The criterion which led us to a separation of the electromagnetic interaction into mass-shell and off mass-shell parts was obtained from physical considerations. The mass-shell part is the customary local Yukawa type interaction; the off mass-shell part is an action-at-a-distance interaction involving no electromagnetic field. This, I believe, is the description which is closest to the experimental situation.

It leads to the following consequences:

(1) The natural gauge is the Coulomb gauge.

(2) The asymptotic free fields $\vec{A}_{in}$ and $\vec{A}_{out}$, being in the Coulomb gauge are nonlocal (commutators do not vanish for spacelike separation).

(3) The total free field $\bar{\vec{A}}$ is transverse and nonlocal; since

$$\bar{\vec{A}} = \tfrac{1}{2}(\vec{A}_{in} + \vec{A}_{out})$$

and $\vec{A}_{in}$ and $\vec{A}_{out}$ are not relatively local, its commutator differs from that of $\vec{A}_{in}$ even for spacelike separations. This is also true in a covariant gauge.

(4) The total interaction consists of a Yukawa type $-\vec{J}\cdot\bar{\vec{A}}$ term for the mass-shell fields and a nonlocal current-current interaction.

(5) The field A^μ, so defined that the basic equations redu-

ce to the conventional theory, has a commutator which is outside the light cone not equal to that of the free field without additional assumptions. This holds true also in a covariant gauge.

(6) The underlying Hilbert space is positive definite.

The physical advantages of this point of view are considerable. The only electromagnetic field which is quantized is the on mass-shell field, eliminating completely the concept of virtual photons. Various phenomena are also understood easier in this fashion. For example, the infrared divergences of the theory have two quite distinct origins: one is the asymptotic behaviour of charged particles, which are incorrectly described by plane waves in the conventional formulation; this is due to the asymptotic presence of the action-at-a-distance forces and is an off mass-shell effect. The other is the emission of soft photons, an on mass-shell effect.

Another example is the lowest order interaction of a bound electron with itself: the level shift is an off mass-shell effect, the line width an on mass-shell effect, being due to the principal part and the pole part, respectively, of the propagator.

The theory is thus formulated with due regard to the mass-shell of physical photons; as a consequence, the nonlocal nature of the theory which is already apparent in the classical formulation is brought out.

In conclusion one can speculate about the importance of the present point of view for the quantization of gravitational interactions. Clearly, the two transverse components of the metric field must be quantized while the remaining parts must be eliminated. The latter are not independent degrees of freedom and must be expressible entirely in terms of the sources. The quantized fields will then be nonlocal.

REFERENCES

1) G. Wentzel, Z. Phys. 86, 479 and 635 (1933); 87, 726 (1933).

2) A. Einstein, Physik. Zeits. 10, 185 (1909).

3) A. Schild, J. Math. Anal. Appl. 1, 127 (1960).

4) P.A.M. Dirac, Proc. Roy. Soc. (London) A 167, 148 (1938).

5) F. Rohrlich, Ann. Phys. (N.Y.) 13, 93 (1961).

6) H. Tetrode, Z. Phys. 10, 317 (1922); A.D. Fokker, ibid., 58, 386 (1929); J.A. Wheeler and R.P. Feynman, Rev. Mod. Phys. 17, 157 (1945) and 21, 425 (1949).

7) F. Rohrlich, Phys. Rev. Lett. 12, 375 (1964).

8) F. Rohrlich, Classical Charged Particles, Addison-Wesley Publishing Company, Reading, 1965.

9) See reference 8, Section 4-8.

10) C. Teitelboim, Phys. Rev. D1, 1572 (1970) and ibid., D3, 297 (1971).

11) F. Strocchi, Phys. Rev. 162, 1429 (1967) and D2, 2334 (1970).

12) C.M. Bender, Phys. Rev. 168, 1809 (1968).

13) F. Hoyle and J.V. Narlikar, Ann. Phys. 54, 207 (1969) and ibid., 62, 44 (1971), Nature, 228, 544 (1970); P.C.W. Davies, J. Phys. A4, 836 (1971) and 5, 1025 (1972), Proc. Camb. Phil. Soc. 68, 751 (1970).

14) F. Rohrlich and J.C. Stoddart, J. Math. Phys. 6, 495 (1965); T.W. Chen, F. Rohrlich, and M. Wilner, J. Math. Phys. 7, 1365 (1966).

LE MODELE DES CHAMPS DE JAUGE UNIFIES

J. LEITE LOPES

Laboratoire de Physique Nucléaire Théorique
Centre de Recherches Nucléaires et
Université Louis Pasteur de Strasbourg

Dans cet article nous nous proposons de discuter les idées physiques qui sont à la base du modèle des champs de jauge unifiés. En dépit des difficultés qu'on trouve actuellement à incorporer, d'une façon naturelle, les muons et hadrons dans ce modèle, on a bien le sentiment qu'on est sur une voie qui semble mener à la construction d'une théorie dans laquelle le champ électromagnétique de Maxwell et le champ des interactions faibles de Fermi ne sont que deux manifestations d'une seule entité physique sous-jacente - les champs de jauge unifiés.

1. INTRODUCTION

Les fondements de la théorie des interactions faibles ont été établis par Fermi[1]. Pour expliquer le spectre continu des électrons émis par les noyaux beta-radioactifs, Pauli avait suggéré que cette émission devait être accompagnée de celle d'une particule neutre, de spin $\frac{1}{2}$, très légère, de telle façon que le processus de la désintégration beta obéirait, lui aussi, aux lois de conservation d'énergie-impulsion, de moment angulaire et de charge. Ce processus consiste en la transformation d'un neutron en un proton et en la création simultanée d'une paire de particules, un électron et un neutrino de Pauli[2]. On sait aujourd'hui qu'il existe deux types de neutrinos : le neutrino-électronique,

Enz/Mehra (eds.), Physical Reality and Mathematical Description. 403–447. *All Rights Reserved.*

ν_e, et le neutrino-muonique, ν_μ; comme toute particule de spin $\frac{1}{2}$, chacun de ces deux neutrinos possède un anti-neutrino associé. C'est l'anti-neutrino $\bar{\nu}_e$, qui accompagne l'électron dans la transformation neutron-proton, le neutrino accompagnant le positron dans la transformation inverse :

$$\begin{aligned} n &\to p + e + \bar{\nu}_e \ , \\ p &\to n + \bar{e} + \nu_e \ . \end{aligned} \tag{1}$$

Fermi suggéra que l'interaction responsable de la réaction (1) était analogue à l'interaction électromagnétique, responsable de l'émission d'un photon, γ, par une particule chargée ε :

$$\varepsilon \to \varepsilon + \gamma \tag{2}$$

Pour la réaction (2), le lagrangien d'interaction s'écrit :

$$L_{(\gamma)} = e j^\mu_{(\gamma)}(x) A_\mu(x) \tag{3}$$

où $A_\mu(x)$ désigne le champ électromagnétique interagissant avec le courant électrique $j^\mu_{(\gamma)}(x)$ de la particule ε au point x. Si $\psi_\varepsilon(x)$ représente le spineur de Dirac qui décrit cette particule, le courant $j^\mu_{(\gamma)}(x)$ est donné par :

$$j^\mu_{(\gamma)}(x) = \bar{\psi}_\varepsilon(x)\gamma^\mu\psi_\varepsilon(x) \ . \tag{3a}$$

A ce courant Fermi a fait correspondre le courant $\bar{\psi}_p(x)\gamma^\mu\psi_n(x)$ associé à la transition d'un neutron en un proton, et au champ $A^\mu(x)$ qui décrit le photon, Fermi a simplement fait correspondre le courant associé à la paire e, $\bar{\nu}_e$, $\bar{\psi}_e(x)\gamma^\mu\psi_{\nu_e}(x)$. Le lagrangien de Fermi s'écrit alors :

$$L = G_F(\bar{\psi}_p(x)\gamma^\mu\psi_n(x)(\bar{\psi}_e(x)\gamma_\mu\psi_{\nu_e}(x)) \ . \tag{4}$$

Dans les années qui suivirent l'article de Fermi, on a cherché la forme du lagrangien mieux adapté à l'expérience puisque, en plus de l'interaction donnée par l'expression (4), d'autres termes étaient a priori possibles, d'après la théorie de Dirac. Comme celle-ci fournit seize formes correspondantes à la base de l'algèbre des matrices gamma, le lagrangien le plus général serait de la forme :

$$L = \sum_a G_a(\bar{\psi}_p(x)\Gamma^a\psi_n(x))(\bar{\psi}_e(x)\Gamma^a\psi_{\nu_e}(x)) \tag{5}$$

où Γ^a désignerait soit l'identité I, soit l'une des quatre matrices formées avec les matrices gamma :

$$\Gamma^a = I;\ \gamma^\mu;\ \frac{i}{2}\left[\gamma^\mu,\gamma^\nu\right];\ \gamma^\mu\gamma^5;\ i\gamma^5; \tag{6}$$

$$\mu,\nu = 0,1,2,3.$$

En 1949 on découvrit le méson π et le muon grâce à de nouvelles réactions suscitées par ces particules[3]. La désintégration du pion positif en un muon :

$$\pi^+ \to \mu^+ + \nu_\mu \tag{7}$$

s'accompagne de l'émission d'un neutrino-muonique ν_μ.
La désintégration du muon s'écrit :

$$\mu^+ \to \bar{\nu}_\mu + \bar{e} + \nu_e \tag{8}$$

et la capture du muon par des noyaux atomiques est due à la réaction :

$$\mu^- + p \to n + \nu_\mu\ . \tag{9}$$

Dans ces réactions nous avons admis que le muon positif était l'antiparticule et le muon négatif, la particule; l'hypothèse inverse est aussi faisable et a été formulée par Konopinski et Mahmoud[4].

Si l'on appelle G_β la constante d'interaction de Fermi correspondant à la réaction (1), G_μ la constante correspondant à la réaction (8) et $G_{\mu n}$ la constante correspondant à la réaction (9), ces réactions étant décrites par un lagrangien analogue au lagrangien (4), on a trouvé de façon qualitative l'universalité des interactions faibles[5] au sens suivant :

$$G_\beta \simeq G_\mu \simeq G_{\mu n} \tag{10}$$

On découvrit[6] en 1956 que les lois des interactions faibles n'étaient pas invariantes par rapport à une réflexion spatiale - la parité étant violée dans ces réactions. Ainsi, le lagrangien (5) devait comprendre non seulement des termes invariants du type $(\bar{\psi}_p\gamma^\mu\psi_n)(\bar{\psi}_e\gamma_\mu\psi_{\nu_e})$ mais aussi une combinaison de termes pseudoscalaires tels que $(\bar{\psi}_p\gamma^\mu\psi_n)(\bar{\psi}_e\gamma_\mu\gamma^5\psi_{\nu_e})$. Ce fut en 1958 que Feynman et Gell-Mann et Marshak et Sudarshan[7] trouvèrent la forme du lagrangien de Fermi décrivant les réactions faibles.

2. LE LAGRANGIEN D'INTERACTION COURANT-COURANT

De nos jours on donne au lagrangien effectif pour les interactions faibles, la forme d'une interaction courant-courant :

$$L = \frac{G_F}{\sqrt{2}} j^{\lambda+}(x) j_\lambda(x) . \qquad (11)$$

G_F est la constante de Fermi et $j^\lambda(x)$, le courant faible total, est la somme des courants faibles leptonique et hadronique :

$$j^\lambda(x) = \ell^\lambda(x) + h^\lambda(x) .$$

Le courant leptonique est réalisé comme somme d'une partie relative à l'électron et d'une partie relative au muon :

$$\ell^\lambda = \ell^\lambda_{(e)} + \ell^\lambda_{(\mu)} =$$
$$= \bar{\psi}_{\nu_e}(x)\gamma^\lambda(1-\gamma^5)\psi_e(x) + \bar{\psi}_{\nu_\mu}(x)\gamma^\lambda(1-\gamma^5)\psi_\mu(x) .$$

Le lagrangien L comprend ainsi une composante pour les réactions purement leptoniques : $L_{\ell\ell}$ (exemple : $\mu^- \to \nu_\mu + e + \bar{\nu}_e$), une composante pour les réactions semi-leptoniques $L_{\ell h}$ (exemple : $n \to p + e + \nu_e$ ou $\pi^+ \to \pi^0 + e + \nu_e$), et une composante pour les réactions ne mettant que les hadrons en jeu L_{hh} (exemple : $\Lambda \to p + \pi^-$)

$$L = L_{\ell\ell} + L_{\ell h} + L_{hh} ,$$

$$L_{\ell\ell} = \frac{G_F}{\sqrt{2}}(\ell^{\lambda+}_{(e)}\ell_{(e)\lambda} + \ell^{\lambda+}_{(e)}\ell_{(\mu)\lambda} + \ell^{\lambda+}_{(\mu)}\ell_{(e)\lambda} + \ell^{\lambda+}_{(\mu)}\ell_{(\mu)\lambda}) ,$$

$$L_{\ell h} = \frac{G_F}{\sqrt{2}}(\ell^{\lambda+}_{(e)}h_\lambda + \ell^{\lambda+}_{(\mu)}h_\lambda + h^{\lambda+}\ell_{(e)\lambda} + h^{\lambda+}\ell_{(\mu)\lambda}) ,$$

$$L_{hh} = \frac{G_F}{\sqrt{2}} h^{\lambda+} h_\lambda .$$

En raison des interactions fortes on ne peut pas donner une expression simple du courant hadronique en fonction d'opérateurs de champ pour les hadrons. On essaie d'obtenir les propriétés des éléments de matrice de ce courant au moyen de ses propriétés algébriques et grâce à des modèles comme le modèle des quarks.

On sait[8)] néanmoins que le courant hadronique a deux composantes - l'une pour les transitions conservant l'étrangeté, $h^\lambda_{(o)} = h^\lambda(\Delta s = 0)$, l'autre pour les transitions qui changent l'étrangeté $h^\lambda_{(1)} = h^\lambda(\Delta s \neq 0)$:

$$h^\lambda(x) = h^\lambda_{(o)}(x)\cos\theta_c + h^\lambda_{(1)}(x)\sin\theta_c$$

θ_c est une constante, l'angle de Cabibbo. On sait aussi que les deux parties du courant hadronique sont une superposition d'une composante vectorielle et d'une composante axiale

$$h^{\lambda}_{(o)}(x) = V^{\lambda}_{(o)}(x) - A^{\lambda}_{(o)}(x) \qquad (\Delta s = 0)$$

$$h^{\lambda}_{(1)}(x) = V^{\lambda}_{(1)}(x) - A^{\lambda}_{(1)}(x) \qquad (\Delta s = 1)$$

comme pour le courant leptonique.

L'angle θ_c a été introduit par Cabibbo pour tenir compte du fait que les réactions avec changement d'étrangeté ont une amplitude plus faible (exemple : $\Lambda \to p + e + \bar{\nu}_e$) que celles qui conservent l'étrangeté (exemple : $n \to p + e + \bar{\nu}_e$). L'universalité des interactions faibles est caractérisée par la première égalité des relations (10) et par le fait que le carré de la constante d'interaction, G_F^2, du courant faible leptonique est égal à la somme des carrés des constantes correspondantes au courant hadronique qui conserve l'étrangeté, $G_F^2\cos^2\theta_c$, et au courant hadronique avec changement d'étrangeté, $G_F^2\sin^2\theta_c$.

On a :

$$G_F = 1{,}02 \times 10^{-5} \, \frac{h^3}{m_P^2 \, c} \, ,$$

$$\theta_c \simeq 0{,}2 \, .$$

Les propriétés du courant hadronique sont étudiées dans le modèle des quarks.

Le modèle des quarks part de l'hypothèse que les hadrons sont des états liés de trois particules fondamentales, les quarks p, n, λ, décrits chacun par un spineur de Dirac et ayant les nombres quantiques-charge Q, hypercharge Y, nombre baryonique B et isospin I- suivants :

	Q	Y	B	I
p	2/3	1/3	1/3	1/2
n	-1/3	1/3	1/3	1/2
λ	-1/3	-2/3	1/3	0

(12)

Le triplet quark q(x)

$$q(x) = \begin{pmatrix} p(x) \\ n(x) \\ \lambda(x) \end{pmatrix} \tag{13}$$

se transforme sous le groupe SU(3) d'après les transformations unitaires infinitésimales (somme sur a de 1 à 8) :

$$q'(x) = (I + i\,\frac{\lambda_a}{2}\,\varepsilon_a)q(x) ,$$

où les matrices λ_a, a = 1,2 ..., 8 sont les matrices hermitiques, sans trace, à trois lignes et trois colonnes, les ε_a étant les paramètres infinitésimaux de la transformation.

Si l'on admet que les masses des quarks sont égales, m_q, l'équation de mouvement du quark libre :

$$(i\gamma^\mu\partial_\mu - m)\, q(x) = 0$$

conduit à la construction du courant conservé :

$$V^\mu_a(x) = \bar{q}(x)\gamma^\mu\,\frac{\lambda_a}{2}\,q(x)$$

$$\partial^\mu V_{\mu a}(x) = 0$$

et du courant axial non-conservé :

$$A^\mu_a(x) = \bar{q}(x)\gamma^\mu\gamma^5\,\frac{\lambda_a}{2}\,q(x) ,$$

$$\partial^\mu A_{\mu a}(x) = m_q i\bar{q}(x)\gamma^5\lambda_a q(x) .$$

Le courant électromagnétique est

$$h^\mu_{(\gamma)}(x) = V^\mu_3(x) + \frac{1}{\sqrt{3}}\,V^\mu_8(x)$$

et le courant faible s'écrit :

$$V^\mu_{(0)} = V^\mu_1 + iV^\mu_2 ,$$

$$V^{\mu+}_{(0)} = V^\mu_1 - iV^\mu_2 ,$$

$$A^\mu_{(0)} = A^\mu_1 + iA^\mu_2 ,$$

$$A^{\mu+}_{(0)} = A^{\mu}_{1} - iA^{\mu}_{2} ,$$
$$V^{\mu}_{(1)} = V^{\mu}_{4} + iV^{\mu}_{5} ,$$
$$V^{\mu+}_{(1)} = V^{\mu}_{4} - iV^{\mu}_{5} ,$$
$$A^{\mu}_{(1)} = A^{\mu}_{4} + iA^{\mu}_{5} ,$$
$$A^{\mu+}_{(1)} = A^{\mu}_{4} - iA^{\mu}_{5} .$$

Le fait que le courant leptonique contienne l'opérateur $1 - \gamma^5$ signifie qu'il existe uniquement des neutrinos polarisés à gauche et des anti-neutrinos polarisés à droite.

Pour distinguer l'électron (et le neutrino électronique) du muon (du neutrino muonique) l'on a introduit un nombre quantique L_e pour le premier, L_μ, pour le deuxième et l'on admet que les leptons ont les nombres quantiques suivants :

	Q	L_e	L_μ	I
ν_e	0	1	0	1/2
e	-1	1	0	1/2
ν_μ	0	0	1	1/2
μ^-	-1	0	1	1/2

Le schéma de Konopinski et Mahmoud remplace μ^- par μ^+ dans cette table. Si l'on introduit les isospineurs (le schéma ci-dessus demande un isovecteur

$\begin{pmatrix} \mu^+ \\ \nu_- \\ e^- \end{pmatrix}$) :
$$\psi_\ell(x) = \begin{pmatrix} \nu_\ell(x) \\ \ell(x) \end{pmatrix} \qquad \text{avec } \ell = e,\mu$$

le courant faible s'écrit :

$$j^\lambda(x) = \ell^\lambda(x) + h^\lambda(x) ,$$
$$\ell^\lambda(x) = \sum_{\ell=e,\mu} \bar{\psi}_\ell(x)\gamma^\lambda(1-\gamma^5)\tau^+\psi_\ell(x), \tag{14}$$
$$h^\lambda(x) = \bar{q}(x)\gamma^\lambda(1-\gamma^5)\tau^+_\theta q(x) ,$$

avec :

$$\tau^+ = \begin{pmatrix} 0 & 1 \\ 0 & 0 \end{pmatrix}$$

et :

$$\tau_\theta^+ = \begin{pmatrix} 0 & \cos\theta_c & \sin\theta_c \\ 0 & 0 & 0 \\ 0 & 0 & 0 \end{pmatrix}$$

3. LES BOSONS INTERMEDIAIRES DES INTERACTIONS FAIBLES

Après la suggestion initiale de Fermi, on avait essayé de décrire les interactions faibles au moyen de composantes scalaire, vectorielle, pseudo-scalaire, axiale et tensorielle du lagrangien, d'après une superposition du type donné par l'équation (5), les combinaisons V-A et S-T+P étant invariantes par rapport au réarrangement de Fierz. Il n'y avait ainsi aucune raison de tenir à l'idée de bosons intermédiaires qui seraient les responsables des interactions faibles - l'idée de postuler un grand nombre de champs intermédiaires et de constantes d'interaction n'est pas satisfaisante.

La conception de Yukawa[9)] d'associer les pions aux interactions faibles aussi bien qu'aux interactions fortes n'avait pas abouti puisque bien que les pions donnent lieu à une interaction faible pseudo-scalaire induite ils ne peuvent pas décrire les interactions de Fermi[10)].

Dès le moment néanmoins où Feynman et Gell-Mann et Marshak et Sudarshan montrèrent que le courant faible était un quadrivecteur, l'analogie avec l'électrodynamique devint plus frappante : on pensa que l'interaction locale de Fermi courant-courant pouvait bien être due à un échange d'un boson vectoriel lourd entre les courants.

Si l'on considère par exemple la désintégration du muon

$$\mu^- \to \nu_\mu + e + \bar{\nu}_e ,$$

le graphe de Feynman pour l'interaction locale courant-courant (fig. 1) :

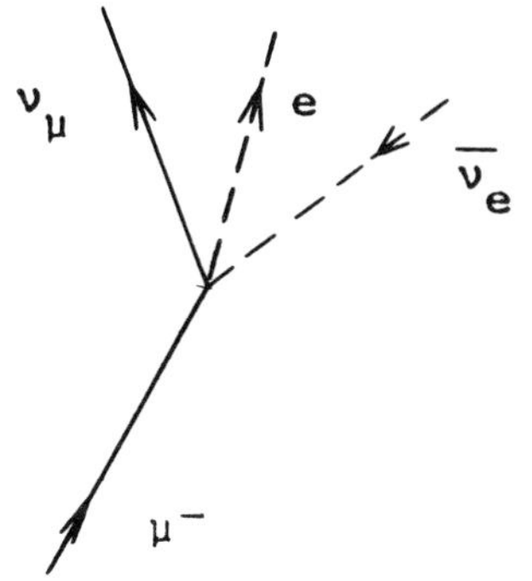

Figure 1

conduit à l'amplitude S de la réaction, qui s'écrit en première approximation, d'après les règles de Feynman[11]

$$S = -\frac{iG_F}{2}\int d^4x(\bar{\nu}_\mu(x)\gamma^\alpha(1-\gamma^5)\mu(x))(\bar{e}(x)\gamma_\alpha(1-\gamma^5)\nu_e(x)) \ .$$

L'idée que l'interaction est le résultat de l'échange d'un méson lourd W, de masse m_W, entre les courants, nous conduit à remplacer ce graphe par le diagramme de la figure 2 :

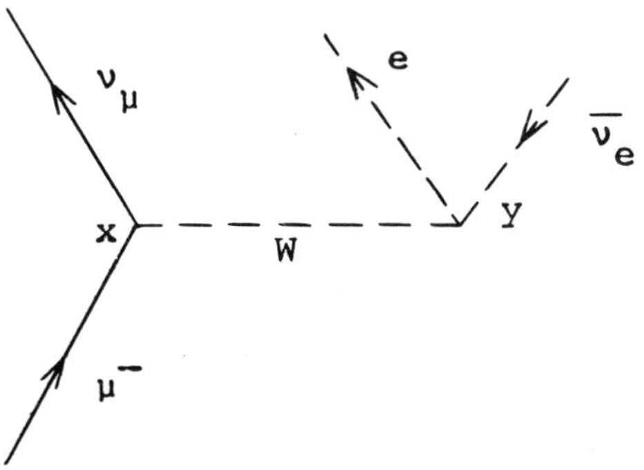

Figure 2

Si l'on désigne par $\Delta_F^{\alpha\beta}(x-y)$ le propagateur de Feynman du champ vectoriel $W^\alpha(x)$ associé aux mésons W, on aura pour l'amplitude correspondant à ce dernier diagramme l'expression

$$S'=-ig_w^2\iint d^4x d^4y(\bar{\nu}_\mu(x)\gamma^\alpha(1-\gamma^5)\mu(x))(\Delta_F(x-y))_{\alpha\beta}(\bar{e}(y)\gamma^\beta(1-\gamma^5)\nu_e(x)).$$

La constante g_w est la constante d'interaction entre les courants et le champ W^α.

Un champ vectoriel $W^\alpha(x)$ doté d'une masse m_w satisfait à l'équation :

$$\partial_\beta G^{\alpha\beta}(x) + m_w^2 W^\alpha(x) = \rho^\alpha(x) \ ,$$

où $\rho^\alpha(x)$ est le courant, source du champ et

$$G^{\alpha\beta}(x) = \partial^\beta W^\alpha(x) - \partial^\alpha W^\beta(x) \ .$$

Cette équation s'écrit

$$P_{\alpha\beta} W^\beta(x) = \rho_\alpha(x) \tag{15}$$

où

$$P_{\alpha\beta} = (\Box + m_w^2) g_{\alpha\beta} - \partial_\alpha \partial_\beta$$

est l'opérateur que nous appellerons opérateur de Proca.

Dans le vide, $\rho_\alpha(x) = 0$, l'équation

$$P_{\alpha\beta} W^\beta(x) = 0$$

doit conduire à l'équation de Klein-Gordon en raison de la relation relativiste entre l'énergie et l'impulsion. Il doit donc exister un opérateur $\pi^{\alpha\beta}$, l'opérateur de Peierls, tel que

$$\pi^{\alpha\lambda} P_{\lambda\beta} = \delta^\alpha{}_\beta (\Box + m_w^2) \ . \tag{16}$$

Une fonction de Green de l'équation de Proca $\Gamma^{\alpha\beta}(x-x')$, obéira à l'équation

$$P_{\alpha\beta} \Gamma^{\beta\lambda}(x-x') = \delta_\alpha{}^{\lambda(4)}(x-x') \ .$$

Posons :

$$\Gamma^{\alpha\beta}(x-x') = \pi^{\alpha\beta}\Gamma(x-x')$$

où $\Gamma(x-x')$ est la fonction de Green correspondante de l'équation de Klein-Gordon; on aura alors encore la relation :

$$P_{\alpha\lambda}\pi^{\lambda\beta} = \delta_\alpha{}^\beta(\Box + m_w^2) \ .$$

De la définition de l'opérateur $P_{\alpha\beta}$ on déduit

$$\partial^\alpha\partial^\beta P_{\beta\lambda} = m_w^2\ \partial^\alpha\partial_\lambda$$

et par conséquent :

$$P_{\alpha\beta} + \frac{1}{m_w^2}\ \partial_\alpha\partial^\lambda P_{\lambda\beta} = (\Box + m_w^2)g_{\alpha\beta} \ . \tag{17}$$

Les équations (16) et (17) montrent que :

$$\pi^{\alpha\beta} = g^{\alpha\beta} + \frac{1}{m_w^2}\ \partial^\alpha\partial^\beta \ .$$

Une fonction de Green du champ de Proca s'obtient donc à partir d'une fonction de Green du champ de Klein-Gordon d'après la relation :

$$\Gamma^{\alpha\beta}(x-x') = (g^{\alpha\beta} + \frac{1}{m_w^2}\ \partial^\alpha\partial^\beta\ \Gamma(x-x') \ .$$

Le propagateur du méson vectoriel est donc :

$$\Delta_F^{\alpha\beta}(x-y) = (g^{\alpha\beta} + \frac{1}{m_w^2}\ \partial^\alpha\partial^\beta)\Delta_F(x-y)$$

où $\Delta_F(x-x')$ est la fonction de Feynman, c'est-à-dire :

$$\Delta_F^{\alpha\beta}(x-y) = -\int\frac{d^4k}{(2\pi)^4}(g^{\alpha\beta} - \frac{1}{m_w^2}\ k^\alpha k^\beta)\ \frac{e^{-ik(x-y)}}{k^2 - m_w^2 + i\varepsilon}$$

Dans l'espace des impulsions on obtiendra ainsi pour l'amplitude S' :

$$S' = (2\pi)^4 \delta^{(4)}(p_{\nu_\mu} + p_e + p_{\bar{\nu}_e} - p_\mu) \frac{(\bar{\nu}(p_{\nu\mu})(-ig_w\gamma^\alpha(1-\gamma^5)\mu(p_\mu))}{V(2E_{\nu_\mu} 2E_\mu)^{\frac{1}{2}}} \cdot$$

$$\cdot \frac{i(g_{\alpha\beta} - \frac{1}{m_w^2} k_\alpha k_\beta)}{k^2 - m_w^2 + i\varepsilon} \frac{(\bar{e}(p_e)(-ig_w\gamma^\beta(1-\gamma^5)\nu(-p_{\nu_e}))}{V(2E_{\nu_e} 2E_e)^{\frac{1}{2}}}$$

L'expression correspondante pour l'amplitude S sera approchée par cette formule de S' si l'on admet que le transfert d'impulsion

$$k = p_{\nu_\mu} - p_\mu$$

est très faible par rapport à la masse m_w :

$$k^2 << m_w^2$$

L'identification de S et S' dans cette approximation conduit à la relation :

$$\frac{g_w^2}{m_w^2} = \frac{G_F}{\sqrt{2}}$$

entre la constante g_w, la masse m_w et la constante de Fermi.

4. INDICATIONS D'UNE POSSIBLE UNIFICATION DES INTERACTIONS FAIBLES ET ELECTROMAGNETIQUES

Pour calculer la masse m_w on doit connaître la constante d'interaction g_w. Puisque le méson W est vectoriel, l'auteur[12)] a suggéré en 1958 que cette constante d'interaction g_w était égale à la charge e, constante d'interaction d'un autre boson vectoriel, le photon, avec la matière :

$$g_w = e$$

Dans ces conditions la masse m_w a une valeur de l'ordre de 40 masses protoniques.

Dans le même article, l'auteur suggérait l'existence de mésons vectoriels chargés $W^\pm$ et de mésons vectoriels neutres W^0 et essayait sans succès d'éliminer les réactions non observées produites par les courants neutres. Par conséquent, l'hypothèse

g_W = e indiquait pour la première fois que les mésons W et les photons appartenaient à la même famille et que les interactions électromagnétiques et faibles avaient la même origine. Nous verrons plus avant le développement élégant et précis de ces idées élaboré par Weinberg et par Salam et Ward[13)].

La théorie de Fermi n'est pas renormalisable : les intégrales divergentes des amplitudes d'ordre supérieur au premier ne sont pas absorbées dans la renormalisation de la masse et de la constante d'interaction. La théorie des mésons vectoriels qui ont une masse n'est pas renormalisable non plus : elle n'a pas d'invariance de jauge et le propagateur de Feynman possède un terme quadratique dans l'impulsion des mésons virtuels qui contribue à la divergence des amplitudes d'ordre supérieur au premier.

L'idée de Salam et Ward et de Weinberg a été de décrire les bosons vectoriels par des champs à masse nulle, ayant une invariance de jauge de Yang-Mills, et donc renormalisables. Et ensuite d'introduire une cassure spontanée de la symétrie pour établir, d'après le mécanisme de Higgs, la masse de ces mésons.

Un progrès important a été accompli par 't Hooft et Veltman[14)] (et Bollini et Giambiagi[15)]), qui ont démontré que la théorie reste renormalisable - si la renormalisabilité d'une théorie n'est pas nécessairement un critère de sa véracité il est, néanmoins, un critère nécessaire à la vérification de la théorie, au moyen du calcul, d'effets observables.

5. L'ALGEBRE DES COURANTS

On peut établir une relation entre les constantes g_W et e si l'on impose au lagrangien des interactions faibles et électromagnétiques de s'écrire sous la forme d'indépendance de charge; soit

$$L = L_W + L_\gamma \tag{18}$$

où :

$$L_\gamma = ej^\mu_{(\gamma)}(x)A_\mu(x)$$

est le lagrangien d'interaction électromagnétique avec :

$$j^\mu_{(\gamma)}(x) = \ell^\mu_{(\gamma)}(x) + h^\mu_{(\gamma)}(x) ,$$

$$\ell^{\lambda}_{(\gamma)}(x) = \overline{\psi}_e(x)\gamma^{\lambda}\psi_e(x) + \overline{\psi}_{\mu}(x)\gamma^{\lambda}\psi_{\mu}(x)$$

où $h^{\mu}_{(\gamma)}(x)$ désigne le courant électromagnétique des hadrons et $\ell^{\mu}_{(\gamma)}(x)$ le courant électromagnétique leptonique;

$$L_w = g_w(j^{\lambda}_w(x)W^{+}_{\lambda}(x) + j^{\lambda +}_w(x)W_{\lambda}(x)) + g_o j^{\lambda}_{(o)}(x)W_{(o)\lambda}(x)$$

où $j^{\lambda}_{(o)}(x)$ indique un possible courant faible neutre et $W^{\lambda}_{(o)}(x)$ un champ vectoriel hermitique.

Ecrire le lagrangien L sous la forme d'indépendance de charge revient à l'écrire

$$L = g\sum_a j^{\lambda}_a(x)\, W_{\lambda a}(x)$$

où g est fonction de g_w, g_o et e et les quatre champs $W_{\lambda a}$ et courants $j_{\lambda a}$ sont fonctions des champs et courants du lagrangien (18). La charge électrique des champs leptoniques est

$$Q^{(\ell)}_{(\gamma)} = -\int d^3x(\psi^{+}_e(x)\psi_e(x) + \psi^{+}_{\mu}(x)\psi_{\mu}(x)) .$$

Par analogie avec cet opérateur on définit la <u>charge faible</u> des champs leptoniques :

$$Q^{(\ell)}_W = \int d^3x(\psi^{+}_{\nu e}(1-\gamma^5)\psi_e + \psi^{+}_{\nu_{\mu}}(1-\gamma^5)\psi_{\mu})$$

et son adjoint :

$$Q^{(\ell)+}_W = \int d^3x(\psi^{+}_e(1-\gamma^5)\psi_{\nu e} + \psi^{+}_{\mu}(1-\gamma^5)\psi_{\nu_{\mu}}) .$$

Ces opérateurs dépendent du temps puisque les courants associés ne se conservent pas. A partir des règles de commutation des champs leptoniques

$$\{\psi_{\ell}(x),\psi_{\ell}(x')\}_o = \{\psi_{\nu_{\ell}}(x),\ \psi_{\nu_{\ell}}(x')\}_o = \delta(\underline{x}-\underline{x}')$$

(où $\ell = e,\mu$ et où $\{\ \}_o$ désigne l'anticommutateur pour $x_o = x'_o$, les anticommutateurs entre les $\psi(x)$ et $\psi(x')$ et entre les $\psi^{+}(x)$ et $\psi^{+}(x')$ étant nuls) et de l'identité :

$$\left[AB,CD\right] = -AC\{D,B\} + A\{C,B\}\, D + \{C,A\}\, DB - C\,\{A,D\}\, B$$

on trouve les commutateurs suivants :

$$\left[\psi^+_{\nu_\ell}(x)\psi_\ell(x),\ \psi^+_\ell(x')\psi_{\nu_\ell}(x')\right]_o =$$

$$= \left[\psi^+_{\nu_\ell}(x)\gamma^5\psi_\ell(x),\ \psi^+_\ell(x')\gamma^5\psi_{\nu_\ell}(x')\right]_o =$$

$$= (\psi^+_{\nu_\ell}(x)\psi_{\nu_\ell}(x') - \psi^+_\ell(x')\psi_\ell(x))_o\ \delta(\underline{x}-\underline{x}') ,$$

$$\left[\psi^+_{\nu_\ell}(x)\psi_\ell(x),\ \psi^+_\ell(x')\gamma^5\psi_{\nu_\ell}(x')\right]_o =$$

$$= (\psi^+_{\nu_\ell}(x)\gamma^5\psi_{\nu_\ell}(x') - \psi^+_\ell(x')\gamma^5\psi_\ell(x'))_o\ \delta(\underline{x}-\underline{x}')$$

avec $\ell = e,\mu$.

On peut alors déterminer aisément le commutateur de la charge faible leptonique $Q_w^{(\ell)}$ avec son adjoint $\overline{Q}_w^{(\ell)}$.

On trouve

$$\left[Q_w^{(\ell)}\ ,\ Q_w^{(\ell)+}\right] = 2Q_{w3}^{(\ell)}$$

où ces opérateurs sont pris au même instant et :

$$Q_{w3}^{(\ell)} = \int d^3x \sum_{\ell=e,\mu} (\psi^+_{\nu_\ell}(x)(1-\gamma^5)\psi_{\nu_\ell}(x) - \psi^+_\ell(x)(1-\gamma^5)\psi_\ell(x)) .$$

On trouve encore :

$$\left[Q_{w3}^{(\ell)}\ ,\ Q_w^{(\ell)}\right] = 4Q_w^{(\ell)} ,$$

$$\left[Q_{w3}^{(\ell)}\ ,\ Q_w^{(\ell)+}\right] = -4Q_w^{(\ell)+} .$$

En posant

$$Q_w^{(\ell)} = 2K_+ ,$$

$$Q_w^{(\ell)+} = 2K_+^+ = 2K_- ,$$

$$Q_{w3}^{(\ell)} = 4K_3$$

ces commutateurs deviennent :

$$\left[K_+ , K_-\right] = 2K_3 ,$$
$$\left[K_3 , K_+\right] = K_+ ,$$
$$\left[K_3 , K_-\right] = -K_- .$$

Ils définissent l'algèbre Su(2), l'algèbre des moments angulaires. Etant donnée l'expression de K_+, K_-, K_3

$$K_+ = \frac{1}{2}\int d^3x \sum_\ell \psi^+_{\nu_\ell} (1-\gamma^5)\psi_\ell ,$$
$$K_- = \frac{1}{2}\int d^3x \sum_\ell \psi^+_\ell (1-\gamma^5)\psi_{\nu_\ell} ,$$
$$K_3 = \frac{1}{4}\int d^3x \sum_\ell \psi^+_{\nu_\ell}(1-\gamma^5) \psi_{\nu_\ell} - \psi^+(1-\gamma^5)\psi_\ell) ,$$

l'on peut utiliser le formalisme de l'isospin et poser

$$\psi_\ell(x) = \begin{pmatrix} \psi_{\nu_\ell}(x) \\ \psi_\ell(x) \end{pmatrix} , \ \ell = e,\mu.$$

Il viendra :

$$K_+ = \frac{1}{2}\int d^3x \sum_\ell \psi^+_\ell(1-\gamma^5) \frac{\tau_+}{2} \psi_\ell ,$$
$$K_- = \frac{1}{2}\int d^3x \sum_\ell \psi^+_\ell(1-\gamma^5) \frac{\tau_-}{2} \psi_\ell ,$$
$$K_3 = \frac{1}{2}\int d^3x \sum_\ell \psi^+_\ell(1-\gamma^5) \frac{\tau_3}{2} \psi_\ell$$

où les matrices d'isospin

$$\tau_+ = \tau_1 + i\tau_2 , \quad \tau_- = \tau_1 - i\tau_2 , \quad \tau_3$$

agissent sur l'isospineur $\psi_\ell(x)$.

Si l'on pose donc

$$K_+ = K_1 + iK_2 \ , \ K_- = K_1 - iK_2$$

on obtient l'isovecteur

$$K_a = \frac{1}{2}\int d^3x \sum_\ell \psi_\ell^+(1-\gamma^5)\ \frac{\tau_a}{2}\ \psi_\ell \qquad a = 1, 2, 3$$

qui satisfait aux commutateurs

$$\left[K_a, K_b\right] = i\,\varepsilon_{abc}\, K_c \tag{19}$$

ε_{abc} est le tenseur totalement antisymétrique.

Pour montrer qu'on a affaire ici, en réalité, à une algèbre $SU(2)\oplus SU(2)$, il faut expliciter les composantes polaire et axiale des K_a

$$K_a = \frac{1}{2}\,(K_a^V - K_a^A)$$

où

$$K_a^V = \sum_\ell \int d^3x\ \psi_\ell^+\ \frac{\tau_a}{2}\ \psi_\ell\ ,$$

$$K_a^A = \sum_\ell \int d^3x\ \psi_\ell \gamma^5\ \frac{\tau_a}{2}\ \psi_\ell\ ,$$

Si l'on introduit alors l'isovecteur :

$$L_a = \frac{1}{2}\,(K_a^V + K_a^A)$$

les règles de commutation (19) seront satisfaites par les commutateurs :

$$\left[K_a^V, K_b^V\right] = \left[K_a^A, K_b^A\right] = i\,\varepsilon_{abc}\, K_c^V\ ,$$

$$\left[K_a^V, K_b^A\right] = i\,\varepsilon_{abc}\, K_c^A$$

et donc :

$$\left[L_a, L_b\right] = i\,\varepsilon_{abc}\, L_c$$

$$\left[K_a, K_b\right] = i\,\varepsilon_{abc}\, K_c$$

$$\left[L_a, K_b\right] = 0\ , \qquad a,b = 1, 2, 3$$

qui définissent bien l'algèbre $SU(2)\oplus SU(2)$.

Ainsi donc les charges faibles $Q_w^{(\ell)}$, $Q_w^{(\ell)+}$, $Q_{w3}^{(\ell)}$ se transforment comme $2K_+$, $2K_-$, $4K_3$ respectivement. Pour voir les propriétés analogues de la charge électrique, définissons les nombres leptoniques

$$N_\ell = \int d^3x\ \psi_\ell^+ \psi_\ell\ , \qquad \ell = e,\mu.$$

La charge $Q_{(\gamma)}^{(\ell)}$ s'écrit :

$$Q_{(\gamma)}^{(\ell)} = -\int d^3x \sum_\ell\ \psi_\ell^+ \frac{1-\tau_3}{2} \psi_\ell\ ,$$

c'est-à-dire

$$Q_{(\gamma)}^{(\ell)} = K_3^V - \frac{1}{2}\sum_\ell N_\ell\ . \tag{20}$$

Comme on le sait, l'expérience suggère que les nombres leptoniques N_ℓ se conservent. Par conséquent, les N_ℓ doivent commuter avec les opérateurs qui décrivent des variables physiques. Sinon, on déduirait de l'hypothèse

$$\left[N_\ell,\ \Omega\right] \neq 0$$

que

$$<\ell|\left[N_\ell,\ \Omega\right]|\ell'> \neq 0$$

où ℓ est un état propre de l'opérateur N_ℓ avec valeur propre n_ℓ

$$(n_\ell - n_{\ell'})\ <\ell|\Omega|\ell'> \neq 0\ .$$

Par conséquent, si $n_\ell \neq n_{\ell'}$, on aurait $<\ell|\Omega|_{\ell'}> \neq 0$ et l'on pourrait produire une transition de l'état ℓ à l'état ℓ' ce qui est impossible si l'on admet la règle de super-sélection pour les nombres leptoniques. On déduit de la propriété de N_ℓ que ces opérateurs sont des multiples de l'identité, que la charge $Q_{(\gamma)}^{(\ell)}$ commute de la manière suivante avec K_+, K_-, K_3 :

$$\begin{aligned}\left[Q_{(\gamma)}^{(\ell)}\ ,\ K_+\right] &= \left[K_3^V,\ K_+\right] = \\ &= \left[K_3 + L_3,\ K_+\right] = \\ &= \left[K_3,\ K_+\right] =\end{aligned}$$

$$= K_+ ,$$

$$\left[Q^{(\ell)}_{(\gamma)}, K_-\right] = -K_-$$

$$\left[Q^{(\ell)}_{(\gamma)}, K_3\right] = 0 .$$

On en conclut que $Q^{(\ell)}_{(\gamma)}$ est de la forme :

$$Q^{(\ell)}_{(\gamma)} = K_o + K_3$$

où K_o est un opérateur qui commute avec K_1, K_2, K_3

$$\left[K_o, K_a\right] = 0 \quad , \quad a = 1, 2, 3$$

et K_o, K_1, K_2, K_3 engendrent l'algèbre $U(1)\oplus SU(2)$.

En effet de l'expression (20) de $Q^{(\ell)}_{(\gamma)}$ on peut déduir

$$Q^{(\ell)}_{(\gamma)} = L_3 + K_3 - \frac{1}{2}\sum_\ell N_\ell$$

d'où

$$K_o = L_3 - \frac{1}{2}\sum_\ell N_\ell$$

6. UN LAGRANGIEN INDEPENDANT DE CHARGE ET LA RELATION ENTRE g_W ET e

Le résultat de ces manipulations est que $Q^{(\ell)}_W$ se transforme comme $2K_+$, $Q^{(\ell)+}_W$ comme $2K_-$ et $Q^{(\ell)}_{(\gamma)}$ comme $K_o + K_3$.

La définition de ces opérateurs nous amène à dire que, par une rotation du spin -K, le courant ℓ^μ_W se transforme comme $2K_+$, $\ell^{\mu+}_W$ comme $2K_-$ et $\ell^\mu_{(\gamma)}$ comme $K_o + K_3$.

On supposera que cette propriété est valable pour les courants hadroniques correspondants et donc pour les courants totaux:

$j^\mu_W(x)$ se transforme comme $2K_+$,

$j^\mu_W(x)$ " " " $2K_-$,

$j^\mu_{(\gamma)}(x)$" " " $K_o + K_3$.

Nous écrivons donc, η étant un nombre :

$$j_w^\mu(x) = \eta(j_1^\mu(x) + i\, j_2^\mu(x)),$$
$$j_w^{\mu+}(x) = \eta(j_1^\mu(x) - i\, j_2^\mu(x)),$$
$$j_{(\gamma)}^\mu(x) = j_3^\mu(x) + j_o^\mu(x),$$

Posons :

$$j_{(o)}^\mu(x) = -j_3^\mu(x) + j_o^\mu(x) .$$

On introduira alors les champs hermitiques $W_1(x)$, $W_2(x)$, $W_3(x)$, $V(x)$ par les relations

$$W^\lambda(x) = \frac{1}{\sqrt{2}}\,(W_1^\lambda(x) + i\, W_2^\lambda(x)),$$
$$W^{\lambda+}(x) = \frac{1}{\sqrt{2}}\,(W_1^\lambda(x) - i\, W_2^\lambda(x)),$$
$$A^\lambda(x) = \frac{1}{2}\,(W_3^\lambda(x) + V^\lambda(x)),$$
$$W_{(o)}^\lambda(x) = \frac{1}{2}\,(W_3^\lambda(x) - V^\lambda(x))$$

et le lagrangien

$$L = e j_{(\gamma)}^\mu A_\mu + g_w(j_w^\lambda W_\lambda^+ + j_w^{\lambda+} W_\lambda) + g_o j_{(o)}^\lambda W_{(o)\lambda}$$

deviendra :

$$L = e\,(j_1^\lambda W_{1\lambda} + j_2^\lambda W_{2\lambda} + j_3^\lambda W_{3\lambda} + j_o^\lambda V_\lambda)$$

dès le moment où les constantes d'interaction satisfont aux relations

$$e = -g_o = \frac{2\eta g_w}{\sqrt{2}} \qquad (21)$$

Pour mettre sur un pied d'égalité les interactions électromagnétiques et les interactions faibles sous la forme d'un lagrangien indépendant de la charge, on devrait donc avoir une relation entre la charge et les constantes d'interaction g_w et g_o, du type indiqué ci-dessus.

Si l'on pose $\eta = \frac{\sqrt{2}}{2}$ dans l'équation (21) on obtient[12) l'égalité $g_w = e$. Pour $\eta = 2$ la relation

$$e = \frac{4g_w}{\sqrt{2}}$$

a été établie par T.D. Lee[16)]. Nous verrons que le modèle de Salam et Weinberg relie les champs W^λ et A^λ à un champ de Yang-Mills $\mathcal{A}_a^\lambda$ et à un champ vectoriel B^λ. La relation (21) sera remplacée par une autre plus générale (voir (33)).

7. LE CHAMP DE YANG-MILLS

Il faut maintenant que nous précisions la nature du champ de jauge de Yang-Mills[17)]. Ce champ est un champ vectoriel avec isospin égal à 1 et à masse nulle, qui interagit avec un courant d'isospin qui se conserve. Si l'on désigne par $\mathcal{A}_a^\mu(x)$, a = 1,2,3, un triplet de champs vectoriels, les équations de ces champs seront

$$\partial_\nu G_a^{\mu\nu}(x) = j_a^\mu(x) \ , \qquad a = 1,2,3$$

où $j_a^\mu(x)$ est le courant d'isospin. Puisque $G_a^{\mu\nu}(x)$ est un tenseur antisymétrique, $j_a^\mu(x)$ sera un courant conservé. L'équation d'un spineur de masse M qui interagit avec $\mathcal{A}_a^\mu(x)$ doit être du type :

$$(i\gamma^\mu\partial_\mu - M)\psi(x) = g\gamma^\lambda \frac{\tau_a}{2} \psi(x)\mathcal{A}_{a\lambda}(x)$$

g étant la constante d'interaction (somme sous-entendue sur a, de 1 à 3).

Les règles de Yang-Mills pour trouver ces équations et la forme du tenseur $G_a^{\mu\nu}(x)$ sont les suivantes :

a) remplacez $i\partial_\mu\psi(x)$ pour un spineur libre

par
$$\left[i\partial_\mu - g\mathcal{A}_{a\mu}(x)\frac{\tau_a}{2}\right]\psi(x) \qquad (22)$$

pour un spineur en interaction avec $\mathcal{A}_a^\mu$;

b) remplacez $i\partial^\mu\mathcal{A}_a^\nu(x)$ par :

$$\left[i\partial^\mu - \frac{1}{2}g\,\mathcal{A}_b^\mu L_b\right]\mathcal{A}_a^\nu(x) \qquad (23)$$

pour le champ vectoriel, où

$$(L_b)_{ac} = i\,\varepsilon_{abc}$$

ε_{abc} est le tenseur totalement antisymétrique.

Le lagrangien de Yang-Mills s'écrit :

$$L = -\frac{1}{4}\,G_a^{\mu\nu}G_{a\mu\nu} + \bar{\psi}\gamma^\mu(i\partial_\mu - g\mathcal{a}_{a\mu}\frac{\tau_a}{a})\psi - M\,\bar{\psi}\,\psi \tag{24}$$

où

$$G_a^{\mu\nu} = \partial^\nu\mathcal{a}_a^\mu - \partial^\mu\mathcal{a}_a^\nu - g\varepsilon_{abc}\,\mathcal{a}_b^\nu\,\mathcal{a}_c^\mu$$

contient une auto-interaction du champ $\mathcal{a}_a^\mu$.

Le courant d'isospin est la somme d'une partie provenant du spineur et d'une autre partie provenant du champ vectoriel :

$$j_a^\mu(x) = g\{\bar{\psi}\gamma^\mu\,\frac{\tau_a}{2}\,\psi - \varepsilon_{abc}\,G_b^{\mu\nu}\,\mathcal{a}_{c\nu}\}\ .$$

Le lagrangien ci-dessus est invariant par rapport aux transformations de jauge suivantes :

$$\begin{aligned} \mathcal{a}_a^{\mu\prime}(x) &= \mathcal{a}_a^\mu(x) - \partial^\mu\Lambda_a(x) - g\varepsilon_{abc}\Lambda_b(x)\,\mathcal{a}_c^\mu(x)\ ,\\ \psi'(x) &= (I + ig\Lambda_a(x)\,\frac{\tau_a}{2})\ \psi(x) \end{aligned} \tag{25}$$

où les fonctions de jauge sont soumises à l'équation

$$\Box\Lambda_a(x) + g\varepsilon_{abc}\ \partial_\mu\ \Lambda_b(x)\cdot\mathcal{a}_c^\mu(x) = 0$$

si le champ vectoriel satisfait à la condition de Lorentz.

Ni le courant ni le tenseur du champ ne sont des invariants de jauge :

$$j_a^{\mu\prime}(x) = j_a^\mu(x) - g\varepsilon_{abc}\{\Lambda_b(x)j_c^\mu(x) + \partial_\nu\Lambda_b(x)\cdot G_c^{\mu\nu}(x)\}\ ,$$

$$G_a^{\mu\nu\prime}(x) = G_a^{\mu\nu}(x) - g\varepsilon_{abc}\Lambda_b(x)\ G_c^{\mu\nu}(x)\ ,$$

mais le lagrangien (24) est invariant.

Le champ de Yang-Mills est donc un champ de jauge plus général que le champ électromagnétique. Les transformations de jauge de l'électrodynamique

$$A'^{\mu}(x) = A^{\mu}(x) - \partial^{\mu}\Lambda(x)$$

$$\psi'(x) = (I + ie\Lambda(x))\psi(x)$$

sont généralisées par les transformations (25). A la dérivée covariante électromagnétique $i\partial_{\mu} - eA_{\mu}$ correspond la dérivée covariante de Yang-Mills (22); pour le champ électromagnétique le produit vectoriel correspondant à celui qui apparaît dans l'opération (23) est évidemment nul.

8. LE MODELE DES CHAMPS DE JAUGE UNIFIES

Tournons-nous à présent vers le modèle des champs de jauge unifiés dans le cas simple de l'interaction faible du champ électron-neutrino électronique avec lui-même.

Tout d'abord remarquons qu'étant donnée l'équation d'un fermion libre de masse m

$$(i\gamma^{\alpha}\partial_{\alpha} - m)\,\psi(x) = 0$$

sa masse peut être considérée comme le résultat de l'interaction d'un tel fermion sans masse avec un champ scalaire $\phi(x)$

$$(i\gamma^{\alpha}\partial_{\alpha} - g\phi(x))\,\psi(x) = 0$$

si on admet que ce champ scalaire a une valeur moyenne dans le vide différente de zéro et réelle :

$$\langle 0|\phi(x)|0\rangle = \lambda \neq 0$$

En effet si l'on pose alors

$$\phi(x) = \lambda + \Phi(x)$$

où

$$\langle 0|\Phi(x)|0\rangle = 0$$

on obtiendra l'équation d'un fermion de masse

$$m = \lambda\, g$$

en interaction avec le champ scalaire $\Phi(x)$:

$$(i\gamma^{\alpha}\partial_{\alpha} - g\lambda - g\Phi(x))\psi(x) = 0$$

Cela n'étant possible que pour un champ scalaire nous aurons besoin d'une interaction de l'électron avec un tel champ si la masse m_e est censée être générée par ce mécanisme, le mécanisme de Higgs[18].

Rappelons que le neutrino est toujours polarisé à gauche; son opérateur sera donc de la forme :

$$\nu_L(x) = \frac{1}{2}(1-\gamma^5)\nu(x) .$$

Comme l'électron réel a une masse, e(x) sera réalisé comme une superposition d'une composante à gauche et d'une composante à droite :

$$e(x) = e_L(x) + e_R(x)$$

Le neutrino étant toujours associé à l'électron il est naturel de considérer l'isospineur :

$$L(x) = \begin{pmatrix} \nu_L(x) \\ e_L(x) \end{pmatrix} = \frac{1}{2}(1-\gamma^5)\begin{pmatrix} \nu(x) \\ e(x) \end{pmatrix}$$

On aura aussi :

$$R(x) = e_R(x) = \frac{1}{2}(1+\gamma^5)e(x)$$

qu'il faudra introduire dans l'expression du lagrangien.

Soit à écrire la partie du lagrangien qui se rapporte à l'électron et au neutrino et à leur interaction avec un champ scalaire de telle sorte qu'on obtienne une masse pour l'électron d'après le mécanisme de Higgs[18] et pas de masse pour le neutrino. Puisque :

$$(1 + \gamma^5)(1 - \gamma^5) \equiv 0$$

on aura :

$$\bar{L}L = \bar{R}R = 0$$

ce qui est une autre manière d'exprimer le fait que les particules polarisées n'ont pas de masse.

Par conséquent le champ $\phi(x)$ devra être complexe et devra apparaître dans l'interaction avec les leptons sous la forme

$$G\{\bar{L}\,\phi\,R + \bar{R}\phi^{+}\,L\} \; . \tag{26}$$

Comme $\bar{L}$ est un isospineur et R un isoscalaire, il faudra que ϕ soit un isospineur pour que cette expression soit isoinvariante :

$$\phi(x) = \begin{pmatrix} \chi(x) \\ \phi_o(x) \end{pmatrix} \; . \tag{26a}$$

On trouve alors :

$$G(\bar{L}\,\phi\,R + \bar{R}\phi^{+}L) =$$

$$= G((\bar{\nu}\;\bar{e})\frac{1}{2}(1+\gamma^5)\begin{pmatrix}\chi \\ \phi_o\end{pmatrix}\frac{1}{2}(1+\gamma^5)e + \bar{e}\,\frac{1}{2}(1-\gamma^5)(\chi^{+}\phi_o^{+})\frac{1}{2}(1-\gamma^5)\begin{pmatrix}\nu \\ e\end{pmatrix}) \; .$$

Si l'on pose

$$\phi_o(x) = \lambda + \frac{1}{\sqrt{2}}(\phi_1 + i\phi_2)$$

le lagrangien de l'électron et du neutrino en interaction avec le scalaire $\phi(x)$ sera :

$$\begin{aligned} L_e &= -\bar{L}i\gamma^{\alpha}\partial_{\alpha}L - \bar{R}i\gamma^{\alpha}\partial_{\alpha}R + G(\bar{L}\phi R + \bar{R}\phi^{+}L) = \\ &= -\bar{e}(i\gamma^{\alpha}\partial_{\alpha} - m_e)e - \bar{\nu}_L i\gamma^{\alpha}\partial_{\alpha}\nu_L + \\ &\quad + G((\bar{\nu}_L e_R)\chi(x) + (\bar{e}_R\nu_L)\chi^{+}(x) + \frac{1}{\sqrt{2}}((\bar{e}_L e_R)(\phi_1 + i\phi_2) + \\ &\quad + (\bar{e}_R e_L)(\phi_1 - i\phi_2)) \end{aligned}$$

où la masse de l'électron est donnée par :

$$m_e = \lambda\; G \tag{26b}$$

ce qui impose à λ la condition d'être réel puisque G, d'après la forme de l'interaction ci-dessus, est réel et que λG est positif strictement.

Comme la théorie doit décrire les interactions faibles, dues à un champ vectoriel complexe W^{μ}, $W^{\mu +}$ de masse m_W, et les interactions électromagnétiques provenant du champ $A^{\mu}(x)$ à masse nulle, la théorie devra contenir au moins trois champs vectoriels.

Associé à l'isospineur L(x) des leptons gauches il y aura un courant

$$j^{\mu}_{a}(x) = \bar{L}(x)\gamma^{\mu} \frac{\tau_a}{2} L(x) , \qquad a = 1,2,3.$$

On introduira alors un champ vectoriel de jauge, de Yang-Mills, qui interagira avec ce courant. Ce sera un champ vectoriel isovecteur, c'est-à-dire un triplet de champs vectoriels $\mathcal{A}^{\lambda}_{a}(x)$, a = 1,2,3.

Or de la même manière que l'opérateur de la charge est l'intégrale spatiale de la composante zéro du courant électrique, l'opérateur d'isospin est l'intégrale spatiale de la composante zéro du courant d'isospin :

$$T_a = \int d^3x \, L^{+}(x) \frac{\tau_a}{2} L(x) \quad .$$

Ces trois opérateurs sont les générateurs du groupe SU(2) que laisse invariante la partie cinématique du lagrangien des leptons

$$- \bar{L}\gamma^{\lambda} i\partial_{\lambda} L - \bar{R}\gamma^{\lambda} i\partial_{\lambda} R .$$

Le groupe plus général de symétrie est le groupe U(1)⊕SU(2), avec en plus le nombre leptonique total $N = N_L + N_R$ comme générateur. Si l'on peut associer le champ $\mathcal{A}^{\lambda}_{a}$ à la symétrie de l'isospin aucun champ vectoriel de masse nulle connu ne peut être associé à la conservation du nombre N[19)]. Weinberg choisit alors l'hypercharge

$$Y = N_R + \frac{1}{2} N_L$$

comme générateur à côté des trois T_a.

Le courant associé à Y est :

$$j^{\mu}_{\gamma}(x) = \bar{R}\gamma^{\mu}R + \tfrac{1}{2}\bar{L}\gamma^{\mu}L$$

et l'on introduira un champ vectoriel B^{μ} qui interagira avec ce courant. La partie du lagrangien correspondant à ces champs vectoriels de masse nulle sera :

$$L_v = -\tfrac{1}{4}G^{\mu\nu}_{a}G_{a\mu\nu} + g\bar{L}\gamma^{\lambda}\frac{\tau_a}{2}\mathcal{A}_{a\lambda}L -$$

$$-\tfrac{1}{4}B^{\mu\nu}B_{\mu\nu} + g'(\bar{R}\gamma^{\mu}R + \tfrac{1}{2}\bar{L}\gamma^{\mu}L)B_{\mu}$$

où

$$G^{\mu\nu}_{a}(x) = \partial^{\nu}\mathcal{A}^{\mu}_{a}(x) - \partial^{\mu}\mathcal{A}^{\nu}_{a}(x) - g\varepsilon_{abc}\mathcal{A}^{\nu}_{b}(x)\mathcal{A}^{\mu}_{c}(x) ,$$

$$B^{\mu\nu}(x) = \partial^{\nu}B^{\mu}(x) - \partial^{\mu}B^{\nu}(x) ,$$

avec une somme sous-entendue sur a, b, c de 1 à 3.

Il nous reste à compléter le lagrangien $L_{e\phi}$ avec l'inclusion de la partie libre du champ scalaire $\phi(x)$ qui sera en interaction avec les champs $\mathcal{A}^{\mu}_{a}$ et B^{μ} et en interaction quartique avec lui-même. Cette dernière interaction est de la forme $f(\phi^{+}(x)\phi(x))^2$ tandis que la première résulte des règles de Yang-Mills pour définir la dérivée d'un champ donné en présence de champs de jauge de Yang-Mills.

Les règles de jauge de Yang-Mills sont les suivantes, pour la différentiation covariante en présence de $\mathcal{A}_a$ et B^{μ} :

d'un spineur singlet R :

$$(i\partial^{\mu} - g'B^{\mu})\,R$$

d'un spineur isospineur L :

$$(i\partial^{\mu} - \tfrac{1}{2}g'B^{\mu} - g\,\frac{\tau_a}{2}\,\mathcal{A}^{\mu}_{a})L ,$$

d'un scalaire isospineur :

$$(i\partial^{\mu} + \tfrac{1}{2}g'B^{\mu} - g\,\frac{\tau_a}{2}\,\mathcal{A}^{\mu}_{a})\phi .$$

Le signe de g' dans la dernière expression est choisi de telle sorte que, comme on le verra, le champ électromagnétique ait une masse nulle ou qu'il n'interagisse pas avec le champ $\phi_o(x)$.

Le lagrangien complet s'écrira donc :

$$L = -\bar{L}\gamma^{\alpha}(i\partial_{\alpha} - \tfrac{1}{2}g'B^{\mu} - g\frac{\tau_a}{2}\mathcal{A}_{a\alpha})L -$$
$$-\bar{R}\gamma^{\alpha}(i\partial_{\alpha} - g'B_{\alpha})R + G(\bar{L}\phi R + \bar{R}\phi^{+}L) -$$
$$-\tfrac{1}{4}G_a^{\mu\nu}G_{a\mu\nu} - \tfrac{1}{4}B^{\mu\nu}B_{\mu\nu} + \tag{26c}$$
$$+(\partial^{\mu}\phi^{+} - ig\phi^{+}\frac{\tau_a}{2}\mathcal{A}_a^{\mu} + i\frac{g'}{2}\phi^{+}B^{\mu})(\partial_{\mu}\phi + ig\frac{\tau_a}{2}\mathcal{A}_{a\mu}\phi - i\frac{g'}{2}B_{\mu}\phi) -$$
$$-M^2\phi^{+}\phi + f(\phi^{+}\phi)^2$$

où M désigne le paramètre de masse du champ ϕ.

Le champ $\phi(x)$ est le doublet (26a) et l'on pose maintenant

$$\phi_o(x) = \lambda\frac{1}{\sqrt{2}}(\phi_1(x) + i\phi_2(x)).$$

La substitution de ϕ_o par cette expression dans L donné ci-dessus, permet d'obtenir, entre autres, des termes linéaires en $\phi_1(x)$:

$$\sqrt{2}(-\lambda M^2 + 2\lambda^3 f)\phi_1 \ ;$$

des termes quadratiques en $\phi_1(x)$:

$$(-\tfrac{1}{2}M^2 + 3f\lambda^2)\,\phi_1^2 \ ; \tag{27a}$$

des termes quadratiques en $\phi_2(x)$:

$$(-\tfrac{1}{2}M^2 + f\lambda^2)\,\phi_2^2 \ ; \tag{27b}$$

des termes en $\chi^{+}(x)\chi(x)$:

$$(-M^2 + 2f\lambda^2)\,\chi^{+}\chi \ . \tag{27c}$$

Si l'on annule les termes linéaires en $\phi_1(x)$ (ces termes impliqueraient des vertices avec absorption d'un méson ϕ_1 dans le vide) on obtient les deux solutions en λ :

$$\lambda = 0 \qquad \text{ou} \qquad \lambda^2 = \frac{M^2}{2f}$$

Comme on impose λ non nul, il restera

$$\lambda = \sqrt{\frac{M^2}{2f}}$$

On obtient ainsi deux possibilités pour λ réel et positif

$$M^2 > 0 \quad , \quad f > 0$$

$$M^2 < 0 \quad , \quad f < 0 \; .$$

Les termes indiqués en (27) deviennent, avec la valeur (26b) de λ (si l'on additionne les termes

$$\tfrac{1}{2}\, \partial^\mu \phi_1 \partial_\mu \phi_1 \; , \; \tfrac{1}{2}\, \partial^\mu \phi_2 \partial_\mu \phi_2 \; , \; \partial^\mu \chi^+ \partial_\mu \chi$$

du lagrangien)

$$\tfrac{1}{2}\, \partial^\mu \phi_1 \partial_\mu \phi_1 + M^2 \phi_1^2 + \tfrac{1}{2}\, \partial^\mu \phi_2 \partial_\mu \phi_2 + \partial^\mu \chi^+ \partial_\mu \chi \; .$$

On obtient ainsi un champ ϕ_2, qui a, comme le champ χ, une masse nulle. Le champ ϕ_1 aura, lui, une masse m_ϕ telle que

$$- \tfrac{1}{2}\, m_\phi^2 = M^2 \qquad \text{avec} \qquad m_\phi^2 > 0 \; .$$

On devra donc imposer

$$M^2 < 0 \quad , \quad f < 0$$

de sorte que

$$m_\phi = \sqrt{2|M^2|} \; .$$

Dans L on trouvera, d'autre part, les termes suivants (une somme est sous-entendue sur a de 1 à 3)

$$-\tfrac{1}{4}\, G_a^{\mu\nu} G_{a\mu\nu} + \frac{\lambda^2 g^2}{4} \mathcal{A}_a^\mu \mathcal{A}_{\mu a} - \tfrac{1}{4}\, B^{\mu\nu} B_{\mu\nu} + \frac{\lambda^2 g'^2}{4} B^\mu B_\mu + \frac{\lambda^2 g g'}{2} \mathcal{A}_3^\mu B_\mu \; .$$

Si l'on introduit à présent les champs W^μ, $W^{\mu +}$

$$W^\mu = \frac{1}{\sqrt{2}} (\mathcal{A}_1^\mu + i \mathcal{A}_2^\mu) \; ,$$
$$W^{\mu +} = \frac{1}{\sqrt{2}} (\mathcal{A}_1^\mu - i \mathcal{A}_2^\mu) \; , \qquad (29)$$

on constate que :

$$\frac{\lambda^2 g^2}{4} (\mathcal{A}_1^\mu \mathcal{A}_{1\mu} + \mathcal{A}_2^\mu \mathcal{A}_{2\mu}) = \tfrac{1}{2}\, \lambda^2 g^2 W^{\mu +} W_\mu \; .$$

Si l'on pose d'autre part :

$$W^{\mu\nu} = \partial^\nu W^\mu - \partial^\mu W^\nu$$

on trouve :

$$\frac{1}{\sqrt{2}}\,(G_1^{\mu\nu} + iG_2^{\mu\nu}) = W^{\mu\nu} - ig(\mathcal{A}_3^{\nu}W^{\mu} - \mathcal{A}_3^{\mu}W^{\nu})$$

d'où

$$-\frac{1}{4}\,(G_1^{\mu\nu}G_{1\mu\nu} + G_2^{\mu\nu}G_{2\mu\nu}) + \frac{1}{4}\,\lambda^2 g^2(\mathcal{A}_1^{\mu}\mathcal{A}_{1\mu} + \mathcal{A}_2^{\mu}\mathcal{A}_{2\mu}) =$$

$$= -\frac{1}{2}\,W^{\mu\nu+}W_{\mu\nu} + m_w^2 W^{\mu}\,W_{\mu} - ig\mathcal{A}_{3\mu}(W^{\mu\nu+}W_{\nu} - W^{\mu\nu}W_{\nu}^{+}) -$$

$$- g^2(\mathcal{A}_3^{\mu}\mathcal{A}_{3\mu}W^{\nu+}W_{\nu} - \mathcal{A}_3^{\nu}W_{\nu}\mathcal{A}_3^{\mu}W_{\mu}^{+}) .$$

Le champ W^{μ} aura donc une masse m_w donnée par

$$m_w^2 = \frac{1}{2}\,\lambda^2\,g^2$$

Les autres termes dans l'expression (28) peuvent s'écrire :

$$-\frac{1}{4}\,G_3^{\mu\nu}G_{3\mu\nu} - \frac{1}{4}\,B^{\mu\nu}B_{\mu\nu} + \frac{1}{4}\,\lambda^2(g\mathcal{A}_3^{\mu} + g'B^{\mu})(g\mathcal{A}_{3\mu} + g'B_{\mu}). \quad (30)$$

D'autre part les termes d'interaction des leptons avec les champs vectoriels sont :

$$\frac{g}{4}\left[(\bar{\nu}\gamma^{\alpha}(1-\gamma^5)e)(\mathcal{A}_{1\alpha} - i\mathcal{A}_{2\alpha}) + (\bar{e}\gamma^{\alpha}(1-\gamma^5)\nu)(\mathcal{A}_{1\alpha} + i\mathcal{A}_{2\alpha})\right] +$$

$$+\frac{1}{4}\,(\bar{\nu}\gamma^{\alpha}(1-\gamma^5)\nu)(g'B_{\alpha} + g\mathcal{A}_{3\alpha}) + (\bar{e}\gamma^{\alpha}e)(\frac{3g'}{4}\,B_{\alpha} - \frac{g}{4}\mathcal{A}_{3\alpha}) +$$

$$+\frac{1}{4}(\bar{e}\gamma^{\alpha}\gamma^5 e)(g'B_{\alpha} + g\mathcal{A}_{3\alpha}) .$$

Les deux premiers termes deviennent, avec la définition (29) du champ W^{μ} :

$$\frac{\sqrt{2}}{4}\,g\,((\bar{e}\gamma^{\alpha}(1-\gamma^5)\nu)W_{\alpha} + (\bar{\nu}\gamma^{\alpha}(1-\gamma^5)e)W_{\alpha}^{+})$$

ce sont les termes de l'interaction faible pour les courants faibles qui transfèrent une charge. On voit qu'apparaissent un courant neutre $(\bar{\nu}\gamma^{\alpha}(1-\gamma^5)\nu)$ et un autre courant $(\bar{e}\gamma^{\alpha}\gamma^5 e)$ qui interagissent avec la combinaison

$$g'B_{\alpha} + g\mathcal{A}_{3\alpha} .$$

Or d'après le terme en λ dans la forme (30) cette combinaison de B_α et $\mathcal{A}_{3\alpha}$ pourrait définir un nouveau champ vectoriel neutre doté d'une masse. Posons en effet :

$$Z^\lambda = \varepsilon(g\mathcal{A}_3^\lambda + g'B^\lambda) \qquad (31)$$

où ε est une certaine constante.

On trouvera pour le terme en λ de l'expression (30)

$$\frac{1}{4}\frac{\lambda^2}{\varepsilon^2} Z^\mu Z_\mu$$

qui suggère une masse :

$$m_z^2 = \frac{1}{2}\frac{\lambda^2}{\varepsilon^2} .$$

Ce champ Z^μ interagit avec les courants neutres $(\bar{\nu}\gamma_\mu(1-\gamma^5)\nu)$ et $(\bar{e}\gamma_\mu\gamma^5 e)$.

Il nous reste le terme

$$(\bar{e}\gamma^\mu e)(\frac{3g'}{4} B_\mu - \frac{g}{4}\mathcal{A}_{3\mu})$$

et l'on serait tenté d'identifier cette combinaison linéaire de B_μ et $\mathcal{A}_{3\mu}$ avec le champ électromagnétique. Ceci n'est cependant pas possible puisque cette identification du champ électromagnétique serait incompatible avec les termes

$$-\frac{1}{4} G_3^{\mu\nu} G_{3\mu\nu} - \frac{1}{4} B^{\mu\nu} B_{\mu\nu} . \qquad (32)$$

Essayons alors de compléter la définition de Z^μ (31) en posant :

$$A^\mu = \varepsilon(-g'\mathcal{A}_3^\mu + gB^\mu) .$$

Cette définition sera bonne pour le champ électromagnétique s'il en résulte une masse nulle pour ce champ.

On obtient alors pour les termes (32) :

$$-\frac{1}{4} G_3^{\mu\nu} G_{3\mu\nu} - \frac{1}{4} B^{\mu\nu} B_{\mu\nu} =$$

$$- \frac{1}{4} \frac{1}{\varepsilon^2(g^2 + g'^2)} (F^{\mu\nu}F_{\mu\nu} + Z^{\mu\nu}Z_{\mu\nu}) +$$

$$+ \frac{ig}{2\varepsilon(g^2 + g'^2)} (gZ_{\mu\nu} - g'F_{\mu\nu})(W^{\nu}W^{\mu+} - W^{\nu+}W^{\mu}) +$$

$$+ \frac{1}{4} g^2(W^{\nu}W^{\mu+} - W^{\nu+}W^{\mu})(W_{\nu}W^{+}_{\mu} - W^{+}_{\nu}W_{\mu})$$

où

$$F^{\mu\nu} = \partial^{\nu}A^{\mu} - \partial^{\mu}A^{\nu} ,$$

$$Z^{\mu\nu} = \partial^{\nu}Z^{\mu} - \partial^{\mu}Z^{\nu} .$$

Pour avoir les termes

$$- \frac{1}{4} F^{\mu\nu}F_{\mu\nu} - \frac{1}{4} Z^{\mu\nu}Z_{\mu\nu}$$

la constante ε sera : $\varepsilon = (g^2 + g'^2)^{-\frac{1}{2}}$.

L'interaction de l'électron avec le champ électromagnétique sera du type :

$$\frac{gg'}{(g^2 + g'^2)^{\frac{1}{2}}} (\bar{e}\gamma^{\mu}e)A_{\mu}(x)$$

et la charge de l'électron sera par conséquent donnée par l'équation

$$e = \frac{gg'}{(g^2 + g'^2)^{\frac{1}{2}}} \tag{33}$$

Il y aura en plus une interaction entre le champ vectoriel neutre Z^{μ} et les courants neutres $\bar{e}\gamma_{\mu}e$, $\bar{e}\gamma^{\mu}\gamma^5 e$, $\bar{\nu}\gamma^{\mu}(1-\gamma^5)\nu$:

$$\frac{1}{4} (g^2 + g'^2)^{\frac{1}{2}}((\bar{\nu}\gamma^{\mu}(1-\gamma^5)\nu) + (\bar{e}\gamma^{\mu}\gamma^5 e) + \frac{3g'^2 - g^2}{g^2+g'^2} (\bar{e}\gamma^{\mu}e))Z_{\mu} .$$

Comme l'interaction des mésons vectoriels chargés W avec les leptons a la forme

$$\frac{\sqrt{2}}{4} g((\bar{e}\gamma^{\mu}(1-\gamma^5)\nu)W_{\mu} \quad + \text{hermitique conjugué}$$

la constante de Fermi sera en relation avec g par

$$\frac{g^2}{8\ m_W^2} = \frac{G_F}{\sqrt{2}}$$

On obtient ainsi

$$G_F = \frac{\sqrt{2}}{4}\frac{1}{\lambda^2}$$

puisque

$$m_W^2 = \frac{1}{2}\lambda^2 g^2 .$$

De

$$m_e = \lambda\ G$$

il résultera :

$$G = 2^{3/4}\sqrt{G_F}\ m_e$$

et finalement :

$$m_Z^2 = \frac{1}{2}\lambda^2(g^2 + g'^2) .$$

9. INVARIANCE DE JAUGE ET ELIMINATION DU CHAMP $\chi(x)$

Le lagrangien (26c) a une propriété importante : il est invariant par rapport à la transformation de jauge suivante :

$$\mathcal{A}_a^{\mu\prime}(x) = \mathcal{A}_a^{\mu}(x) - \partial^\mu\theta_a(x) - g\varepsilon_{abc}\theta_b(x)\mathcal{A}_c^{\mu}(x) ,$$

$$B^{\mu\prime}(x) = B^\mu(x) - \partial^\mu\Lambda(x) ,$$

$$\phi'(x) = \left[\exp(ig\frac{\tau_a}{2}\theta_a(x) - i\frac{g'}{2}\Lambda(x))\right]\phi(x) ,$$

$$L'(x) = \left[\exp(ig\frac{\tau_a}{2}\theta_a(x) + i\frac{g'}{2}\Lambda(x)\right]L(x) ,$$

$$R'(x) = \left[\exp(ig'\Lambda(x))\right]R(x) ,$$

où $\theta_a(x)$, $a = 1,2,3$, $\Lambda(x)$ sont les fonctions de jauge.

On peut se servir de cette invariance comme un instrument

pour éliminer le champ scalaire $\chi(x)$ (voir éq. (26a)). En effet, puisque $\phi(x)$ est un isospineur, il est équivalent à quatre fonctions réelles. Nous pouvons donc écrire pour le champ $\phi(x)$

$$\phi(x) = \left[\exp(ig \frac{\tau_a}{2} \theta_a(x))\right] \begin{pmatrix} 0 \\ \lambda + \frac{1}{\sqrt{2}} \Phi(x) \end{pmatrix}$$

qui est équivalent à l'équation (26a) puisqu'on a ici quatre fonctions réelles $\theta_a(x)$, $a = 1,2,3$ et $\Phi(x)$.

Comme le lagrangien initial est un invariant de jauge, les fonctions $\theta_a(x)$ n'apparaissent pas dans l'expression du lagrangien et par conséquent nous pouvons remplacer $\phi(x)$ par

$$\begin{pmatrix} 0 \\ \lambda + \frac{1}{\sqrt{2}} \Phi(x) \end{pmatrix}$$

ce qui revient à poser $\chi(x) = \phi_2(x) = 0$ dans l'expression précédente de L est $\phi_1(x) = \Phi(x)$.

Le lagrangien transformé a à présent l'expression suivante :

$$\begin{aligned} L = & \frac{1}{2} \partial^\mu\Phi\partial_\mu \Phi - \frac{1}{2} m_\phi^2\Phi^2 - \frac{1}{4} F^{\mu\nu}F_{\mu\nu} - \frac{1}{4} Z^{\mu\nu}Z_{\mu\nu} + \\ & + \frac{1}{2} m_z^2 Z^\mu Z_\mu - \frac{1}{2} W^{\mu\nu +}W_{\mu\nu} + \frac{1}{2} m_W^2 W^{\mu +} W_\mu - \\ & - \bar{e}(i\gamma^\alpha\partial_\alpha - m_e)e - \bar{\nu}\, i\gamma^\mu\partial_\mu \frac{1}{2}(1-\gamma^5)\nu + \\ & + \frac{1}{\sqrt{2}} \frac{m_e}{\lambda} (\bar{e}e)\,\Phi + \frac{\sqrt{2}}{4} g((\bar{\nu}\gamma^\mu(1-\gamma^5)e)W_\mu + \\ & + (\bar{e}\gamma^\mu(1-\gamma^5)\nu)W_\mu^+ + e(\bar{e}\gamma^\mu e)A_\mu + \\ & + \frac{1}{4}(g^2 + g'^2)^{\frac{1}{2}} ((\bar{\nu}\gamma^\mu(1-\gamma^5)\nu) + (\bar{e}\gamma^\mu\gamma^5 e) + \\ & + \frac{3g'^2 - g^2}{g^2 + g'^2} (\bar{e}\gamma^\mu e)\, Z_\mu - \\ & - \frac{ig}{(g^2 + g'^2)^{\frac{1}{2}}} (gZ_\mu - g'A_\mu)(W^{\mu\nu +}W_\nu - W^{\mu\nu}W_\nu^+) + \end{aligned}$$

$$+ \frac{ig}{2(g^2 + g'^2)^{\frac{1}{2}}} (gZ_{\mu\nu} - g'F_{\mu\nu})(W^\nu W^{\mu+} - W^{\nu+}W^\mu) -$$

$$- \frac{g^2}{g^2 + g'^2} (gZ_\alpha - g'A_\alpha)(gZ_\beta - g'A_\beta)W_\mu^+ W_\nu \cdot$$

$$\cdot (g^{\alpha\beta}g^{\mu\nu} - g^{\alpha\nu}g^{\mu\beta}) +$$

$$+ \frac{1}{4} g^2(W^\nu W^{\mu+} - W^{\nu+}W^\mu)(W_\nu W_\mu^+ - W_\nu^+ W_\mu) +$$

$$+ (\frac{1}{2} \Phi^2 + \lambda\sqrt{2}\Phi)(\frac{g^2}{2} W^{\mu+}W_\mu + \frac{1}{4}(g^2 + g'^2)Z^\mu Z_\mu) +$$

$$+ f(\lambda^2 + \frac{1}{2} \Phi^2 + \lambda\sqrt{2}\Phi)^2$$

avec :

$$m_\phi^2 = 2|M^2| ,$$

$$\lambda^2 = \frac{-m_\phi^2}{4 f} , \ f < 0, \ \lambda > 0$$

$$m_e = \lambda G , \ m_W = \frac{1}{\sqrt{2}}\lambda g ,$$

$$m_Z^2 = \frac{1}{2} \lambda^2(g^2 + g'^2) ,$$

$$e = \frac{gg'}{(g^2 + g'^2)^{\frac{1}{2}}} , \ \frac{g^2}{8 m_W^2} = \frac{G_F}{\sqrt{2}} ,$$

$$G = 2^{3/4}\sqrt{G_F} m_e .$$

Si l'on pose :

$$\tan\theta_W = \frac{g'}{g}$$

les relations entre le champ vectoriel neutre Z et le champ électromagnétique A avec les champs de jauge $\mathcal{A}$ et B s'écriront :

$$Z^\mu = \mathcal{A}_3^\mu\cos\theta_W + B^\mu\sin\theta_W ,$$

$$A^\mu = -\mathcal{A}_3^\mu\sin\theta_W + B^\mu\cos\theta_W ,$$

θ_W est l'angle de Weinberg. La charge e et la constante g satisfont aux relations :

$$e = g \sin\theta_W \leqslant g$$

Les masses des mésons W et Z seront données par les relations :

$$\frac{m_W}{m_Z} = |\cos\theta_W|$$

$$\left(\frac{m_W}{m_\phi}\right)^2 = \frac{e^2\sqrt{2}}{8\ G_F} \operatorname{cosec}^2\theta_W \ .$$

On aura donc la limitation suivante sur la masse m_W :

$$m_W \geqslant 40\ m_p$$

où m_p est la masse du proton.

La constante d'interaction entre leptons et le champ scalaire est très faible :

$$G = \frac{m_e}{m_W}\frac{g}{\sqrt{2}} \ .$$

La beauté du modèle est l'unification des forces électromagnétiques et des forces d'interaction faible : à partir de l'hypothèse qu'un triplet de champs vectoriels à masse nulle interagit avec le courant d'isospin et qu'un champ vectoriel singlet à masse nulle interagit avec le courant d'hypercharge avec une invariance de jauge, on déduit, avec le postulat de la cassure spontanée de la symétrie - un champ scalaire à valeur moyenne dans le vide non nulle étant introduit pour attribuer une masse à l'électron -,les interactions électromagnétiques et les interactions faibles avec échange de charge. Le champ électromagnétique et le champ des bosons vectoriels chargés sont déduits des champs vectoriels donnés mais un autre champ, qui décrit des mésons vectoriels neutres, s'impose. La théorie a donc des conséquences expérimentales qui sont à vérifier puisqu'elle prédit des interactions faibles de courants neutres telles que le processus de diffusion élastique

$$\nu_\mu + e \to \nu_\mu + e \ .$$

En effet le processus

$$\nu_e + e \to \nu_e + e$$

peut avoir lieu si l'on n'invoque que des mésons chargés dans le modèle (figure 3)

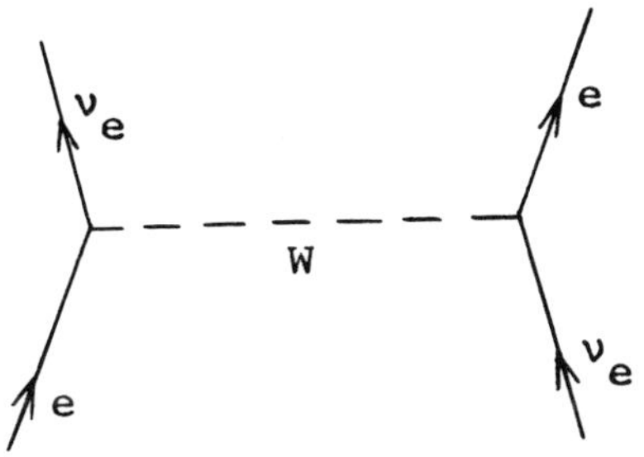

Figure 3

tandis que la réaction $\nu_\mu + e \to \nu_\mu + e$ ne peut avoir lieu, en premier ordre, que si le modèle admet des mésons vectoriels neutres (figure 4)

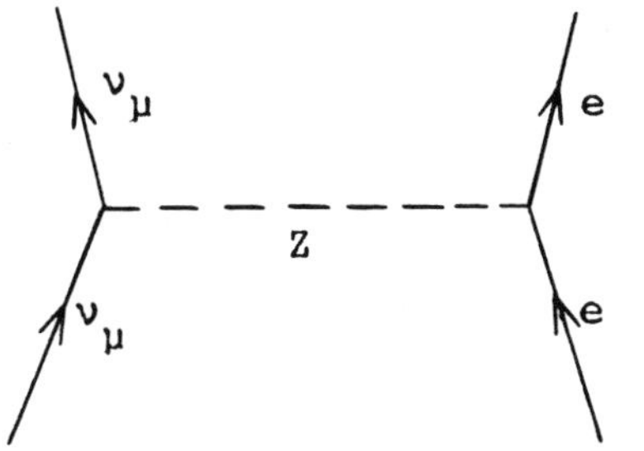

Figure 4

Des processus électromagnétiques auront, en principe, des corrections provenant de l'interaction faible avec les champs neutre Z et Φ. Les diagrammes de Feynman pour le moment magnétique, par exemple, seront maintenant jusqu'au deuxième ordre (figure 5) :

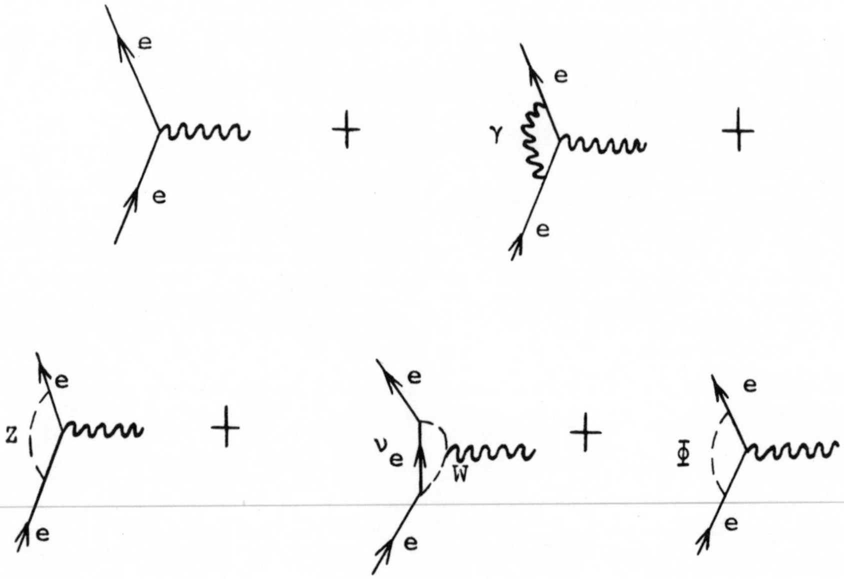

Figure 5

10. LE MODELE DES CHAMPS DE JAUGE UNIFIES ET LE MUON

Le modèle des champs de jauge unifiés ne peut pas se restreindre au seul traitement de l'interaction entre électrons et neutrinos électroniques. Il faut le généraliser de sorte que la théorie comprenne les muons et les hadrons.

De prime abord, il suffirait, pour introduire les muons, d'écrire des termes additionnels dans le lagrangien, obtenus en remplaçant partout :

$$e \to \mu \ , \ \nu_e \to \nu_\mu \ .$$

Ainsi, au lieu de l'interaction (26) on aurait

$$G(\bar{L}_e \ \phi \ R_e + \bar{R}_e \ \phi^+ L_e + \bar{L}_\mu \ \phi \ R_\mu + \bar{R}_\mu \phi^+ L_\mu) \ . \tag{34}$$

pour l'interaction entre leptons et le champ scalaire.

Cette règle n'est pas acceptable si la masse du muon doit provenir de la valeur moyenne de $\phi(x)$ dans le vide. En effet, pour un seul isospineur $\phi(x)$ et la même constante G, l'interaction (33) donnerait lieu à une masse m_μ égale à m_e.

L'hypothèse la plus simple est donc de remplacer l'interaction (26) par la suivante :

$$G_e(\bar{L}_e \phi R_e + \bar{R}_e \phi^+ L_e) + G_\mu(\bar{L}_\mu \phi R_\mu + \bar{R}_\mu \phi^+ L_\mu)$$

avec :

$$L_e = \frac{1}{2}(1-\gamma^5)\binom{\nu_e}{e} , \quad R_e = \frac{1}{2}(1+\gamma^5)e ,$$

$$L_\mu = \frac{1}{2}(1-\gamma^5)\binom{\nu_\mu}{\mu} , \quad R_\mu = \frac{1}{2}(1+\gamma^5)\mu .$$

On maintient dans ce modèle un seul champ scalaire isospineur $\phi(x)$ mais on est obligé de détruire l'universalité de l'interaction de ce champ avec les composantes des leptons, en introduisant deux constantes telles que :

$$m_e = \lambda G_e \quad , \quad m_\mu = \lambda G_\mu$$

et donc :

$$G_\mu \simeq 206\ G_e .$$

Une autre possibilité que nous proposons[20] est d'introduire un angle α et d'écrire l'interaction ϕ-leptons comme il suit (ce qui est équivalent à la formulation précédente, les champs e et μ ne pouvant pas se mélanger si les nombres leptoniques se conservent) :

$$G\{\bar{L}_e R_e \cos\alpha + \bar{L}_\mu R_\mu \sin\alpha\}\phi + \text{herm. conj.} \tag{35}$$

On maintiendrait ainsi une seule constante G et la valeur de l'angle α expliquerait la différence de masse électron-muon :

$$\begin{aligned} m_e &= G\ \lambda\ \cos\alpha \quad , \\ m_\mu &= G\ \lambda\ \sin\alpha \quad . \end{aligned} \tag{36}$$

La valeur de α est : $\alpha = 89^o 43'22''$ et on obtient

$$G = m_\mu (2\sqrt{2}\, G_F)^{\frac{1}{2}} (1 + \frac{m_e^2}{m_\mu^2})^{\frac{1}{2}} .$$

Il résulte de l'expression (35) que l'interaction entre le champ scalaire réel $\Phi(x)$ et les leptons a la forme suivante :

$$\frac{1}{\sqrt{2}} (\frac{m_e}{\lambda} (\bar{e}e) + \frac{m_\mu}{\lambda} (\bar{\mu}\mu)) \Phi(x) . \qquad (37)$$

Ainsi la théorie des champs de jauge unifiés pour les leptons admettra le lagrangien (26c) avec la convention de faire la somme des termes leptoniques pour électrons et muons et de remplacer l'interaction électron-champ scalaire par le terme (35).

11. L'INCORPORATION DES HADRONS DANS LE MODELE DE JAUGE

Une généralisation de ce modèle de sorte à inclure les hadrons est au premier abord simple : il faudrait tout simplement introduire le courant de quarks, $h^\alpha(x)$ donnée par les équations (13 et (14) :

$$h^\alpha(x) = \bar{p}(x)\gamma^\alpha(1-\gamma^5)\{n(x)\cos\theta_c + \lambda(x)\sin\theta_c\} .$$

Le lagrangien devrait donc avoir des termes d'interactions entre les champs $\mathcal{A}_a^\mu(x)$ et $B^\mu(x)$, d'une part, et les courants d'isospin et d'hypercharge des quarks d'autre part, de manière à contenir, après la cassure de la symétrie, des termes du type :

$$\frac{\sqrt{2}}{4}\, g\{\bar{p}(x)\gamma^\lambda(1-\gamma^5)(n(x)\cos\theta_c + \lambda(x)\sin\theta_c)W_\lambda + h.c.\}.$$

La grande difficulté de ce modèle est qu'il prédit des courants neutres avec changement d'étrangeté ($\Delta S = 1$) qui interagissent avec le champ vectoriel neutre Z^0 (en plus des courants neutres avec $\Delta S = 0$). Or les résultats expérimentaux sont très rigoureux à ce sujet et semblent exclure de tels courants neutres avec $\Delta S = 1$.

Pour éviter ces difficultés, on peut admettre l'existence d'un quatrième quark, p', et considérer les deux doublets de quarks suivants :

$$\begin{pmatrix} p \\ n\cos\theta_c + \lambda\sin\theta_c \end{pmatrix} \quad \text{et} \quad \begin{pmatrix} p' \\ -n\sin\theta_c + \lambda\cos\theta_c \end{pmatrix}$$

Alors le courant neutre avec $\Delta S = 1$ qui serait contenu dans le terme :

$$(\bar{n}\cos\theta_c + \bar{\lambda}\sin\theta_c)\gamma^\alpha(1-\gamma^5)(n\cos\theta_c + \lambda\sin\theta_c)$$

dans le modèle antérieur, serait dans ce modèle-ci du type :

$$j^\alpha_{(o)} = (\bar{n}\cos\theta_c + \bar{\lambda}\sin\theta_c)\gamma^\alpha(1-\gamma^5)(n\cos\theta_c + \lambda\sin\theta_c) +$$

$$+ (-\bar{n}\sin\theta_c + \bar{\lambda}\cos\theta_c)\gamma^\alpha(1-\gamma^5)(-n\sin\theta_c + \lambda\cos\theta_c)$$

c'est-à-dire

$$j^\alpha_{(o)} = \bar{n}\gamma^\alpha(1-\gamma^5)n + \bar{\lambda}\gamma^\alpha(1-\gamma^5)$$

qui ne contient pas des termes avec $\Delta S = 1$. S'il y a des termes de masse pour les quarks il sera nécessaire de faire restreindre la valeur de ces masses pour que les effets de courant neutre indésirables soient supprimés. Les diagrammes indésirables sont ceux qui entraînent une transition de λ en n et qui sont donc responsables de réactions telles que :

$$K_L \longrightarrow \mu^+ + \mu^- \quad ,$$

$$K^+ \longrightarrow \pi^+ + \nu + \bar{\nu} \ ,$$

$$\Sigma^+ \longrightarrow p + e + \bar{e} \quad , \text{etc.}$$

La première de ces réactions présente un problème[21)] qui n'est pas encore résolu. Cette réaction peut se produire d'après le diagramme de la figure 6

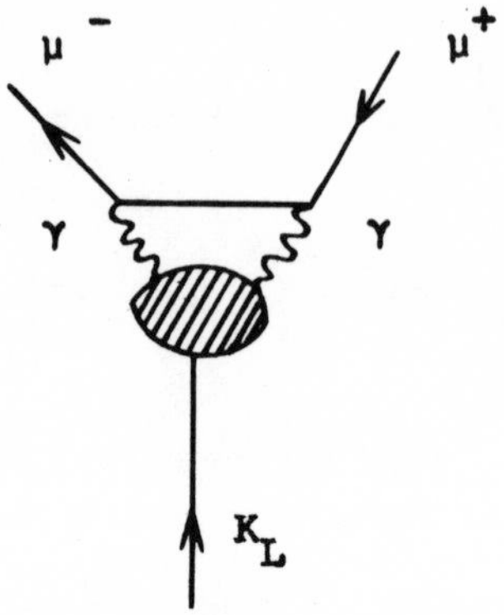

Figure 6

c'est-à-dire le méson K_L donne lieu à deux photons virtuels qui produisent, d'après les lois de l'électrodynamique quantique, une paire de muons. On espère donc avoir grossièrement le rapport :

$$\frac{\text{ampl.}(K_L \to \mu^+ + \mu^-)}{\text{ampl.}(K_L \to \gamma + \gamma)} \simeq \left(\frac{1}{137}\right)^2$$

Comme l'expérience indique que :

$$\frac{\text{ampl.}(K_L \to \gamma + \gamma)}{\text{ampl.}(K_L \to \text{somme tous processus})} \simeq 4{,}5 \times 10^{-4}$$

on doit avoir :

$$\frac{\text{ampl.}(K_L \to \mu^+ + \mu^-)}{\text{ampl.}(K_K \to \text{somme})} \simeq 2 \times 10^{-8}$$

Un calcul plus rigoureux basé sur l'unitarité de la matrice S donne une borne inférieure :

$$\frac{\text{ampl.}(K_L \to \mu^+ + \mu^-)}{\text{ampl.}(K_L \to \text{somme})} > 6 \times 10^{-9} .$$

Or l'expérience indique jusqu'à aujourd'hui la relation suivante :

$$\frac{\text{ampl.}(K_L \to \mu^+ + \mu^-)}{\text{ampl.}(K_L \to \text{somme})_{\text{exp.}}} < 1.8 \times 10^{-9}$$

ce qui contredit la valeur théorique ci-dessus.

Il faudra donc que les modèles de jauge qui admettant l'existence de bosons neutres ne donnent pas une contribution nouvelle à ce processus qui augmente le désaccord avec la donnée expérimentale. Nous n'examinerons pas ici cette question ni les modèles qui postulent des leptons lourds.

On est encore loin d'une synthèse simple et élégante entre les champs de jauge unifiés et les hadrons. Les modèles se multiplient[22)], une indication que l'on ne sait pas encore comment faire cette synthèse entre leptons et hadrons - surtout les particules étranges - pour avoir une théorie des champs de jauge unifiés.

REFERENCES

1) E. FERMI, Zs.f. Physik 88, 161 (1934) Voir: The development of Weak Interaction Theory, P.K. Kabir Editor, Gordon and Breach 1963, New York.

2) W. PAULI, Collected Scientific Papers, vol. 2, 1313, Interscience 1964, New York.

3) C.M.G. LATTES, G. OCCHIALINI and C. POWELL, Nature 160, 453, 486 (1947).

4) E. KONOPINSKI and H.M. MAHMOUD, Phys. Rev. 92, 1054 (1953).

5) G. PUPPI, Nuovo Cim. 5, 587 (1948); O. KLEIN, Nature 161, 897 (1948); J. TIOMNO and J.A. WHEELER, Rev. Mod. Phys. 21, 153 (1949).

6) T.D. LEE and C.N. YANG, Phys. Rev. 104, 254 (1956).

7) R.P. FEYNMAN and M. GELL-MANN, Phys. Rev. 109, 193 (1958); E.C.G. SUDARSHAN and R.E. MARSHAK, Phys. Rev. 109, 1860 (1958).

8) M. GELL-MANN and M. LEVY, Nuovo Cimento 16, 705 (1958) (voir note added in proof page 708); N. CABIBBO, Phys. Rev. Letters 10, 531 (1963).

9) Cf. H. YUKAWA, Rev. Mod. Phys. 21, 474 (1949).

10) J. LEITE LOPES, Phys. Rev. 109, 509 (1958).
L. WOLFENSTEIN, Nuovo Cimento 8, 882 (1958).

11) Cf. R.P. FEYNMAN, Quantum Electrodynamics, Benjamin New-York (1962); J.M. JAUCH and F. ROHRLICH, The theory of photons and electrons, Addison-Wesley, Cambridge, Mass. 1955; J.D. BJORKEN and S.D. DRELL, Relativistic Quantum Mechanics I; Relativistic Quantum Field II, Mc Graw-Hill, New York, 1965.

12) J. LEITE LOPES, Nucl. Phys. 8, 234 (1958); voir aussi S.A. BLUDMAN, Nuovo Cim. 9, 433 (1958) pour l'hypothèse des courants neutres.

13) S. WEINBERG, Phys. Rev. Letters 19, 1264 (1967); 27, 1688 (1971).
A. SALAM and J. WARD, Phys. Lett. 13, 168 (1964).

14) G.t' HOOFT and M. VELTMAN, Proc. Colloquium on Renormalization of Yang-Mills Fields, Marseille, 1972.

15) C.G. BOLLINI and J.J. GIAMBIAGI, Phys. Lett. 40B, 566 (1972).

16) T.D. LEE, Phys. Rev. Letters 26, 801 (1971).
J. LEITE LOPES, Nucl. Phys. B38, 555 (1972). Dans ce dernier article, le modèle de la dominance vectorielle est généralisée au cas des interactions faibles.

17) C.N. YANG and R.L. MILLS, Phys. Rev. 96, 191 (1954).

18) P.W. HIGGS, Phys. Lett. 12, 132 (1964); Phys. Rev. 145, 1156 (1966).
F. ENGLERT and R. BROUT, Phys. Rev. Letters 13, 321 (1964).
G.S. GURALNIK, C.R. HAGEN and T.W.B. KIBBLE, Phys. Rev. Lett. 13, 585 (1964).

19) T.D. LEE and C.N. YANG, Phys. Rev. 98, 1501 (1955).

20) J. LEITE LOPES, The muon-mass and the unified gauge field theory for leptons, Strasbourg, preprint CRN/HE 73-11 (1973).

21) H. STERN and M.K. GAILLARD, Ann. Phys. (N.Y.) 76, 580 (1973).

22) Cf. J.D. BJORKEN and C.H. LLEWELLYN SMITH, Phys. Rev. D7, 887 (1973).

IS ANTI-GRAVITATION POSSIBLE?

E.C.G. Stueckelberg de Breidenbach

Département de Physique théorique
Université de Genève

It is a pleasure for me to dedicate these few lines to my friend J.M. Jauch who has contributed largely to the expansion of the Geneva University's "Ecole de Physique".

* * *

We introduce the following terminology: we call "matter" all sources of the gravitational field in Einstein's equation

$$R^{(\mu\nu)} - \frac{1}{2} g^{(\mu\nu)} R = \kappa_E \theta^{(\mu\nu)}(x) \tag{1}$$

where $\kappa_E > 0$ is Einstein's gravitational constant. With respect to "matter", we distinguish between "substance" and "no-substance". "Substance" obeys a conservation law for the particle number $N_A = N_{A+} - N_{\bar{A}-}$, and A (= particle) $\neq \bar{A}$ (= anti-particle). "No-substance" obeys $A = \bar{A}$. We shall base our considerations on the Newton-Nordström scalar theory of gravitation, in which we distinguish between inert mass $M > 0$ and passive (P) and active (A) gravitational charge.

We show that

$$\frac{A}{P} = \frac{\bar{A}}{\bar{P}} = \text{const} > 0 .$$

The constant may be chosen equal to one because all numerical factors may be absorbed into Newton's gravitational constant $\kappa_N > 0$.

Enz/Mehra (eds.), Physical Reality and Mathematical Description. 448–452.

Thus substance attracts substance and anti-substance attracts anti-substance. However, this theory leaves open the sign of

$$\frac{\bar{A}}{P} = \frac{A}{\bar{P}} = \overline{\text{const}} \gtrless 0 \ .$$

Thus between substance and anti-substance there may be gravitational attraction or anti-gravitational repulsion.

As is well known, tensor potentials of <u>even tensor degree</u> (scalar Newton-Nordström potential ϕ, Einstein's potential $g_{\alpha\beta}$ (the metric)) give <u>attraction</u> between <u>charges of equal sign</u>, while <u>vector potentials</u> (the electromagnetic potential A_α) and potentials of <u>odd tensor degree</u> give <u>repulsion</u> between <u>charges of equal sign</u>.

The Nordström theory is Lorentz-covariant and gives (for zero graviton mass) propagation of gravitational waves with light velocity.

From the thermodynamical point of view, M, P and A are extensive quantities. Furthermore, and for convenience, they are measured in the same units (g).

We assume physical space to have d dimensions

$$\vec{x} = \{x^i\} \ ; \ ik.. = 12..d \tag{2}$$

and

$$dV(\vec{x}) = \underset{(+1)}{g} d^d x; \ d^d x = dx^1..dx^i..dx^d$$
$$dx^i > 0 \tag{3}$$

$\underset{(+1)}{g}$ is a scalar tensor density of the first kind and of order + 1. In Euclidian space it may be assumed = + 1.

Now for any extensive quantity A a density a is defined according to

$$\frac{dA(\vec{x})}{dV(\vec{x})} = a(\vec{x}) \tag{4}$$

and, analogously, $m(\vec{x})$ and $p(\vec{x})$ in terms of $dM(\vec{x})$ and $dP(\vec{x})$.

The Newton-Nordström equation in gravistatics,

$$\text{div}\,\vec{G}(\vec{x}) = \partial_i G^i(\vec{x}) = 4\pi\kappa_N a(\vec{x}) \ , \tag{5}$$

is analogous to Maxwell's electrostatic law

$$\text{div}\,\vec{D}(\vec{x}) = \partial_i D^i(\vec{x}) = 4\pi q(\vec{x}) \tag{6}$$

where $\kappa_N > 0$ is Newton's (positive) gravitational constant and

$$q(\vec{x}) = \frac{dQ(\vec{x})}{dV(\vec{x})}$$

the density of <u>active</u> electric charge. Furthermore, we have in the electrostatic isotropic case

$$\vec{D}(\vec{x}) = \varepsilon(\vec{x})\overleftarrow{E}(\vec{x});\ \varepsilon(\vec{x}) \geq \varepsilon_o = 1 \tag{7}$$

ε_o being the (positive) dielectric constant of the vacuum. Deriving electrostatics from a variational principle one finds

$$\overleftarrow{E}(\vec{x}) = -\overleftarrow{\text{grad}}\phi(\vec{x}) \rightarrow \overset{\circlearrowleft}{\text{rot}}\ \overleftarrow{E}(\vec{x}) = 0 \tag{8}$$

and the force is (active and passive electric charges are equal)

$$d\overleftarrow{K}(\vec{x}) = (dVq\overleftarrow{E})\ (\vec{x}) = (dQ.\overleftarrow{E})(\vec{x}) \ . \tag{9}$$

In gravitation there seems to be no difference between "gravitational displacement $\vec{G}(\vec{x})$" and the covariant "gravitational field $\overleftarrow{G}(\vec{x})$". However, the sign of the force changes:

$$d\overleftarrow{K}(\vec{x}) = -dP(\vec{x})\overleftarrow{G}(\vec{x}) \tag{10}$$

where $-dP(\vec{x})\overleftarrow{G}(\vec{x})$ is the <u>weight</u> (minus passive gravitational charge times gravitational field!) of the substance contained in $dV(\vec{x})$. For spherically symmetric bodies at distances large compared to their radius, neglecting self-forces which are compensated by a positive internal pressure (see my forthcoming theory of the point electron), we have ($\overleftarrow{K}_{12}$ is the force produced by body 2 on 1)

$$\overleftarrow{K}_{12} = -\left[\vec{r}_{12}\right] \kappa_N \frac{P_1 A_2}{r_{12}^{\ d-1}} \tag{11}$$

Now we turn to the inert mass. The momentum ("quantité de mouvement") $\vec{\pi}_\alpha$ of a body, α, is at time t

$$\vec{\pi}_\alpha(t) = M_\alpha\left[v_\alpha{}^2\right]\vec{v}(t) \tag{12}$$

where $\vec{v}(t)$ is the d(=3)-velocity of body α. Now the "quantity of motion" $\vec{\pi}$ must evidently be parallel (and not anti-parallel) to the d(=3)-velocity. Otherwise the French expression "quantité de mouvement" would not make sense, hence

$$M_\alpha\left[v_\alpha^2\right] > 0 . \tag{13}$$

Thus we have, to first approximation $M(v^2) = M(o)$,

$$\dot{\vec{\pi}}_\alpha(t) = M(o)\dot{\vec{v}}_\alpha = \sum_\beta{}' \vec{K}_{\alpha\beta} = -\kappa_N \sum_\beta{}' \left[r_{\alpha\beta}\right] \frac{P_\alpha A_\beta}{r_{\alpha\beta}^{d-1}} \tag{14}$$

where Σ_β' is the sum over all β, excluding α. Thus, as I. Kant has already shown, Newton's law of gravitational attraction gives $d = 3$.

Furthermore $\vec{\pi}(t) = \Sigma_\alpha \vec{\pi}_\alpha(t) = \vec{\pi}'$ must be a constant, (according to F. Hund) at least when $\vec{v}_\alpha(t)$ is measured with respect to the galaxy.

Thus, we have

$$P_\alpha A_\beta = P_\beta A_\alpha ; \quad \alpha \neq \beta \text{ or}$$

$$\frac{P_\beta}{A_\beta} = \frac{P_\alpha}{A_\alpha} = \frac{p(\vec{x})}{a(\vec{x})} = \text{const} > 0 .$$

The same holds for $\bar{P}_\alpha/\bar{A}_\beta$, on account of the CPT-theorem which itself follows from Lorentz-Poincaré invariance. However, the sign of the $\overline{\text{const}}$ in

$$\frac{\bar{P}_\alpha}{A_\alpha} = \frac{P_\alpha}{\bar{A}_\alpha} = \frac{\bar{p}(\vec{x})}{a(\vec{x})} = \frac{p(\vec{x})}{\bar{a}(\vec{x})} = \overline{\text{const}} \gtrless 0$$

is still open.

Apparently it has been shown at SLAC that $e_+ = \bar{e}_-$ (positron) falls downward in the terrestrial gravitational field. It seems to us still worthwhile to confirm this experiment further with anti-protons.

In closing we remind the reader that W. Thirring jr. came to the conclusion that anti-gravitation does not exist which is because he does not distinguish between M, P and A. We intend to look further into this question (following suggestions by L. van Hove and R. Hagedorn) in relation to the $K^{\underline{o}}$ problem, where anti-substance gets sort of mixed with substance.

* * *

I thank my friends at CERN, Jean V. Augsburger, Rolf Hagedorn, V. Weisskopf, for conversations and W. Heisenberg and F. Hund for an encouraging correspondence which is largely responsible for the results described in these lines.

Part IV

QUANTUM THEORY AND STATISTICAL MECHANICS

ON A NEW DEFINITION OF QUANTAL STATES

B. Misra

Department of Physics and Astrophysics
University of Colorado
Boulder, Colorado

1. INTRODUCTION

The concept of states of a physical system is, perhaps, the most fundamental of all the concepts of physics. In fact, it seems possible to defend the thesis that the history of physics is the same as the history of the concept of state; that the emergence of a new physical theory has often been consequent upon the emergence of a new conception of state. Thus, Galilei - Newtonian Physics emerged by rejecting the preceeding Aristotlian conception of state and adopting a new one in its place. Thermodynamics was built upon a newer non-mechanical conception of state. Gibbs-Boltzmann statistical mechanics attempting to reduce the concept of state in thermodynamics to the older mechanical conception of state required the enlarged concept of the Mixture states or statistical ensembles. The old (Bohr-Sommerfeld) quantum theory signaled a drastic departure from the classical conception of state. And lastly, quantum theory as we know it today has emerged as a result of a most profound criticism and revision of the classical concept of state. It is therefore, not unexpected, that the concept of state in quantum mechanics should be, in its turn, the subject of vigorous critical analysis as it has been in recent years.

The motivation for this renewed critical analysis of the concept of states is two-fold: one is the desire to obtain a conceptually coherent interpretation of quantum mechanics. For, the usual ensemble interpretation of quantal states and the usual ma-

Enz/Mehra (eds.), Physical Reality and Mathematical Description. 455–476.

thematical formulation in terms of the unit vectors (and the density operators) of the Hilbert space do lead to conclusions that seem to conflict with some of our deeply held convictions about the physical reality. To convince oneself of this, one need only recall the celebrated Einstein - Podolsky - Rosen Paradox and its more recent discussions [1, 2].

The second motivation is a more pragmatic one. While quantum theory of non-relativistic systems with finite degrees of freedoms is a mathematically complete and beautiful theory with amazingly unerring predictive power, the same can not be held of any of the attempted theories for relativistic systems with infinite degrees of freedoms. This has lead to a critical re-examination of the foundations of quantum theory with the hope that such a critical re-examination might suggest the necessary generalizations in the conceptual and formal structure of quantum theory that are needed for its successful application to relativistic systems with infinite degrees of freedoms.

These efforts have resulted in a variety of proposals suggesting modifications in the conceptual or formal structure of quantum theory. The, so-called, hidden variables theories, the axiomatics of quantum theory based on the, so-called, propositional system and the algebraic formulation of quantum theory are among the most well-known of such attempts. And correspondingly, the concept of state has been defined variously in terms of a variety of mathematical concepts; such as : generalized probability measures on Non Boolean Lattice [3,4,5], positive linear functionals on Jordan algebras or C*-algebras [6,7], (not necessarily linear) monotone positive functionals on Jordan or C*-algebras [8] and certain special subsets of orthocomplemented, weakly modular lattice [9,10] etc.

To survey these various proposals and discuss their relative merits (however fascinating a task) is beyond the scope of the present article. Its aim is the more modest one of examining yet another possible definition of (quantal) states that does not seem to have been considered explicitly in the literature.

The definition of state that we wish to examine is directly suggested by the corresponding definition in classical mechanics. As it is well-known, the general states (i.e. pure states as well as mixture states) of a classical system are represented by probability measures on the phase space of the system in question. It is, then, natural to try defining the states of quantal systems as the probability measures on the appropriate quantal analogue of the classical phase space. And it is this definition that we discuss in this note.

In order to avoid misunderstanding and possible disappointment it should be stated here that this is <u>not</u> related to the past attempts to represent quantal states in terms of distribution functions on the <u>classical</u> phase space, or its more recent revivals. [11]. And we need not discuss here the merits and blemishes of this approach.

The motivations for studying the definition of states proposed here derive from two sources : one is the desire to formulate the corresponding concepts of classical theory and quantum theory in as close an analogy to each other as possible.

The other is the hope, based on certain indications resulting from this article, that the definition of states as probability measures on the quantal phase space would, in fact, be better suited for discussion of certain fundamental questions than the usually accepted definition of states.

It has been argued recently that the description of (genuine) <u>mixtures</u> in terms of the Von Newmann density operators is incomplete and even misleading in two respects.

(1) It fails to distinguish between certain ensembles that are objectively different.

(2) To attribute density operators to the so-called, <u>improper mixtures</u> while useful in certain practical context would, under critical examination, lead to inconsistencies. It is suggested in section (4) of this article that these objections do not apply when ensembles are represented by the probability measures on the quantal phase space rather than the density operators.

Moreover, since probability measures on the quantal phase space can be interpreted as representing quantal states the vast body of concepts and results (for instance, "Ergodicity," "mixing", "Entropy," "Conditional Entropy" and so on) that have been developed in probability theory can be interpreted as referring to quantal states. It would be interesting to explicate in detail these "quantal interpretations" of the concepts and results of classical probability theory. In particular, it seems that the foundational questions of quantum statistical mechanics (of finite systems at least) can be discussed in much closer analogy to be the corresponding discussion of classical statistical mechanics.

Finally, from a technical-mathematical point of view the present article is a (preliminary) study of Borel structures and measures on the projective Hilbert space; i.e., the set of all rays of a complex and separable Hilbert space. The main technical results are contained in the two theorems: theorem (3) and theorem (4).

2. THE QUANTAL ANALOGUE OF THE CLASSICAL PHASE SPACE

This section is devoted to motivating the choice of an appropriate quantal analogue of the classical phase space. It consists, largely, of a sketchy review of an axiomatic scheme whose formulation runs along exactly parallel lines for both classical as well as quantal systems [3,4,5,10].

The objects of this axiomatics are the, so-called « yes-no experiments » (or more strictly « equivalence classes» of yes-no experiments), called the Propositions, pertaining to the system under consideration. These are (equivalence classes) of possible experiments on the system that have only two possible outcomes, labelled "yes" and "no" (or 1 and 0). The object of the axiomatics is to formulate, on the basis of physical interpretations, a set of postulates that would exhibit the essential structure of the set of all propositions.

To this end, one introduces in the set of all propositions a relation of partial ordering which is denoted by the symbol $\subset$, and whose physical interpretation is given by the following : Two propositions p and q stand in the relation $p \subset q$ if and only if, whenever p is « true» of the system q is « true » too. In this context a proposition p is said to be « true» of the system if and only if the system is such that the outcome of any experiment in the equivalence class of p is "yes" with certainty.

With the aid of this relation one can then define (and interpret physically) new propositions (called the greatest lower bounds) from given families $\{p_i\}$ of propositions. The greatest lower bound of a given family $\{p_i\}$ of propositions is denoted by $\bigcap_i p_i$ and it is defined as the proposition for which the following holds :

$$a \subset \bigcap_i p_i \Leftrightarrow a \subset p_i \quad \text{for all } p_i \text{ in the given}$$

family.

In order to equip the set of propositions with additional structure it is further assumed that it admits of an ortho-complementation. Explicitly, it is postulated that to every proposition p there corresponda a proposition p', called the (ortho) complement of p, such that:

(i) $p \cap p' = \Phi$

(ii) $(p')' = p$

(iii) $p \subset q \Rightarrow q' \subset p'$

Here Φ denotes the so-called absurd proposition which is always « false» for the system. It may be formally defined as the greatest lower bound of all the propositions of the system. The physical justification for this postulate flows from the interpretation of p' as the « negation » of p. The equivalence class of yes-no experiments of p' consists of exactly the same equipments that belong to the class of p except that the pointer readings labelled "yes" and "no" are reversed so that whenever an experiment in the class of p yields the result "yes" an experiment from the class of p' would yield the result "no" and vice-versa.

The existence of an orthocomplementation and of greatest lower bounds now implies the existence of the least upper bounds $\bigcup_i a_i$ of families $\{a_i\}$ of proposition.

$$\bigcup_i a_i = (\bigcap_i a'_i)'$$

There is one additional property which is generally postulated of the set of propositions of any physical system. It is the postulate of "atomicity" which, roughly speaking, asserts the existence of sufficiently many "atoms" or minimal propositions. We need not formulate and discuss this postulate explicitly here but only recall the definition of a minimal proposition: a proposition p is, by definition, minimal if and only if

$$q \subset p \Rightarrow \text{either } q = \Phi \text{ or } q = p$$

If thus turns out that the set of propositions of a physical system has the structure of an Atomic, Orthocomplemented, Lattice. And this is true of both classical as well as quantal systems.

The characteristic property that distinguishes classical from quantal systems is the property that the Lattice of propositions of a classical system if distributive (Boolean); that is,

$$a \cap (b \cup c) = (a \cap b) \cup (a \cap c), \text{ and}$$

$$a \cup (b \cap c) = (a \cup b) \cap (a \cup c)$$

for all propositions a, b, c of classical system, where as the same is not true of quantal systems.

The appropriate weakening of the distributive law that holds for the propositions of quantal systems was found and its physical justification was discussed in [5]. It may be formulated as

axiom P

If two propositions p and q of a quantal system stand in the relation $p \subset q$ to each other, then the sublattice generated by {p, q, p', q'} is distributive.

The preceeding brief account of the axiomatics of the systems of propositions of physical systems may be summarized in the two following statements : (1) The set of all propositions of a classical system has the structure of a (complete) Boolean, atomic, orthocomplemented Lattice; and (2) the set of all propositions of a quantal system, on the other hand, possesses the structure of a (complete) non-Boolean, atomic, orthocomplemented lattice satisfying the axiom P. For brevity, lattices of this type will be referred to in the future as (atomic) CROC.

The statement (1) about classical systems is well-known to be an abstract formalization of the somewhat obvious but basic fact that a classical system can always be associated with a, so-called, phase-space which is defined in terms of the possible simultaneous values of the generalized co-ordinates and the conjugate momenta of the system and that the yes-no experiments of the classical system may be identified with (Borel) subsets of its phase space. Under this identification, one easily recognizes that the relation of partial ordering $\subset$ among the propositions is the relation of set theoretic inclusion among (Borel) subsets of the phase space, greatest lower bound corresponds to set theoretic intersection, orthocomplementation is the operation of taking set theoretic complements and the minimal propositions correspond to the points of the phase space. The truth of the statement (1) then follows immediately.

Thus, although statement (1) is seen to be a restatement of some obvious facts about classical systems it is a useful restatement for the purpose of motivating the choice of an appropriate quantal analogue of the classical phase space. For, the classical phase can be defined now as the set of all minimal propositions pertaining to the system under consideration. And this definition, unlike the usual definition in terms of the possible simultaneous values of the set of generalized co-ordinates and conjugate momenta, can obviously be extended to quantal systems as well.

We conclude, then, that the natural quantal analogue of the classical phase space is the set of all minimal propositions pertaining to the system under consideration. And its mathematical representation is provided by the set of all "atoms" of an (atomic) CROC.

While it is desirable to endow this set with a "natural" Borel structure and to discuss the implications of defining quantal states as probability measures on this Borel structure we shall, actually, discuss this problem in a slightly more specialized and less abstract setting. This; because one is confronted with, as yet, unresolved technical difficulties when one tries to pose the above mentioned problem in the general and abstract setting of the set of all "atoms" in an (atomic) CROC. The specific specialization we have in mind is the case when the set of "atoms" in the CROC is realized as the set of all one-dimensional projections of a separable, Hilbert space over the complex field which, for brevity, will be referred to as the projective Hilbert space.

In order to see the restrictions implied by such a specialization we recall a well-known theorem [5]. According to this theorem the (atomic) CROC representing the propositions of a quantal system without superselection rules can always be realized as the set of all projections of a Hilbert space, with the "atoms" of the CROC being represented by the one-dimensional projections. Moreover, the Hilbert space would be separable if the CROC satisfies certain mild and physically reasonable separability condition. The only thing that is left open by this theorem is the field of the Hilbert space. It need not be the complex field. Thus, modulo this open possibilities as regards the field of the Hilbert space, there is no essential restriction involved in identifying the phase spaces of quantal systems without super selection rules with the projective Hilbert space.

We conclude this section with a comment on the preceeding discussion.

It might be thought that the preceeding motivating discussion leading up to the definition of the quantal phase space as the projective Hilbert space could have been substantially curtailed by simply observing that the classical phase space could also be defined as the set of all pure states of the system and by noticing further that the corresponding object for quantal systems, according to the usual postulates of quantum mechanics, is the set of all rays of a complex, separable, Hilbert space. But taking such a short cut would involve logical inconsistency as regard physical interpretations. For, the concept of quantal phase space is intended to be used for defining the concept of quantal states and it would be, clearly, in consistent to invoke a prior conception of state in defining quantal phase space. Therefore, we have chosen to interpret the quantal phase space as the set of all minimal propositions and not as the set of all pure states. Whether the points of the phase space are also associated with pure states or not should be a conclusion flowing from the subsequently adopted definition of states.

Moreover, we wanted to keep in sight the more general possibility of defining phase space as the set of all "atoms" in an (atomic) CROC.

3. THE NATURAL BOREL STRUCTURE IN THE QUANTAL PHASE SPACE

Definition

The quantal phase space (of quantal systems without super selection rules) is, by definition, the set of all one-dimensional projections in a separable, complex Hilbert space H, or more briefly, the projective Hilbert space. It will be denoted as $\mathcal{P}(H)$.

The preceeding section was devoted to motivating this definition. In the following section we shall develop the natural suggestion of defining quantal states as probability measures on the quantal phase space. For this, however, we need to single out, first, a natural class of subsets of the quantal phase space which will be considered as its "Borel sets" and will serve as the natural arena for defining probability measures. The present section is devoted to this end.

Since we shall need in our discussion some of the standard concepts and results pertaining to the concept of Borel structures in a set it will be useful to review them briefly [12].

A Borel structure in a set S is, by definition, a collection B of subsets of S which is closed with respect to complementation and taking countable intersection :

(i) $\Delta \in B \Rightarrow S - \Delta \in B$

(ii) $\Delta_i \in B_i \ (i=1, 2, \ldots) \Rightarrow \bigcap_{i=1}^{\infty} \Delta_i \in B$

It follows from this definition that a Borel structure is also closed with respect to taking countable unions. A set S together with a given Borel structure B is said to form a Borel space which we denote as S, B. And the subsets in B are called the Borel Sets of the Borel space S, B.

It is clear that the intersection of any collection of Borel structures in a set S is again a Borel structure in S. This permits the definition of the Borel structure generated by a given family F of subsets of S as the intersection of all these Borel structures containing F. It is clear that for any family F of subsets of S the Borel structure generated by F exists and it is the smallest Borel structure containing F.

When a set X is not only an abstract set but a topological space as well there is a distinguished Borel structure in X which is referred to as The Borel Structure of X. It is the Borel structure generated by the closed subsets (or equivatley the open subsets) of X.

A family F of subsets of a set S is said to be a separating family if for any two elements x,y in S with $x \neq y$ there exists a subset A belonging to F such that $x \in A$ but $y \notin A$. The Borel space S, B is said to be separated if the Borel structure B is a separating family in S; and S, B is called a countably separated Borel space if the Borel structure B contains a countable subfamily, say F, which is separating. It is trivial to verify that if B is a countably separating Borel structure in S then the family B contains the entire set S as a member. Hence, B, in this case, is also a , so-called, σ-algebra of subsets.

Let S, B be a Borel space and Δ an arbitrary given subset of B. Then it is obvious that the family B_Δ consisting of subsets of the form

$$A \cap \Delta \text{ with } A \in B$$

is a Borel structure in Δ.

The Borel space Δ, B_Δ is then referred to as the (Borel) subspace based on Δ of the Borel space S, B.

We now introduce the important concepts of Borel maps and Borel isomorphism. Let S_1, B_1 and S_2, B_2 be two Borel spaces. A map T from S_1 into S_2 is called a Borel (measurable) map if and only if the inverse image $T^{-1}(\Delta)$ of every set Δ in B_2 belongs to B_1. This definition of Borel maps obviously includes, as a special case, the definition of (Borel) measurable (numerically valued) functions on Borel spaces. The Borel spaces S_1, B_1 and S_2, B_2 are said to be (Borel) isomorphic if there exists a one - one map T form S_1 on to S_2 such that T as well as the inverse map T^{-1} are both Borel maps. And maps T with the above properties are called Borel isomorphisms.

Let T be a Borel isomorphism of the Borel space S_1, B_1 to the Borel space S_2, B_2. Then it is evident that Δ belongs to B_1, if and only if $T(\Delta)$ belongs to B_2. In fact the correspondence $\Delta \to T(\Delta)$ is a lattice -isomorphism between Boolean σ - lattices B_1 and B_2.

Owing to the above remark any countably additive (Borel) measure defined on a given Borel space can always be transferred to any other Borel space that is isomorphic to the original one. Thus, from the point of view of general theory of measures, the distinction between isomorphic Borel spaces is inessential.

Of all Borel spaces the, so-called, standard Borel spaces are the "best-behaved", mathematically. They are defined as follows. A Borel space is said to be standard if and only if it is isomorphic with the subspace that is based on some Borel set of the Borel space of a complete, separable metric space. In particular the Borel space associated with any separable, complete metric space is, itself, standard. Recall, in this connection, that by the Borel space of a given topological space X we mean the Borel space X, B, with B the Borel structure generated by the closed sets of X.

We conclude this brief review of Borel structures by stating two fundamental theorems that will be used in our subsequent discussion. The first of these expresses the surprising fact that although there are a bewildering variety of (topologically) non isomorphic metric spaces the possibilities for non-isomorphic standard Borel spaces is very severly limited.

Theorem 1

Let S_1, B_1 and S_2, B_2 be two standard Borel spaces. Then they are (Borel) isomorphic if and only if the cardinal numbers of the sets S_1 and S_2 are the same. Moreover, the cardinal numbers of S_1 and S_2 will be the same as soon as they both are non denumerable.

The following corollary is essentially a restatement of theorem 1 in more explicit terms.

Corollary 1

Every standard Borel space S, B is isomorphic to one of the

following : (i) the natural Borel space associate with the unit interval $[0,1]$ of the real line. (ii) a countable set C equipped with the Borel structure consisting of all subsets of C.

Theorem 2

Every standard Borel space S, B is contabley separated. Moreover, if F is any countable separating family of subsets in B then F is a generating family as well.

After this disgression on the general properties of Borel structures we turn now to the particular problem of equipping the projective Hilbert space $\mathcal{P}(H)$ with a suitable Borel structure. It is obvious, of course, that in a given abstract set there are in general, many distinct Borel structures. And none of these can be singled out as being more suitable than the others without invoking some ulterior objective. Usually, the ulterior objective that guides the choice of a Borel structure on a given set is the objective of obtaining a mathematically well-behaved and rich theory of measures and integration.

Now, it is clear from theorem 1 and its corollary that standard Borel spaces are the most well-behaved of all Borel spaces from the point of view of the theory of measures and integration. Measures and integrations on such spaces are no more forbidding than that on the unit interval. Thus, in our search for a suitable Borel structure on $\mathcal{P}(H)$ it is natural to look for one that is standard.

But the requirement that the Borel structure on $\mathcal{P}(H)$ be standard is matter of mathematical convenience alone. While it is a desirable property it is not absolutely essential. For our purpose, there is yet another property which is even of greater importance.

Our interest in equipping $\mathcal{P}(H)$ with a Borel structure is not the purely mathematical interest of defining measures and integration on $\mathcal{P}(H)$. We want also to interpret the (probability) measures on $\mathcal{P}(H)$ as quantal states. It is this second objective which determines the important criterion that will guide our choice of Borel structure in $\mathcal{P}(H)$.

In order to formulate this criterion we observe first that every projection operator E of H determines a function $\Phi_E(p)$ $(p \in \mathcal{P}(H))$ which is defined by $\Phi_E(p) = \mathrm{Tr}pE$ (1)
It will be seen in the next section that the interpretation of probability measures on $\mathcal{P}(H)$ as quantal states requires that all the functions $\Phi_E(p)$ be integrable with respect the probability measu-

res and hence, a fortiori, measurable. Thus, our primary requirement on the Borel structure of $\mathcal{P}(H)$, is that it renders all the functions $\Phi_E(p)$ (Borel) measurable. And it is, of course, natural to choose the smallest of such Borel structure as the natural Borel structure of $\mathcal{P}(H)$. This motivates the

Definition

By the natural Borel structure in $\mathcal{P}(H)$ we mean the smallest Borel structure that renders all the functions Φ_E defined by the relation (1) (with E a projection operator of H) (Borel) measurable. And the Borel space defined by this Borel structure will be called the natural Borel Space associated with $\mathcal{P}(H)$.

Evidently, the existence of the natural Borel structure in $\mathcal{P}(H)$ as defined above is not in doubt. In fact, it is the Borel structure generater by all the sets of the form $\Phi_E^{-1}(\Delta)$ (with Δ borel sets of the real line). We shall denote this Borel structure by the symbol B_n.

What is in doubt, however, is whether the Borel structure B_n is mathematically well-behaved. The following theorem resolves this doubt.

Theorem 3

The natural Borel space $\mathcal{P}(H)$, B_n associated with the projective Hilbert space $\mathcal{P}(H)$ is standard.

Since the set $\mathcal{P}(H)$ is, evidently, non-denumerable it then follows from combining theorem 3 with theorem 1 that $\mathcal{P}(H)$, B_n is (Borel) isomorphic to the Borel space of the unit interval.

The proof of theorem 3 is based on the following three lemmata.

Lemma 1

For any two elements p_1, p_2 of $\mathcal{P}(H)$ define $\rho_t\,(p_1,p_2)$ by

$$\rho_t\,(p_1\,,\,p_2) = \| p_1 - p_2 \|_1,$$

where $\|\ \ \|_1$, denotes the trace-norm of nuclear (trace-class) operators in the Hilbert space H. Then ρ_t is a metric function on $\mathcal{P}(H)$ and $\mathcal{P}(H)$ is complete and separable in metric topology of ρ_t.

The properties of various possible metric structures on $\mathcal{P}(H)$, or more generally on the set of all, so-called, density operators, have been discussed else where in a different physical context [13].

And a proof of lemma (1) is contained in [13], at least implicitly. Nevertheless, we supply a proof here for the sake of completeness.

Proof of Lemma 1

The proof that ρ_t is a metric function on $\mathcal{P}(H)$ is trivial.

To verify that $\mathcal{P}(H)$ is complete with respect to the metric ρ_t consider an arbitrary cauchy sequence $p_1, \ldots p_n, \ldots$ of elements in $\mathcal{P}(H)$:

(A) $\rho_t(p_n, p_m) \equiv \| p_n - p_m \|_1 \to 0$ as $n, m \to \infty$.

We desire to prove the existence of an element, say p, of $\mathcal{P}(H)$ such that $\rho_t(p_n, p) \to 0$ as $n \to \infty$.

Since the set of all nuclear operators in H is known to be complete in the trace-norm topology it follows (from the hypothesis (A)) that there exists a nuclear operator p with the property that $\| p_n - p \|_1 \to 0$ as $n \to \infty$.

The point, now, is to show that this nuclear operator p is in fact an element of $\mathcal{P}(H)$ (i.e. a one-dimensional projection operator). But this latter statement follows easily from the following well-known properties of trace-norm limits.

Let A be the limit in the trace-norm topology of the sequence $A_1, \ldots A_n, \ldots$ of nuclear operators :

$$\| A_n - A \|_1 \to 0 \quad \text{as } n \to \infty.$$

Then:

(i) A is self adjoint if each A_n in the sequence is so.

(ii) The sequence $A_1^2, \ldots A_n^2, \ldots$ is also convergent in the trace-norm topology and its trace-norm limit is the operator A^2.

(iii) $\mathrm{Tr}\, A_n \to \mathrm{Tr}\, A$ as $n \to \infty$.

Now, the sequence $p_1, \ldots p_n$ is a sequence of one-dimensional projection. That is:

$$\text{(i)} \quad p_n^* = p_n$$

$$\text{(ii)} \quad p_n^2 = p_n$$

$$\text{(iii)} \quad \mathrm{Tr}\, p_n = 1$$

Thus, owing to the above mentioned properties of trace-norm

topology, the trace-norm limit p of the sequence $p_1, \ldots p_n, \ldots$ must also satisfy the preceeding set of conditions. In other words, p would be a one-dimensional projection.

To verify the separability of $\mathcal{P}(H)$ in the metric topology of ρ_t we need to exhibit a countable subset, say D, of $\mathcal{P}(H)$ such that for every given $p \in \mathcal{P}(H)$ and any $\varepsilon>0$ there exists an element q of D for which $\rho_t(q,p) \leqslant \varepsilon$. Such a set D is readily at hand. In fact let $\mathcal{S}$ be a countable set of unit vectors in H which is dense on the unit sphere of H (such a set exists since H is assumed to be separable). And let D be the set of all one-dimensional projections determined by the vectors in $\mathcal{S}$. It is now easy to verify that the countable set D is dense in $\mathcal{P}(H)$ with respect to the metric topology of ρ_t.

In fact, let p be an arbitrary element in $\mathcal{P}(H)$ and let Ψ denote a unit vector in the range of p. Since the set $\mathcal{S}$ is dense on the unit sphere there exists a vector ϕ in $\mathcal{S}$ such that $|(\Psi,\phi)|^2 \geqslant 1-4\,\varepsilon^2$ for any given $\varepsilon>0$. Let q be the one-dimensional projection defined by ϕ. Evidently $q \in D$ and it can be easily verified that

$$\rho_t(q,p) = \| q - p \|_1 \leqslant \varepsilon.$$

This latter statement follows from the known fact that for any two one-dimensional projections p and q

$$\| p - q \|_1 = 2 \left[1 - |(\Psi,\phi)|^2 \right]^{\frac{1}{2}}$$

when Ψ and ϕ are two unit vectors in the ranges of p and q respectively [13].

Before proceeding to the next lemma we may remark that the completeness and separability of $\mathcal{P}(H)$ seems not to depend critically on the choice of ρ_t as the metric function. It seems any other "natural" choice would do. In fact two other metric structures that suggest themselves naturally are the metrics ρ_u and ρ_v defined as follows:

$$\rho_u(p_1,p_2) = \| p_1 - p_2 \|, \text{ the operator norm of } p_1 - p_2;$$

$$\rho_v(p_1,p_2) = \sup_{E = E^* = E^2} \operatorname{Tr} \left[(p_1 - p_2)\, E \right]$$

And it is found that ρ_t, ρ_u and ρ_v are all topologically equivalent metrics. This is because for any two one-dimensional projections p_1, p_2.

$$\| p_1 - p_2 \|_1 = 2 \| P_1 - P_2 \|$$

and due to the fact that

$$\rho_u(p_1,p_2) \leqslant \rho_v(p_1,p_2) \leqslant \rho_t(p_1,p_2)$$

as can be verified easily.

Lemma 2

Consider the separable, complete metric space $\mathcal{P}(H)$ with the metric ρ_t. Then the Borel space defined by the Borel structure B_T that is generated by the closed sets of $\mathcal{P}(H)$ in the metric topology of ρ_t is Borel isomorphic to the Borel space of the unit interval $[0,1]$. Moreover, all functions $\phi_E(p) = \mathrm{Tr}\, pE$ (E a projection) are Borel measurable with respect to the Borel structure B_T.

Proof of Lemma 2

The Borel space $\mathcal{P}(H)$, B_T is, of course, standard by definition. And theorem 1 (with the evident fact that $\mathcal{P}(H)$ is nondenumerable) then implies its Borel isomorphism with the Borel space of $[0,1]$.

That all functions $\phi_E(p)$ are (Borel) measurable with respect to B_T follows from direct verification. In fact, the functions $\phi_E(p)$ are continuous in the topology of the metric ρ_t. For if

$$\rho_t(p_n,p) = \| p_n - p \|_1 \to 0 \quad \text{as } n \to \infty$$

then $\| E\, p_n - E\, p \|_1 \leq \| p_n - p \|_1 \to 0 \quad \text{as } n \to \infty$.

Thus $\phi_E(p_n) \equiv \mathrm{Tr}\, Ep_n \to \mathrm{Tr}Ep = \phi_E(p) \quad \text{as } n \to \infty$.

Hence, the functions $\phi_E(p)$ are, a fortiori, (Borel) measurable with respect to the Borel structure B_T.

With lemma 2 we have established the existence of a standard Borel structure B_T in $\mathcal{P}(H)$ which has also the desired property of rendering all functions ϕ_E measurable. What is not known is whether it is the smallest of such structures. The next lemma answers this question.

Lemma 3

The Borel structure B_T of lemma 2 is identical with the natural Borel structure B_n of $\mathcal{P}(H)$.

Proof

It follows from the definition of B_n and the preceeding lemma that $B_n \subseteq B_T$.

To prove that they are in fact identical we consider another Borel structure B_o which is defined as follows :

Let D $\{p_1, \ldots p_n, \ldots\}$ be a countable (indexed) subset of $\mathcal{P}(H)$ which is also dense in $\mathcal{P}(H)$ with respect to the metric topology of ρ_T (cf. lemma 1). Denote by ϕ_i the functions $\phi_i(p) = \mathrm{Tr}\, p_i\, p$ (with $p_i \in D$) on $\mathcal{P}(H)$. And let F be the collection of all subsets of $\mathcal{P}(H)$ that are of the form $\phi_i^{-1}(\Delta)$ when ϕ_i is a function as described above and Δ a sub interval of $[0,1]$ with rational end points. The Borel structure B_o is now defined to be the Borel structure generated by F.

It follows from the definition of B_o that $B_o \subseteq B_n \subseteq B_T$. Thus to show $B_n = B_T$ it suffices to show that $B_o = B_T$.

It follows from its construction that the generating family F of B_o is a countable separating family of subsets in $\mathcal{P}(H)$. The fact that F is countable is obvious. That it is also separating needs a little argument for its proof. But we omit this argument and mention only that it follows essentially from the fact that D is chosen to be dense in $\mathcal{P}(H)$.

An application of Theorem (2) now immediately yields the identity $B_o = B_T$. In fact F is a countable, separating subfamily of the standard Borel structure B_T. Thus according to Theorem (2) B_T is generated by F and thus coincides with B_o.

Theorem (3) is established with the completion of the proof of lemma 3. In fact, a little more than what was asserted in Theorem (3) has been proved. We have not only shown that B_n is standard but have shown also that B_n coincides with the Borel structure generated by the closed sets of the metric space $\mathcal{P}(H)$ with any one of the metric functions ρ_T, ρ_n, ρ_v as its metric.

Finally we mention that it will be of interest to carry through an analysis similar to the preceeding in the abstract setting of the sets of all "atoms" in an (atomic) CROC.

4. QUANTAL STATES AS PROBABILITY MEASURES ON $\mathcal{P}(H)$

In the preceding section we have introduced a natural Borel structure B_n in the projective Hilbert space $\mathcal{P}(H)$. The Borel space defined by this Borel structure was shown to be (Borel) isomorphic to the natural Borel space of the unit interval $[0,1]$. Thus the definitions and properties of measures an integration in the Borel space $\mathcal{P}(H)$, B_n is not essentially different from that in the unit interval and we are now prepared to discuss the interpretation of probability measures on $\mathcal{P}(H)$ as quantal states. For the sake of completeness we recall, first, the well-known definition of probability measures.

Definition

By a probability measure μ on $\mathcal{P}(H)$ we mean a function μ which assigns a non-negative number $\mu(E)$ to every $E \in B_n$ in such a manner that :

(i) $$\mu\left(\bigcup_{n=1}^{\infty} E_n\right) = \sum_{n=1}^{\infty} \mu(E_n)$$ for every countable

family $\{E_n\}$ of pair-wise disjoint sets in B_n; and

(ii) $\mu(\mathcal{P}(H)) = 1$

The interpretations of probability measures on $\mathcal{P}(H)$ as quantal states is based upon the following theorem and its corollary.

Theorem 4

To every probability measure μ on $\mathcal{P}(H)$ there corresponds a unique density operator (i.e. a positive nuclear operator with unit trace), say W_μ, such that

$$\operatorname{Tr} W_\mu A = \int_{\mathcal{P}(H)} \operatorname{Tr}(pA)\, d\mu(p) \qquad (1)$$

for every bounded operator A of H.

Corollary to Theorem 4

The mapping $\mu \to W_\mu$ from the set of all probability measures on $\mathcal{P}(H)$ to the set of density operators has the following additional properties :

(i) It is on to the entire set of all density operators.

It is not one-one, however, but many to one.

(ii) If $\mu = \lambda\mu_1 + (1-\lambda)\mu_2$ (with $0 \leqslant \lambda \leqslant 1$) then

$$W_\mu = \lambda W_{\mu 1} + (1-\lambda) W_{\mu 2}.$$

(iii) W_μ is a one-dimensional projection if and only if μ is <u>extremal</u> in the sense that $\mu = \lambda\mu_1 + (1-\lambda)\mu_2$ with $0 < \lambda < 1$ implies that $\mu_1 = \mu_2 = \mu$.

(iv) The mapping $\mu \to W_\mu$ restricted to the extremal probability measures is <u>one-one</u> and <u>on to</u> the set of all one-dimensional projections of H.

<u>Proof of Theorem 1 and its Corollary</u>

The function Tr pA (for every given bounded operator A) is continuous in the metric topology of ρ_T on $\mathcal{P}(H)$. Thus it is (Borel) measurable with respect to the Borel structure B_n. Moreover, $|\mathrm{Tr}\, pA| \leqslant \|A\|$. Hence

$$\phi_\mu(A) \equiv \int_{\mathcal{P}(H)} \mathrm{Tr}(pA)\, d\mu(p) < \infty.$$

It is obvious that the map $A \to \phi_\mu(A)$ defines a positive linear functional on the algebra of all bounded operators of H. We have to verify that this functional is induced by a density operator W_μ. It is known that every positive linear functional $A \to \phi(A)$ on the <u>algebra of all compact operators</u> of a Hilbert space H determines a (unique) density operator such that

$$\mathrm{Tr} WA = \phi(A) \quad (A \text{ compact}) \quad [\text{cf. } 14]$$

Thus there exists a unique density operator W_μ such that

$$\int_{\mathcal{P}(H)} \mathrm{Tr}(pA) \quad d\mu(p) = \mathrm{Tr} W_\mu A \quad \text{for } \underline{\text{every}}$$

<u>compact operator A</u>. It remains only to verify that the above equality holds not only for compact operators but for any bounded operator B as well. To this end consider a sequence Pn of <u>finite-dimensional</u> projections such that

$$\operatorname*{s\,lim}_{n\to\infty} Pn = I$$

Then $\mathrm{Tr}pBPn \to \mathrm{Tr}pB$ for every $p \in \mathcal{P}(H)$. Moreover $|\mathrm{Tr}pB| \leqslant \|B\|$ and the constant function $\|B\|$ on $\mathcal{P}(H)$ is integrable with respect to the probability measure μ. Thus according to the Lebesgue dominated convergence theorem.

$$\int_{\mathcal{P}(H)} \mathrm{Tr}\ (pBPn)\ d\mu(p) \to \int_{\mathcal{P}(H)} \mathrm{Tr}\ (pB)\ d\mu(p) \text{ as } n \to \infty.$$

The operators BP_n are, however, compact and hence,

$\int \mathrm{Tr}\ (pBPn)\ d\mu(p) = \mathrm{Tr}\ W_\mu BPn$. And theorem 4

is established by noting that

$$\mathrm{Tr}\ W_\mu BPn \to \mathrm{Tr}\ W_\mu B \text{ as } n \to \infty.$$

The proofs of the statements of the corollary are easy and will be omitted here.

It is apparent from the preceeding theorem and its corollary that every probability measure μ on $\mathcal{P}(H)$ can be interpreted as representing a quantal state; the expectation value $\langle A\rangle_\mu$ of the observable represented by the operator A of H being given by

$$\langle A\rangle_\mu = \int_{\mathcal{P}(H)} \mathrm{Tr}(pA)\ d\mu(p) \qquad \text{(E)}.$$

Since the correspondence between extremal measures and one-dimensional projections is one-one the representation of pure states (≡ homogeneous ensembles) by extremal probability measures is entirely in accord with the usual representation of pure states in terms of the unit rays of the Hilbert space. The representation of <u>mixtures</u> (≡ non Homogeneous ensembles) by probability measures, however, is <u>not</u> completely equivalent to the usual representation of mixtures in terms of the density operators. In general there are several distinct measures. (Some of which may be even <u>continuous measures</u> in the sense that they assign zero "probability" to every discrete subset of $\mathcal{P}(H)$) that correspond to the same density operator W (when $W^2 \neq W$). Thus, it appears as if the description of mixture states in terms of probability measures on $\mathcal{P}(H)$ contains unobservable redundancies and hence, is unsatisfactory to this extent.

This conclusion would follow, however, only if it is assumed that the density operators W (with $W^2 \neq W$) provide an adequate and complete description of the nonhomogeneous ensembles. But it is just this assumption that is found wanting, under critical examination, in two respects [2].

1. It is known that if $W \neq W^2$ then not one but many proper mixtures can be thought of that have the same operator W as their density operator. For instance, if the density operator W has two distinct representations $W = \lambda_1 p_1 + \lambda_2 p_2$ and $W = \lambda_3 p_3 + \lambda_4 p_4$ in terms of the four distinct one-dimensional projections p_1, p_2, p_3, p_4 then the mixture S_1 obtained by combining (with the appropriate relative "weights") the homogeneous ensembles corresponding to p_1 and p_2 as well as the ensemble S_2 obtained from combining the homogeneous ensembles corresponding to p_3 and p_4 will both have the same density operator W. But the ensembles S_1 and S_2 are prepared, in very different ways and it has been persuasively argued in [2] that such ensembles are in general objectively distinguishable too.

2. It is misleading also, on two counts, to represent the so-called, improper mixtures by density operators.

(a) It is shown in [2] that such mixtures cannot be consistently conceived of as consisting of a number of homogeneous subensembles. Thus, if attributing a density operator to an ensemble is meant to carry the implication that the ensemble can be "split" in to a number of homogeneous subensembles then it is clearly inconsistent to attribute density operators to improper mixtures.

(b) Moreover, the density operator associated with an improper mixture provides, at the best, an incomplete description since it does not contain informations about the correlations of the constituent systems of the improper mixture and the systems with which they interacted in the past.

Thus it seems more plausible that the formalism in which mixtures are represented by probability measures, instead of containing physically irrelevant redundancies, does in fact furnish a more complete and consistent description of mixtures than that provided by the usual formalism of density operators. To make this suggestion more specific, it would be necessary, of course, to provide a more direct interpretation of the correspondence between probability measures and the ensembles they represent than the indirect interpretation given in (E). Without going into an explicit formulation of such a direct interpretation we shall merely assume here that the following statement follows from it as a consequence.

Let S_μ, $S_{\mu 1}$ and $S_{\mu 2}$ be the ensembles corresponding to probability measures μ, μ_1 and μ_2 (respectively). Then the ensemble S_μ can be obtained by combining the ensembles $S_{\mu 1}$ and $S_{\mu 2}$ (with appropriate relative weights) if and only if

$$\mu = \lambda\mu + (1- \lambda)\mu_2 \quad \text{with } 0 < \lambda < 1.$$

If the preceding statement is accepted then it is clear that proper mixtures that are "prepared" by combining different sets of homogeneous ensembles will be represented by different (discrete) probability measures even though they may all have the same density operator. For instance, the measure representing the ensemble of an unpolarized beam of spin 1/2 particles that is prepared by mixing two completely polarized beams along ±z direction will be different from the probability measure corresponding to the beam that is prepared by mixing polarized beams along ±x direction. As it has been argued in [2], although these two beams have the same density operator they can be objectively distinguished by the fluctuations in the values of

$$\sum_{n=1}^{N} (\sigma_z)_n$$

measured on these two beams. And thus, the description of these beams in terms of probability measures is more complete than the description provided by the density operator.

Again, it is clear from the preceding statement that the ensemble S_μ corresponding to a <u>continuous</u> probability measure μ cannot be conceived of as resulting from mixing a homogeneous ensemble with another (suitable) ensemble. (The "probability" or "weight" of occurence of any given homogeneous ensemble as a subensemble of S_μ is always zero.) As argued in [2] this is the characteristic property of the, so-called, improper mixtures. It is, thus, tempting to represent the <u>improper mixtures</u> by <u>continuous probability measures</u> on $\mathcal{P}(H)$. It should be emphasized also that the representation of improper mixtures in terms of continuous probability measures is not bereft of observable consequences but would lead to predictions as regards the correlations between the elements of the improper mixtures and the systems with which they interacted in the past. This last point has been argued, in the setting of a specific example, and from a slightly different point of view in [9].

The preceeding discussion thus indicates that representing mixtures in terms of probability measures as proposed here would avoid some of the objection raised against the attribution of density operators to mixtures. Moreover, as remarked in the introduction, the formalism of quantum mechanics in which states are represented by probability measures on the quantal phase space seems to be better suited than the usual formalism for the discussion of foundational questions of quantum statistical mechanics. We plan to return to these suggestions in a subsequent communication.

REFERENCES*

1. A.Einstein, B.Podolsky, N.Rosen: Phy.Rev. 47, 777 (1935)
2. B.d'Espgnat: Conceptual Foundations of Quantum Mechanics (W.A.Benjamin, Inc.1971)
3. J.M.Jauch: Foundations of Quantum Mechanics (Addison-Wesley, Reading, Mass. 1968)
4. G.W.Mackey: Mathematical Foundations of Quantum Mechanics (Benjamin, New York,1963)
5. C.Piron: Helv.Phys.Acta, 37, 439 (1963)
6. I.E.Segal: Mathematical Problems of Relativistic Physics (AMS; Providence, Rhode Island, 1963)
7. R.Haag and D.Kastler: J.Math.Phys., 5, 848 (1964)
8. B.Misra: Nuovo Cimento, 47, 841 (1967)
9. J.M.Jauch in Proceedings of the International School of Physics; Varenna 1971 ed. B.d'Espagnat.
10. C.Piron: Foundations of Physics, 2, 287 (1972)
11. E.P.Wigner: Phys.Rev. 40, 749 (1932)
12. C.Kuratowski: Topologie (Volume 1) (Warszwa, 1958)
13. J.M.Jauch, B.Misra and A.Gibson: Helv.Phys.Acta, 41, 513 (1968)
14. J.Dixmier: Les C*-Algebres (Gauthier-Villars, Paris, 1964)p.83.

* This bibliography is, obviously, incomplete and it does not attempt to trace the ideas to their original source. We are content with mentioning only some readily available references where the subjects in question are discussed.

THE MINIMAL K-FLOW ASSOCIATED TO A QUANTUM DIFFUSION PROCESS

Gérard G. Emch

Laboratoire de Physique Théorique, Ecole
Polytechnique Fédérale, Lausanne (Switzerland)

The existence of a minimal dynamical extension of a stochastic quantum diffusion process is proven. This extension is shown to satisfy all the properties of a non-abelian special K-flow.

INTRODUCTION AND CONCLUSION

The aim of this paper is to show that there exists an intimate connection between a natural quantum generalization of the Kolmogorov-Sinai concept of classical K-flow, and a canonical, mechanical model for the diffusion of a quantum harmonic oscillator embedded in a thermal bath.

In section I, we describe the stochastic, markovian process corresponding to the diffusion equations for a quantum harmonic oscillator.

In section II, we show that this stochastic process can be embedded canonically in a minimal, deterministic quantum dynamical system.

In section III, we examine the ergodic properties of the dynamical system obtained in section II. We show in particular that this system satisfies all the axioms of a non-abelian special K-flow[1)].

Mathematically, one interest of this result is that it provides a concrete example of the abstract structure described in [1)],

Enz/Mehra (eds.), Physical Reality and Mathematical Description. 477–493.

thus showing that its axioms are self-consistent.

At the interpretative level, this model points towards a connection between, on the one hand, the mathematical notion of "dissipativity" associated to a K-system, and on the other hand, the physical notion of "dissipativity" encountered in the study of transport processes such as that described by the diffusion equations.

I. DIFFUSION PROCESS

In this section we describe the quantum stochastic process associated to the diffusion of a one-dimensional harmonic oscillator of frequency ω. This is done at a purely phenomenological level, i.e. within the framework of a kind of "quantum thermodynamics"; we thus postpone to the next section all the questions relative to the possible understanding of this process within the framework of quantum statistical mechanics.

To establish the notation, we first review the usual description of a one-dimensional quantum oscillator. Its position and momentum are described by two (unbounded) selfadjoint operators Q and P, acting in the Hilbert space $\mathcal{H}_o = \mathcal{L}^2(\mathbb{R}, dx)$ and defined by:

$$(Qf)(x) = xf(x) \tag{I.1}$$

$$(Pf)(x) = -id_x f(x) \; ; \tag{I.2}$$

they satisfy the canonical commutation relation

$$[P,Q] \subseteq -iI \tag{I.3}$$

The Hamiltonian H_o of the system is given as usual by:

$$H_o = \frac{1}{2}(P^2 + \omega^2 Q^2) \tag{I.4}$$

From H_o we define two operators

$$U_o(t) = \exp\{-i\, H_o t\} \tag{I.5}$$

$$\rho^\beta = \exp\{-\beta H_o\} / \mathrm{Tr} \exp\{-\beta H_o\} \tag{I.6}$$

which we use to describe the time-evolution $\alpha_o(t)$ in the Heisenberg picture, and the canonical equilibrium state ϕ_o^β for the natural temperature $\beta = (kT)^{-1}$ via the expressions:

$$\alpha_o(t)[A] = U_o(t)\, A\, U_o(-t) \tag{I.7}$$

$$<\phi_o^\beta ; A> = \mathrm{Tr}\, \rho^\beta\, A \tag{I.8}$$

for every (bounded) observable A on our system.

In order to avoid the superfluous technicalities encountered when dealing directly with the unbounded observables Q and P, we will use instead in the sequel the Weyl form of the canonical commutation relations. Specifically we define for every a and b in $\mathbb{R}$:

$$z = \omega^{\frac{1}{2}}\, a = i\, \omega^{-\frac{1}{2}}\, b \tag{I.9}$$

$$W_o(z) = \exp\{-i(aP + bQ)\} \tag{I.10}$$

We notice that for every z_1, z_2 and z in $\mathbb{C}$

$$\left.\begin{aligned} &W_o(z_1)W_o(z_2) = W_o(z_1+z_2)\exp\{i\mathrm{Im}(z_1^* z_2)/2\} \\ &W_o(z)^* = W_o(-z) \end{aligned}\right\} \tag{I.11}$$

which in turn completely characterize the canonical commutation relation (I.3). Moreover, since $W_o(\mathbb{C})' \equiv \{W_o(z) | z \in \mathbb{C}\}' \equiv \{A \in \mathcal{B}(\mathcal{H}_o) \mid [A, W_o(z)] = 0\ \forall\ z \in \mathbb{C}\} = \mathbb{C}\, I$, every operator A in $\mathcal{B}(\mathcal{H}_o)$ is the limit (in the weak operator topology) of operators of the form $\Sigma_n\, a_n\, W_o(z_n)$, so that we do not loose anything by considering only $W_o(\mathbb{C}) = \{W_o(z) | z \in \mathbb{C}\}$. In particular the time evolution $\alpha_o(\mathbb{R}) = \{\alpha_o(t) | t \in \mathbb{R}\}$ and the equilibrium state ϕ_o^β are completely characterized by their restriction to $W_o(\mathbb{C})$, which moreover takes on a particularly simple form (see for instance [2]), or [3])), namely:

$$\alpha_o(t)\left[W_o(z)\right] = W_o(e^{-i\omega t)} z) \tag{I.12}$$

and with $\Phi_o^\beta(z) \equiv <\phi_o^\beta ; W_o(z)>$

$$\left.\begin{aligned} &\Phi_o^\beta(z) = \exp\{-\tfrac{1}{4}\,\Theta\, |z|^2\} \\ &\text{with } \Theta = \coth(\tfrac{1}{2}\beta\omega) \end{aligned}\right\} \tag{I.13}$$

It follows from the general theory (see for instance Theorem III.1.7 in [3])) of the representations of the canonical commutation relations, that there exists a unique (up to unitary equivalence) triple $(\mathfrak{H}_o^\beta, W_o^\beta(\mathbb{C}), \Phi_o^\beta)$, where $\mathfrak{H}_o^\beta$ is a separable Hilbert space, $W_o^\beta(\mathbb{C}) \equiv \{W_o^\beta(z) \mid z \in \mathbb{C}\}$ is a family of unitarity operators acting in $\mathfrak{H}_o^\beta$ and satisfying the relations (I.11), and Φ_o^β is a unit vector in $\mathfrak{H}_o^\beta$ satisfying: (i) $\hat{\phi}_o^\beta(z) = (\Phi_o^\beta, W_o^\beta(z)\,\Phi_o^\beta)$ for all z in $\mathbb{C}$, and (ii) Φ_o^β cyclic in $\mathfrak{H}_o^\beta$ with respect to $W_o^\beta(\mathbb{C})$, i.e. Ψ in $\mathfrak{H}_o^\beta$ and $(\Psi, W_o^\beta(z)\,\Phi_o^\beta) = 0$ for all z in $\mathbb{C}$ imply $\Psi = 0$. We denote by $\mathfrak{N}_o$ the von Neumann algebra $W_o^\beta(\mathbb{C})''$ generated by $\{W_o(z) \mid z \in \mathbb{C}\}$. The element N of $\mathfrak{N}_o$ will play, in the quantum system considered here, the role of the "stochastic variables" for the stochastic process to be now defined.

For a fixed positive constant λ and every t in $\mathbb{R}^+$ we define the linear transformation $\gamma(t)$ of $\mathfrak{N}_o$ by:

$$\gamma(t)\left[W_o^\beta(z)\right] = W_o^\beta(e^{-\lambda t} z)\,\exp\{-\tfrac{1}{4}\,\Theta|z|^2(1-e^{-2\lambda t})\} \tag{I.14}$$

and the transformation $\gamma_*(t) : \psi \to \psi_t$ on the predual $\mathfrak{N}_*$ of $\mathfrak{N}_o$ by:

$$<\gamma_*(t)[\psi];N> = <\psi;\gamma(t)[N]> \quad \forall\ N \in \mathfrak{N}_o \tag{I.15}$$

We notice immediately that $\gamma(\mathbb{R}^+) = \{\gamma(t) \mid t \in \mathbb{R}^+\}$ defines a Markovian process on $\mathfrak{N}_o$, i.e. in particular $\gamma(t_1)\gamma(t_2) = \gamma(t_1+t_2)$ for all t_1 and t_2 in $\mathbb{R}^+$. It is also trivial to verify from (I.13) and (I.14) that

$$<\phi_o^\beta;\gamma(t)\left[W_o^\beta(z)\right]> = <\phi_o^\beta;W_o^\beta(z)> \tag{I.16}$$

for all z in $\mathbb{C}$ and all t in $\mathbb{R}^+$ so that ϕ_o^β is an invariant state for the process $\gamma(\mathbb{R}^+)$. One actually gets from the same definitions an even stronger (ergodic) result namely that for every normal state ψ on $\mathfrak{N}_o$:

$$\lim_{t\to\infty}\ <\psi;\gamma(t)\left[W_o^\beta(z)\right]> = <\phi_o^\beta;W_o^\beta(z)> \tag{I.17}$$

for all z in $\mathbb{C}$, i.e. ϕ_o^β is the only normal state on $\mathfrak{N}_o$ which is invariant for the process $\gamma(\mathbb{R}^+)$, <u>and</u> every other normal state approaches ϕ_o^β as t tends to $+\infty$, a situation which already indicates some sort of "return to equilibrium" which we now want to investigate more closely.

To obtain the physical interpretation of the semi-group $\gamma(\mathbb{R}^+)$ we first remark that if for any fixed z_o in $\mathbb{C}$ we denote by $\mathfrak{N}_o(z_o)$ the von Neumann algebra $\{W_o^\beta(az_o)|a \in \mathbb{R}\}''$, then $\mathfrak{N}_o(z_o)$ is abelian and stable under $\gamma(\mathbb{R}^+)$. We now analyze the case $z_o = i\omega^{-\frac{1}{2}}$ for which we have simply $W_o(kz_o) = \exp\{-ikQ\}$ so that $W_o(\mathbb{R}z_o)$ generates a von Neumann algebra which is isomorphic to $\mathcal{L}^\infty(\mathbb{R},dx)$. This implies in particular that for every normal state ψ on $\mathfrak{N}_o$ there exists a configuration space density ρ_ψ such that

$$\langle\psi;N\rangle = \int\rho_\psi(x)\ N(x)\ dx \qquad \forall\ N\in\mathfrak{N}_o(z_o) \tag{I.18}$$

The Fourier transform $\tilde{\rho}_\psi$ of this density takes on a particularly simple form in the formalism which we have used up to this point; we have indeed:

$$\begin{aligned}\tilde{\rho}_\psi(k) &= \int dx\ \rho_\psi(x)\ \exp\{-ikx\} \\ &= \langle\psi;\ W_o^\beta(kz_o)\rangle\end{aligned} \tag{I.19}$$

We thus get, for instance, that the density ρ^β corresponding to ϕ_o^β satisfies

$$\tilde{\rho}^\beta(k) = \exp\{-\frac{1}{4}\ \Theta\ \frac{1}{\omega}\ k^2\} \tag{I.20}$$

i.e.

$$\rho^\beta(x) = \exp\{-\beta V(x)\} \Big/ \int dx\ \exp\{-\beta V(x)\} \tag{I.21}$$

with

$$\left.\begin{aligned} V(x) &= \frac{1}{2}\ \Omega^2\ x^2 \\ \Omega^2 &= \omega^2(\tfrac{1}{2}\beta\omega)^{-1}\ \tanh(\tfrac{1}{2}\beta\omega)\end{aligned}\right\} \tag{I.22}$$

a well-known result which we shall use in the sequel.

We now use (I.19) to derive the differential equation satisfied by the distribution $\rho_\psi(\cdot,t)$ corresponding to ψ_t. We have:

$$\tilde{\rho}_\psi(k,t) = \langle\psi;\ \gamma(t)\left[W_o^\beta(kz_o)\right]\rangle \tag{I.23}$$

and thus, upon using the definition (I.14)

$$\tilde{\rho}_\psi(k,t) =$$

$$\tilde{\rho}_\psi(e^{-\lambda t}k)\exp\{-\frac{1}{4}\,\Theta\,\frac{1}{\omega}\,k^2(1-e^{-2\lambda t})\} \qquad (I.24)$$

From this explicit form, we verify immediately that

$$\{\partial_t - \lambda\left[1-\partial_k\cdot k - \frac{1}{2}\,\Theta\,\frac{1}{\omega}\,k^2\right]\}\tilde{\rho}_\psi\,(k,t) = 0 \qquad (I.25)$$

i.e.

$$\{\partial_t - D\left[\partial_x^{\,2}+\beta V'(x)\partial_x+\beta V''(x)\right]\}\,\rho_\psi(x,t) = 0 \qquad (I.26)$$

where $D = \lambda/\beta\Omega^2$. We can therefore conclude that $\rho_\psi(x,t)$ satisfies the phenomenological diffusion equation for a particle moving in the potential $V(x)$ given by (I.22), which is the effective classical potential at the natural temperature β for our quantum harmonic oscillator. Clearly, the above derivation is not limited to the algebra of the position observables and can be transferred immediately to any algebra $W_o^\beta(\mathbb{R}z_o)''$. For instance with $z_o = \omega^{\frac{1}{2}}$ we obtain the algebra of the momentum observables, and an equation similar to (I.26) holds, with the only modification that Ω^2 now is replaced by $\Lambda^2 = \omega^{-2}\Omega^2$, as to be expected.

We thus arrived now at a complete interpretation of the stochastic process $\gamma(\mathbb{R}^+)$ as a diffusion process for the quantum harmonic oscillator considered in this section.

It will be useful, for the analysis to be carried on in the next section, to notice that this process can be described by a contractive semi-group of self-adjoint operators acting in the Hilbert space $\mathfrak{K}_o^\beta$ and defined by:

$$S(t)W_o^\beta(z)\Phi_o^\beta = \gamma(t)\left[W_o^\beta(z)\right]\Phi_o^\beta \qquad (I.27)$$

where z runs over $\mathbb{C}$ and t over $\mathbb{R}^+$.

II. STATISTICAL MECHANICS

The first aim of this section is to show that the dissipative process of the preceding section can be described within the framework of quantum statistical mechanics. The second aim will then be to show that among the many such descriptions which are indeed possible, there exists one which is minimal in a sense to be made precise below.

We shall admit here that our first aim will be reached if we can produce: (i) a von Neumann algebra $\mathfrak{N}_1$, (ii) a (normal) state ϕ_1 on $\mathfrak{N}_1$, and (iii) a continuous one-parameter group $\alpha(\mathbb{R})$ of automorphisms of the von Neumann algebra $\mathfrak{N} = \mathfrak{N}_o \otimes \mathfrak{N}_1$ such that for every normal state ψ on $\mathfrak{N}_o$ we have

$$<\psi;\gamma(t)[N_o]> = <\psi\otimes\phi_1;\alpha(t)\left[N_o\otimes I\right]> \tag{II.1}$$

for all N_o in $\mathfrak{N}_o$ and all t in $\mathbb{R}^+$. Physically $(\mathfrak{N}_1,\phi_1)$ will thus play the role of a thermal bath, and $\alpha(\mathbb{R})$ will describe the total time-evolution of the whole system consisting of the "system-of interest" ($\mathfrak{N}_o$ and its normal states) in interaction with the "thermal bath" $(\mathfrak{N}_1,\phi_1)$.

We first construct the algebra $\mathfrak{N}$. Let $\mathcal{T}$ denote the Hilbert space $\mathcal{L}^2(\mathbb{R},dx)$. With Θ defined as in (I.13) we define for every f in $\mathcal{T}$:

$$\hat{\phi}(f) = \exp \{- \frac{1}{4} \Theta \ ||f||^2 \} \tag{II.2}$$

We know (see for instance Theorem III.1.7 in [3]) that there exists: (i) a separable Hilbert space $\mathfrak{H}$; (ii) a mapping W^β from $\mathcal{T}$ into $\mathcal{U}(\mathfrak{H})$, the unitary operators on $\mathfrak{H}$, such that for every f and g in $\mathcal{T}$

$$\left.\begin{aligned} &W^\beta(f)W^\beta(g) = W^\beta(f+g) \exp\{i \text{ Im } (f,g)/2\} \\ &W^\beta(f)^* = W^\beta(-f) \end{aligned}\right\} \tag{II.3}$$

and (iii) a normalized vector Φ in $\mathfrak{H}$ such that

$$\hat{\phi}(f) = (\Phi,W^\beta(f)\ \Phi) \tag{II.4}$$

for all f in $\mathcal{T}$, and the linear manifold generated by $W^\beta(f)\ \Phi$, when f runs over $\mathcal{T}$, is dense in $\mathfrak{H}$. This representation, which is uniquely determined (up to unitary equivalence) by the above conditions can be explicitly presented as follows. Let $(\mathfrak{H}_F,W_F(\mathcal{T}),\Phi_F)$ be the usual Fock space representation of the canonical commutation relations. Let now $\mathfrak{H} = \mathfrak{H}_F\otimes\mathfrak{H}_F$ and $\Phi = \Phi_F\otimes\Phi_F$; one verifies that the above conditions are satisfied with

$$\left.\begin{aligned} &W^\beta(f) = W_F(\xi_+f) \otimes W_F(\xi_-f) \\ &\text{with } \xi_\pm = \left[\tfrac{1}{2}\ (\Theta \pm 1)\right]^{\frac{1}{2}} \end{aligned}\right\} \tag{II.5}$$

We now take for $\mathfrak{N}$ the von Neumann algebra $W^{\beta}(\mathcal{T})''$ generated by $\{W^{\beta}(f) \mid f \in \mathcal{T}\}$, and for ϕ the normal state on $\mathfrak{N}$ defined by $\langle\phi;N\rangle = (\Phi, N\Phi)$ for all N in $\mathfrak{N}$. To identify the von Neumann algebras $\mathfrak{N}_o$ and $\mathfrak{N}_1$, to be respectively associated to the "system of interest" and to the "thermal bath", we single out the vector f_o in $\mathcal{T}$ with:

$$f_o(x) = \left[\frac{1}{\pi}\left(\frac{\lambda}{\lambda^2 + x^2}\right)\right]^{\frac{1}{2}} \tag{II.6}$$

Let $\mathcal{T}_o$ be the one-dimensional subspace of $\mathcal{T}$ containing f_o, and $\mathcal{T}_1$ be its orthogonal complement in $\mathcal{T}$. We then identify $\mathfrak{N}_o$ with $W^{\beta}(\mathcal{T}_o)''$, $\mathfrak{N}_1$ with $W^{\beta}(\mathcal{T}_1)''$ and ϕ_1 with the restriction of ϕ to $\mathfrak{N}_1$; we also notice that the restriction of ϕ to $\mathfrak{N}_o$ coïncides with ϕ_o^{β}, the equilibrium state characterized by (I.13).

The next step is to define the time-evolution of the composite system. To this end, we consider the mapping $u : \mathbb{R} \to \mathcal{U}(\mathcal{T})$ defined by:

$$\Big(u(t)f\Big)(x) = \exp(-ixt)\, f(x) \tag{II.7}$$

For every t in $\mathbb{R}$ and every f in   we define

$$\alpha(t)\left[W^{\beta}(f)\right] = W^{\beta}\Big(u(t)f\Big) \tag{II.8}$$

and denote by $\alpha : \mathbb{R} \to \mathcal{A}ut(\mathfrak{N})$ its extension to $\mathfrak{N}$. Clearly $\alpha(\mathbb{R})$ is a continuous group of automorphisms of $\mathfrak{N}$ and its dual $\alpha^*(\mathbb{R})$ leaves ϕ invariant since $\hat{\phi}(f) = \hat{\phi}\Big(u(t)f\Big)$ for all t in $\mathbb{R}$ and all f in $\mathcal{T}$. Therefore there exists a representation $U : \mathbb{R} \to \mathcal{U}(\mathcal{H})$ such that

$$\left.\begin{aligned} &\alpha(t)\left[N\right] = U(t)\, N\, U(-t) \\ &U(t)\Phi = \Phi \end{aligned}\right\} \tag{II.9}$$

for all t in $\mathbb{R}$ and all N in $\mathfrak{N}$.

To prove the fundamental embedding formula (II.1) we make the following two preliminary remarks: first,

$$\Big(f_o\,,\, u(t)\, f_o\Big) = \exp(-\lambda t) \tag{II.10}$$

for all t in $\mathbb{R}^+$; and second, with P_o (resp. P_1) denoting the projector from $\mathcal{T}$ to $\mathcal{T}_o$ (resp. $\mathcal{T}_1$) we have, for every f in $\mathcal{T}$:

$$W^{\beta}(f) = W^{\beta}(P_o f)\, W^{\beta}(P_1 f) \tag{II.11}$$

with $W^{\beta}(P_o f)$ in $\mathfrak{N}_o$ and $W^{\beta}(P_1 f)$ in $\mathfrak{N}_1$. We can now compute for every normal state ψ on $\mathfrak{N}_o$ and every f in $\mathcal{T}_o$:

$$<\psi\otimes\phi_1;\alpha(t)\left[W^{\beta}(f)\right]> =$$

$$<\psi;W^{\beta}(P_o u(t)f)><\phi_1;W^{\beta}(P_1\ u(t)\ f)$$

$$<\psi;W_o^{\beta}(e^{-\lambda t}f)>\ \exp\{-\tfrac{1}{4}\Theta\|u(t)f-(f_o,u(t)f)f_o\|^2\}$$

$$<\psi;W_o^{\beta}(e^{-\lambda t}f)>\ \exp\{-\tfrac{1}{4}\Theta\|f\|^2\ (1-e^{-2\lambda t})\}$$

$$<\psi;\gamma(t)\left[W_o^{\beta}(f)\right]>,$$

from which (II.1) follows by continuity, thus completing the proof of the assertion that the dissipative process of the preceding section can be described within the framework of statistical mechanics, in the sense described in the beginning of the present section. It should furthermore be added that the above model is essentially equivalent to that obtained by Davies[4] as a limit of Hamiltonian systems, specifically the van Hove "long-time-weak-coupling" limit of an infinite one-dimensional chain of harmonic oscillators interacting via a quadratic, translation invariant interaction of the Ford-Kac-Mazur variety.

We now want to point out that this "mechanical" model of the "diffusion process" considered in the first section, is minimal in a precise sense. We first remark that f_o is cyclic in $\mathcal{T}$ with respect to $u(\mathbb{R})$. To see this, suppose indeed that g in $\mathcal{T}$ satisfies $(g,u(t)f_o) = 0$ for all t in $\mathbb{R}$; this is equivalent to saying that the Fourier transform $(g^*f_o)^{\sim}(k)$ of g^*f_o vanishes for all k, i.e. that g^*f_o is identically zero; since f_o is strictly positive, this is equivalent to $g \equiv 0$, thus proving our assertion. From this and from the fact that Φ is cyclic in $\mathcal{H}$ for $W^{\beta}(\mathcal{T})$, follows that $\mathcal{H}_o^{\beta}$ (which is the close linear span of $W^{\beta}(\mathcal{T}_o)\Phi$) is cyclic in $\mathcal{H}$ for $U(\mathbb{R})$ defined by (II.9). We therefore conclude that $(\mathcal{H}, U(\mathbb{R}),P)$ is <u>the</u> minimal Nagy extension[5] of $(\mathcal{H}_o^{\beta}, S(\mathbb{R}^+))$, where P denotes the projector from $\mathcal{H}$ onto $\mathcal{H}_o^{\beta}$, and $S(\mathbb{R}^+)$ is the contractive semi-group defined by (I.27). Clearly then, every dynamical system $(\bar{\mathfrak{N}}, \bar{\Phi}, \alpha(\mathbb{R}))$ containing $(\mathfrak{N}_o, \Phi_o^{\beta}, \gamma(\mathbb{R}^+))$ in the sense described in the beginning of this section will contain our model $(\mathfrak{N},\Phi, \alpha(\mathbb{R}))$ which is therefore minimal.

III. ERGODIC PROPERTIES

The purpose of this section is to examine the ergodic properties of the system $(\mathfrak{N}, \Phi, \alpha(\mathbb{R}))$ constructed in section II. Asides from the didactic interest of this simple system where most things can easily be computed explicitly, the canonical way in which this model is linked to the diffusion process of section I seems to indicate that some of these properties might indeed belong intrinsically to the microscopic theory of quantum transport phenomena.

We first recall that the normal state ϕ on $\mathfrak{N}$ given by $<\phi;N> = (\Phi,N\Phi)$ for all N in $\mathfrak{N}$ is invariant under the automorphism group $\alpha(\mathbb{R})$. Since the action of $\alpha(\mathbb{R})$ is isomorphic to that of the translation group (albeit in momentum space), we have that $\alpha(\mathbb{R})$ acts in a "norm-asymptotically abelian" manner (for a precise definition, see [3]) on the algebra of (quasi-local) observables. In particular, we can easily verify here, as a consequence of the Riemann-Lebesgue lemma, that

$$\lim_{t\to\infty} \left\| \left[W^\beta(f) \, , \alpha(t) \left[W^\beta(g) \right] \right] \right\| = 0 \tag{III.1}$$

for all f and g in $\mathcal{S}$. From this we conclude in particular that $\alpha(\mathbb{R})$ acts in a "η-asymptotically abelian" manner (see for instance [3]) on our algebra $\mathfrak{N}$, i.e. for every invariant mean η on $\mathbb{R}$ we have:

$$\eta<\phi;N_1 \left[N_2, \alpha(t)[N_3] \right] N_4> = 0 \tag{III.2}$$

for all N_1, N_2, N_3, N_4 in $\mathfrak{N}$.

Both for the physical interpretation of the theory and for its mathematical development, it is useful to consider the automorphism group $\alpha^\beta(\mathbb{R})$ of $\mathfrak{N}$ defined as follows: for every t in $\mathbb{R}$ and every f in $\mathcal{S}$ we write:

$$\left.\begin{aligned} &\alpha^\beta(t) \left[W^\beta(f) \right] = W^\beta\big(u^\beta(t)f\big) \\ &\text{with } u^\beta(t)f = \exp(-i\omega t)f \end{aligned}\right\} \tag{III.3}$$

From (II.2), (II.3) and (II.3) one checks by a straight-forward computation that for every f and g in $\mathcal{S}$ there exists a function $\phi^\beta_{f,g}(z)$ holomorphic in the strip $0 \leq \text{Im } z \leq \beta$, and such that for every t in $\mathbb{R}$

$$\left.\begin{aligned} \phi^{\beta}_{f,g}(t) &= \langle\phi;W^{\beta}(g)\alpha^{\beta}(t)\left[W^{\beta}(f)\right]\rangle \\ \phi^{\beta}_{f,g}(t+i\beta) &= \langle\phi;\alpha^{\beta}(t)\left[W^{\beta}(f)\right]W^{\beta}(g)\rangle \end{aligned}\right\} \qquad (III.4)$$

i.e. 3) that ϕ is a KMS state with respect to $\alpha^{\beta}(\mathbb{R})$. From the explicit form (II.5) of $W^{\beta}(\mathcal{Y})$ one checks that $\mathfrak{N}$ is a factor, i.e. $\mathfrak{N}\cap\mathfrak{N}' = I$. Consequently 3), ϕ cannot be decomposed as a mixture of other states satisfying (III.4), a fact which we refer to by saying that:

$$\phi \text{ is extremal KMS with respect to } \alpha^{\beta}(\mathbb{R}) \qquad (III.5)$$

Since clearly $\alpha^{\beta}(\mathbb{R}) \neq$ id, the identity automorphism of $\mathfrak{N}$, we can conclude (see for instance Theorem II.2.14 in 3)) that:

$$\mathfrak{N} \text{ is a type III - factor.} \qquad (III.6)$$

Moreover, from the explicit form (II.5) of $W^{\beta}(\mathcal{Y})$, one can see that Φ is also cyclic in $\mathcal{H}$ with respect to $W^{\beta}(\mathcal{Y})'$, so that for any N in $\mathfrak{N}$, $N\Phi = 0$ if and only if $N = 0$, a fact which is referred to by the assertion that

$$\Phi \text{ is separating for } \mathfrak{N}. \qquad (III.7)$$

We now come back to $\alpha(\mathbb{R})$ and notice that the definitions (II.7-8) and (III.3) immediately imply that

$$\alpha^{\beta}(\mathbb{R}) \text{ and } \alpha(\mathbb{R}) \text{ commute.} \qquad (III.8)$$

From (III.2) and (III.5) we conclude (see for instance Cor. 3 to Theorem II.2.12 in 3)) that the following ergodic properties hold for the automorphism group $\alpha(\mathbb{R})$ of $\mathfrak{N}$:

$$\phi \text{ is extremal } \alpha(\mathbb{R})\text{-invariant;} \qquad (III.9)$$

$$\phi \text{ is the only normal } \alpha(\mathbb{R})\text{-invariant state on } \mathfrak{N}; \qquad (III.10)$$

with $U(\mathbb{R})$ defined as in (II.9) and E_o denoting the projector on $\{\Psi\in\mathcal{H} \mid U(t)\Psi = \Psi \quad \forall\, t \in\mathbb{R}\}$:

$$E_o \text{ is one-dimensional} \qquad (III.11)$$

and thus coïncides with the projector E_{Φ} on $\mathbb{C}\Phi$; for any invariant mean η on $\mathbb{R}$, the following clustering property hold

$$\eta\langle\phi;N_1\alpha(t)[N_2]\rangle = \langle\phi;N_1\rangle\langle\phi;N_2\rangle \tag{III.12}$$

for all N_1 and N_2 in $\mathfrak{N}$; moreover

$$\mathfrak{N}' \cap U(\mathbb{R})' = \mathbb{C}\, I \tag{III.13}$$

Furthermore, we get (see for instance Cor. 2 to Theorem II.2.7 in 3)):

$$\mathfrak{N} \cap U(\mathbb{R})' = \mathbb{C}\, I \tag{III.14}$$

i.e. the only observables N in $\mathfrak{N}$ which are invariant under $\alpha(\mathbb{R})$ are the multiples of the identity. If in addition we use, instead of (III.2), the stronger property (III.1) the clustering property (III.12) can be strengthened (see for instance Theorem III.1.4 in 3)) to:

$$\lim_{t\to\infty} \langle\phi;N_1\alpha(t)[N_2]N_3\rangle = \langle\phi;N_2\rangle\langle\phi;N_1N_3\rangle \tag{III.15}$$

for all N_1, N_2 and N_3 in $\mathfrak{N}$. With $N_3 = I$ this relation reduces to the "strong mixing" property:

$$\lim_{t\to\infty} \langle\phi;N_1\alpha(t)\,[N_2]\rangle = \langle\phi;N_1\rangle\langle\phi;N_2\rangle \tag{III.16}$$

Following the scheme encountered in classical ergodic theory 6), we now try to climb up the ladder towards stronger ergodic properties.

The next natural question therefore concerns the nature of the spectrum of the generator H of the unitary group $U(\mathbb{R})$ describing the time evolution via equation (II.9). To answer this question we first notice that the spectrum of the generator h of $u(\mathbb{R})$ in $\mathcal{T}$ extends over $\mathbb{R}$, is absolutely continuous with respect to Lebesgue measure, and is simple. We next define, in the Fock space $\mathcal{H}_F$, the unitary group $U_F(\mathbb{R})$ by:

$$\left.\begin{aligned} W_F\big(u(t)f\big) &= U_F(t)\; W_F(f)\; U_F(-t) \\ U_F(t)\;\Phi_F &= \Phi_F. \end{aligned}\right\} \tag{III.17}$$

Since $U_F(\mathbb{R})$ commutes with the number operator, the spectrum of its generator H_F is the union of a discrete, non-degenerate eigenvalue {0} (corresponding to the vacuum vector Φ_F), and $\mathbb{R}$ over which the spectrum is absolutely continuous with respect to Lebesgue measure, homogeneous, and countably infinitely degenerated. Following the usage in classical ergodic theory[6], we characterize this situation by saying that H_F has "Lebesgue spectrum". To transfer this information to the generator H of the unitary group defined in (II.9) we notice that the conjugation $f \to f^*$ appearing in the explicit form (II.5) of the representation $W^\beta(\mathcal{S})$ of the canonical commutation relations is still unspecified up to unitary operator. A particular choice of that operation will clearly not affect the spectrum of H; we use this freedom to define now $f \to f^*$ as the usual complex conjugation in the momentum representation of $\mathcal{S}$, namely $f^*(k) = f(k)^*$. In this representation (II.7) becomes

$$\big(u(t)f\big)(k) = f(k-t) \qquad \text{(III.18)}$$

so that

$$\big(u(t)f\big)^* = u(t)f^* \qquad \text{(III.19)}$$

We therefore have, upon comparing (II.5), (II.8), (III.19):

$$U(t) = U_F(t) \otimes U_F(t) \qquad \text{(III.20)}$$

i.e.

$$H = H_F \otimes I + I \otimes H_F \qquad \text{(III.21)}$$

from which it clearly follows that

$$H \text{ has Lebesgue spectrum .} \qquad \text{(III.22)}$$

As in the case of classical ergodic theory, this is a rather strong property which implies in particular: ϕ extremal $\alpha(\mathbb{R})$-invariant (III.9); (III.11); and the strong mixing property (III.16).

It is interesting at this point to compare the evolution $\alpha(\mathbb{R})$ and the evolution $\alpha^\beta(\mathbb{R})$ with respect to which ϕ is an extremal KMS state. In Fock space, we define the unitary group $U_F^\beta(\mathbb{R})$ by:

$$\left.\begin{aligned} &W_F\big(u^\beta(t)f\big) = U_F^\beta(t)\, W_F(f)\, U_F^\beta(-t) \\ &U_F^\beta(t)\, \Phi_F = \Phi_F \end{aligned}\right\} \qquad \text{(III.23)}$$

Its generator H_F^β clearly has discrete spectrum, namely $\omega\mathbb{Z}^+ = \{n\omega | n = 0,1,2,\ldots\}$. In $\mathcal{H}$ itself the unitary group $U^\beta(\mathbb{R})$ is defined by:

$$\left.\begin{aligned} W^{\beta}\big(u^{\beta}(t)f\big) &= U^{\beta}(t)W^{\beta}(f)U^{\beta}(-t) \\ U^{\beta}(t)\,\Phi &= \Phi \end{aligned}\right\} \qquad \text{(III.24)}$$

Since now (III.19) is replaced by

$$\big(u^{\beta}(t)f\big)^{*} = u^{\beta}(-t)\, f^{*} \qquad \text{(III.25)}$$

we get, by a reasoning similar to that done on $U(\mathbb{R})$, that:

$$U^{\beta}(t) = U_F^{\beta}(t) \otimes U_F^{\beta}(-t) \qquad \text{(III.26)}$$

so that the generator H^{β} of $U^{\beta}(\mathbb{R})$ is:

$$H^{\beta} = H_F^{\beta} \otimes I - I \otimes H_F^{\beta} \qquad \text{(III.27)}$$

Consequently:

$$\left.\begin{array}{l}\text{the spectrum of } H^{\beta} \text{ is discrete and coïncides} \\ \text{with } \omega\mathbb{Z} = \{n\omega \mid n = 0, \pm 1, \pm 2, \dots\}.\end{array}\right\} \qquad \text{(III.28)}$$

The above analysis of the behaviour of $(\mathfrak{N},\phi)$ under $\alpha(\mathbb{R})$ and $\alpha^{\beta}(\mathbb{R})$ shows that:

$$\phi \text{ is a homogeneous periodic state on } \mathfrak{N} \qquad \text{(III.29)}$$

i.e. [7] (i) with $G(\phi)$ denoting the group of all automorphisms of $\mathfrak{N}$ the adjoints of which leave ϕ invariant, and $\mathfrak{N}^{G(\phi)}$ denoting the set of all elements of $\mathfrak{N}$ which are left invariant under the automorphisms belonging to $G(\phi)$, then $\mathfrak{N}^{G(\phi)} = \mathbb{C}\, I$; and (ii) there exists a smallest $T \neq 0$ (namely here $T = 2\pi/\omega$) such that the modular group $\alpha^{\beta}(\mathbb{R})$ associated to ϕ satisfies $\alpha^{\beta}(T) = id$, the identity automorphism of $\mathfrak{N}$.

Our last step is to verify that the dynamical system $(\mathfrak{N},\phi,\alpha)$ is a non-abelian special K-flow in the sense of [1], i.e. satisfies certain properties which make it a natural quantum generalization of the classical systems known [6] as Kolmogorov-Sinai K-flows.

We define for every t in $\mathbb{R}$ the closed subspace $\mathcal{Y}(t)$ of $\mathcal{Y}$ by:

$$\mathcal{Y}(t) = \overline{\mathrm{Sp}}\,\{u(s)f_o \mid s \leq t\}\,. \qquad \text{(III.30)}$$

Its orthocomplement $\mathcal{Y}^{\perp}(t)$ in $\mathcal{Y}$ is characterized as the subspace of all g in $\mathcal{Y}$ such that the Fourier transform $(g^* f_o)^{\sim}$ of $g^* f_o$ has support in the semi-infinite interval $[s,\infty)$. We have:

$$\left.\begin{array}{lll} \text{(i)} & s \leq t & \text{implies } \mathcal{Y}^{\perp}(t) \subseteq \mathcal{Y}^{\perp}(s) \\ \text{(ii)} & \cap_{t \in \mathbb{R}} \mathcal{Y}^{\perp}(t) = \{0\} & \\ \text{(iii)} & \cup_{t \in \mathbb{R}} \mathcal{Y}^{\perp}(t) = \mathcal{Y} & \end{array}\right\} \qquad \text{(III.31)}$$

from which we conclude

$$\left.\begin{array}{lll} \text{(i)} & s \leq t & \text{implies } \mathcal{Y}(s) \subseteq \mathcal{Y}(t) \\ \text{(ii)} & \cup_{t \in \mathbb{R}} \mathcal{Y}(t) = \mathcal{Y} & \\ \text{(iii)} & \cap_{t \in \mathbb{R}} \mathcal{Y}(t) = \{0\} & \end{array}\right\} \qquad \text{(III.32)}$$

We next define for every t in $\mathbb{R}$ the von Neumann algebra

$$\mathcal{A}(t) = W^{\beta}(\mathcal{Y}(t))'' \qquad \text{(III.33)}$$

and we notice that the closed linear span of (t)Φ is

$$[\mathcal{A}(t)\Phi] = \mathcal{H}_F(\mathcal{Y}(t)) \otimes \mathcal{H}_F(\mathcal{Y}(t)) \qquad \text{(III.34)}$$

where $\mathcal{H}_F(\mathcal{Y}(t))$ is the Fock space constructed over $\mathcal{Y}(t)$. Upon denoting simply by $\mathcal{A}$ the von Neumann algebra $\mathcal{A}(t=0)$ we have:

$$\mathcal{A}(t) = \alpha(t)[\mathcal{A}] \qquad \text{(III.35)}$$

$$\left.\begin{array}{ll} \mathcal{A}(-t) \subseteq \mathcal{A} & \text{for all } t \geq 0 \\ \cap_{t \in \mathbb{R}}[\mathcal{A}(t)\Phi] = \mathbb{C}\,\Phi & \\ \cup_{t \in \mathbb{R}} \mathcal{A}(t) = \mathcal{N} & \end{array}\right\} \qquad \text{(III.36)}$$

(where, in the last equality, the left-hand side is meant to be the von Neumann algebra generated by all $\mathcal{A}(t)$ as t runs over $\mathbb{R}$).

The relations (III.36) being satisfied, the system $(\mathcal{N},\Phi,\alpha,\mathcal{A})$ is then (by definition [1]) a "completely self-refining dynamical system". Since, in addition it satisfies: (III.7);

$$\left.\begin{array}{l}\alpha(t)[Z] = Z \quad \forall\, t \in \mathbb{R},\ \forall\, Z \in \mathfrak{N} \cap \mathfrak{N}' \\ \phi \text{ is not a trace on } \mathfrak{N}\end{array}\right\} \qquad \text{(III.37)}$$

the latter conditions being equivalent [1] to (III.6); and

$$\alpha^{\beta}(t)\mathcal{A} \subseteq \mathcal{A} \qquad \forall\ t \in \mathbb{R} \qquad \text{(III.38)}$$

$(\mathfrak{N},\phi,\alpha,\mathcal{A})$ is by definition [1] a "non-abelian special K-flow". This property is a very strong ergodic property, comparable to that encountered in the study of K-flows in classical ergodic theory. Its strength is illustrated in particular by the fact that from this property alone one can conclude [1] that all the following properties hold: (III.2); (III.4) (including the existence and uniqueness of the group of automorphisms $\alpha^{\beta}(\mathbb{R})$ entering there, and in the formulation of (III.38)); (III.5); (III.8); (III.9); (III.10); (III.11); (III.12); (III.13); (III.14); (III.16); (III.22); (III.28); and (III.29). We also noted in [1] that the condition (III.38), satisfied by the model constructed here, implies that there exists a conditional expectation ξ from $\mathfrak{N}$ to $\mathcal{A}$ with respect to ϕ. It might be added here that in this specific model we also have:

$$\alpha^{\beta}(t)\left[\mathfrak{N}_o\right] \subseteq \mathfrak{N}_o \qquad \forall\ t \in \mathbb{R} \qquad \text{(III.39)}$$

and that there thus exists, for the same general reason, a conditional expectation $\mathfrak{D}$ from $\mathfrak{N}$ to $\mathfrak{N}_o$ with respect to ϕ.

We actually have for every N_o in $\mathfrak{N}_o$ and every N_1 in $\mathfrak{N}_1$:

$$\mathfrak{D}(N_o \otimes N_1) = \langle\phi_1;N_1\rangle\, N_o \qquad \text{(III.40)}$$

From the fact that ϕ is the product state $\phi_o^{\beta}\otimes\phi_1$ follows indeed that the σ-weakly continuous, faithful projector $\mathfrak{D}:\mathfrak{N}\to\mathfrak{N}_o$ satisfies the defining condition for a conditional expectation, namely (taking into account the fact that ϕ_o^{β} is the restriction of ϕ to $\mathfrak{N}_o$):

$$\langle\phi;N_o^{*}\, N\, N_o\rangle = \langle\phi_o^{\beta};N_o^{*}\mathfrak{D}(N)N_o\rangle \qquad \text{(III.41)}$$

for all N_o in $\mathfrak{N}_o$, and all N in $\mathfrak{N}$. With the help of $\mathfrak{D}$ our embedding condition (III.1) can be rewritten as:

$$\mathfrak{D}\,\alpha(t)\mathfrak{D} = \gamma(t)\mathfrak{D} \quad \text{for all} \quad t \geq 0 \qquad \text{(III.42)}$$

$\mathfrak{D}$ is thus to be interpreted physically as a "coarse-graining" operation with respect to which the dissipative system $(\mathfrak{N}_o,\gamma)$

appears as a "reduced description" of the conservative system $(\mathfrak{N},\alpha)$. The consistency between the irreversibility characteristic of $(\mathfrak{N}_0,\gamma)$ and the reversibility of $(\mathfrak{N},\alpha)$ is underlined here by the fact that (III.42) is actually a particular case of:

$$\mathfrak{D}\,\alpha(t)\mathfrak{D} = \gamma(|t|)\mathfrak{D} \qquad \text{(III.43)}$$

which holds for all t in $\mathbb{R}$.

REFERENCES

1) G.G. Emch, Non-abelian special K-flows, preprint, Rochester, 1973.

2) A. Messiah, Quantum Mechanics, Vol. 1, North-Holland, Amsterdam, 1969.

3) G.G. Emch, Algebraic Methods in Statistical Mechanics and Quantum Field Theory, J. Wiley-Interscience, New York, 1972.

4) E.B. Davies, Diffusion for Weakly Coupled Quantum Oscillators, Commun. Math. Phys. 27 (1972), 309-325.

5) B. Sz.-Nagy, Appendice in F. Riesz et B. Sz. Nagy, Leçons d'Analyse Fonctionnelle, Gauthier-Villars, Paris, 1955.

6) V.I. Arnold and A. Avez, Ergodic Problems of Classical Mechanics, W.A. Benjamin, New York, 1968.

7) M. Takesaki, Periodic and Homogeneous States on a von Neumann Algebra I, Bull. Amer. Math. Soc. 79 (1973), 202-206; II, ibid. 79 (1973), 416-402; III, ibid. 79 (1973), 559-563; The Structure of a von Neumann Algebra with Homogeneous Periodic State, Acta Math. 131 (1973), 79-121.

COMPOSITE PARTICLES IN MANY-BODY THEORY

Wesley E. Brittin* and Arthur Y. Sakakura†

* University of Colorado, Boulder, Colorado
† Colorado School of Mines, Golden, Colorado

The stability of free atoms and molecules (and even interacting atoms and molecules, if the interactions are not too large) allows us to consider them as "particles" with structure. The successes of early kinetic theory and spectroscopy bear this out. We know in fact that atoms, molecules, nuclei, etc. may be thought of as being composed of particles more elementary than themselves. Of course the nuclei which are "elementary" for atoms and molecules are "composite" in terms of protons and neutrons. Composite structures when far enough apart or not interacting too strongly obey Fermi (Bose) statistics if they are themselves made of odd (even) numbers of fermions. This is just a statement of the famous Ehrenfest-Oppenheimer theorem[1]. We present a method for treating interacting composites as though they were particles obeying simple Bose or Fermi commutation relations. Our method appears to generalize and unify a number of previous methods[2-6].

For simplicity we consider a system of hydrogen-like atoms, and let $|\gamma\rangle$ be a complete orthonormal set of states for free atoms. The lable γ includes translation of the center of mass and refers to continuum as well as bound states for the relative motion. We have orthogonality,

$$\langle\gamma|\gamma'\rangle = \delta(\gamma,\gamma') \qquad (1.1)$$

and completeness

$$\sum_{\gamma} |\gamma\rangle\langle\gamma| = 1 \qquad (1.2)$$

Enz/Mehra (eds.), Physical Reality and Mathematical Description. 494–500.

where 1 is the unit operator for 2-particle states.

A basis for the neutral N-atom system is given by

$$|\Psi_N^F(\gamma_1 \dots \gamma_N)\rangle = |\gamma_1\rangle\ |\gamma_2\rangle\ \dots |\gamma_N\rangle. \tag{1.3}$$

Orthogonality and completeness for these states are expressed by

$$\langle\Psi_N^F(\gamma_1 .. \gamma_N)|\Psi_N^F(\gamma_1' .. \gamma_N')\rangle = \delta(\gamma_1,\gamma_1')\delta(\gamma_2,\gamma_2') .. \delta(\gamma_N,\gamma_N') \tag{1.4}$$

and

$$\sum_{\gamma_1 .. \gamma_N} |\Psi_N^F(\gamma_1 \dots \gamma_N)\rangle\ \langle\Psi_N^F(\gamma_1 .. \gamma_N)| = \hat{P}_N^F \tag{1.5}$$

The label F refers to Fock and $\hat{P}_N^F$ is the projector onto the subspace of Fock space having precisely N-atoms. Our Fock space has all types of symmetry present, so the physical states form the subspace corresponding to complete antisymmetry of both protons and electrons. We denote our Fock space by H^F.

We introduce an ideal[2] space H^G which corresponds to "bound" atoms, and "free" protons and electrons. H^G is constructed from the vacuum through the application of a^*_α's, a^*_i's, and a^*_j's where a^*_α creates a bound atom in the state $|\alpha\rangle$, a^*_i creates a proton in the state $|i\rangle$ and a^*_j an electron in the state $|j\rangle$. These operators satisfy the usual elementary commutation (anticommutation) relations

$$\left[a_\alpha,\ a^*_\beta\right]_- = \delta(\alpha,\beta) \tag{1.6}$$

$$\left[a_i,\ a^*_{i'}\right]_+ = \delta(i,i') \tag{1.7}$$

$$\left[a_i,\ a^*_{j'}\right]_+ = \delta(j,j'),\dots,. \tag{1.8}$$

The vectors

$$|\Psi^G(\alpha_1 \dots \alpha_N\ i_1 \dots i_P, j_1 .. j_E\rangle \equiv |\Psi^G(\alpha i j)\rangle = \frac{1}{\sqrt{M!P!E!}}\ a^*_{\alpha_1} \dots a^*_{\alpha_N}\ a^*_{i_1} \dots$$

$$a^*_{i_P}\ a^*_{j_1}\ \dots\ a^*_{j_E}|0\rangle^G \tag{1.9}$$

are normalized such that

$$\langle\Psi^G(\alpha i j)\rangle|\Psi^G(\alpha i j)\rangle = \delta^S(\alpha,\alpha')\ \delta^A(i,i')\ \delta^A(j,j') \tag{1.10}$$

where δ^S is completely symmetric in its indices and δ^A is completely antisymmetric in its indices. The states (1.9) are complete in the sense that

$$\sum_{\alpha,i,j} |\Psi^G(\alpha ij)\rangle \langle\Psi^G(\alpha ij)| = \hat{P}^G(M,P,E) \qquad (1.11)$$

where $\hat{P}^G(M,P,E)$ is the projector for that subspace of H^G corresponding to M bound atoms, P protons and E electrons.

We wish to transform H^F into H^G in order to treat interacting atoms in terms of "bound" atoms and "free" protons and electrons. To do this we first would like $|\gamma\rangle$ to correspond to $a^*_\alpha|0\rangle^G$ if γ corresponds to a bound atom and to correspond to some $a^*_i\ a^*_j|0\rangle^G$ if γ corresponds to an ionized atom. To make this correspondence involves some choice which hopefully can be made on physical grounds. At this stage we note that the interacting continuum 2-particle states can always be obtained from the free 2-particle states by means of an isometry $\hat{\Omega}$

$$\hat{\Omega}|i\rangle\ |j\rangle \equiv |ij\rangle \quad \text{continuum} \qquad (1.12)$$

The isometry $\hat{\Omega}$ satisfies the relations

$$\hat{\Omega}^*\ \hat{\Omega} = 1 \qquad (1.13)$$

$$\hat{\Omega}\ \hat{\Omega}^* = 1 - \hat{P}_B \qquad (1.14)$$

where $\hat{P}_B$ is the projector onto the 2-particle bound states,

$$\hat{P}_B = \Sigma_\alpha|\alpha\rangle\langle\alpha| \qquad (1.15)$$

The isometry $\hat{\Omega}$ may be used to lable the continuum states

$$|ij\rangle \equiv \hat{\Omega}|i\rangle\ |j\rangle \qquad (1.16)$$

and in this manner we may make the correspondence

$$|\gamma\rangle \to a_\gamma^*|0\rangle^G = \begin{cases} a^*_\alpha|0\rangle^G, & \text{if } \gamma = \alpha \\ a^*_i\ a^*_j|0\rangle^G, & \text{if } \gamma = ij \end{cases} \qquad (1.17)$$

The operator $\hat{\Omega}$ has the well-known intertwining property

$$\hat{h}\ \hat{\Omega} = \hat{\Omega}\ \hat{h}_o \qquad (1.18)$$

where $\hat{h}$ is the full two-particle hamiltonian and $\hat{h}_o$ is the free two-particle hamiltonian.

The Mapping $\hat{V}^*$

In H^G we now introduce the states

$$|\Psi_N^G(\gamma)> \equiv |\Psi_N^G(\gamma_1 \dots \gamma_N)> = N(\gamma)\ a^*_{\gamma_1} \dots a^*_{\gamma_N} |0>^G \tag{2.1}$$

where $N(\gamma) = 1/\sqrt{N!\,(N-M)!}$, M being the number of α's appearing among the N γ's. The $|\Psi^G{}_N>$ are thus normalized so that

$$\sum_{\{\gamma\}} |\Psi^G{}_N(\gamma)><\Psi^G{}_N(\gamma)| = \hat{P}^G{}_N \tag{2.2}$$

is the projector onto the subspace of H^G having N free proton-electron pairs and no bound atoms, or N-1 free proton-electron pairs and one bound atom,.... That is

$$\hat{P}^G_N = \sum_{M=0}^{N} \hat{P}^G_{M,\ N-M,\ N-M} \tag{2.3}$$

Because combinatorics can become complicated we introduce operators $\tilde{a}^*_\gamma$:

$$\tilde{a}^*_\gamma = \begin{cases} a^*_\alpha & \text{if } \gamma = \alpha \\ a^*_i\ a^*_j \,/\, \sqrt{\hat{N}_e+1} & \text{if } \gamma = ij \end{cases} \tag{2.4}$$

where $\hat{N}_e = \sum_j \hat{a}^*_j\, \hat{a}_j$ is the electron number operator. The $\tilde{a}^*_\gamma$'s take care of the $1/\sqrt{(N-M)!}$ in the normalization of $|\Psi^G{}_N>$, and we may write

$$|\Psi^G{}_N(\gamma)> = \frac{1}{\sqrt{N!}}\ \tilde{a}^*_{\gamma_1} \dots \tilde{a}^*_{\gamma_N} |0>^G \tag{2.5}$$

We now introduce the mapping $\hat{V}^* : H^F \longrightarrow H^G$ defined by

$$V^*_N\ |\Psi^F_N\ (\gamma_1 \dots \gamma_N)> = |\Psi^G_N(\gamma_1 \dots \gamma_N)> \tag{2.6}$$

or

$$\hat{V}^*_N = \sum_{\{\gamma\}} |\Psi^G_N\ (\gamma)><\Psi^F_N\ (\gamma)| \tag{2.7}$$

We have

$$\hat{V}^*_N \hat{V}_N = \sum_{\{\gamma\}\{\gamma'\}} |\Psi^G_N(\gamma)\rangle \langle\Psi^F_N(\gamma)|\Psi^F_N(\gamma')\rangle \langle\Psi^G_N(\gamma')|$$

$$= \sum_{\{\gamma\}} |\Psi^G_N(\gamma)\rangle \langle\Psi^G_N(\gamma)| = \hat{P}_N{}^G \qquad (2.8)$$

which means that $\hat{V}_N$ is an isometry. It follows that $\hat{V}_N{}^*$ is also an isometry and

$$\hat{V}_N \hat{V}^*_N = \hat{Q}^F_N \qquad (2.9)$$

is a projector on H^F. $\hat{Q}^F_N$ projects onto the subspace Q^F_N of H^F having the symmetry corresponding to H^G.

Physics in H^G

We may use $\hat{V}$, $\hat{V}^*$ to transform from H^F to H^G.

$$H^F \ni |F\rangle \rightarrow |G\rangle = \hat{V}^* |F\rangle \in H^G \qquad (3.1)$$

Operators $\hat{O}^F$ on H^F are transformed to operators $\hat{O}^G$ on H^G by

$$\hat{O}^F \rightarrow \hat{O}^G = \hat{V}^* \hat{O}^F \hat{V} \qquad (3.2)$$

In this manner relations in H^F are transformed into corresponding relations in H^G.

Physical states $|F\rangle_A$ in H^F are those which satisfy

$$\hat{A}^F|F\rangle_A = |F\rangle_A \qquad (3.3)$$

where $\hat{A}^F$ is the antisymmetrizer for protons and electrons. Since $\hat{A}^F < \hat{Q}^F$

$$\hat{W}^* \equiv \hat{V}^*\hat{A}^F \qquad (3.4)$$

satisfies $\hat{W}\,\hat{W}^* = \hat{A}^F \hat{V} \hat{V}^* \hat{A}^F$

$$= \hat{A}^F \hat{Q}^F \hat{A}^F = \hat{A}^{F^2} = \hat{A}^F. \qquad (3.5)$$

Therefore we have also

$$\hat{W}^* \hat{W} \equiv \hat{A}^G$$

is a projector on H^G, and physical states in H^G must satisfy

$$\hat{A}^G \, |G>_A = |G>_A. \qquad (3.6)$$

Thus if we want to describe our system in terms of "bound" atoms and "free" protons and electrons, we must impose the subsidiary conditions (3.6) on states in H^G. The mapping $\hat{W}^* : |F> \rightarrow |G> = \hat{W}^* \, |F>$ should be ideally the mapping to use since arbitrary states in H^F will automatically be mapped into physical states in H^G. However $\hat{W}^*$ is quite complicated and it is therefore better to map with $\hat{V}^*$ and then impose (3.6). Actually for problems where exchange is very important, our description would not appear to be very useful. It would be better to use the description in terms of elementary particles and not composite particles. However, there are many situations such as in the kinetic theory of ionization processes where the composite bases may prove useful.

The Transformed Hamiltonian

We wish to compute the transform by $\hat{V}$, $\hat{V}^*$ of a two-particle hamiltonian

$$H_N^F = \sum_{k=1}^{N} h_1(k) + \sum_{k>l} h_2(k,l) \qquad (4.1)$$

where $h_1(k)$ acts on the kth pair of particles and $h_2(k,l)$ acts between the kth and lst pairs of particles, eg. in configuration space the lst term of H_N^F is

$$\sum_{i=1}^{N} t_e(i) + \sum_{i=1}^{N} t_p(i) + \sum_{i=1}^{N} V(i,i) \qquad (4.2)$$

where $t_e(i)$ is the kinetic energy of the ith electron, $t_p(i)$ the kinetic energy of the ith proton and $V(i,i)$ is the interaction between the ith electron and the ith proton. The calculation of $\hat{H}_N^G = \hat{V}_N^* \, \hat{H}_N^F \, \hat{V}_N$ is straight-forward (if one uses the $\tilde{a}$'s !) and we find

$$\hat{H}_N^G = \hat{P}_N^G \{ \sum_{\gamma} \tilde{a}^*_{\gamma} <\gamma|h_1|\gamma'> \tilde{a}_{\gamma'} + \frac{1}{2} \Sigma_{\gamma_1\gamma_2\gamma_1'\gamma_2'} \tilde{a}^*_{\gamma_1} \tilde{a}^*_{\gamma_2} <\gamma_1\gamma_2|h_2|\gamma_1'\gamma_2'> \tilde{a}_{\gamma_2'} \tilde{a}_{\gamma_1'} \} \qquad (4.3)$$

The first term $\hat{H}_O^G$ of $\hat{H}^G$ may be written

$$\hat{H}_{NO}^G = \hat{P}_N^G \left\{ \sum_{\alpha\ \alpha'} a_\alpha^* \langle\alpha|\hat{h}_1|\alpha'\rangle a_\alpha \right. \tag{4.4}$$

$$\left. + \sum_{ij\ i'j'} a_i^* a_j^* \frac{1}{\sqrt{\hat{N}_e+1}} \langle ij|h_1|i'j'\rangle \frac{1}{\sqrt{\hat{N}_e+1}} a_{j'} a_{i'} \right\}$$

$$= \hat{P}_N^G \left\{ \sum_\alpha \epsilon_\alpha \hat{n}_\alpha \right.$$

$$\left. + \sum_{iji'j'} a_i^* a_j^* \frac{1}{\sqrt{\hat{N}_e+1}} \langle i|\langle j|\hat{\Omega}^*\hat{h}_1\hat{\Omega}|i'\rangle|j'\rangle \frac{1}{\sqrt{\hat{N}_e+1}} a_{j'} a_{i'} \right\} \tag{4.5}$$

since the cross terms involving $\langle\alpha|ij\rangle$ are zero. But $\hat{\Omega}^*\hat{h}_1\ \hat{\Omega} = \hat{\Omega}^*\hat{\Omega}\ \hat{h}_o = \hat{h}_o$ and $\hat{h}_o|i'\rangle|j'\rangle = (\epsilon_{i'}+\epsilon_{j'})|i'\rangle|j'\rangle$, and

$$H_{ON}^G = \hat{P}_N^G \left\{ \sum_\alpha \epsilon_\alpha \hat{n}_\alpha + \sum_i \epsilon_i \hat{n}_i + \sum_j \epsilon_j \hat{n}_j \right\} . \tag{4.6}$$

So this part of the hamiltonian corresponds to non-interacting bound atoms, protons, and electrons. The interaction part of the hamiltonian is more complicated since it corresponds to interacting pairs of pairs. The N dependence of the interaction hamiltonian is not surprising since we have folded into H_o essentially the fraction 1/N of the proton-electron interaction which corresponds to the complete solution of just the two-body problem.

REFERENCES

1. P.Ehrenfest and J.R.Oppenheimer, Phys.Rev.37, 333 (1931).
2. F.J.Dyson, Phys.Rev.102, 1217 (1956).
3. M.Girardeau, J.Math.Phys.4 1096 (1963).
4. R.H.Stolt and W.E.Brittin, Phys.Rev.Lett. 27, 616 (1971).
5. A.Y.Sakakura, Phys.Rev.Lett. 27, 822 (1971).
6. M.Girardeau, Phys.Rev.Lett. 27, 1416 (1971).

ON THE QUANTUM ANALOGUE OF THE LEVY DISTRIBUTION

Elliot W. Montroll*

Institute for Fundamental Studies
Department of Physics and Astronomy
University of Rochester
Rochester, New York

I. INTRODUCTION

Josef Jauch has made important contributions to the extension of the algebraic formulation of quantum mechanical ideas through the use of quaternions.

This is a short note to indicate how an extension can be made in another direction through a generalization of the basic probability model used in quantum theory. The extension given here has been motivated by the observation that an increasing number of practical stochastic processes seem to be better characterized by Lévy distributions[1] than by Gaussian distributions. A quantum analogue of the Lévy distribution will be presented here.

Certain analogies between quantum processes and classical stochastic processes have been recognized for many years[2]. For example, if $P(x_2,x_0;t)$ is the probability of a transition of a variable x from x_0 to x_2 in time t, and if $K(x_2,x_0;t)$ is the quantum mechanical propagator for the transition $x_0 \to x_2$ in time t, both of these functions satisfy chain conditions:

$$P(x_2,x_0;t) = \int P(x_2,x_1;t-\tau)\, P(x_1,x_0;\tau)dx_1 \tag{1a}$$

$$K(x_2,x_0;t) = \int K(x_2,x_1;t-\tau)\, K(x_1,x_0;\tau)dx_1 \tag{1b}$$

---

* This research was partially supported by the U.S. National Science Foundation.

Enz/Mehra (eds.), Physical Reality and Mathematical Description. 501–508.

Of course P and K have different probabilistic interpretations. Let $p(x_0,0)dx_0$ be the probability that the variable x, whose variation is generated by a classical stochastic process, has a value between x_0 and x_0+dx_0 at time t=0. Then the probability distribution of x at time t is

$$p(x,t) = \int P(x,x_0;t)p(x_0,0)\, dx_0 \tag{2a}$$

If the variation of x is generated by a quantum process the wave function $\psi(x,t)$ is obtained from $\psi(x_0,0)$ by

$$\psi(x,t) = \int K(x,x_0;t)\psi(x_0,0)dx_0 \tag{2b}$$

while

$$p(x,t) = |\psi(x,t)|^2 \tag{2c}$$

Furthermore the transition probability $P(x_2,x_0;t)$ is always potitive

$$P(x_2,x_0;t) \geq 0 \tag{3}$$

while the propagator $K(x_2,x_0;t)$ is generally a complex unitary kernel[2] satisfying the relation

$$\int_{-\infty}^{\infty} K(x_2,x_1;t)\, K^*(x_2,x_0;t)dx_2 = \delta(x_1-x_0) \tag{4}$$

The remainder of this section as well as the next section will be concerned with models describing the propagation of packets in a one dimensional translationally invariant infinite medium which has the properties that

$$P(x_2,x,t) \equiv P(x_2-x;t) \tag{5a}$$

$$K(x_2,x;t) \equiv K(x_2-x;t) \tag{5b}$$

The chain conditions (1a) and (1b) then become

$$P(x_2-x_0;t) = \int_{-\infty}^{\infty} P(x_2-x_1;t-\tau)\, P(x_1-x_0;\tau)\, dx_1 \tag{6a}$$

$$K(x_2-x_0;t) = \int_{-\infty}^{\infty} K(x_2-x_1;t-\tau)\, K(x_1-x_0;\tau)dx_2 \tag{6b}$$

It is well known and easy to verify that these equations are satisfied by the Gaussian Packets

$$P(x,t) = \begin{cases} (4\pi Dt)^{-\frac{1}{2}} \exp(-x^2/4Dt) & t \geq 0 \\ 0 & t<0 \end{cases} \tag{7a}$$

$$K(x,t) = \begin{cases} (4\pi\lambda i)^{-\frac{1}{2}} \exp(-ix^2/4\lambda t) & t \geq 0 \\ 0 & t<0 \end{cases} \tag{7b}$$

The Gaussian distribution (7a) is the Green's function of the diffusion equation

$$P_t - DP_{xx} = \delta(x)\delta(t) \tag{8a}$$

and the complex Gaussian (7b) is the Green's function of the free particle Schrödinger equation

$$K_t - (\lambda/i)K_{xx} = \delta(x)\delta(t) \quad \text{with} \quad \lambda = \hbar/2m \tag{8b}$$

The connection between the free particle Schrödinger equation and the diffusion equation is sometimes made by noting that the Schrödinger equation is the diffusion equation with the imaginary diffusion constant $D=(\hbar/2mi)$

We now consider a more general class of solutions of the chain equations (6).

II. THE LEVY DISTRIBUTION AND ITS QUANTUM ANALOGUE

The chain conditions are best discussed through the fourier transforms

$$p(k,t) = \int_{-\infty}^{\infty} P(x,t)\, e^{ikx}\, dx \tag{9a}$$

$$\kappa(k,t) = \int_{-\infty}^{\infty} K(x,t)\, e^{ikx}\, dx \tag{9b}$$

An application of the convolution theorem to (6) yields

$$p(k,t) = p(k,t-\tau)p(k,\tau) \tag{10a}$$

$$\kappa(k,t) = \kappa(k,t-\tau)\kappa(k,\tau) \tag{10b}$$

The general solution of (10a) whose fourier transform has properties suitable for a probability distribution function is the characteristic function of a Lévy distribution

$$p(k,t) = \exp\{-\beta t|k|^{\alpha}\left[1+i\gamma(\alpha)\operatorname{sign}k\right]\} \tag{11}$$

where

$$\gamma(\alpha) = \begin{cases} \tan\frac{1}{2}\alpha\pi & \text{if } 0<\alpha<1 \text{ or } 1<\alpha\leq 2 \\ \log|k| & \text{if } \alpha=1 \end{cases} \tag{12}$$

The case $\alpha=2$ and $\beta=D$ corresponds to the Gaussian diffusion form of $P(x,t)$. Generally

$$P(x,t) = \frac{1}{2\pi}\int_{-\infty}^{\infty} \exp\{-ikx-\beta t|k|^{\alpha}\left[1+i\gamma(\alpha)\operatorname{sign} k\right]\} \tag{13}$$

This function is properly normalized, since

$$\delta(k) = \frac{1}{2\pi}\int_{-\infty}^{\infty} e^{-ikx}\,dx \tag{14}$$

The three cases for which the fourier integral in (13) can be evaluated in closed form are

a) $\alpha=2$, which yields the Gaussian form

b) $\alpha=0$, $\gamma=0$ which yields the Cauchy (or Lorentz) distribution

$$P(x,t) = (2\beta t/\pi)(x^2+\beta^2t^2)^{-1}$$

c) $\alpha=\frac{1}{2}$, $\alpha=1$ which yields the Smirnov distribution

If $\alpha\neq 2$ the asymptotic form for large t is

$$P(x,t) \sim (\beta/\pi)t(1+\gamma)x^{-\alpha-1}\,\Gamma(\alpha+1)\sin\tfrac{1}{2}\pi\alpha \quad \text{if } x^{\alpha+1}\gg\beta t \tag{15a}$$

$$P(x,t) \sim (\beta/\pi)t(1-\gamma)|x|^{-\alpha-1}\Gamma(\alpha+1)\sin\tfrac{1}{2}\pi\alpha \quad \text{if } |x|^{\alpha+1}\gg\beta t \text{ and } x<0 \tag{15b}$$

It is natural to consider the quantum mechanical analogue to the Lévy distribution to be (we choose $\gamma=0$ in the remainder of our discussion) the propagator

$$K(x,t) = \frac{1}{2\pi}\int_{-\infty}^{\infty} \exp -i\{kx+\beta|k|^{\alpha}t\}dk \tag{16}$$

This clearly satisfies the chain condition since the integrand satisfies (10b). That K(x,t) is a unitary kernel can be verified by substituting it into the left hand side of (4) to obtain

$$\frac{1}{(2\pi)}\iint_{-\infty}^{\infty} dk_1\, dk_2 \exp i\{(k_2x_0-k_1x_1)-\beta(|k_1|^{\alpha}-|k_2|^{\alpha})t\} \times$$

$$\frac{1}{2\pi}\int_{-\infty}^{\infty} \exp i\, x_2(k_1-k_2)dx_2$$

$$= \frac{1}{2\pi}\int_{-\infty}^{\infty} dk_1 \exp i\, k_1(x_0-x_1) = \delta(x_0-x_1) \qquad (17)$$

The above results are appropriate for a packet which spreads but which is non-propagating. If the packet moves with velocity v, the generalization of (16) is

$$K(x,t) = \frac{1}{2\pi}\int_{-\infty}^{\infty} \exp -\{ik(x-vt)+i\beta t|k|^{\alpha}\}dk \qquad (18)$$

It is easy to generalize (17) to show that this kernel is also unitary.

III. PERIODIC BOUNDARY CONDITIONS

Except for some special values of α (α=2 being one of the exceptional cases) one cannot construct a simple differential equation for the propagator K(x,t). Hence boundary value problems involving the propagator cannot be solved by the usual methods of theory of differential equations. We employ a method which is closely related to the theory of images. To show how it works we discuss periodic boundary conditions

Consider a function f(x) defined on a ring of circumference L. Then

$$f(x) = f(x\pm L) = f(x\pm 2L) = \ldots. \qquad (19)$$

Such a function has a fourier series expansion

$$f(x) = \sum_{s=\infty}^{\infty} c(s) \exp 2\pi is/L \qquad (20)$$

where the c(s) are given by

$$c(s) = L^{-1}\int_0^L f(x)\exp(-2\pi isx/L)\,dx \tag{21}$$

There is an alternative to (20) which follows from the Poisson summation formula

$$\sum_{s=-\infty}^{\infty} g(k) = \sum_{n=-\infty}^{\infty} G(n) \tag{22}$$

where

$$G(\gamma) \equiv \int_{-\infty}^{\infty} g(y)\exp -2\pi iy\gamma\,dy \tag{23}$$

Now suppose that

$$C(\gamma) = \int_{-\infty}^{\infty} c(y)e^{-2\pi i\gamma y}\,dy \tag{24}$$

Then we find the fourier transform of

$$c(y)\exp 2\pi ixy/L$$

to be

$$\int_{-\infty}^{\infty} c(y)\exp -2\pi iy(\gamma - x/L)dy = C(\gamma - x/L) \tag{25}$$

Hence from (20) and (22)

$$f(x) = \sum_{s=-\infty}^{\infty} c(s)\exp(2\pi ixs/L)$$

$$= \sum_{n=-\infty}^{\infty} C(n-[x/L]). \tag{26}$$

The propagator K_L of a free particle with periodic boundary conditions has the properties listed above plus the analogue of the chain equation (16)

$$K_L(x,t) = \int_0^L K_L(x_1,\tau)K_L(x-x_1,t-\tau)dx_1 \tag{27}$$

If $K_L(x,t)$ has the fourier series expansion

$$K_L(x,t) = \sum_{s=-\infty}^{\infty} c(k,t)\ \exp(-2\pi ixs/L) \qquad (28)$$

then

$$\kappa_L(s,t) = \kappa_L(s,t-\tau)\kappa_L(s,\tau) \qquad (29)$$

where

$$\kappa_L(s,t) \equiv Lc(s,t) \qquad (30)$$

the functional equation (29) is exactly the same as (10b). We choose the solution to depend on a parameter Γ so that

$$K_L(x,t) = \frac{1}{L} \sum_{s=-\infty}^{\infty} e^{-i\Gamma t|s|^{\alpha}}\ e^{-2\pi isx/L} \qquad (31)$$

reduces to (16) in the limit as $L\to\infty$. Let $k=2\pi s/L$ and take the limit of (31) as $L\to\infty$. To be consistent with (16) we must set

$$\Gamma = \beta(2\pi/L)^{\alpha} \qquad (32)$$

A packet which propagates with velocity v can be obtained from (31) by replacing x by (x-vt). One can proceed in a similar manner to construct other propagators which satisfy boundary conditions.

IV. A "FUNDAMENTAL LENGTH" WHEN $\alpha\neq 2$

When $\alpha=2$ our propagators are the tradition ones of non-relativistic quantum theory and the value of the parameter $\beta(\alpha)$ of eqs. (16) and (18) is

$$\beta(2) = \hbar/2m$$

so that the combination

$$\beta(2)|k|^2t = \hbar t|k|^2/2m$$

is dimensionless as required.

If we wish to connect eq. (16) to some quantum mechanical process for arbitrary α, $\beta(\alpha)$ should depend on $\hbar$ in some way, on the mass of the "particle" whose motion it describes, and on the medium through which it propagates. If it is assumed that the medium is characterized by a fundamental length, which we call b, then the dimensions of $\beta(\alpha)$ would be

$$\beta(\alpha) = b^{\lambda_1} \hbar^{\lambda_2} m^{\lambda_3}$$

To make the combination $\hbar\beta(\alpha)|k|^{\alpha}$ dimensionless the values of λ_1, λ_2, and λ_3 must be

$$\lambda_1 = \alpha-2;\ \lambda_2 = 1;\ \text{and}\ \lambda_3 = -1$$

Since

$$\beta(\alpha) = b^{\alpha-2}\hbar/m,$$

the case $\alpha=2$, of standard quantum theory, is singular, being the only one which does not enjoy a fundamental length.

REFERENCES

1. P.Lévy, Processus Stochastique et Mouvement Brownien (Gauthier-Villars, Paris 1948).

2. cf. E.W.Montroll, Comm. on Pure and App. Math, 5, 415 (1952).

EXISTENCE AND BOUNDS FOR CRITICAL ENERGIES OF THE HARTREE OPERATOR

N. Bazley and R. Seydel

Mathematisches Institut
Universität zu Köln

A method is presented for proving the existence and calculating lower bounds for critical energies of the Hartree equation for the helium atom. Our idea is to introduce an "energy" scalar product and use it to approximate the fourth order term in the potential by a smaller second order term. The rigorous lower bounds are then obtained from an associated linear operator containing a trial vector. By a suitable choice of the trial vector we are able to show that the minimal solution exists and is in fact a pointwise positive solution.

Recently M. Reeken[1] demonstrated the existence of a pointwise positive eigenfunction for the radial Hartree equation of the helium atom. Part of his analysis extends the theory in[2] to unbounded regions. Several interesting articles on the mathematical properties of the Hartree equation have already appeared since Reeken's work [3,4,5,6]. All of these results are primarily concerned with the properties of eigenfunctions in terms of the eigenvalue parameter. However, it is well known that in nonlinear problems the eigenvalue is usually not equal to the associated critical energy (see, for example, [7]). In this article we are concerned with the existence of the lowest critical energy of the Hartree equation for helium and the determination of upper and lower bounds to it, a problem which goes back to Wilson and

Enz/Mehra (eds.), Physical Reality and Mathematical Description. 509–515. *All Rights Reserved.*

Lindsay(8). Once the existence has been established, upper bounds are obtained by a nonlinear analogue of the Rayleigh-Ritz procedure. However, the more difficult problem of lower bounds was first considered in (9). Here we introduce a new method to obtain lower bounds which makes essential use of the energy scalar product and the triangle inequality. We are able to reduce the original nonlinear problem to that of finding lower bounds for eigenvalues of an associated linear eigenvalue problem containing a trial vector. We then show by a special choice of this vector that Reeken's solution minimizes the energy. Our analysis was suggested by the chapter on variational methods in Stakgold's book (10) and the results can be extended to more general operators. We emphasize that our motivation stems from the mathematical behavior of the Hartree operator as a nonlinear problem.

In atomic units the Hartree operator A(u) for helium is

$$A(u) = -\tfrac{1}{2}\nabla^2 u - \frac{2u}{|\vec{r}|} + u(\vec{r}) \int_{\mathbb{R}^3} \frac{u^2(\vec{r}\,')}{|\vec{r}-\vec{r}\,'|}\, d\vec{r}\,' , \tag{1}$$

where $\mathbb{R}^3$ denotes three-dimensional real Euclidian space. Each solution to the eigenvalue problem $A(u) = \lambda u$ is required to satisfy the normalization condition

$$\int_{\mathbb{R}^3} u^2(\vec{r})\, d\vec{r} = 1 . \tag{2}$$

The potential energy corresponding to the operator A(u) is given by

$$\Phi(u) = \tfrac{1}{2} \int_{\mathbb{R}^3} |\text{grad } u|^2 d\vec{r} - 2 \int_{\mathbb{R}^3} \frac{u^2(\vec{r})}{|\vec{r}|} d\vec{r} + \tfrac{1}{2} \iint_{\mathbb{R}^3} \frac{u^2(\vec{r})u^2(\vec{r}\,')}{|\vec{r}-\vec{r}\,'|} d\vec{r}\, d\vec{r}\,' \tag{3}$$

and the minimal critical energy E_1 is determined*) by

$$E_1 = \underset{\int_{\mathbb{R}^3} u^2(\vec{r})d\vec{r}=1}{\text{minimum}} \Phi(u) . \tag{4}$$

*) Since in our approximation the electrons are in identical states, the total energy of the atom is simply $2E_1$. This is the best upper bound to the ground state of the linear helium problem obtainable by product test functions.

Let $u_1(\vec{r})$ denote the minimizing function. By the principle of Euler-Lagrange it satisfies the eigenvalue equation

$$-\tfrac{1}{2}\nabla^2 u_1 - \frac{2u_1}{|\vec{r}|} + u_1 \int_{\mathbb{R}^3} \frac{u_1^2(\vec{r}\,')d\vec{r}\,'}{|\vec{r}-\vec{r}\,'|} = \lambda u_1 \quad , \tag{5}$$

where it follows at once from (4) and (5) that λ is related to E_1 by

$$\lambda = E_1 + \tfrac{1}{2} \iint_{\mathbb{R}^3} \frac{u_1^2(\vec{r})u_1^2(\vec{r}\,')}{|\vec{r}-\vec{r}\,'|} d\vec{r}d\vec{r}\,' \quad . \tag{6}$$

It will be shown at the end of this article that this minimum exists and is in fact the positive solution found by Reeken; here, however, we restrict ourselves to the problem of obtaining bounds for E_1.

The variational characterization (4) of E_1 immediately yields the upper bound

$$E_1 \leq \Phi(\tilde{u}) \quad , \tag{7}$$

where $\tilde{u}$ is any admissible trial vector. Our method for a lower bound to E_1 is also based on the use of a trial vector. First we consider the "energy" scalar product $[v,w]$ defined as

$$\left[v,w\right] = \iint_{\mathbb{R}^3} \frac{v(\vec{r})w(\vec{r}\,')}{|\vec{r}-\vec{r}\,'|} d\vec{r}d\vec{r}\,' \tag{8}$$

and remark that the last term in the expression (3) for Φ is simply $\tfrac{1}{2}\,[u^2,u^2]$. The inequality $[v-w,v-w] \geq 0$ implies that for any two functions v and w we have

$$\left[v,v\right] \geq 2\left[v,w\right] - \left[w,w\right] \quad . \tag{9}$$

If we choose $v = u^2$ and $w = z^2$, where $z(\vec{r})$ is a trial vector, (9) becomes

$$\left[u^2,u^2\right] \geq 2\left[u^2,z^2\right] - \left[z^2,z^2\right] \quad . \tag{10}$$

Taking (10) together with (3) we obtain for each u and z the inequalities

$$\Phi(u) \geq \psi(u,z) \quad , \tag{11}$$

where

$$\psi(u,z) = \tfrac{1}{2}\int_{\mathbb{R}^3} |\text{grad } u|^2 d\vec{r} - 2\int_{\mathbb{R}^3} \frac{u^2(\vec{r})}{|\vec{r}|} d\vec{r} + \left[u^2, z^2\right] - \tfrac{1}{2}\left[z^2, z^2\right]$$

$$= \tfrac{1}{2}\int_{\mathbb{R}^3} |\text{grad } u|^2 d\vec{r} - 2\int_{\mathbb{R}^3} \frac{u^2(\vec{r})}{|\vec{r}|} d\vec{r} \tag{12}$$

$$+ \iint_{\mathbb{R}^3} \frac{u^2(\vec{r})z^2(\vec{r}\,')}{|\vec{r}-\vec{r}\,'|} d\vec{r}d\vec{r}\,' - \tfrac{1}{2}\iint_{\mathbb{R}^3} \frac{z^2(\vec{r})z^2(\vec{r}\,')}{|\vec{r}-\vec{r}\,'|} d\vec{r}d\vec{r}\,' \quad .$$

Thus for a fixed trial vector z we have

$$E_1 = \underset{\int_{\mathbb{R}^3} u^2(\vec{r})d\vec{r}=1}{\text{minimum}} \Phi(u) \geq \underset{\int_{\mathbb{R}^3} u^2(\vec{r})d\vec{r}=1}{\text{minimum}} \psi(u,z) \quad . \tag{13}$$

But the first three terms in $\psi(u,z)$ are quadratic in u while the last term is independent of u , so that

$$\underset{\int_{\mathbb{R}^3} u^2(\vec{r})d\vec{r}=1}{\text{minimum}} \psi(u,z) = \underset{\int_{\mathbb{R}^3} u^2(\vec{r})d\vec{r}=1}{\text{minimum}} \left\{ \tfrac{1}{2}\int_{\mathbb{R}^3} |\text{grad } u|^2 d\vec{r} - 2\int_{\mathbb{R}^3} \frac{u^2(\vec{r})d\vec{r}}{|\vec{r}|} \right.$$

$$\left. + \iint_{\mathbb{R}^3} \frac{u^2(\vec{r})z^2(\vec{r}\,')}{|\vec{r}-\vec{r}\,'|} d\vec{r}d\vec{r}\,' \right\} - \tfrac{1}{2}\iint_{\mathbb{R}^3} \frac{z^2(\vec{r})z^2(\vec{r}\,')}{|\vec{r}-\vec{r}\,'|} d\vec{r}d\vec{r}\,'$$

$$= \tilde{\lambda}_1 - \tfrac{1}{2}\iint_{\mathbb{R}^3} \frac{z^2(\vec{r})z^2(\vec{r}\,')}{|\vec{r}-\vec{r}\,'|} d\vec{r}d\vec{r}\,' \quad . \tag{14}$$

Here $\tilde{\lambda}_1$ is the lowest eigenvalue of the <u>linear</u> operator $\tilde{A}$ given by

$$\tilde{A}u = -\tfrac{1}{2}\nabla^2 u - \frac{2u}{|\vec{r}|} + q_z(\vec{r})u \quad , \tag{15}$$

where

$$q_z(\vec{r}) = \int_{\mathbb{R}^3} \frac{z^2(\vec{r}\,')}{|\vec{r}-\vec{r}\,'|} d\vec{r}\,' \tag{16}$$

can be considered as a known potential for fixed $z(\vec{r})$. The eigenvalues of $\tilde{A}$ are not generally known and must be approximated from below. However this can be done by the A. Weinstein theory of intermediate operators (see, for example, [11,12]). Thus we have the final inequality

$$\tilde{\lambda}_1^{\ell} - \tfrac{1}{2} \iint_{\mathbb{R}^3} \frac{z^2(\vec{r})z^2(\vec{r}\,')}{|\vec{r}-\vec{r}\,'|} d\vec{r}d\vec{r}\,' \leq E_1 \leq \Phi(\tilde{u}) \quad , \tag{17}$$

where $\tilde{\lambda}_1^{\ell}$ is a lower bound to the first eigenvalue of $\tilde{A}$.

We observe without proof that another operator which gives lower bounds to the critical values by (17) is

$$\tilde{A}u = -\tfrac{1}{2}\nabla^2 u - \frac{2u}{|\vec{r}|} + z(\vec{r}) \int_{\mathbb{R}^3} \frac{z(\vec{r}\,')u(\vec{r}\,')}{|\vec{r}-\vec{r}\,'|} d\vec{r}\,' \tag{18}$$

This operator is especially suitable when the method of "special choice" is employed to obtain lower bounds to eigenvalues, since

$$A'u = z(\vec{r}) \int_{\mathbb{R}^3} \frac{z(\vec{r}\,')u(\vec{r}\,')}{|\vec{r}-\vec{r}\,'|} d\vec{r}\,'$$

has the easily calculated inverse $(A')^{-1}u = -(1/z)\nabla^2(u/z)$. Numerical estimations of E_1 are presently being carried out.

We remark that in applications of our method we try to choose $z(\vec{r})$ as close to $u_1(\vec{r})$ as possible. Thus reasonable candidates for z might be the Rayleigh-Ritz trial vector $\tilde{u}$ or the nth Hartree iteration vector u_n obtained from the iteration scheme

$$-\tfrac{1}{2}\nabla^2 u_n - \frac{2u_n}{|\vec{r}|} + u_n \int_{\mathbb{R}^3} \frac{u_{n-1}^2(\vec{r})}{|\vec{r}-\vec{r}\,'|} d\vec{r}\,' = \lambda u_n \quad . \tag{19}$$

Until now we have assumed the existence of a minimizing function. Here we use a special choice of z in (13) to prove that the positive solution of M. Reeken[1] uniquely minimizes E_1. First we recall that in his article Reeken proves the existence of a pointwise positive radially symmetric solution $\bar{u}(\vec{r})$ of (1) and (2) with a corresponding eigenvalue $\bar{\lambda}$. Thus $\bar{\lambda}$ and $\bar{u}(\vec{r})$ are respectively the first eigenvalue $\xi_1 = \bar{\lambda}$ and eigenvector $v_1 = \bar{u}$ of the problem

$$-\tfrac{1}{2}\nabla^2 v - \frac{2v}{|\vec{r}|} + q_{\bar{u}}(\vec{r})v = \xi v \quad , \tag{20}$$

where

$$q_{\bar{u}}(\vec{r}) = \int_{\mathbb{R}^3} \frac{\bar{u}^2(\vec{r}\,')d\vec{r}\,'}{|\vec{r}-\vec{r}\,'|} \quad . \tag{21}$$

Hence by the minimum characterization of the first eigenvalue,

$$\bar{\lambda} = \underset{\int_{\mathbb{R}^3} u^2(\vec{r})d\vec{r}=1}{\text{minimum}} \left\{ \tfrac{1}{2}\int_{\mathbb{R}^3} |\text{grad } u|^2 \, d\vec{r} - 2\int_{\mathbb{R}^3} \frac{u^2(\vec{r})}{|\vec{r}|}d\vec{r} + \int_{\mathbb{R}^3} q_{\bar{u}}(\vec{r})u^2(\vec{r})d\vec{r} \right\} \tag{22}$$

is uniquely minimized by the vector $u = \bar{u}(\vec{r})$.

Our object is to prove that the infinum is a minimum in the inequality

$$E_1 = \underset{\int_{\mathbb{R}^3} u^2(\vec{r})d\vec{r}=1}{\text{infinum}} \Phi(u) \geq \underset{\int_{\mathbb{R}^3} u^2(\vec{r})d\vec{r}=1}{\text{minimum}} \psi(u,z) \quad , \tag{23}$$

where $\Phi(u)$ and $\psi(u,z)$ are respectively given by (3) and (12). To do this choose $z = \bar{u}$, the positive solution of Reeken, and use (21) to write $\psi(u,\bar{u})$ as

$$\underset{\int_{\mathbb{R}^3} u^2(\vec{r})d\vec{r}=1}{\text{minimum}} \psi(u,\bar{u}) = \underset{\int_{\mathbb{R}^3} u^2(\vec{r})d\vec{r}=1}{\text{minimum}} \left\{ \tfrac{1}{2}\int_{\mathbb{R}^3} |\text{grad } u|^2 d\vec{r} \right.$$

$$- 2\int_{\mathbb{R}^3} \frac{u^2(\vec{r})d\vec{r}}{|\vec{r}|} + \int q_{\bar{u}}(\vec{r})u^2(\vec{r})d\vec{r} \Bigg\} - \frac{1}{2}\iint_{\mathbb{R}^3} \frac{\bar{u}^2(\vec{r})\bar{u}^2(\vec{r}\,')}{|\vec{r}-\vec{r}\,'|}d\vec{r}d\vec{r}\,' \quad . \qquad (24)$$

Therefore by (22) and (3)

$$\underset{\int_{\mathbb{R}^3} u^2(\vec{r})d\vec{r}=1}{\text{minimum}} \quad \psi(u,\bar{u}) = \bar{\lambda} - \frac{1}{2}\iint_{\mathbb{R}^3} \frac{\bar{u}^2(\vec{r})\bar{u}^2(\vec{r}\,')}{|\vec{r}-\vec{r}\,'|}d\vec{r}d\vec{r}\,' = \Phi(\bar{u}) \ . \qquad (25)$$

Thus (23) now reads

$$E_1 = \underset{\int_{\mathbb{R}^3} u^2(\vec{r})d\vec{r}=1}{\text{infininum}} \quad \Phi(u) \geq \Phi(\bar{u}) \ . \qquad (26)$$

which proves that Reeken's positive solution minimizes Φ . Uniqueness of the minimum easily follows from nodal line properties.

REFERENCES

(1) M. Reeken, J. Math. Phys. 11 (1970) 2505.

(2) N. Bazley, B. Zwahlen, Manuscripta Math. 2 (1970) 365.

(3) K. Gustafson, D. Sather, Rendiconti di Matematica, Vol. IV, Serie VI, (1971) 723.

(4) J. Wolkowisky, Indiana Univ. Math. J. 22 (1972) 551.

(5) C. Stuart, Arch. Rat. Mech. Anal. 51 (1973) 60.

(6) W. Quirmbach, Diplomarbeit, Universität zu Köln.

(7) N. Bazley, M. Reeken, B. Zwahlen, Math. Z. 123 (1971) 301.

(8) W. Wilson, R. Lindsay, Phys. Rev. 47 (1935) 681.

(9) N. Bazley, M. Reeken, Applicable Anal. (to appear).

(10) I. Stakgold, Boundary Value Problems of Mathematical Physics, Vol. II (MacMillan, New York, 1968).

(11) N. Bazley, D. Fox, Intern. J. Quantum Chem. 3 (1969) 587.

(12) J. Gay, Phys. Rev. 135 (1964) A 1220.

LONG RANGE ORDERING IN ONE-COMPONENT COULOMB SYSTEMS

Ph. Choquard

Laboratoire de Physique Théorique
Ecole Polytechnique Fédérale
Lausanne

Coulomb systems are known to play an important role in the non-relativistic physics of matter. One way to classify these systems is by the number of their components : one, two or more. In this paper, we consider in particular the one-component Coulomb systems (O.C.C.S,) and we propose a mechanism leading to the onset of long range order (density oscillations) in those systems. The exploratory character of the third part of this paper should be stressed beforehand.

****

The O.C.C.S. are characterized by an assembly of particles, with charge $\sigma = \pm |\sigma|$ enclosed in a box of volume Λ and immersed in a uniform background of opposite charge. Either rigid walls or periodic boundary conditions are imposed on the system. If the particles are electrons, treated as fermions, then the system is called jellium. In three dimensions, it is used as a model of simple metals where the ionic charge density is represented by a uniform background and, needless to say, this model has been one of the most extensively studied example of the Many-Body Problem [1]. If the particles are ions (with $\sigma = Z|e|$), treated classically and if the uniform background is supposed to represent the electron cloud of the fully ionized matter, then the system is called one-component plasma (O.C.P.), as a model equally extensively studied [2].

The equilibrium properties of the O.C.P. and of the jellium at zero temperature can be analyzed in terms of a single parameter which yields an effective ratio of potential to kinetic energy.

Enz/Mehra (eds.), Physical Reality and Mathematical Description. 516–532. *All Rights Reserved.*

In the three dimensional case, with a designating the mean interparticle distance, ρ the particle density ($a = (3/4\pi\rho)^{1/3}$), β the natural temperature $= 1/k_BT$, the relevant parameter of the O.C.P. model is $\beta\sigma^2/a \equiv \Gamma$. For the jellium k_BT is suitably replaced by $\hbar^2/ma^2$ and the corresponding parameter becomes $a/a_o \equiv r_s$ where $a_o = \hbar^2/m\sigma^2$ is Bohr's radius. At finite temperature and apart from the question of statistics applying to the particles of a given O.C.C.S. we have two parameters. This means that any comprehensive description of the equilibrium properties of an O.C.C.S. in the thermodynamic domain ($\rho \geqslant 0$, $T \geqslant 0$) will have to be based on a two-parameter theory. However there are regions in the (ρ,T)-plane where it is meaningful to study the properties of one-parameter models.

Consider, for example, the following properties of the O.C.P. and of the jellium models. In the small coupling limit the O.C.P. and jellium are well described by their respective Random Phase Approximations, which both start from spatially homogeneous equilibrium densities. In the very stong coupling limit, the equilibrium properties of both models have been investigated in starting from an ordered configuration of the particles, a crystal lattice, the Wigner lattice. Since all improvements made to cover intermediate Γ and r_s values are based on either one of the two starting points, it suffices here to refer to the literature specialized on these subjects [1,2] and to attempt at elucidating the fundamental questions of why can, how and when do density oscillations occur in the O.C.C. systems.

****

The first question : why ? can be dealt with by inspecting the Hamiltonian of the system and by invoking a general argument valid for repulsive system. Let us begin with the Hamiltonian. Let $w(|\underset{\sim}{x}-\underset{\sim}{x}'|)$ be the Coulomb potential in ν dimensions*; let Λ be the volume of a ν dimensional box centered at the origin of the euclidean space R^ν, let Z=1, N the number of particles and ρ, the bath density equal to the particle density N/Λ. Then, disregarding the kinetic energy not relevant to the point, we have :

$$H = H_{pp} + H_{pb} + H_{bb} \tag{1}$$

where

$$H_{pp} = \frac{1}{2}\sum_{i\neq j} \sigma^2 w(|\underset{\sim}{x}_j-\underset{\sim}{x}_i|)$$

* $w(|\underset{\sim}{x}|)$ is defined as the elementary solution of $\Delta w = -\omega_\nu \delta(\underset{\sim}{x})$ with $\omega_1=2$, $\omega_2=2$ and $\omega_3=4$.

$$H_{pb} = -\sum_i \sigma^2 \int_\Lambda d^\nu x w(|\underset{\sim}{x}-\underset{\sim}{x}_i|)$$

$$H_{bb} = \frac{1}{2}\sigma^2\rho^2 \int_\Lambda d^\nu x d^\nu x' w(|\underset{\sim}{x}-\underset{\sim}{x}'|)$$

It is clear that H_{pb}, which expresses the coupling of the system of particles with the bath breaks the translational invariance of the particle system which H_{pp} possesses, but not the rotational invariance, and that the souvenir of the chosen origin in R^ν persists in the thermodynamic limit irrespective of the boundary conditions imposed on the initial box.

The reader familiar with the procedure usually followed to eliminate this feature and to combine the three contributions to H in order to obtain, in three dimensions, for instance, the well-known euclidean invariant Hamiltonian, designated here as standard Hamiltonian, namely

$$H_{st} = \frac{1}{2\Lambda}\sum_{\underset{\sim}{k}\neq 0} \frac{4\pi\sigma^2}{\underset{\sim}{k}^2}(\rho_{\underset{\sim}{k}},\ \rho_{-\underset{\sim}{k}} - N) \tag{2}$$

where

$$\rho_{\underset{\sim}{k}} = \sum_j e^{-i\underset{\sim}{k}\,\underset{\sim}{x}j} \quad ,$$

will have noticed that a key point for going from H to H_{st} is precisely an assumption* of translational invariance, assumption permitting then to shift the integration variables $\{\underset{\sim}{x}-\underset{\sim}{x}_j \rightarrow \underset{\sim}{x}\ \forall j\}$ in H_{pb}. An elementary counter-example showing the persistence of the chosen origin in R^ν is the case where Λ is a sphere of radius R. Writing H of eq.(1) as $H_{pp}-H_{bb} + (H_{pb}+2H_{bb})$ results in

$$H_\odot = H_{pp} - H_{bb} + \sum_j \frac{4\pi\sigma^2\rho}{6}\left(\underset{\sim}{x}_j^2 - \frac{3}{5}R^2\right)$$

then, with the identity

$$\sum_j \underset{\sim}{x}_i^2 \equiv \frac{1}{2N}\sum_{i,j}(\underset{\sim}{x}_j-\underset{\sim}{x}_i)^2 + \frac{1}{N}\left(\sum_j \underset{\sim}{x}_j\right)^2$$

---

* See f.i. [1], p.22 for a careful derivation of eq.(2) where all the steps are explicitly given.

it is seen how the center of mass coordinate $\underset{\sim}{C} = \frac{1}{N}\sum_j \underset{\sim}{x}_j$ occurs and H becomes, with $w^*(|\underset{\sim}{x}_j-\underset{\sim}{x}_i|) = w(|\underset{\sim}{x}_j-\underset{\sim}{x}_i|)+ \frac{2\pi}{3N}\rho(\underset{\sim}{x}_j-\underset{\sim}{x}_i)^2$ and $\omega_p^2 = 4\pi\sigma^2\rho/m$,

$$H_\odot = \frac{1}{2}\sum_{j\neq i}\sigma^2 w^*(\ \underset{\sim}{x}_j-\underset{\sim}{x}_i\ ) + H'$$

with

$$H' = \frac{N}{6} m\omega_p^2 \left(\underset{\sim}{C}^2 - \frac{3}{5}R^2\right),$$

$H_\odot - H'$ being manifestly translational invariant and thus corresponds to H_{st}. For ordinary cubic boxes with either rigid walls or periodic boundary conditions* applied to the system, the problem is more complicated. The key point is to perform a double Fourier series expansion of $w(|\underset{\sim}{x}-\underset{\sim}{x}'|)$ over the domain Λ. One thing is certain, H' will always occur, in addition to other contributions arising from the non-diagonal part of $w_{\underset{\sim}{k},\underset{\sim}{k}'}$. Since the diagonal part $w_{\underset{\sim}{k},\underset{\sim}{k}}$ always contains $4\pi/k^2$ we might just consider a model Hamiltonian $\widetilde{H} = H_{st} + H'$, the factor $3/5\ R^2$ being replaced by $\Lambda^{-1}\int_\Lambda d^3C\ \underset{\sim}{C}^2$. In any case, the breaking of the translation invariance of the particle system is a built-in feature of O.C.C.S. Hamiltonians and such a condition is necessary for permitting spacially inhomogeneous equilibrium densities to occur.

A neat example illustrating the points made above is the one-dimensional case, recently published [3]. In this case, rigid walls and periodic boundary conditions give rise to the same Hamiltonian (since $w(|x-x'|) = -|x-x'|$) and it is found that H of eq.(1), with Λ=L can be written in terms of the ρ_k as

$$H = \frac{1}{L}\sum_{k\neq 0}\frac{\sigma^2}{k^2}\rho_k\,\rho_{-k} + \frac{1}{L}\sum_{\substack{k\neq 0\\ k'\neq 0}}\frac{\sigma^2}{kk'}\rho_k\,\rho_{-k'} \tag{3}$$

The first part is useful for the one-dimensional analogue of eq.(2). As to the second part, in permuting the $\sum_K$ and $\sum_J$ it is found to be exactly $\sigma^2\rho NC^2$ with $C = \frac{1}{N}\sum_j x_j$ and $-L/2 \leqslant \{x_j\} \leqslant L/2$.

---

* Here the p.b.c. are understood in the sense that $\rho(\underset{\sim}{x}+\underset{\sim}{n}L)=\rho(\underset{\sim}{x})$ with $\rho(\underset{\sim}{x}) = \sum_J \delta(\underset{\sim}{x}-\underset{\sim}{x}_j)-\rho$. The corresponding instantanous electric potential $\psi(x)$ is, in general, not periodic over Λ, whereas $\psi'(x)$ is. At ν=1, f.i. $\psi(L/2)-\psi(-L/2)=L\psi'(L/2) + \sigma NC$.

In this form we see that H reduces to H_{st} provided that we freeze the center of mass of the particles. The third useful form for H is obtained in terms of the particle coordinates, once the latter have been ordered : $-L/2 \leqslant x_1 \leqslant x_2 \ldots \leqslant x_N \leqslant L/2$. In this case, we have :

$$H = N \frac{\sigma^2}{12\rho} + \sum_j \sigma^2 \rho (x_j - x_j^o)^2 \tag{4}$$

with

$$x_j^o = \frac{1}{\rho}\left(j - \frac{N+1}{2}\right) \qquad j = 1, \ldots N$$

The first term represents the Madelung energy and the second shows how the ordered particles are harmonically coupled to the bath around their static equilibrium position x_j^o (notice that $C^o = 0$).

In the next section we shall need a slightly generalized version of (4) namely for the case where the particle density ρ_p is varied independently from the bath density ρ. For this purpose let M be the number of particles ($1 \leqslant M \leqslant \rho L = N$) immersed in the bath centered at the origin. In this case, H reads, with $n = \rho L.M$

$$H = M \frac{\sigma^2}{12\rho}(1 - 3n^2) - \frac{\sigma^2}{6\rho} n^3 + \sum_{1 \leqslant j \leqslant M} \sigma^2 \rho \left(x_j - \frac{1}{\rho}\left(j - \frac{M+1}{2}\right)\right)^2 \tag{5}$$

We notice that the static equilibrium positions have not changed, their nearest neighbour spacing being $1/\rho$ $\forall j$ and $M > 1$!

Finally the general argument invoked to favor density oscillations in O.C.C. systems is the same which applies to all repulsive systems, namely a tendency for the particles to repel each other. In summary, intrinsic symmetry breaking due to the attraction of the particles with the bath plus repulsion between the particles are the ingredients answering the first question.

****

The next question : how does long range ordering occur ? will be treated in several steps. In a first one, intended to unveil the underlying mechanism, we want to develop an approximate, strong coupling theory for the one-dimensional O.C.P. In this case the coupling parameter is $\lambda = \beta\sigma^2/\rho$, ρ being the bath density. Examples of approximate strong coupling theories are, typically, Mean-Field Approximations. So, and despite the well-known weakness of such approximations, we look for an appropriate M.F.T. of the

O.C.P. model developed first in the canonical ensemble specified by M, $\rho L = N$, T with $1 \leqslant M \leqslant N$, and then in the thermodynamic limit $M = N \to \infty$, $L \to \infty$, $\rho = N/L$ fixed.

The order parameter of the model is clearly the equilibrium expectation value of the density fluctuation $\rho_p(x) - \rho$, of one particle distribution function

$$\mathcal{P}_M(x) = < \rho_p(x) - \rho >_{M,L,T} = < \sum_{i \leqslant j \leqslant M} \delta(x-x_j) - \rho > \tag{6}$$

A natural candidate for the field variable conjugated to $\mathcal{P}_M(x)$ is the electrostatic potential $\psi_M(x)$, produced by $\mathcal{P}_M(x)$ on a test particle with charge σ^* moving through the system. Since $_M(x)$ will later on play the role of Weiss'molecular field, we choose $\sigma^* = \sigma_p = \sigma$. Thus

$$\psi_M(x) = \sigma \int_{-L/2}^{+L/2} dx' \; w(|x-x'|) \, \mathcal{P}_M(x')$$

where $w(|x-x'|) = - |x-x'|$ in one dimension. This potential function satisfies Poisson's equation

$$\Delta\psi_M(x) = - 2\sigma \, \mathcal{P}_M(x) \tag{7}$$

which gives us a first equation relating ψ_M to $\mathcal{P}_M$. We need a second equation relating the order parameter to ψ_M. But ψ_M is the equilibrium expectation value of the instantaneous electrostatic potential $u_M(x)$ namely

$$u_M(x) = \sigma \sum_{i \leqslant j \leqslant M} (-1) \; |x-x_j| + \sigma\rho \int_{-L/2}^{+L/2} dx' \; |x-x'|$$

and thus, in evaluating the R.H.S. of eq.(7), namely,

$$\mathcal{P}_M(x) = \frac{\int_\Lambda d^M x \sum_j \delta(x-x_j) \; e^{-\beta H_M(x_1,..x_M)}}{\int_\Lambda d^M x \; e^{-\beta H_M(x_1...x_M)}} = \rho$$

we take advantage of the fact that for any one of the M possibilities to have $x_j = x$, the Hamiltonian (1), symmetric in all coordinates, can be written, in setting $(x_1,\ldots,x,..x_M \equiv x,y)$, as

$$H_M(x,y) = u_{M-1}(x,y) + H_{M-1}(y),$$

in such a way that

$$\Delta\langle u_M(x)\rangle_M = 2\sigma\ (\rho - M\ \frac{\int_\Lambda d^{M-1}y\ e^{-\beta u_{M-1}(x,y) - \beta H_{M-1}(y)}}{\int_\Lambda dx\ (\text{numerator})}) \quad (8)$$

This equation is still exact. We have only expressed $\mathcal{P}_M(x)$ as an expectation value of exp $-\beta u_{M-1}(x,y)$ over a canonical ensemble of M-1 particles in the same bath and at the same temperature and thus have prepared the ground for invoking the Mean-Field concept. Before doing that, however, it is enlightening to consider the infinite coupling limit, i.e., the classical ground state of the system. In this case

$$\mathcal{P}^o_M(x) = \sum_j \delta(x - x^o_j) - \rho$$

with

$$x^o_j = \frac{1}{\rho}\ (j - \frac{M+1}{2})$$

and thus $\psi^o_M(x)$ is made up of a sequence of pieces of parabolae having their minima in the midpoint between two neighbouring equilibrium positions and adjusting their slopes so as to accommodate the δ singularities. More precisely, for j=2,.. M-1 and

$$\frac{1}{\rho}\ (j - \frac{M+1}{2}) \leqslant x \leqslant \frac{1}{\rho}\ (j - \frac{M-1}{2})$$

the ground state potential reads

$$\psi^o_M(x) = \sigma\rho\ (x - \frac{1}{\rho}(j - \frac{M}{2}))^2$$

whereas for j = M and $\frac{1}{2\rho}\ (M-1) \leqslant x \leqslant L/2$

$$\psi^o_M(x) = \sigma\rho\ (x - \frac{M}{2\rho})^2$$

and for j = 1 and $-L/2 \leqslant x \leqslant \frac{1}{2\rho}\ (1-M)$

$$\psi^o_M(x) = \sigma\rho\ (x + \frac{1}{\rho}\ (\frac{M}{2} - 1))^2$$

The important remark is now made that if we add one particle to a system of M particles $\psi_M^o(x)$ is bodily shifted by $\rho/2$ to produce $\psi_{M+1}^o(x)$ (apart from an additional branch of parabola) and this in a way which keeps the center of mass frozen at the origin. A corollary is that, apart from edge effects, $\psi_{M+2}^o(x) = \psi_M^o(x)$. If we now let M = N = Lρ the equations defining $\psi_M^o(x)$ remain valid but if we proceed to the thermodynamic limit we must distinguish between N even and odd. Indeed for $N = 2N'$, $N' \to \infty$, $L \to \infty$, $\rho=2N'/L$ fixed, $\psi_{2N'}^o(x) \to \psi_e^o(x) = \psi_e^o(x+1/\rho)$ with $\psi_e^o(x) = \sigma\rho\, x^2$ for $-1/2\rho \leqslant x \leqslant 1/2\rho$ whereas for $N = 2N'+1$, $\psi_{2N'+1}^o(x) \to \psi_o^o(x) = \psi_e^o(x+1/2\rho)$ with $\psi_o^o(x) = \sigma\rho\, (x-1/2\rho)^L$ for $0 \leqslant x \leqslant 1/\rho$.

In summary we find two fixed points for the ground state potential function in the thermodynamic limit,

$$\psi_e^o(x) = \psi_o^o(x+\tfrac{1}{2}a)$$

and (9)

$$\psi_{e,o}^o(x) = \psi_{e,o}^o(x+a)$$

where $a = 1/\rho$ is the lattice constant. It is from this observation that the central idea of this paper stemmed, namely that of a double fixed point theory for the onset of long range order in O.C.C. systems.

What about finite temperatures ? The first remark to make is that Landau-Peierls' argument concerning the absence of long range ordering in one-dimensional crystal lattices at $T > 0$, absence due to the divergence of the mean square fluctuation $<(x_j - x_j^o)^2>$, does not apply here simply because there are no acoustic phonons in the one-dimensional O.C.P. We have indeed an optical branch only with $\omega^2(q) \to \omega_p^2 = 2\sigma^2\rho/m$ as $q \to 0$, where ω_p is the well-known plasma frequency. Therefore the even and odd one-dimensional O.C.P. crystals should persist at finite temperature and since no phase transition is expected to occur in this case, they should always persist. This is precisely what H.Kunz* has recently proved [4].

---

* The first of his results concerning this subject, namely that

$$\lim\,(N'\to\infty,\ L\to\infty,\ \frac{2N'}{L} = \rho)\ \mathcal{P}_{2N'}(x,\lambda) = \mathcal{P}_e(x,\lambda) = \mathcal{P}_e(x+a,\lambda) \neq \text{const.}$$

for all values of $\lambda = \sigma^2\beta a$ was in fact communicated to us before the development of the M.F.T. presented below.

It followed that the search for an approximate strong coupling theory was greatly stimulated. So we come back to eq.(8) and proceed to obtain its Mean-Field approximation. The procedure is standard : if $A_M(\xi, x_1 \ldots x_M)$ is an integrable function of the coordinates, ξ a parameter, then

$$< e^{-\beta A_M(S,x)} >_M = \frac{\int d^M x \, e^{-\beta A_M - \beta H_M}}{\int d^M x \, e^{-\beta H_M}} \xrightarrow{\text{M.F.A.}} e^{-\beta < A_M(\xi) >_M}, \quad (10)$$

the effect of the fluctuations being neglected. Application of this procedure to the R.H.S. of eq.(8), divided by the partition function in the numberator and denominator and with $A_M(\xi,x)=u_{M-1}(x,y)$, results in the following approximate recurrence relation for the Mean Field

$$\phi_M(x) = < u_M(x) >_M \quad \text{(M.F.A.)}$$

namely

$$\Delta \, \phi_M(x) = 2\sigma \left(\rho - \frac{e^{-\beta\sigma\phi_{M-1}(x)}}{\frac{1}{M}\int_\Lambda dx \, e^{-\beta\sigma\phi_{M-1}(x)}}\right) \quad (11)$$

with M = 1, 2, .. N and

$$\phi_o(x) = \sigma\rho x^2 \qquad -L/2 \leqslant x \leqslant L/2$$

Before proceeding to the thermodynamic limit it is worthwhile inspecting some properties of eq.(11) considered per se. To do so, it is convenient to introduce $\lambda = \sigma^2\beta / \rho$, $N = \rho L$,

$$x = (1/\rho) y$$

$$\phi(x,\beta,\tau,\rho) = \sigma \frac{1}{\rho} \varphi(y,\lambda)$$

and obtain

$$\varphi''_M(y) = 2 \left(1 - \frac{e^{-\lambda\varphi_{M-1}(y)}}{\frac{1}{M}\int_{-N/2}^{+N/2} dy \, e^{-\lambda\varphi_{M-1}(y)}}\right)$$

with

$$\varphi_o(y) = y^2 \qquad -N/2 \leqslant y \leqslant N/2$$

An interesting feature of this rather simple recurrence equation is indeed that it contains a mechanism producing density oscillations for sufficiently large λ. The physical mechanism is that if a particle is injected into the system, it is submitted to two opposite tendencies, one to be attracted towards the center of the bath, the other to be repelled by the other particles already present, the new assembly having to collectively re-adjust itself in order to keep its center of mass at the origin, in the mean. These two tendencies are of equal strength and if this strength is sufficient to overcome the thermal agitation, density oscillations occur. As to the mathematical mechanism it is unveiled by the inspection of the zeros, if any, of the r.h.s. of $\varphi''_M(y)$. For this purpose, let $\mathcal{P}_M = \rho p_M$, $q_M = 1-p_M$ so that $\varphi''_M = 2q_M\{\lambda\varphi_{M-1}\}$; then the following occurs : if p_1 maximum at the origin is > 1, q_1 has two zeros symmetric about the origin; since φ''_1 is successively +, -, +, this entails that φ_1 acquires one maximum at the origin and two satellite minima, thus producing a p_2 with one minimum at the origin and two satellite maxima. The argument can then be repeated : if p_M edge maxima are > 1 (the maxima in the interior being larger) then q_M, i.e. φ''_M possesses 2M zeros, producing M+1 maxima of p_{M+1} within which those of p_M are intercalated and so on until M=N. If on the contrary, p_1 maximum is < 1, q_1 is always > 0, i.e. φ''_1 as well, so that $\varphi_1(x)$ will be flatter than $\varphi_o(x)$ around the origin and the flattening mechanism expands <u>without</u> oscillations since $0 \leqslant \varphi''_M \leqslant 2$ for $-N/2 \leqslant x \leqslant N/2$ and M=1,..N. It is clear that the above situations are intimately connected with the strength of the coupling parameter λ, a question which is taken up in the next section. What a numerical analysis of eq.(11) carried out for N = 6, 8, 10 and M = 1 to 4 indicates, is that for $\lambda = 25$ for example, density oscillations are present and that the spacing between the maxima of $p_M(y)$ is, within 5% equal to 1.0 for $M \geqslant 2$ already !

What we have learned from the approximate recurrence relation permits us to write down the equations in the thermodynamic limit <u>provided</u> that we proceed through even (odd) subsequences of N values, i.e. :

$$M = N = 2N' \ ; \quad \text{th.lim. } \phi_{2N'}(x) = \phi_e(x)$$

$$M = N = 2N'+1; \quad \text{th.lim. } \phi_{2N'+1}(x) = \phi_o(x)$$

with the properties,

$$\phi_o(x) = \phi_e(x+b) \qquad\qquad b = \frac{1}{2}a$$

and

$$\phi_{o,e}(x+a) = \phi_{o,e}(x)$$

the trivial case $\phi_o = \text{const} = \phi_e$ being included. Then, and since

$$\text{th. lim. } \frac{1}{N} \int_{-L/2}^{+L/2} dx\, e^{-\beta\sigma\phi_N(x)} = \rho \int_{-b}^{+b} dx\, e^{-\beta\sigma\phi_{e,o}(x)} ,$$

the resulting Mean Field equation, written for $\phi = \phi_e$ or ϕ_o finally reads :

$$\phi''(x) = 2\sigma\rho\left(1 - \frac{e^{-\beta\sigma\phi(x+\frac{1}{2}a)}}{\rho \int_0^a dx\, e^{-\beta\sigma\phi(x)}}\right) \tag{12}$$

It is a non-local, normalized version, of the non-linear Debye-Hückel equation.

In dimensionless form it becomes :

$$\varphi''(y) = 2\left(1 - \frac{e^{-\lambda\varphi(y+\frac{1}{2})}}{\int_0^1 dy\, e^{-\lambda\varphi(y)}}\right)$$

and the question is to study its solutions in the domain $[-\frac{1}{2}, +\frac{1}{2}]$ with the condition $\varphi(y+1) = \varphi(y)$. The mathematical analysis of this equation is under way. The numerical analysis performed in injecting the ground state potential $\varphi^o = y^2 \,(-1/2 \leqslant y \leqslant 1/2)$ and in finding out for which values of λ the solution saturates toward a constant or toward a periodic function reveals that for $\lambda > 20$ a periodic solution symmetric around $y = 0$ persists, whereas for $\lambda < 20$ the trivial solution is reached after the first few iterations. Again the question of exactly when oscillations occur will be discussed in the next section. So far we can say that the approximate strong coupling theory does what we expected. Like any M.F.A. it is invalid in the small coupling limit owing to the effects of the fluctuations of the order parameter having been neglected.

However, it enables us to envisage generalizations in two and three dimensions where M.F.T. is usually meaningful as first order approximation, and this in being guided by the multiple fixed-point concept associated with frozen center of mass. Consider for example the three-dimensional case where the order parameter might assume, among many others, simple cubic symmetry, this case being treated as illustration. Let $N = N_1 N_2 N_3$ be the number of simple cubic cells in Λ with a lattice constant a_L related to the mean interparticle distance defined before through $(4\pi/3)\, a^3 = \rho^{-1} = a_L^3$, i.e. $a_L = (4\pi/3)^{1/3}\, a$. A natural generalization of the one-dimensional situation $\mathcal{P}_{N+2}(x) = \mathcal{P}_N(x)$, apart from edge effects, is that $\mathcal{P}_{N_1+2,N_2+2,N_3+2}(x) = \mathcal{P}_{N_1,N_2,N_3}(x)$, apart from surface effects. Between these equivalent fixed points of the order parameter we have in particular $\mathcal{P}_{N_1+1,N_2+1,N_3+1}(x)$. If we now invoke the principle of frozen center of mass, we are led to make the Ansatz, for the interior of the crystal,

$$\mathcal{P}_{N_1+1,N_2+1,N_3+1}(x_1,x_2,x_3) = \mathcal{P}_{N_1,N_2,N_3}(x+\tfrac{1}{2}a, y+\tfrac{1}{2}a, z+\tfrac{1}{2}a).$$

Then a ν dimensional generalization of the Mean-Field eq.(12) for an O.C.P. with long range order possessing the symmetries generated by the basis vectors $\underset{\sim}{a}_1$, $\underset{\sim}{a}_2$, $\underset{\sim}{a}_3$, of a primitive, electrically neutral, cell, may be conjectured to read

$$\Delta\, \phi(\underset{\sim}{x}) = \omega_\nu \sigma\rho \left(1 - \frac{e^{-\beta\sigma\phi(\underset{\sim}{x}+\underset{\sim}{b})}}{\rho_v \int d^\nu x\; e^{-\beta\sigma\phi(\underset{\sim}{x})}} \right) \tag{13}$$

where v is the volume of a cell and $\underset{\sim}{b}$ a vector being chosen after inspection of the minima found in calculating the potential ψ^* produced by the supposed crystal on a test particle with a small charge σ^* having the same sign as the particle charge. As an illustration of this point, consider the primitive cell of a face-centered cubic lattice : Since there are two tetrahedral holes and one octahedral hole in this cell and since the minimum of ψ^* is normally deeper in the center of the octahedral ones, we shall have $\underset{\sim}{b} = \frac{1}{2}(\underset{\sim}{a}_1 + \underset{\sim}{a}_2 + \underset{\sim}{a}_3)$. Now it is clear that the above equation may also describe states of higher symmetries than the ν dimensional crystalline one. This feature will naturally belong to the analysis presented in the next section. In summary, a double fixed-point mechanism has been found which can produce long range order.

****

The last question : when do density oscillations set in ? can in principe be treated in the framework of the theory of bifurcations. However, the non-local character of the proposed equations is not a harmless feature. We shall therefore follow here a pedestrian way to explore the situation. We are primarily interested in finding out the smallest values of the coupling parameters at which a non trivial solution for the periodic potentials can emerge. To do so, we inspect the linearized eqs.(13) studied here for simple cubic symmetry only. Defining

$$K_y^2 = \omega_\nu \beta\sigma^2\rho$$

where we recall that $\omega_1=2$, $\omega_2=2\pi$, and $\omega_3=4\pi$, we find the non-local linear Debye-Hückel equation

$$\Delta\phi(\underset{\sim}{x}) = K_\nu^2 \, \phi(\underset{\sim}{x}+\underset{\sim}{b}) \qquad \underset{\sim}{b}=\frac{1}{2}(a,a,a) \tag{14}$$

Making the Ansatz

$$\phi(\underset{\sim}{x}) = \phi_{\underset{\sim}{Q}} \, e^{i\,\underset{\sim}{Q}\cdot\underset{\sim}{x}} + c.c$$

where $\underset{\sim}{Q}$ is a reciprocal lattice vector, we find :

$$(-\underset{\sim}{Q}^2 - e^{i\,\underset{\sim}{Q}\cdot\underset{\sim}{b}} K_\nu^2)\,\phi_{\underset{\sim}{Q}} = 0$$

Clearly if $\underset{\sim}{b}$ were absent, the parenthesis would be the dielectric function $\varepsilon(Q)$ in the Debye-Hückel approximation and there would be no real critical value K_ν^2 making the parenthesis zero. The bifurcation equation thus reads

$$e^{i(\pi - \underset{\sim}{Q}\cdot\underset{\sim}{b})}\,\underset{\sim}{Q}^2 = K_{\nu c}^2$$

It follows that the smallest $K_{\nu c}^2$ is associated with the smallest reciprocal lattice vectors compatible with the condition $\underset{\sim}{Q}\cdot\underset{\sim}{b} = \pi$. These $\underset{\sim}{Q}'_s$ are the three basis vectors $(\frac{2\pi}{a_L},0,0;\ 0,\frac{2\pi}{a_L},0;\ 0,0,\frac{2\pi}{a_L})$, since $\underset{\sim}{b} = (\frac{1}{2}\,a_L,\ \frac{1}{2}\,a_L,\ \frac{1}{2}\,a_L)$. Introducing the coupling parameters :

$$\lambda_\nu = \beta\sigma^2\rho a_L^2$$

we find the $\lambda_{\nu c}$ to be

$$\lambda_{\nu c} = (2\pi)^2/\omega_\nu \quad , \tag{15}$$

i.e. $\lambda_{1c} = 2\pi^2$, $\lambda_{2c} = 2\pi$ and $\lambda_{3c} = \pi$. Remarkably enough, λ_{1c}=19,8 accurately corresponds to the results of the numerical analysis reported in the preceding section. In the two- and three-dimensional cases we have new predictions : the onset of uni-axial long range ordering at critical values of λ_ν which are not very large. These predictions call for the following comments. i) According to Landau's argument, such states should not occur basically for the same reason which preludes one-dimensional crystal lattices above T = 0. Here again this argument does not apply because of the frequency gap ω_p at q = 0. Therefore we venture to suggest the possibility for one-component plasmas in 2 (3) dimensions to admit equilibrium states in the form of fluid rows (sheets) periodically arranged in the system and this within a certain range of the coupling parameters λ_ν. (ii) While ours is a purely classical and exploratory theory, the possibility for such equilibrium states to exist for quantum systems has already been advanced and rigorousl formulated by G. Emch and al. [5]. In fact such states belong to the class SP_3 in these author's classification. We note in passing, however, that the states characterized by $\underset{\sim}{Q} = (\frac{2\pi}{a_L}, \frac{2\pi}{a_L})$ at ν = 2 and $\underset{\sim}{Q} = (\frac{2\pi}{a_L}, \frac{2\pi}{a_L}, 0)$ at ν = 3 are forbidden according to the bifurcation equation. In three dimensions the next possible state of simple cubic symmetry is $\underset{\sim}{Q} = (\frac{2\pi}{a_L}, \frac{2\pi}{a_L}, \frac{2\pi}{a_L})$ and in this case $\lambda_c = 3\pi$.(iii) Considering the body of knowledge accumulated, so far, for the three dimensional O.C.P. through computer simulation as well as numerical and theoretical studies, the prediction of the onset of uni-axial long range ordering at λ_c(M.F.A) = π, i.e. since $\Gamma = \lambda\frac{a_L}{a}$, at Γ_c(M.F.A) = $\pi(4\pi/3)^{1/3}$, is certainly the most interesting one. In a few words the equilibrium properties of the O.C.P. are indeed understood as follows : for Γ values up to 1 the system is well represented by its non-linear Debye-Hückel approximation typically described by its monotonous increasing pair distribution function $g(x,r) = \exp\left[-(\Gamma/x)\exp\left[-(3\Gamma)^{\frac{1}{2}}x\right]\right]$, x=r/a. For Γ between 1 and 2, short range order sets in, as indicated by damped oscillations of g(x). The theoretical origin of this feature is well understood [6]. For $\Gamma \simeq 3$ (!) all data collected so far indicate that the system becomes unstable (negative compressibility). For very large Γ values we have already said that the system had been studied in starting from an ordered configuration of the particles, a crystal lattice. The lattice model is, however, not stable either because of the dominant, negative, Madelung compressibility. Thus for very large Γ values alterations of the model have to be made to provide stability, as for instance by smearing out the point charges, by quantizing the system of particles or by replacing the background by a real Fermi sea, all changes introducing an essential second parameter in the sense discussed in the introduction. It remains, however, that the O.C.P. model as such, seems to be stable in a certain domain of the ρ, T plane bounded by a curve $r(\rho,T) = \overline{r}$ and that long range ordering may further stabilize the system as preliminary calculations seem to indicate. We have to recall, however, that no proof for this limited stability of the C.O.P. model is, at present, available. It thus would be of great interest to "prepare" the system for computer simulation in single "domain" form in this range of Γ values and thus try to improve our understanding of the equilibrium properties of this model. If this proposal makes sense then it is imperative to improve our estimate of Γ_c.

As a last subject fitting within this section, we take up the question of improving the bifurcation equation so as to incorporate the effects of the fluctuations of the order parameter. A tempting asssumption to make here is the applicability of linear response of the order parameter $\mathcal{P}(x)$ linear in the selfconsistent field, not in an external field. The reason for being careful about this assumption is that in general $\mathcal{P}(\phi)$may be non-analytic in the amplitude ϕ_Q, ϕ^*_Q. A typical mechanism which might occur here and produce non-analyticity could be the parallel of the well-known breakdown of perturbative band theory near Bragg reflexions, a problem solved by Peierls' method which led to the gap energy being $= 2\,|V_{\underset{\sim}{Q}}|$ where $V_{\underset{\sim}{Q}}$ is a Fourier component of the periodic potential $V(\underset{\sim}{x})$. While keeping in mind this restriction, nothing prevents us to explore the potentialities offered by linear response theory. The tremendous advantage is clearly that we could exploit all the knowledge accumulated concerning the properties of equilibrium density - density correlation functions of spatially homogeneous systems. So we may expect non-trivial applications of the generalized bifurcation equation now proposed.

We start again from Poisson's equation, in Q space, with the displaced selfconsistent field on the left, i.e.

$$- \underset{\sim}{Q}^2 \, e^{-i\underset{\sim}{Q}\cdot\underset{\sim}{b}} \phi(\underset{\sim}{Q}) = - \omega_\nu \, \sigma \mathcal{P}_{\underset{\sim}{Q}}$$

or

$$e^{i(\pi-\underset{\sim}{Q}\cdot\underset{\sim}{b})} \phi(\underset{\sim}{Q}) = -\omega_\nu \frac{\sigma}{\underset{\sim}{Q}^2} \mathcal{P}_{\underset{\sim}{Q}}$$

Now the linear response of $\mathcal{P}(\phi) \sim - \sigma\, \phi_Q$, which is the selfconsistent field, directly produces, on the r.h.s. of the last equation, the polarizability $+ \omega_\nu\, \alpha(\underset{\sim}{Q})\, \phi_{\underset{\sim}{Q}}$*. The generalized bifurcation equation would then read :

$$(e^{i(\pi - \underset{\sim}{Q}.\underset{\sim}{b})} - \omega_\nu\, \alpha(Q))\, \phi_{\underset{\sim}{Q}} = 0 \qquad (15)$$

Again it is clear that if $\underset{\sim}{b} = 0$ the parenthesis would become $-(1 + \omega_\nu\, \alpha(\underset{\sim}{Q})) = - \varepsilon(Q)$, $\varepsilon(Q)$ being the dielectric function. A similar equation can be written down for quantum systems*. Here the dielectric function becomes the so-called retarded $\varepsilon^R(Q,0)$ at zero frequency. A first inspection of the jellium model indicates that since $2\, K_F/Q_{min} = 1/2,\ (2/\pi)^{1/2},\ (3/\pi)^{1/3}$ in one, two and three dimensions, respectively, eq.(15) is confortably applicable in one dimension, roughly in two and not in three dimensions, precisely for the reason anticipated above. In the last case Peierls' technique will have to be employed. An estimate of the critical value $\lambda_{1,c}$, based on the Random Phase Approximation results in $\lambda_{1,c}(R.P.A.) = m\sigma^2 a^3/\hbar^2 = 2\pi^4/en^3$, a value which is to be considered as purely indicative. In summary, a careful application of the theory of bifurcation will enable us to answer the third question.

As a conclusion to this paper, the following perspectives may be mentioned : if the multiple fixed-point idea proves right, then the study of long range ordering in other systems such as the two-component Coulomb systems and even the hard sphere model could be envisaged. However, considerable cooperative efforts will still be needed to make progress towards a real theory of crystallization. Here, we wish to acknowledge invaluable discussions with H.Kunz, with J.Lebowitz in September 1973 and lately with G.Emch.

---

* See ref. [1] eq.(14.4), p.175 for the quantum case : the r.h.s. is $= \dfrac{\omega_\nu \sigma}{\underset{\sim}{Q}^2} (\pi^* U)^R$ where π^* is the "proper" polarization and U the selfconsistent field.

REFERENCES

1. Quantum Theory of Many-Particles Systems, sections 3, 10, 11, 12, 14.
A.L.Fetter, J.D.Walecka, I.S.P.A.P., Mc Graw-Hill, 1971.

2. Statistical Mechanics of Dense Ionized Matter, I, II.
J.P.Hansen, Phys. Rev. A6, 3096, Dec. 1973.
J.P.Hansen, E.L.Pollock, Phys. Rev. A6, 3107, Dec. 1973.

3. L'Hamiltonien du Plasma à une Composante et une Dimension.
Ph.Choquard et R.R.Sari, H.P.A. 46, 464, 1973.

4. Equilibrium Properties of the One-Dimensional Classical Electron Gas.
H.Kunz, submitted to Annals of Physics.

5. Breaking of Euclidean Symmetry with an Application to the Theory of Crystallization.
G.G.Emch, H.J.Knops and E.J.Verboven, J.M.P. 11, 1655, 1970.

6. Onset of Short Range Order in a One-Component Plasma.
Ph.Choquard and R.R.Sari, Phys. Lett. 40A, 109, 1972.

A SCALE GROUP FOR BOLTZMANN-TYPE EQUATIONS

Erdal Inönü

Department of Physics, Middel East Technical University, Ankara, Turkey

For problems of multiple scattering a group of scale transformations on the scattering kernel is introduced. The group is shown to be valid in general for both linear and non-linear Boltzmann-type equations and expresses the fact that the effect of collisions which do not change the state of the motion can be taken care of by a change in the mean free path.

To look for new symmetries in old and familiar equations is a fascinating, although rarely fruitful game. Thinking that Josef Jauch would understand well the temptation to take part in this game, I wish to indicate here the existence of a scale group for multiple scattering problems and make some comments on it.

* * *

The group is defined on the scattering kernel in a rather trivial manner and simply expresses the fact that intuitively one can arbitrarily change the effective mean free path between successive collisions by changing the proportion of collisions which do not disturb the state of the system. Whether this group may have any special meaning in quantum theory remains to be seen. Here it is considered only for classical statistical problems.

Let us start with a linear problem of multiple scattering. Such a problem consists in general of a repetition of elementary events, the combined result appearing in a linear way. Denote by $f_n(s)$ the probability distribution function for the system under consideration after n elementary events, where s stands for all

Enz/Mehra (eds.), Physical Reality and Mathematical Description. 533–540.

the variables which are needed to define the state of the system, and let $P(s,t)$ be the probability that an elementary event carries the state s into a unit volume element at t. Then, the characteristic equation defining the problem of multiple scattering is

$$f_{n+1}(t) = \int f_n(s)\, P(s,t)ds \quad (1)$$

where the integral is taken over all possible values of s and $P(s,t)$ is normalized as

$$\int P(s,t)ds = 1 \ . \quad (2)$$

The total distribution function $f(t)$ defined as

$$f(t) = \sum_{n=o}^{\infty} f_n(t) \quad (3)$$

obeys the following Boltzmann-type integral equation:

$$f(t) = f_o(t) + \int f(s)\, P(s,t)ds \quad (4)$$

where $f_o(t)$ is the given initial distribution.

In a classic paper written in 1954, Wigner[1)] has shown that if the elementary probability law $P(s,t)$ remains invariant under a group of transformations, one can immediately obtain from Eqs.(1-4) the appropriate moments of the distribution function which, if the group is large enough, suffices to determine $f(s)$ itself. The groups considered by Wigner and his followers[2)] are the displacements and rotations in three dimensional space and time displacements for scattering without energy change, the multiplicative group of real numbers (a scale group) for energy changes due to elastic collisions with nuclei at rest and a group of linear transformations in time and energy with appropriate conditions on the cross sections. In all these cases, the linear Boltzmann equation of the problem remains invariant under the group considered and the methods of solution based on taking the convenient group representations coincide essentially with the well known integral transform methods.

We have noticed during a study of extremely anisotropic scattering in neutron transport theory[3)] that one can also consider symmetry transformations valid only for Eq.(4) but not for (1). Consider, for instance, the following transformation:

$$P(s,t) \rightarrow \alpha P(s,t) + (1-\alpha)\ \delta(s,t) \tag{5a}$$

$$f_o(t) \rightarrow \alpha f_o(t) \tag{5b}$$

where α is a real, positive parameter and, by definition,

$$\int f(s)\ \delta(s,t)ds = f(t)\ . \tag{6}$$

Eq.(4) remains unchanged under the transformation (5a-b). Note also that the integral of $P(s,t)$ remains equal to unity. Combining two successive transformations we have

$$P(s,t) \rightarrow \alpha_1\{\alpha_2 P(s,t) + (1-\alpha_2)\ \delta(s,t)\} + (1-\alpha_1)\ \delta(s,t)$$

or $$P(s,t) \rightarrow \alpha_1\alpha_2 P(s,t) + (1-\alpha_1\alpha_2)\ \delta(s,t) \tag{7}$$

and $$f_o(t) \rightarrow \alpha_1\alpha_2 f_o(t)\ .$$

Thus the transformations (5a-b) form a group isomorphic to the multiplicative group of real, positive numbers, i.e. a kind of scale group.

Let us consider as an example the linear homogeneous Boltzmann equation for neutron transport which reads (using the notation of Ref. 4)

$$\frac{\partial\psi(\vec{r},\vec{v},t)}{\partial t} + \vec{v}\cdot\vec{\nabla}\psi(\vec{r},\vec{v},t) + v\sigma(\vec{r},\vec{v})\psi(\vec{r},\vec{v},t)$$

$$= \int d^3\vec{v}'\ \sigma(\vec{v}'\rightarrow\vec{v},\vec{r})v'\psi(\vec{r},\vec{v}',t)\ . \tag{8}$$

Here $\psi(\vec{r},\vec{v},t)$ is the distribution function giving the average number of neutrons around the point $\vec{r}$, moving with velocities around $\vec{v}$ at time t. $\sigma(\vec{r},\vec{v})$ is the macroscopic total cross section at $\vec{r},\vec{v}$ and $\sigma(\vec{v}'\rightarrow\vec{v},\vec{r})$ is the macroscopic differential cross section at $\vec{r}$ for a scattering from $\vec{v}'$ to $\vec{v}$. Defining the mean number of secondary neutrons coming out of a collision at $\vec{r}$ with velocity $\vec{v}$ as

$$c(\vec{r},\vec{v}) = \frac{\sigma_s(\vec{r},\vec{v})}{\sigma(\vec{r},\vec{v})} \tag{9}$$

where

$$\sigma_s(\vec{r},\vec{v}) = \int \sigma(\vec{v}'\rightarrow\vec{v},\vec{r})d^3\vec{v}' \tag{10}$$

and writing

$$f(\vec{v}'\rightarrow\vec{v},\vec{r}) = \frac{\sigma(\vec{v}'\rightarrow\vec{v},\vec{v})}{\sigma_s(\vec{r},\vec{v})} \tag{11}$$

Eq.(8) becomes

$$\frac{\partial\psi}{\partial t} + \vec{v}\cdot\nabla\psi + v\sigma\psi = c\sigma\int d^3\vec{v}'f(\vec{v}'\rightarrow\vec{v},\vec{r})\ v'\psi(\vec{r},\vec{v}',t) + Q(\vec{r},\vec{v}) \quad . \tag{12}$$

Consider first the special case of $c(\vec{r},\vec{v}) = 1$. We see that under the transformation

$$f(\vec{v}'\rightarrow\vec{v},\vec{r}) = \alpha f'(\vec{v}'\rightarrow\vec{v},\vec{r}) + (1-\alpha)\ \delta(\vec{v}'-\vec{v}) \tag{13a}$$

$$\vec{r} = \frac{\vec{r}'}{\alpha} \tag{13b}$$

$$t = \frac{t'}{\alpha} \tag{13c}$$

with

$$\psi(\vec{r},\vec{v},t) = \psi'(\vec{r}',\vec{v},t') \tag{13d}$$

Eq.(12) remains unchanged if all quantities (except $\vec{v}$ and σ) are replaced by primed ones.

If $c(\vec{r},\vec{v}) \neq 1$, then (13a) must be replaced by

$$c(\vec{r},\vec{v})f(\vec{v}'\rightarrow\vec{v},\vec{r}) = \alpha c'f'(\vec{v}'\rightarrow\vec{v},\vec{r}) + (1-\alpha)\ \delta(\vec{v}'-\vec{v}) \tag{14a}$$

and

$$c(\vec{r},\vec{v}) = 1-\alpha + \alpha c'(\vec{r},\vec{v}) \ . \tag{14b}$$

Another way to express this invariance is to say that under the transformation (14) the linear Boltzmann operator

$$B \equiv \frac{\partial}{\partial t} + \vec{v}\cdot\nabla + v\sigma(\vec{r},\vec{v}) - c(\vec{r},\vec{v})\sigma(\vec{r},\vec{v})\int v'f(\vec{v}'\rightarrow\vec{v},\vec{r})d^3\vec{v}' \tag{15}$$

is simply multiplied by α. If at the same time ψ is divided by α the inhomogeneous equation will also remain unchanged.

A physical interpretation for this invariance may be found in the fact that collisions which do not change the state of the motion (except for a possible absorption) only serve to change the mean free path; hence their effect can simply be taken care of by a suitable scale change.

This interpretation suggests that the same group should also be valid for non-linear Boltzmann equations. Consider for instance the following equation of kinetic theory for a system of gas molecules in which only binary collisions occur[4]:

$$\frac{\partial\psi(\vec{r},\vec{v},t)}{\partial t} = -\vec{v}\cdot\vec{\nabla}_r\psi - \vec{a}\cdot\vec{\nabla}_v\psi + 2\pi\int d^3v_1 \int d\mu|\vec{v}-\vec{v}_1|\sigma(|\vec{v}-\vec{v}_1|,\mu) (\tilde{\psi}\tilde{\psi}_1-\psi\psi_1) \ . \tag{16}$$

Here $\psi(\vec{r},\vec{v},t)$ is the angular density of molecules, $\vec{a}$ the acceleration at the position $\vec{r}$ due to any external forces and the subscripts on the gradient operators indicate the relevant variables. The collision integral represents a binary collision in which two molecules of velocity $\vec{v}$ and $\vec{v}_1$ collide at $\vec{r}$ with the relative velocity vector $\vec{v}-\vec{v}_1$, changing through an angle $\theta = \cos^{-1}\mu$ and the velocities after the collision being $\tilde{\vec{v}}$ and $\tilde{\vec{v}}_1$. The differential macroscopic cross section for this process is $\sigma(|\vec{v}-\vec{v}_1|,\mu)$. Note that for $\mu = 1$ we shall have by definition $\tilde{\psi}\tilde{\psi}_1 = \psi\psi_1$.

Under the transformation

$$\sigma(|\vec{v}-\vec{v}_1|,\mu) = \alpha\sigma'(|\vec{v}-\vec{v}_1|,\mu) + (1-\alpha)\sigma(|\vec{v}-\vec{v}_1|)\delta(\mu-1)$$

$$\vec{r} = \frac{\vec{r}'}{\alpha}$$

$$t = \frac{t'}{\alpha} \tag{17}$$

with

$$\psi(\vec{r},\vec{v},t) = \psi'(\vec{r}',\vec{v},t')$$

where

$$\sigma(|\vec{v}-\vec{v}_1|) = \int\sigma(|\vec{v}-\vec{v}_1|,\mu)d\mu \ ,$$

Eq.(16) keeps its form, all quantities (except $\vec{v}$ and μ) being replaced by primed ones. Thus the existence of the scale group considered is not restricted to the linear equation.

At this point one wonders whether the transformation introduced essentially by (5a) cannot be used to induce, on the coordinates, transformations more general than scaling. It seems plausible to expect that this will also be possible. Let us look at the simplest linear Boltzmann equation obtained in the one-velocity, time-independent, plane-symmetric model in neutron transport theory. It may be written as

$$\mu\frac{\partial\psi(x,\mu)}{\partial x} + \sigma(x)\psi(x,\mu) = c(x)\sigma(x)\int\psi(x,\vec{\Omega}')f(\vec{\Omega}'\to\vec{\Omega})d\vec{\Omega}' + Q(x,\mu) \tag{18}$$

where $\vec{\Omega}$ is the unit vector in the direction of motion of the neutron ($\vec{v}=v\vec{\Omega}$, but v does not appear as it is constant), μ is the cosine of the angle between $\vec{\Omega}$ and the x-axis, (which is the symmetry axis), c is the mean number of secondary neutrons coming out of a collision and $f(\vec{\Omega}'\to\vec{\Omega})$ is the probability distribution for a scattering from $\vec{\Omega}'$ into $\vec{\Omega}$. We have $\int f(\vec{\Omega}'\to\vec{\Omega})d\vec{\Omega}' = 1$.

Now consider a general transformation group given by

$$x = \phi(x') \quad \text{with} \quad \frac{dx}{dx'} \neq 0 \text{ or } \infty \ . \tag{19}$$

It is easy to see that Eq.(18) will keep its form under the transformation (all quantities except μ being replaced by primed ones)

$$x = \phi(x')$$

$$cf(\vec{\Omega}'\to\vec{\Omega}) = \frac{dx'}{dx}\, c'f'(\vec{\Omega}'\to\vec{\Omega}) + (1-\frac{dx'}{dx})\delta(\vec{\Omega}'-\vec{\Omega}) \tag{20}$$

$$c(x) = (1-\frac{dx'}{dx}) + \frac{dx'}{dx}\, c'(x)$$

with

$$Q(x,\mu) = Q'(x',\mu)\,\frac{dx'}{dx}$$

$$\psi(x,\mu) = \psi'(x',\mu)$$

$$\sigma(x) = \sigma'(x') \ .$$

Let me not pursue further possible generalizations but conclude instead by pointing out a connection with known solution.

To solve the plane-symmetric, one-velocity equation (18) with constant σ and c in practice, one expands the probability distribution $f(\vec{\Omega}'\to\vec{\Omega})$ into a Legendre series:

$$f(\vec{\Omega}'\to\vec{\Omega}) = f(\vec{\Omega}'\cdot\vec{\Omega}) = \frac{1}{4\pi}\sum_{\ell=0}^{\infty}(2\ell+1)f_\ell P_\ell(\vec{\Omega}'\cdot\vec{\Omega}) \tag{21}$$

where $f_o = 1$.

A well-known method then expresses the solution of the homogeneous form of (18) as

$$\psi(x,\mu) = e^{-\frac{x}{\nu}} \sum_{\ell=0}^{\infty} \left(\frac{2\ell+1}{2}\right) P_\ell(\mu)\, h_\ell(\nu) \tag{22}$$

where ν is a real parameter, $P_\ell(\mu)$ are the Legendre polynomials and $h_\ell(\nu)$ are polynomials which satisfy the recurrence relation

$$(\ell+1)h_{\ell+1}(\nu) + \ell h_{\ell-1}(\nu) - (2\ell+1)(1-cf_\ell)\nu h_\ell(\nu) = 0 \tag{23}$$

with $h_o = 1$, $h_{-1} = 0$.

Now it follows from the existence of the group (14) that the polynomials $h_\ell(\nu)$ remain invariant under the group

$$\begin{aligned} \nu &\to \frac{\nu}{\alpha} \\ cf_\ell &\to c\alpha f_\ell + (1-\alpha) \end{aligned} \tag{24}$$

which may directly be checked on the relation (23). In fact, the product $\nu(1-cf_\ell)$ remains invariant under this group. Thus the well-known polynomials $h_\ell(\nu)$ of neutron transport theory come out as the polynomials which are invariant under the scale group considered in this article.

REFERENCES

1) E.P. Wigner, Phys. Rev. 94, 17 (1954).

2) E. Guth and E. Inönü, J. Math. Phys. 2, 451 (1961).

3) A. Eris, E. Inönü, O. Öztunali and I. Usseli, IAEA Project Report 1228-RB (1974).

4) K.M. Case and P.F. Zweifel, Linear Transport Theory (Addison-Wesley, Reading, Mass., 1967).

EFFECT OF A NON-RESONANT ELECTROMAGNETIC FIELD ON THE FREQUENCIES OF A NUCLEAR MAGNETIC MOMENT SYSTEM

G. Béné, B. Borcard, M. Guenin[+], E. Hiltbrand, C. Piron[+] and R. Séchehaye

Département de physique de la matière condensée
+ Département de physique théorique
Université de Genève

1. Introduction

Such an effect was pointed out for the first time by Bloch and Siegert[1)], if we agree that only the component of an alternating field which rotates in the same sense as the Larmor precession is a resonant one. The other component (rotating in the opposite sense) gives a displacement of the resonance location proportional to the ratio $(H_1/H_o)^2$ in which H_1 is the amplitude of the magnetic component of the RF field and H_o the amplitude of the constant magnetic field which gives the splitting of the magnetic levels. H_γ is the gyromagnetic ratio of the nuclei involved. We can write H_1 (amplitude of the applied RF field, resonant or not) and H_o, in frequency units by the relation $\gamma H_o = 2\pi\nu_o$; $\gamma H_1 = 2\pi\nu_1$. While the Bloch-Siegert effect was known a long time ago, there has been recent progress in its analysis, mainly in the case of atomic magnetic moments and of transitions involving many quanta[2)].

More recently, Cohen - Tannoudji et al.[3)] studied the direct effect of a non-resonant RF magnetic field $H_1 \cos 2\pi \nu t$ on the magnetic moments of a mono-atomic paramagnetic vapour. These authors measured the Larmor frequency of the system atom plus RF field - the dressed atom - after a $\pi/2$ pulse, by its effects in the optical range. The special case $\nu >> \nu_o$ was analysed experimentally and theoretically. Some other new effects - anisotropy of the g factor of the dressed atom - were also observed.

Enz/Mehra (eds.), Physical Reality and Mathematical Description. 541–552.

A. Theoretical determination of the frequency shift.

The dressed atom concept led Cohen-Tannoudji and his coworkers to give a quantitative determination of the frequency shift observed when $\nu >> \nu_o$.

An atom is able to absorb a non-resonant photon, provided it reemits it again before a time interval

$$\delta t \leq \hbar\,(\delta E)^{-1} \quad : \quad h\nu = E_2 - E_1 + \delta E \qquad (1.1)$$

Here $E_2 - E_1$ is the energy interval between two discret energy states E_2 and E_1 of the atom.

When such a process happens spontaneously, we are led, for example, to a slight variation of the electron magnetic moment, as was shown by P. Kush et al.[4] and to a shift of $2^2S_{\frac{1}{2}}$ and $2^2P_{\frac{1}{2}}$ states of the atomic hydrogen, the Lamb-Retherford effect[5].

If an atom is in a radiation field[6] such a process of absorption and reemission of photons produces analogous effects. If the frequency of the induced field is high by reference to the Larmor frequency, we have a variation of the g factor of the atom given by

$$g = g_o\, J_o\left(\frac{\nu_1}{\nu}\right) \qquad (1.2)$$

where J_o is the Bessel function of zero order. Such a relation is well confirmed by experiments.

The main consequences of the Bloch-Siegert effect may be simply interpreted by pure geometrical consideration[7]. The dressed atom concept, or a semi-classical interpretation[8] lead to a very precise approach of complex cases and particularly coherence resonances.

B. Guidelines of this work.

As we see, in the anterior works, the effect of a non-resonant RF field was analysed in two special cases :

- the component of the resonant radiation which rotates in the opposite sense

- a non-resonant RF field with a frequency higher than the Larmor frequency.

In this paper we intend to study the general problem of the action of a RF field - in the whole range of amplitude and frequencies - on a system of nuclear moments submitted to a constant magnetic field.

More precisely, in the experiments described and analysed here after, the influence of a non-resonant RF field on protons of liquid water (or methanol) in the earth field, is shown either by a transient method (free precession of nuclei after switching out a prepolarizing field (9) normal to the earth field) or by a permanent one (NMR of protons, magnified by Overhauser effect[10]).

In such situations, when the RF resonant field is cut off, the Larmor frequency ν_o of the protons is in the range of 2 KHz. Measurements were made with ν in the range 0 - 6 KHz and amplitudes (given by ν_1 in the same frequency units) between 0 and 2,5 KHz.

The approximations involved in the theoretical analysis limit the comparison with the experiments to the range 0 - 500 Hz for ν_1.

This low frequency range of EM field used is particularly convenient to have the three frequencies involved (ν_o, ν, ν_1) in the same range 0 - 5 KHz.

We note briefly here that we were able to observe analogous effects in the current range of NMR spectrometers (proton resonances between 10 and 100 MHz) on similar samples (protons in liquid systems) in collaboration with Descout's Group of our department and with Briguet of the "Laboratoire de Spectroscopie et de Luminescence" (University of Lyon - France). These works are in progress and a full account will be published in the near future.

In fact, the analysis of the results in the range $\nu < \nu_o$, with $\nu_1 < \nu$, ν_o gives a good agreement between theory and experiment, by using a vectorial model in which only the component of the alternating field rotating in the same sense than the Larmor precession is taken into account.

In the range $\nu > \nu_o$, always with $\nu_1 < \nu$, ν_o we noted[11] in our first papers a good agreement with the expression (1.2) used by Cohen-Tannoudji et al. in this work.

We shall give in this paper succesively :

1) The experimental methods used and the main results observed.

2) A theoretical analysis of the phenomenon

- vectorial model (with and without rotating field approximation)
- new quantum derivation of the frequency spectrum of this problem
- a discussion of the results and some concluding remarks.

2. Experimental arrangements

We shall briefly describe the two kinds of experimental arrangement used in order to observe the effect of a non-resonant RF field pointed out in the introduction.

A) Free precession

A sample of condensed matter (water, methanol, ...) is acted upon by a steady magnetic field $\vec{H}_p$ produced by a prepolarisation coil. After a time longer than T_1 a macroscopic nuclear magnetization $\vec{M}_p$ arises whose magnitude is given by

$$M_p = \chi H_p \tag{2.1}$$

where χ is the nuclear susceptibility for the particular nucleus we are interested in.

Generally $\vec{H}_p$ makes a 90° angle with the earth magnetic field.

In a time sufficiently short compared with T_1, $\vec{H}_p$ is removed and then we observe the free precession of $\vec{M}_p$ around $\vec{H}_o$ with a frequency $\nu_o = \gamma H_o$ through the induced voltage in the prepolarisation coil used now as a pick-up coil[13].

In our experiments it is very useful to get a magnetization $|\vec{M}_p|$ as big as possible, because the amplitude of the induced signal depends directly on it. Besides the possibility to raise up $|\vec{H}_p|$, see formula (2.1), we get a considerably greater magnetization using the dynamical polarization or Overhauser effect[10]. Electronic transitions of a free radical (tanone[14]) diluted in the sample, are induced by a RF field $H_y = H_{DP} \cos\Omega t$ producing an electronic polarization[15),16)]. This electronic polarization is transferred to the nuclear spins, which in turn are polarized via the dipolar interaction. The net result is a nuclear magnetization enhanced by a factor up to 1000. The field H_{DP} cosΩt is supplied by a coil directly wired around the sample. If the order of magnitude of $|\vec{H}_o|$ is near the earth magnetic field (0,5 G), for tanone in water we have an Ω of about 67 MHz[11)]. In Table I a summary of the method and technical data are given.

To the free precession with dynamical polarization just described, is added a RF field H_1 cosΩt : the "dressing" field, perpendicular to H_o, generated by a system of Helmoltz coils. With this supplementary field we observe a shift of the Larmor frequency ν_o[12)]. This shift depends on both frequencies ν and ν_1, the latter being related to the magnitude of the "dressing" field H_1 by $\nu = \gamma H_1$[8)].

B) Double coil spectrometer

The basic apparatus we use, is the well known Bloch crossed coils system[17)] summarized in Figure 1.

For the same raason as in the free precession method it is suitable to magnify the nuclear magnetization M_o, using the dynamical polarization property.

Now here again we put an additional "dressing" field H_1cosΩt along the Ox axis supplied by Helmoltz coils. As expected, we see a shift of the nuclear resonance frequency ν_o. Owing to the fact that the receiving frequency is fixed by the quartz oscillator at 1965 Hz[17)] we have to fit the $\vec{H}_o$ field according to the shift. Fig. 2 shows the signals shifted in terms of the amplitude of the "dressing" field.

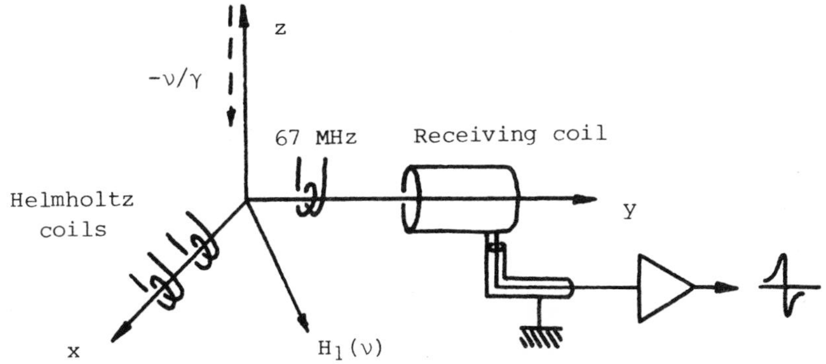

Fig. 1

C) Results

The results appear in Table I below.

On Figure 3, besides the experimental points, we plotted the theoretical curves for several approximations which will be discussed in the following paragraph.

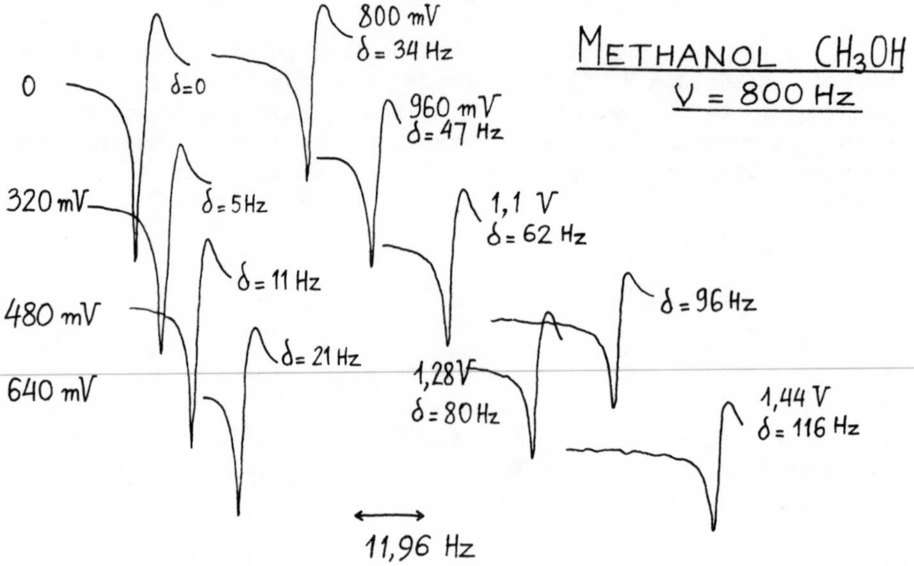

Fig. 2

3. Theoretical considerations.

To study this problem, we shall consider an ensemble of spin $\frac{1}{2}$, all in the same state, and which precess in the magnetic field $\vec{H}_o$ of the earth, taken in the direction z, and in a magnetic field $\vec{H}_1$ cosωt, in the direction x.

The corresponding Schrödinger equation is

$$i\partial_t\psi_t = -\frac{\gamma}{2}\,(\sigma_z H_o + \sigma_x\,H_1\cos\omega t) \qquad (3.1)$$

The initial state corresponding to a spin orthogonal to $\vec{H}_o$ is given by

$$\psi_o = \begin{pmatrix} e^{i\,\phi/2} \\ e^{-i\,\phi/2} \end{pmatrix} \qquad (3.2)$$

where ϕ is the usual polar angle in the plane orthogonal to $\vec{H}_o$.

The measured frequency in the pick-up coil is nothing else than the fundamental of

$$M_y = \langle \bar{\psi}_t^{+} \sigma_y \psi_t \rangle \tag{3.3}$$

Unfortunately, the exact solution of (3.1) is not known. We must therefore resort to approximations.

The first idea would be to change the problem, and thus the equation, in replacing the linear field by a rotating field. That is, to write

$$i\partial_t \psi_t = -\frac{\gamma}{2} (\sigma_z H_o + \sigma_x H_1 \cos\omega t + \sigma_y H_1 \sin\omega t) \tag{3.4}$$

The sense for the rotation is chosen so as to give the best possible approximation. The solution is then obtained in going over to the so-called rotating frame, this amounts to making the unitary transformation

$$\psi_t^r = U_t^{-1} \psi_t \quad ; \quad U_t = e^{-i \frac{1}{2} \sigma_z \omega t} \tag{3.5}$$

In this new representation, the Schrödinger equation reduces to

$$\begin{aligned} i\partial_t \psi_t^r &= -\frac{\gamma}{2} \left[(H_o - \frac{\omega}{\gamma}) \sigma_z + H_1 \sigma_x \right] \psi_t^r \\ &= -\frac{1}{2} \left[(\omega_o - \omega) \sigma_z + \omega_1 \sigma_x \right] \psi_t^r \end{aligned} \tag{3.6}$$

In this "rotating frame", the motion of ψ_t^r is easily found. It is a rotation around the direction defined by $(\omega_1, 0, \omega_o - \omega)$ and with the Rabi angular velocity

$$\left[(\omega_o - \omega) + \omega_1^2 \right]^{\frac{1}{2}} \tag{3.7}$$

The observed frequency shift is therefore going to be

$$\Delta\nu_o = \left[(\nu_o - \nu)^2 + \nu_1^2 \right]^{\frac{1}{2}} - \nu_o \tag{3.8}$$

Comparing perturbative expansions of (3.3) based on (3.1) or on (3.4), one can expect that the formula (3.8) is going to give good results for small values of ν as compared with ν_o.

A second method[3] is to treat the radiofrequency field not as an external, classical field, but as a quantized field. The state to consider is then that of a coherent state, but we don't have to consider it explicitely. As our radiofrequency field is linearly polarized and of sharp frequency, we only have to introduce one type of creation operator and the Hamiltonian of our system becomes

$$H = \frac{1}{2}\,\omega_o\sigma_z + \omega a^* a + \frac{\lambda}{2}\,\sigma_x(a+a^*) \;;\quad \lambda = \frac{\omega_1}{2\sqrt{N}} \tag{3.9}$$

N = mean number of photons

In this new setting, what is observed are the energy difference between the levels of this Hamiltonian. This levels, however, are fantastically numerous. Indeed, in the case $\lambda = 0$, the eigenvalues are simply

$$n\omega \pm \tfrac{1}{2}\,\omega_o$$

The difference, however, does not depend upon n if we consider the transition between $n\omega + \frac{1}{2}\,\omega_o$ and $n\omega - \frac{1}{2}\,\omega_o$ which gives of course only ω_o. One can prove in complete mathematical rigour that this remains true of the exact solution for the difference which is nearest to ω_o, in the limit $n\to\infty$ (and hence $N\to\infty$, since asymptotically $n \sim 2N$). The fact that we are going to make the limit $n\to\infty$ in the computations is in itself no restriction to the accuracy of the method, since typical numerical values for N are in the range of 10^{20}. The limitations arise much more from the fact that ω should not be too small if we want to describe our radiofrequency field in the cavity by a coherent state. It therefore just turns out that the cases for which our second method fails are those for which the first was good.

The computation of the eigenvalues and thus finally of the frequency shift are completely straightforward, but somehow tedious. The formulas for the perturbation expansion can be taken out of the book of Kato[18], the final result being

$$\Delta\nu_o = \frac{\nu_1^2}{4\pi}\cdot\frac{\nu_o}{\nu_o^2-\nu^2} - \frac{\nu_1^4}{16\pi}\cdot\frac{\nu_o(\nu_o^2+3\nu^2)}{(\nu_o^2-\nu^2)^3}$$

Apart from the small values of ω, the agreement with the experimental values is good (Figure 3). We remark that in the vicinity of the resonance, the fit given by the second order term only is much better than the fit obtained in taking the 4th order term into account. This is not astonishing. In the vicinity of

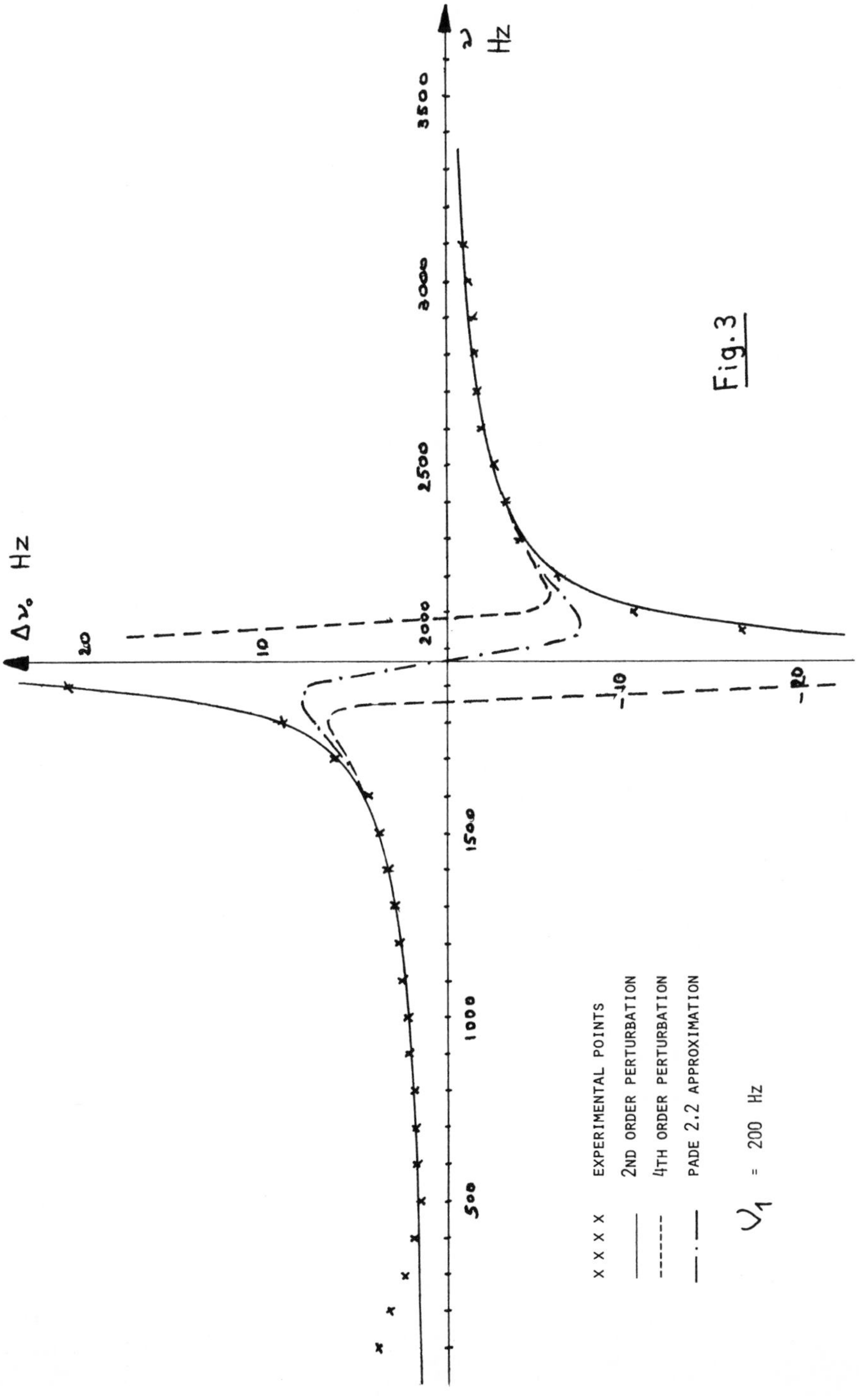

Δν₀ Hz
ν Hz
20
10
-10
-20
500
1000
1500
2000
2500
3000
3500
x x x x EXPERIMENTAL POINTS
2ND ORDER PERTURBATION
4TH ORDER PERTURBATION
PADE 2.2 APPROXIMATION
ν₁ = 200 Hz

Fig. 3

the resonance ($\omega_o = 1971.3$) the coefficient of ω_1^4 is much larger than the coefficient of ω_1^2 and the series is clearly asymptotic. Under such circumstances, the sequence of Padé approximants usually gives a much better result. The formula for the diagonal (2,2) Padé approximant corresponding to the 4th order Taylor expansion is

$$\Delta\nu_o = \frac{\nu_1^2}{4\pi} \cdot \frac{\nu_o}{\nu_o^2-\nu^2}\left[1 + \frac{\nu_1^2}{4}\,\frac{(\nu_o^2 + 3\nu^2)}{(\nu_o^2-\nu^2)^2}\right]^{-1}$$

Indeed, the (2,2) diagonal Padé approximant gives a much better fit than the 4th order Taylor expansion, giving even nearly exactly the behavior in the immediate vicinity of the resonance; further away, however, the fit given by the second order is much better, up to the point where all approximations coincide.

As a conclusion, we can say that the correspondence with theory is good, even astonishingly good if we remember the fairly simple model which we are describing theoretically.

Acknowledgments

The authors are indebted to Mr. Richard Gavaggio (Laboratory of Jussy) who constructed most of the experimental apparatus.

References

1) F. Bloch and A. Siegert, Phys. Rev. 57, 522 (1940).

2) G.W. Series, *Quantum Optics*, St Andrews Summer School, Academic Press, London (1970).

3) C. Cohen-Tannoudji and S. Haroche, J. Phys. 30, 125,153 (1969).

4) P. Kusch, Phys. Rev. 101, 627 (1956).

5) W.E. Lamb and R.C. Refherford, Phys. Rev. 72, 241 (1947).

6) A. Kastler, Proc. Ampère International Summer School, Basko Polje (1971).

7) A. Abragam, *The Principles of Nuclear Magnetism*, Oxford, Clarendon Press (1961).

8) D.T. Pegg and G.W. Series, Proc. Roy. Soc. London A 332, 281 (1973).

9) M. Packard and R. Varian, Phys. Rev. 93, 941 (1954).

10) A.W. Overhauser, Phys. Rev. 89, 689 (1953); 92, 411 (1953).

11) R. Séchehaye et al., Helv. Phys. Acta 45, 842 (1972); 46, 408 (1973).

12) E. Hiltbrand et al., C.R. Acad. Sc. (Paris) 277, 531 (1973): 278, 243 (1974).

13) Magnetic Resonance and Relaxation, Proc. of the XIV Coll. Ampère, Ljubljana, North-Holland (1967).

14) R. Brière, H. Lemaire et A. Rassat, Bull. Soc. Chim. France 11, 3273 (1965).

15) G. Breit and I.I. Rabi, Phys. Rev. 38, 2082 (1931).

16) A. Landesman, J. Phys. Rad. 20B, 937 (1959).

17) R. Séchehaye et P.A. Dreyfuss, Z.A.M.P. 21, 660 (1970).

18) T. Kato, *Perturbation Theory for Linear Operators*, Springer Verlag, Berlin, (1966).

TABLE I

ν	ν_1 = 100 Hz $\Delta\nu_o$	ν_1 = 200 Hz $\Delta\nu_o$	ν_1 = 300 Hz $\Delta\nu_o$	ν_1 = 400 Hz $\Delta\nu_o$	ν_1 = 500 Hz $\Delta\nu_o$
10	1.91	1.98	1.67	6.16	-
30	1.91	7.61	12.77	5.26	-
50	1.79	6.94	15.86	30.47	-
70	1.60	6.20	14.36	27.99	-
100	1.36	5.22	11.92	24.40	32.87
150	1.05	3.97	9.01	18.16	25.07
200	0.86	3.23	7.29	14.49	19.82
300	0.66	2.45	5.49	10.86	14.99
400	0.58	2.10	4.79	9.41	12.31
500	0.58	2.02	4.48	8.74	12.12
600	0.54	1.95	4.36	8.59	11.81
700	0.51	1.95	4.40	8.55	11.92
800	0.51	1.98	4.44	8.86	12.16
900	0.54	2.02	4.59	9.02	12.55
1000	0.58	2.30	4.83	9.45	13.15

ν	ν_1 = 100 Hz $\Delta\nu_o$	ν_1 = 200 Hz $\Delta\nu_o$	ν_1 = 300 Hz $\Delta\nu_o$	ν_1 = 400 Hz $\Delta\nu_o$	ν_1 = 500 Hz $\Delta\nu_o$
1100	0.62	2.45	5.22	10.04	14.05
1200	0.62	2.57	5.69	10.94	15.47
1300	0.74	2.88	6.23	12.36	16.85
1400	0.78	3.23	7.02	13.86	18.91
1500	1.01	3.74	8.35	16.26	22.24
1600	1.13	4.40	10.19	19.7	26.87
1700	1.52	6.32	13.26	25.83	35.04
1800	2.30	9.18	19.69	37.63	49.48
1900	5.14	21.29	38.94	67.37	-
1920	6.71	25.87	47.21	76.85	-
1940	12.01	33.12	58.02	-	-
1960	17.77	54.77	72.74	127.24	-
1970	- 1.63	- 1.48	-	-	-
1980	-22.10	-38.00	-73.38	-103.31	-
2000	-10.90	-32.23	-59.01	- 85.97	-
2020	- 7.05	-24.11	-46.61	- 72.32	-
2050	- 4.38	-16.50	-34.84	- 57.55	-
2100	- 2.76	-10.52	-23.60	- 39.75	-58.82
2200	- 1.63	- 6.16	-13.62	- 24.39	-36.55
2300	- 1.09	- 4.23	- 9.39	- 18.67	-26.23
2400	- 0.78	- 3.30	- 7.01	- 13.93	-19.78
2500	- 0.62	- 2.64	- 5.50	- 11.02	-15.81
2600	- 0.47	- 2.21	- 4.57	- 9.21	-13.05
2700	- 0.43	- 1.86	- 3.76	- 7.82	-11.08
2800	- 0.36	- 1.63	- 3.22	- 6.74	- 9.48
2900	- 0.31	- 1.48	- 2.75	- 5.93	- 8.24
3000	- 0.27	- 1.20	- 2.48	- 5.16	- 7.32
3500	- 0.16	- 0.85	- 1.55	- 3.10	- 4.73
4000	- 0.12	- 0.66	- 1.01	- 2.21	- 3.26
4500	- 0.12	- 0.51	- 0.74	- 1.97	- 2.45
5000	- 0.08	- 0.43	- 0.54	- 1.28	- 1.98